PHOTOBIOLOGY OF MICROORGANISMS

Photobiology of Microorganisms

Edited by
PER HALLDAL
Professor of Plant Physiology, Umeå University, Sweden

WILEY - INTERSCIENCE
a division of John Wiley & Sons Ltd
London New York Sydney Toronto

Library of Congress Catalog card number 78-116159

ISBN 0 471 34340 4

Printed in Great Britain by
Adlard & Son Ltd, Bartholomew Press, Dorking

Preface

Photobiology is a very diverse research field, even if we restrict the consideration only to microorganisms. Experiments on this subject have been performed by investigators with very different backgrounds and it is possible to find pure physicists who have minor training in biology and biologists who have little knowledge of rather elementary physics. With the highly specialized training that is required to perform top quality research today, this division is unfortunately unavoidable. Many physicists who are working on light quanta problems and excitation energy transfer have discovered that biological materials serve their need for suitable models. In a similar way, biologists have discovered how their material has been affected by radiation and may use pure physical models to explain their results. In this way many physicists and biologists meet on a common ground. Both may be dealing with the related effect of radiation on the same biological material, but according to their scientific training the approach to the problem will naturally be quite different. They may also find it very difficult to agree on a language that is understandable to both.

As the editor of this book is trained mainly as a biologist, the approach is biological. Many biologists often have difficulties with proper physical design of photophysiological experiments and are also uncertain about the way obtained data should be handled and interpreted. Therefore, Professor Duysens has emphasized some of the fundamental photobiological principles in the first chapter of this book. It is hoped that this article will serve as a guide for those who feel uncertain on this particular point.

The reader of this book will find that photosynthesis as a main subject is treated in only two chapters. This may seem unjust as photosynthetic bacteria and algae have contributed tremendously to our present knowledge of this process. However, photosynthesis has recently been covered extensively in several books (for references see Chapters 2 and 3). It is hoped that these two chapters, in addition to summing up the present status of the process, will also supplement the reading of other articles where the photosynthetic process has been treated peripherally.

The reader will further note that several photobiological topics, for

v

example photoinactivation, photoreactivation and mutagenic effects, are not dealt with separately. With a few exceptions the book thus deals only with the physiological effects of visible and near infrared radiation, and not with the lesion processes in the far ultraviolet spectral region. These missing subjects, however, are distinctly independent and clearly distinguished from the other articles. They have a marked chemical and biophysical character, in contrast to the biological approach of the book, and would therefore have formed a separate subdivision. These topics have been covered extensively in several recent reviews.

PER HALLDAL

Contributing Authors

R. L. Airth   *Department of Botany and The Cell Research Institute, The University of Texas, Austin, Texas, USA*

M. J. Carlile   *Department of Biochemistry, Imperial College, London, England*

D. Davenport   *Department of Biological Sciences, University of California, Santa Barbara, California, USA*

M. J. Dring   *Department of Botany, The Queen's University of Belfast, Northern Ireland*

L. N. M. Duysens   *Biophysical Laboratory of the State University, Leiden, The Netherlands*

C. F. Ehret   *Division of Biological and Medical Research, Argonne National Laboratory, Argonne, Illinois, USA*

G. E. Foerster   *Department of Botany and The Cell Research Institute, The University of Texas, Austin, Texas, USA*

P. Halldal   *Department of Biology, University of Umeå, Umeå, Sweden*

W. G. Hand   *Department of Biology, Occidental College, Los Angeles, California, USA*

W. Haupt   *Department of Biology, University of Erlangen-Nürnberg, Germany*

R. Hinde   *Department of Botany and The Cell Research Institute, The University of Texas, Austin, Texas, USA*

W. Kowallik   *Department of Botany, University of Cologne, Germany*

H. Lorenzen   *Department of Plant Physiology, University of Göttingen, Germany*

W. Nultsch   *Department of Botany, University of Marburg, Germany*

E. Schönbohm   *Department of Biology, University of Marburg, Germany*

Chr. Sybesma   *Department of Physiology and Biophysics, University of Illinois, Urbana, Illinois, USA*

vii

W. Wiessner   *Department of Plant Physiology, University of Göttingen, Germany*

J. J. Wille   *Department of Biological Sciences, University of Cincinnati, Cincinnati, Ohio, USA*

Contents

CHAPTER 1

Photobiological principles and methods

L. N. M. DUYSENS

Biophysical Laboratory of the State University, Leiden, The Netherlands

Introduction

As a starting point for discussion let us take the presently best known
and probably most elaborate process for absorption and transformation
of light energy, namely photosynthesis.

Light is a very dilute source of energy: even in direct sunlight a strongly
absorbing pigment molecule whose average molar absorption coefficient
for visible light is of the order of 3×10^4 will absorb about twelve quanta

1

every second (Rabinowitch, 1951). A monomolecular layer of molecules such as chlorophyll *a* would absorb only about 1 per cent of the incident radiation. In order to make possible the absorption of an appreciable fraction of the radiation in thin photosynthetic cells, a high concentration (about 10^{-2} M) of photosynthetic pigment is required.

Conversion of light energy into a chemical form takes place only in reaction centres which possess an enzymatic apparatus (Duysens, 1964). These reaction centres are present in a concentration in the order of 1 per cent of that of the total pigment. Light energy absorbed by the pigments is transferred via various pigment molecules to (and thus channelled into) the reaction centres in the form of electronic excitation energy. In the reaction centres the conversion of the short-lived electronically excited state into long-lived 'chemical' states takes place. This is the primary photochemical reaction. Secondary reactions result in the reduction of carbon dioxide and the oxidation of a hydrogen donor, which is water in algae and a sulphur compound such as thiosulphate or an organic substance such as succinate in purple bacteria.

In vision the light-absorbing protein pigment complex, rhodopsin, itself changes in shape. Secondary reactions lead to excitation of a nerve cell and other secondary reactions restore the original shape of the rhodopsin.

Because of their diversity, the secondary reactions cannot be the subject of this paper. In this paper we will describe the fundamentals of processes leading up to the primary photochemical reactions, as well as the reverse processes as exemplified by bioluminescence and delayed fluorescence. Methods will be mentioned or described for identifying the components and studying the processes occurring in reaction centres, and for determining the spectra of pigments active in photoreactions.

For more extensive treatment and for aspects not discussed here, we refer the reader to Murrell (1963), Seliger and McElroy (1965), and Thomas (1965).

Absorption and emission of light energy; transfer and other conversions of electronic excitation

Introduction

As far as is known, electromagnetic radiation is only active in biological systems (if heating effects are disregarded) through excitation of essentially electronic energy levels of molecules, which absorb in the visible and adjacent ultraviolet and infrared regions.

Molecular reactions including photoreactions can be described in principle by quantum mechanics. In practice a mixture of empirical and approximate quantum mechanical methods has been used with good results. Further rapid development may be expected, on the one hand by the entry of theoretical chemists, physicists and mathematicians into the field using improved approximative methods and larger and more efficient computer systems, and on the other hand by improved experimental techniques furthered by the dramatic improvement of electronic and electrooptical components.

The following approximations apply with good precision to probably most photobiologically active molecules (see, for example, Seliger and Morton, 1968).

(1) There are a number of electronic energy states in a molecule. Each of these states corresponds with a number of vibrational (and rotational) levels of the system of nuclei.

(2) These electronic states can be divided into two groups; one group with paired electron spins (singlet electronic states), another group with unpaired spins (triplet states).

Absorption and emission of radiation

After a dark period the molecule is in the lowest electronic state, S_0, which is a singlet stage. Light absorption (by light we mean radiation in the visible and adjacent regions) causes a transition to a higher electronic singlet state; simultaneously one of the vibrational levels is excited. Visible excitation in organic molecules usually takes place in the system of π electrons. These are the delocalized electrons associated with the system of conjugated bonds present in all organic molecules absorbing in the visible region. To each singlet state except S_0 corresponds a triplet state with lower energy. Transitions involving light absorption or emission can occur with high transition rate only between singlet states or between triplet states, but not between singlet and triplet states.

Figure 1.1 shows an energy level diagram with typical rates (i.e. the number of transitions per second per excited molecule) for some transitions. Processes following light absorption may be discussed in relation to this diagram. Light absorption in the longest wave absorption maximum occurs, for example, by the transition S_{00} to S_{11}. The vibrational energy of S_{11} is converted within about 10^{-12} seconds into heat, resulting in the transition S_{11} to S_{10}. The rate of internal conversion, k_{ic}, from S_{10} to one of the vibrational states of S_0 is in general much lower than from higher electronic levels to lower ones, or from higher to lower vibrational levels.

If k_{ic} is smaller than or of the same order of magnitude as the fluorescence rate (which in strongly absorbing molecules is about 10^8), and if the transition rates to the states P and T are relatively small, then fluorescence will be observed. The fluorescence in the maximum of the emission spectrum is, for example, represented by the transition S_{10} to S_{01}. Absorption into a shorter wave band, represented by the transition S_{00} to S_{21}, leads to S_{20} in a very short time. From S_{20} short wave fluorescence may occur (transi-

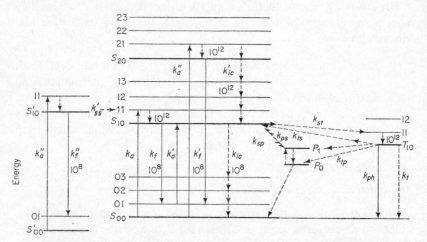

Figure 1.1 Energy level diagram. S are singlet levels, T triplet levels of the reaction centre pigment, and P 'levels' or products of the photochemical reaction. S_{10}, S_{01}, etc., are vibrational levels corresponding to the lowest excited state of the reaction centre complex. The S' levels are singlet levels of a molecule transferring energy to the reaction centre. The solid vertical lines represent transitions involving emission or absorption of radiation and the broken lines represent radiationless transitions, in which an appreciable part of the energy is converted into heat. The ks are rate constants; k_a is a rate constant for absorption, k_f for fluorescence and k_{ic} for internal conversion. The order of magnitude of some rate constants in number of transitions per second per excited molecule is indicated for strongly absorbing molecules

tion S_{20} to S_{01}) or internal conversion k'_{ic} to S_{10}. Since the internal conversion process between higher excited states is much more efficient than light emission, the short wave fluorescence is extremely weak and so far has not been detected. Thus light absorption exciting a higher excited state has the same result as excitation of the lowest excited state and leads to the same fluorescence emission. This explains the rule that the shape of the emission (fluorescence) spectrum and the quantum yield of fluorescence (number (≤ 1) of quanta emitted per quantum absorbed) are independent of the wavelength of the exciting light. Furthermore, the emission occurs at or beyond the long wave absorption band (Stokes' rule).

If the rate k_{st} is of the same order of magnitude or larger than k_f and k_{ic}, which often is the case, an appreciable fraction of the transitions occurs to the lowest triplet level. If in addition k_{tp} is small, and if the rate k_t of radiationless transitions to S_{00} is minimized by lowering the diffusion rate of 'quenching' molecules by solidification (e.g. by low temperature) or by viscous media, an appreciable amount of molecules may accumulate in the triplet state T_{10}. Even though k_{ph} is small, appreciable light emission may then occur from T_{10}. This emission, which lasts for some time after the light exciting a singlet state has been turned off, is called phosphorescence. In addition to or instead of phosphorescence, a long lasting emission will occur from the state S_{10}, due to the transition from T_{10} to S_{10}, if k_{ts} is not negligible compared to k_{tp}, k_{ph} or k_t. This emission, which has the same spectrum as the fluorescence spectrum is called delayed fluorescence. Delayed fluorescence may also occur by a transition from a product, P_0, of the photochemical reaction to S_{10}, if k_{sp} is not negligible. The latter phenomenon probably occurs in the so-called delayed light emission from chlorophyll *a* in photosynthesizing cells or chloroplasts.

Transfer of electronic excitation energy

An electronically excited pigment molecule will transfer its electronic excitation energy to another pigment molecule with high efficiency under conditions described in the following. The lifetime of the excited state should be relatively long. The state S'_{10} (see the left part of Figure 1.1) has the longest lifetime. At a distance smaller than 100 Å the electromagnetic field of the excited molecule can be described as the coulomb field of an electric dipole vibrating with the frequency of emission corresponding to the transition from S'_{10} to S'_{01}. Another molecule in this field will be excited, if this frequency is equal to a frequency at which the energy-accepting molecule shows appreciable absorption. In Figure 1.1 this is assumed to be the case for the frequency of the transition from S_{00} to S_{10}. Since the frequencies of energy donor and energy acceptor are equal, the transfer is said to occur by induced resonance. Förster (1948, 1951) has shown that the transfer efficiency is proportional to

$$(1/R^6) \times c \times \int \phi(\lambda)\, \alpha(\lambda)\, f(\lambda)\, \mathrm{d}\lambda$$

The transfer efficiency increases rapidly with decreasing distance R and is proportional to the overlap integral between the spectral fluorescence yield $\phi(\lambda)$ of the donor and the specific absorption $\alpha(\lambda)$ of the acceptor; $f(\lambda)$ is a universal function of λ and c a constant. The equation also applies to two identical molecules. In photosynthetic cells, in which the distance between the chlorophyll molecules is of the order of 20 Å, the transfer frequency, $k_{s's}$, between these molecules is much greater than k_f or k_{ic}, so the energy is transferred between many molecules until it is trapped in a reaction centre (Duysens, 1964; Knox, 1968).

Primary photochemical reactions

The primary photochemical reactions proceed with good efficiency if k_{sp} or k_{st} and k_{tp} are greater than k_{ic}, or frequencies of other processes leading to S_{00}. As an example we may discuss the primary photochemical reaction in purple bacteria (Beugeling, 1968; Clayton, 1967; Duysens, 1965; Duysens and coworkers, 1956; Ke, 1969; Parson, 1968; Vredenberg and Duysens, 1963; Zankel, Reed and Clayton, 1968). The relevant part of the reaction centre may be written as cyt PXQ. P is reactive bacteriochlorophyll and has a low intrinsic fluorescence yield. Cyt is a cytochrome, Q possibly ubiquinone and X an unknown intermediate. In darkness cyt and P are largely reduced, X and Q largely oxidized. Light energy absorbed by one of the bulk pigment molecules is transferred by induced resonance through a matrix of bacteriochlorophyll molecules in which the reaction centres are imbedded. Finally the excitation excites a P molecule, for example, to the state S_{11} from which it returns to S_{10}. A rapid reaction occurs with rate k_{sp} (see Figure 1.1), probably the transfer of an electron from P to X. The low fluorescence yield of PX indicates that k_{sp} is much larger than $k_f + k_{ic}$. Subsequently cyt is oxidized by P and Q is reduced by X^- within less than 10 microseconds. The redox reactions of P, cyt and Q can be followed by difference absorption spectrophotometry. Very rapid reactions are measured after excitation with an electronic flash lamp or a laser. Subsequent states of the reaction centre are P_1 : cyt P^+X^-Q and P_0 : cyt$^+$ PXQ^-. Only one quantum absorbed by or transferred to P is needed to oxidize a cytochrome molecule.

Subsequent reactions which have half times longer than 10 milliseconds finally result, for example, in oxidation of the hydrogen donor and reduction of carbon dioxide. These reactions restore the reaction centre to its original state. We have omitted, for the sake of simplicity, the intermediate states cyt$^+$ P X^- Q and cyt P^+ X Q^-. If we had taken P X as the reaction centre in the state S_{00}, P_1 would have been P^+ X^-. Other states such as P X^- and P^+ X would then have no place in the energy diagram.

Other photoreactions, such as the change in form of rhodopsin (see the introduction, page 2), can be described analogously.

In photosynthetic cells after illumination, delayed fluorescence occurs (Fleischman and Clayton, 1968; Strehler and Arnold, 1951), which may result from the transition from P_1 to S_{10}. Treatments enhancing the concentration of P_1 will stimulate the delayed fluorescence, and provide information concerning reactions, for example, leading from P_0 to P_1 and vice versa.

If the triplet state T_{10} is an intermediate between S_{10} and P_1, then the fluorescence yield of P will in principle be independent of the redox state of X, since the excited singlet state of P would not react directly with X. The fluorescence yield of P is increased upon reduction of X in purple bacteria (Clayton, 1967) and the fluorescence yield of the chlorophyll a of the second reaction centre in algae is probably enhanced by the reduction of Q in the second photoreaction of algae (Duysens and Sweers, 1963). Both these observations suggest that the excited singlet states of these reaction centres react directly with the electron acceptor, and that these reactions do not proceed via the triplet state.

In bioluminescence long-lasting fluorescence is emitted due to a chemical reaction. Here again chemical reactions leading to P_1 and from P_1 to S_{10} may occur.

Identification and reaction sequences of compounds in the reaction centres

Photochemical reactions in general cause changes in light absorption or fluorescence of molecules in the reaction centre. In principle it is possible to measure, by means of sensitive differential spectrophotometers, the spectral dependence and the kinetics of these changes, and to determine the identity of participants and the sequence and rate of reactions. Rapid excitation has become possible through the use of special electronic flash lamps and of lasers. Rapid developments occur in this field. Precise recording (through averaging) and rapid and detailed analysis are feasible through the use of hybrid analogue–digital systems, interfaced with the apparatus. These systems are also being developed.

Further accessory techniques are the cooling of the photosystems by which the reactions are slowed down, some to a larger extent than others. Some photoreactions proceed at the temperature of liquid helium. Other more conventional techniques are the use of inhibitors and other substances selectively affecting the processes under study, and of biochemical disintegration. Part of these techniques can of course also be used for the study of secondary reactions.

Thermodynamical considerations

Maximum efficiency of photochemical reactions at constant temperature

Stated somewhat oversimply, it can be shown on the basis of the second law of thermodynamics (Duysens, 1958; Knox, 1969; Ross and Calvin, 1967) that the maximum efficiency with which absorbed light energy can be converted by any photoprocess into Gibbs free energy or other 'high grade' energy is lower than 100 per cent, and decreases to zero with decreasing intensity. Application of the equations obtained for photosynthesis led to the result that at a certain weak intensity, at which the efficiency can still be measured with good precision, the calculated *maximum* efficiency is about 70 per cent. It is a well-established observation that the relative photosynthetic efficiency in algae does not further increase with light intensity, so that a theoretical efficiency limit of 70 per cent is established for algal photosynthesis.

For the reverse process, the conversion of chemical or other high grade energy into light, such as occurs in bioluminescence, analogous conclusions follow from thermodynamic considerations. The *maximum* efficiency with which in principle high grade energy can be converted into light is greater than 100 per cent and increases with decreasing light intensity. The first law of thermodynamics is not violated since heat is taken up from the surroundings.

To my knowledge, so far no applications of this theory have been made outside the field of photosynthesis.

Relationship between absorption and the fluorescence spectrum

Consider a black body at temperature T (e.g. room temperature) with one pigment molecule (say chlorophyll a) in heat contact with the inside wall. The second law of thermodynamics requires that the radiation emitted at a certain wavelength per unit of wavelength interval, $E(\lambda, T)$, is equal to the radiation absorbed, or

$$E(\lambda,T) = cP(\lambda,T)\,\alpha(\lambda,T) \tag{1.1}$$

in which $P(\lambda,T)$ is the well-known Planck distribution of the radiation in the black body, $\alpha(\lambda, T)$ is the specific absorption of the molecule, and c a constant only depending upon the units used.

Compare $E(\lambda,T)$ with the conventional fluorescence from the first excited singlet state S_1. If the lifetime of this state is of the order of 10^{-9} seconds, which is long compared to the nuclear vibration time of about 10^{-12} seconds, during most of this lifetime a temperature equilibrium will probably exist between the vibrational states S_{10}, S_{11}, S_{12}, ... The occupation of these levels, which determines the shape of the fluorescence spectrum, will then be proportional to the occupation of the corresponding levels of S_1 which determines $E(\lambda,T)$: in other words, the fluorescence spectrum $E(\lambda,T)$ will be proportional to $E(\lambda,T)$. Substitution in equation (1.1) gives:

$$F(\lambda,T) = c_1 P(\lambda,T)\,\alpha(\lambda,T) \qquad (1.2)$$

This so-called Stepanov relation gives an *exact relation* between *shapes* of absorption and the fluorescence spectrum of any molecule which shows fluorescence and in which equilibrium is obtained between the various vibrational states of S_1; c_1 depends on the internal conversion of the molecule and will be large for molecules with high fluorescence yield.

Applications to photosynthesis have been made by Szalay and others (1967) and by Zankel, Reed and Clayton (1968). Perhaps the most useful application, similar to that of Zankel, is the following. If it is known that in a mixture of pigments only one fluoresces, then it is possible by means of equation (1.2) to calculate the absorption spectrum for those wavelengths for which the fluorescence spectrum of the pigment is known with sufficient precision. Attempts to check equation (1.2) with a solution of one substance have probably not always been satisfactory, because small amounts of absorbing impurities may cause errors at longer wavelengths where $\alpha(\lambda,T)$ is small and $P(\lambda,T)$ very large.

Another relation between the fluorescence spectrum and the absorption spectrum is of much less general validity and exactness, namely that of mirror symmetry (Förster, 1951). Whether conditions for mirror symmetry exist can at present only be discovered experimentally.

Action spectra and quantum yield

Introduction

An action spectrum should provide information for determining the identity and efficiency of pigments which bring about a certain photoreaction. Such photoreactions are, for example, phototaxis or the rate of uptake of carbon dioxide in photosynthesis. An action spectrum is most useful if it is defined and determined in such a way that it is proportional to the sum of the absorption spectra of the pigments that bring about the photoreaction, each multiplied with an efficiency factor. A pigment may participate directly in the primary reaction or it may transfer the excitation energy to the reaction centre. Many 'action spectra' reported in the literature are not proportional to the absorption spectrum of the active pigments, because certain corrections are omitted, or because the activity is influenced in different ways by two or more primary photoreactions with different action spectra. Such action spectra are at best of only qualitative significance. Precise action spectra can be determined by relatively simple and general procedures if the system is weakly absorbing or if the response is linear with intensity over a certain range of intensities, and if only one primary reaction influences the effect. If more than one photoreaction are present the disturbing reaction(s) may be eliminated, for example, by means of inhibitors; or by using a background light the disturbing reactions may be run at such a high rate that only the action spectrum of one rate limiting photoreaction is measured. Only these cases, for which the conditions mentioned in this paragraph are fulfilled, will be considered in the following.

Weakly absorbing systems

The light intensity, $I(\lambda)$ in this case, does not vary with the depth of penetration of the incident light. Let the effect or action be $a[I(\lambda)]$. This action is not assumed to be linear with intensity. The action a is, for example, the amount of oxygen produced per unit of time by a dilute suspension of algae upon illumination with light of intensity I of wavelength λ. The total action a_t can then be expressed as

$$a_t[I(\lambda)] = Vf[\alpha(\lambda) . I(\lambda)] \tag{1.3}$$

V is the volume of the cuvet and $\alpha(\lambda)$ is the absorption coefficient for the actively absorbed quanta; the number of actively absorbed quanta is then equal to $\alpha(\lambda) . I(\lambda)$, and the action is a function of this number, as expressed by the equation. If we now adjust the intensities $I(\lambda_1)$, $I(\lambda_2)$ and $I(\lambda_j)$ in such a way that

$$a_0 = a[I(\lambda_1)] = a[I(\lambda_2)] = a[I(\lambda_j)]$$

then it follows, if $f[u]$ is an increasing function of u, that the corresponding values of u are also equal, or

$$\alpha(\lambda_1) . I(\lambda_1) = \alpha(\lambda_j) . I(\lambda_j) = K$$

and

$$\alpha(\lambda_j) = K/I(\lambda_j) \tag{1.4}$$

In these equalities K is a proportionality constant.

If we define the action spectrum as a spectrum obtained by plotting $1/I(\lambda_j)$ against wavelength, then this action spectrum is, according to the foregoing equations, proportional to the active absorption spectrum $\alpha(\lambda)$, but only when the suspension is weakly absorbing. If only one active pigment is present with the absorption coefficient $\beta_1(\lambda)$, then

$$\alpha(\lambda) = \phi_1\beta_1(\lambda)$$

in which ϕ_1 is an efficiency factor. If more active pigments are present, transferring excitation energy to the photoreaction, then

$$\alpha(\lambda) = \sum_k\phi_k\beta_k(\lambda)$$

in which the β_ks are the absorption coefficients of the pigments and the ϕ_ks the efficiency factors. We will describe how the ϕ_ks can be determined in the text below.

In practice it is desirable to determine at least part of the intensity curve for certain critical wavelengths. The intensity curve is obtained by plotting the action $a_t(\lambda)$, e.g. the rate of oxygen production, against the intensity of the actinic light of wavelength λ. From equation (1.3) it follows that the intensity curve at another wavelength is obtainable by multiplying all abscissa values of the first curve with a certain number. We may say that the intensity curves are proportional to each other (in the direction of the abscissa). In Figure 1.2(A), in which we have given an example of

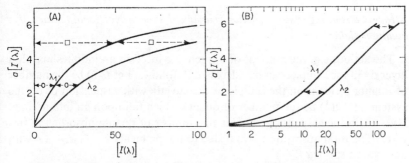

Figure 1.2 Hypothetical non-linear intensity curves of a weakly absorbing sample, giving the activity of a photoreaction as a function of intensity at two wavelengths, for which the ratio of activities is 2

intensity curves at two wavelengths, the proportionality factor is 2. If we can assume that at another wavelength, λ_3, the intensity curve is proportional to that at λ_1, then it suffices to determine the action at λ_3 for only one intensity. The ratio $\alpha(\lambda_3)/\alpha(\lambda_1)$ is then equal to $I(\lambda_1)/I(\lambda_3)$, in which $I(\lambda_1)$ is the intensity read from the intensity curve for λ_1 at which the action is the same as that measured at λ_3.

If the action is the result of two photochemical reactions, or if the suspension is strongly absorbing, then the measured curves in general will not be proportional to each other. The shape of the action spectrum would then depend upon the shapes of the action spectra of the two photo-reactions and would be a function of intensity; in the case of a strongly absorbing suspension it would be a function of the concentration of the suspension. Proportionality of intensity curves can be checked by plotting the action against ln $I(\lambda)$, or by plotting $I(\lambda)$ on graph paper with logarithmic divisions along the abscissa (see Figure 1.2(B)). The curves for various wavelengths should then be superimposable by moving the curves along the abscissa. This can be checked, and averaging to determine the proportionality constant or action ratios $\alpha(\lambda_j)/\alpha(\lambda_1)$ can be done more precisely by drawing the curves on transparent paper, or more quantitatively by using methods of numerical analysis and a computer.

A simpler procedure is possible if the action is a linear function of the intensity: then straight lines through the origin are found if the action is plotted against intensity.

If the action is an all or none phenomenon, such as seeing a light flash as a function of its intensity and wavelength, we may experimentally determine the probability of the occurrence of the phenomenon, which is taken as the action, and use similar procedures as above.

Strongly absorbing systems; action is assumed to be proportional to intensity; quantum yield

The action, e.g. the amount of oxygen produced in an infinitesimal thin layer dx in the sample, can be calculated as follows. Let $I_0(\lambda)$ be the number of quanta incident on the sample or the cuvette with suspension or on the system, and $I(\lambda)$ be the number of quanta which fall upon an infinitesimal layer dx in the suspension; $-\mathrm{d}I$ is the number of quanta absorbed in this layer. The total absorption $\beta(\lambda)$ is due to the pigments $1, \ldots k, \ldots n$ with absorption spectra $\beta_k(\lambda)$

$$\beta(\lambda) = \Sigma k \beta_k(\lambda)$$

Pigment k absorbs $\{\beta_k(\lambda)/\beta(\lambda)\}$ $(-\mathrm{d}I)$ quanta, since β_k/β is the fraction absorbed by pigment k. The action da_k produced by the quanta absorbed by pigment k is

$$\mathrm{d}a_k = \phi_k \beta_k(\lambda) \, (-\mathrm{d}I)/\beta(\lambda)$$

in which ϕ_k is a proportionality constant, which is independent of λ and of I. We will call ϕ_k the *quantum yield* for light absorbed by pigment k. It is indeed the yield or action produced per quantum absorbed by pigment k. If pigment k would be the only pigment, then this definition for quantum yield is the same as that usually given. The total action a_t produced by all pigments and all layers is found by summing over all pigments and layers:

$$a_t[I_0(\lambda)] = \{\Sigma\phi_k \, \beta_k(\lambda)\} \, \Sigma(-dI)/B(\lambda) \tag{1.5}$$

$\Sigma(-dI)$ is the total amount of light absorbed by the suspension or system. If the light incident on the suspension is $I_0(\lambda)$, and $A(\lambda)$ is the absorptance, we can write

$$(\Sigma - dI) = I_0(\lambda) \, . \, A(\lambda) \tag{1.6}$$

Substituting equation (1.6) in equation (1.5) and rearranging, we get

$$\Sigma\phi_k\beta_k(\lambda) = \frac{a_t[I_0(\lambda)] \, . \, \beta(\lambda)}{I_0(\lambda) \quad . \, A(\lambda)} \tag{1.7}$$

If only the light absorbed by pigment k is active, then

$$\alpha(\lambda) = \phi_k\beta_k(\lambda)$$

would be the action spectrum as usually defined. If more than one pigment is active, the expression on the left-hand side of equation (1.7) is the appropriate generalization of the definition, since it is equal to the sum of the absorption spectra of the active pigments weighted with an efficiency factor ϕ_k:

$$\alpha(\lambda) = \Sigma\phi_k\beta_k(\lambda) \tag{1.8}$$

The action spectrum $\alpha(\lambda)$ is proportional to the concentration of the suspension, since the β_ks are proportional to this concentration. From equations (1.7) and (1.8) it follows that

$$\alpha(\lambda) = \frac{a_t(\lambda) \, . \, \beta(\lambda)}{I_0(\lambda) \, . \, A(\lambda)} \tag{1.9}$$

Equation (1.9) makes it possible to calculate the action spectrum $\alpha(\lambda)$ from the experimental data. If $\alpha(\lambda)$ is known, and if the spectrum of the suspension, $\beta(\lambda)$, can be analysed in terms of the absorption spectra, $\beta_k(\lambda)$, of its pigments ($\beta(\lambda) = \Sigma\beta_k(\lambda)$, in other words, if the $\beta_k(\lambda)$ are known), then the factors ϕ_k can be calculated by means of equation (1.8). If there are n unknown factors ϕ_k, n values of λ are substituted in equation (1.8), and the n resulting equations can be solved for the n unknown ϕ_ks. The quantum yields or efficiencies of the pigments are the most important data obtainable from the action spectrum. If $I(\lambda)$ and/or $\alpha_t(\lambda)$ are measured in relative units, then the ϕ_ks are relative efficiencies, which are proportional to the quantum efficiencies.

Simplifications are possible in the following cases:

(1) Weakly absorbing suspension; $\beta(\lambda)$ and $A(\lambda)$ small. Then, if $\beta(\lambda)$ is defined as $\beta(\lambda) = \ln I_0(\lambda)/I(\lambda)$,

$$\beta(\lambda)/A(\lambda) = \beta(\lambda)/\{1 - \exp[-\beta(\lambda)]\} = \beta(\lambda)/\{1 - [1 - \beta(\lambda) + \ldots]\} = 1$$

or in first approximation

$$\alpha(\lambda) = a_t(\lambda)/I_0(\lambda) \tag{1.10}$$

This is a special case of that treated on page 10.

(2) Light absorption is complete. Then $A(\lambda) = 1$ and

$$\alpha(\lambda) = a_t(\lambda) \cdot \beta(\lambda)/I_0(\lambda) \tag{1.11}$$

The methods to be used to determine the total action a_t are different for different types of action. For example, if oxygen is measured by means of an oxygen electrode, the total amount may be measured by stirring the suspension, so that the oxygen is homogeneously distributed. If the reaction has to be measured rapidly, stirring may not be possible, and special experimental and mathematical methods may have to be devised to determine the total action.

Highly absorbing systems with non-linear response

Summarizing the last two sections, we can state that if the action is a linear function of intensity, or $a_t(\lambda) = c(\lambda) I_0(\lambda)$, equation (1.9) can be used to determine the action spectrum. If the action is not linear with intensity, the method for weakly absorbing systems can be used, if the system studied has little absorption.

When the response is not linear, $\alpha(\lambda_1)$ at a wavelength λ_1 of high absorbancy may be determined from the measured absorbancy $\beta(\lambda_1)$ and the total action $a_t(I_0, \lambda_1)$ as follows. In a layer dx at distance x from the illuminated surface, the action $a(I_0, \lambda_1)$ is a function of the light intensity and the active absorption:

$$a(I_0, \lambda_1) = F[I_0(\lambda_1) \exp\{-\beta(\lambda_1)x\} \cdot \alpha(\lambda_1)]$$

Integrating over all layers we obtain the total action:

$$a_t(I_0, \lambda_1) = \int a(I_0, \lambda_1) \, dx = \int F[I_0 \exp\{-\beta(\lambda)x\} \cdot \alpha(\lambda_1)] \, dx \tag{1.12}$$

The function, $F[u]$, is the intensity curve, which, except for a proportionality factor in u, can be found from measurements at wavelengths of low absorption. We assume that such measurements are possible. We calculate the integral of equation (1.12) for a number of values $\alpha(\lambda_1)$. The $\alpha(\lambda_1)$ required is that which gives the same value of $a_t(I_0, \lambda_1)$ as that which is measured. The $\alpha(\lambda)$'s for various wavelengths can be found in this way except for a proportionality constant. Preliminary calculations suggest that a more general procedure is possible, but this has not been worked out.

In the foregoing we have assumed that the pigments in the sample or suspension are homogeneously distributed, or more precisely that no highly absorbing particles are present. If that is not the case, corrections for flattening of the absorption spectra should be applied (Duysens, 1956). Additional corrections are needed if scattering occurs (Amesz, Duysens and Brandt, 1961; Latimer, Moore and Dudley Bryant, 1968).

Final remark

Sometimes it is asked whether action and absorption spectra should be plotted as absorptance (in per cent absorption), or absorbancy ($\ln I_0/I$). In this paper absorbancy has been used. For all cases, but especially for suspensions with non-negligible absorption, the equations are simplest, and the figures are compared most readily by plotting against absorbancy. All spectra then are the sum of absorbancy spectra of the pigments present multiplied by certain factors, which is not the case for absorptance spectra. If the absorption is negligible, the absorptance is approximately proportional to the absorbance, and plotting as for absorptance would be acceptable.

REFERENCES

Amesz, J., L. N. M. Duysens, and D. C. Brandt (1961) Methods for measuring and correcting the absorption spectrum of scattering suspensions. *J. Theoret. Biol.*, **1**, 59.

Beugeling, T. (1968). Photochemical activities of $K_3Fe(CN)_6$-treated chromatophores from *Rhodospirillum rubrum*. *Biochim. Biophys. Acta*, **153**, 143.

Clayton, R. K. (1967). The bacterial photosynthetic reaction center. In *Energy Conversion by the Photosynthetic Apparatus*, Brookhaven Symposia in Biology Number 19, Brookhaven National Laboratory, Upton, New York. pp. 62–70.

Duysens, L. N. M. (1956). The flattening of the absorption spectrum of suspensions, as compared to that of solutions. *Biochim. Biophys. Acta*, **19**, 1.

Duysens, L. N. M. (1958). In *The Photochemical Apparatus; Its Structure and Function*, Brookhaven Symposia in Biology Number 11, Brookhaven National Laboratory, Upton, New York. pp. 10–25.

Duysens, L. N. M. (1964). Photosynthesis. *Progr. Biophys. Molecul. Biol.*, **14**, 1.

Duysens, L. N. M. (1965). On the structure and function of the primary reaction centers of photosynthesis. *Arch. Biol. (Liège)*, **76**, 251.

Duysens, L. N. M., W. J. Huiskamp, J. J. Vos, and J. M. van der Hart (1956). Reversible changes in bacteriochlorophyll in purple bacteria upon illumination. *Biochim. Biophys. Acta*, **19**, 188–190.

Duysens, L. N. M., and H. E. Sweers (1963). In Jap. Soc. Plant Physiologists (Ed.), *Studies on Microalgae and Photosynthetic Bacteria*. Special issue of *Plant and Cell Physiology*, University of Tokyo Press, Tokyo. pp. 353–372.

Fleischman, D. E. (1968). The effect of phosphorylation uncouplers and electron transport inhibitors upon spectral shifts and delayed light emission of photosynthetic bacteria. *Photochem. Photobiol.*, **8**, 287.

Förster, Th. (1948). Zwischenmolekulare Energiewanderung und Fluoreszenz. *Ann. Physik*, **55**, 55.

Förster, Th. (1951). *Fluoreszenz Organischer Verbindungen*, Vandenhoeck and Ruprecht, Göttingen.

Ke, B. (1969). Nature of the primary electron acceptor in bacterial photosynthesis. *Biochim. Biophys. Acta*, **172**, 583.

Knox, R. S. (1968). On the theory of trapping of excitation in the photosynthetic unit. *J. Theoret. Biol.*, **21**, 244.

Knox, R. S. (1969). Thermodynamics and the primary processes of photosynthesis. *Biophys. J.*, **11**, 1351.

Latimer, P., D. M. Moore, and F. Dudley Bryant (1968). Changes in total light scattering and absorption caused by changes in particle conformation. *J. Theoret. Biol.*, **21**, 348.

Murrell. J. N. (1963). *The Theory of the Electronic Spectra of Organic Molecules*, Methuen and Co. Ltd., London and Wiley and Sons Inc., New York.

Parson W. W. (1968). The role of P870 in bacterial photosynthesis. *Biochim. Biophys. Acta*, **153**, 248.

Rabinowitch, E. (1951). Photosynthesis. *Ann. Rev. Phys. Chem.*, **2**, 361.

Ross, R. T., and M. Calvin (1967). Thermodynamics of light emission and free energy storage in photosynthesis. *Biophys J.*, **7**, 595.

Seliger, H. H., and W. D. McElroy (1965). *Light: Physical and Biological Action*, Academic Press, New York, London.

Seliger, H. H., and R. A. Morton (1968). In A. C. Giese (Ed.), *Photophysiology. Current topics*, Vol. IV. Academic Press, New York, London. pp. 253–314.

Strehler, B. L., and W. Arnold (1951). Light production by green plants. *J. Gen. Physiol.*, **34**, 809.

Szalay, L., E. Rabinowitch, N. R. Murty, and Govindjee (1967). Relationship between the absorption and emission spectra and the 'red drop' in the action spectra of fluorescence *in vivo*. *Biophys. J.*, **7**, 137.

Thomas, J. B. (1965). *Primary photoprocesses in biology*. North Holl. Publ. Cy., Amsterdam, and Wiley and Sons Inc., New York.

Vredenberg, W. J., and L. N. M. Duysens (1963). Transfer of energy from bacteriochlorophyll to a reaction center during bacterial photosynthesis. *Nature*, **197**, 355.

Zankel, K. L., D. W. Reed, and R. K. Clayton (1968). Fluorescence and photochemical quenching in photosynthetic reaction centers. *Proc. Natl Acad. Sci. U.S.*, **61**, 1243.

CHAPTER 2

The photosynthetic apparatus of microalgae and its adaptation to environmental factors

PER HALLDAL

Department of Biology, University of Umeå, Sweden

Introduction

In the ingenious experiments by Engelman, the results of which were published in 1881 and 1882, aerotactic bacteria were used to demonstrate oxygen evolution in filiform algae. By exposing different parts of a cell of the green alga *Spirogyra* to a tiny light spot, Engelman found that photosynthesis took place only in the chloroplast. A projected spectrum of light from a prism gave distinct accumulation of bacteria around 450 and 680 nm in green algae. The conclusion was drawn that the light which drives photosynthetic oxygen evolution was absorbed by chlorophyll. The minimum in green and yellow light observed for green algae was absent in diatoms and blue-greens, indicating for the first time the participation of accessory pigments (the carotenoid fucoxanthin for the diatoms, and phycobilins for the blue-green algae). These simple experiments, which gave such important and fundamental information, did not have the high precision that is required today for analyses of action spectra of photosynthesis. The coloured light was impure, and the projected spectrum had different intensities at different wavelengths and was also biased in wavelength dispersion. However, the experiments of Engelman have been repeated on many occasions with more refined methods and instruments, and his general conclusions have been confirmed.

In experiments aimed at the detection and further analysis of the effectiveness of photosynthetic active pigments, it is compulsory to perform some sort of action spectra determinations and/or quantum yield measurements. The determinations of *in vivo* absorption spectra of chloroplasts or photosynthetic lamella systems are as a rule not sufficient. In many cases *in vivo* absorption characteristics of living algae may be used as a means of predicting the photosynthetic spectral response with a fair degree of precision. This is notably true for green and for brown algae, and for the diatoms and dinoflagellates (Haxo, 1960; Haxo and Blinks, 1950; Tanda, 1951). For red and for blue-green algae, this is not the case (Duysens, 1952; Haxo, 1960; Haxo and Blinks, 1950). For these groups of algae the absorption characteristics greatly deviate from the photosynthetic spectral response. Chlorophyll seems to have little effect, and the carotenoids are evidently completely inactive in photosynthesis of red and of blue-green algae, while the phycobilins are of main importance.

In living plants the photosynthetic active pigments are localized in membrane or lamellar systems whose building elements are orderly arranged, the photosynthetic pigments being closely associated with proteins and phospholipids. It has not been possible by standard enzymatic preparative methods to isolate a distinct chlorophyll–protein complex; one is therefore inclined to assume that the pigment–macromolecule association in the lamellae is of a different nature from that in enzymes. Several models of aggregations of pigment molecules in lipid–protein membranes have been proposed.

When extracted from the thylakoid membrane with organic solvents chlorophylls and some carotenoids undergo pronounced absorption changes. For chlorophyll *a* the *in vivo* absorption peak is around 675 nm, while the maximum absorption in ether occurs at 662 nm. During the last 10 to 15 years it has been evident through a number of experiments that several *in vivo* chlorophyll *a* forms exist. It has also been established that these specific forms serve different functions in the process that captures light and transforms quantum energy into high energy bonds in organic substances. The presence of at least six such *in vivo* forms of chlorophyll *a* has been demonstrated; they can be identified and characterized through their different red absorption maxima. When extracted, the differences disappear. The different *in vivo* chlorophyll *a* forms are distinguished through their red absorption maximum (see page 24). One should, however, be somewhat cautious to fix a certain form to a distinct function in photosynthesis, as it has been demonstrated that in the red region of the spectrum both the absorption characteristics and the photosynthetic response of algae may be altered through different environmental conditions (Halldal, 1968; Öquist, 1969). Different forms of chlorophyll characterized by their red absorption maxima may thus represent substances which serve the same function.

In the structural analyses of photosynthetic organelles and the membrane systems, chemical and biophysical investigations should be combined and coordinated. The value of this experimental approach is demonstrated, for instance, by the work of Moudrianakis, Howell and Karu (1968).

The chemical properties of the extractable pigments of the photosynthetic apparatus are fairly well known. The structures of chlorophyll *a* and *b* have been known for many years. We also know that different chlorophyll *c* can be extracted from certain algae (Dougherty and coworkers, 1966; Jeffrey, 1968b). The chemical structure of the accessory carotenoid fucoxanthin has been known for several years (Jensen, 1964), while that of peridinin is still unknown. We also know a good deal about the chromophore groups of different biliproteins (Ó hEocha, 1966). However, we are only at the very beginning of the process of identification and elucidation of the arrangement of different elements in the photosynthetic membranes. The synthesis of chlorophyll *a* was completed in 1960 by Woodward and coworkers (for references also see Lwowski, 1966). It is a long and difficult way from this most elegant synthesis of a very complex chemical molecule to the 'construction' of the membrane system which will function in photosynthesis. Today we see only the beginning of this process as we attempt to take apart and understand the function of the different components. Tomorrow it may be possible to put these components back into place or even to synthesize new ones.

During the planning of this book it was decided, with a few exceptions, to exclude any extensive treatment of photosynthesis. This may seem somewhat unfair because photosynthetic bacteria and microalgae have contributed so much to our present knowledge on the photosynthetic process. Consequently space has been reserved for brief summaries on the photosynthetic process of these groups. They should be considered as a general survey. For the microalgae, emphasis will be given to the flexibility of the photosynthetic apparatus, which has not been treated very well in recent reviews of photosynthesis. We feel that this approach is justified, as excellent survey articles written by some of the most competent workers in the field, have recently appeared, (Goodwin (Ed.), 1965, 1966, 1967; Shibata and coworkers (Eds.), 1968; Vernon and Seely (Eds.), 1966).

Morphology and structure of the photosynthetic apparatus

A recent review of the structure of the photosynthetic apparatus has been given by Branton (1968). Textbooks on botany carefully describe and emphasize the great variability of chloroplast morphology in algae. In addition the distinction between bacterial 'chromatophore', the 'chromato-plasm' of the blue-green algae and the chloroplasts of algae and higher plants is clearly pointed out. The terms 'chromatophore' and 'chromato-plasm' seem to have disappeared in more recent literature. Today one is more inclined to describe different types of arrangements of photosynthetic lamellae. The description of chloroplasts has unfortunately been biased to some degree, due to the extensive analyses of spinach chloroplast and related types. When reading modern literature on photosynthesis one is often left with the impression that the spinach chloroplast is a universal

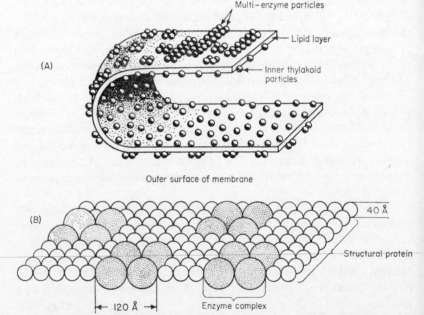

Figure 2.1 (A) Model of the thylakoid membrane. After Mühlethaler, 1966. (B) Detail of the membrane. After Mühlethaler, 1970

model for this subcellular photosynthetic light receptor in all plants, which is by no means true.

In all algal groups and in higher plants, the photosynthetic apparatus contains a lamellar system built up of units called thylakoids (sack-like), a term introduced by Menke (1962). A thylakoid consists of two parallel membranes enclosed in themselves (Figure 2.1A). One view is that the membranes are formed by regularly arranged protein molecules, termed 'structural protein' by Menke. These structural protein units form regular sheets in which large enzyme complexes are inserted (Figure 2.1B). The lipid portion of the membranes is assumed to be packed in cavities within the structural protein (Mühlethaler, 1970). A different opinion is advocated by Heslop-Harrison (1962, 1963), Wehrmeyer (1963, 1964a, 1964b) and Weier (1961). These authors do not consider the granum as a pile of closed discs adhering by their faces. Their view is that the whole lamellar system within the chloroplast constitutes a single elaborated complex membrane-bounded cavity embedded in the stroma (Figure 2.2).

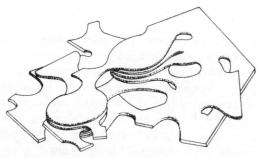

Figure 2.2 A model presenting the lamellar system in chloroplasts as a single complex membrane-bounded cavity embedded in the stroma. After Wehrmeyer, 1964b

Park and Pon (1961) described distinct morphological structures containing four subunits, which they called 'quantasomes'. Similar structures were described by Mühlethaler though they were not identical in size (Mühlethaler, 1966; Park and Biggins, 1964). Park and collaborators consider that the 'quantasomes' are embedded in the thylakoid membrane, and that they represent the 'photosynthetic unit' of Emerson and Arnold (1932), that is, the smallest unit that can complete the photosynthetic process. Mühlethaler favours the opinion that they represent subdivisions of the thylakoid membrane which have been modified in structure in the course of preparation. In later analyses Park and Shumway (1968) claim additional support for the Park–Branton model (Park and Branton, 1966). Particles bound to chloroplast lamellae which appear to contain four subunits and presumably are identical with the 'quantasomes' were analysed by Moudrianakis, Howell and Karu (1968). These authors, however, could not find any evidence for their function as the 'photosynthetic unit'; neither did they appear to participate in the electron transport chain in photosynthesis. Moudrianakis, Howell and Karu came to the conclusion that the photochemical reaction in photosynthesis appeared to be a total membrane phenomenon, and that the membrane-bound particles were the site of carboxydismutase, and of calcium- and magnesium-activated ATPases and photophosphorylase. Such analyses support the view of Mühlethaler (1966) that these units are multienzyme particles.

Ji, Hess and Benson (1968) and Weier and Benson (1966) give an extensive literature analysis of the molecular structure of thylakoid membranes. Through experiments with polarized light, X-ray diffraction analyses and studies with the electron microscope, they deduced a model including the site and orientation of photosynthetic pigments and electron carriers. This model is much more advanced and sophisticated than any of the previous ones.

The thylakoids, with the presumably multienzymatic particles embedded in the membranes, are arranged differently in different classes of algae. It was formerly assumed that the photosynthetic pigments of prokaryotic organisms (e.g. blue-green algae) were dispersed throughout the 'chromatoplasm'. When the microstructure of prokaryotic organisms was studied in the electron microscope it was found that the 'chromatoplasm' contained an extensive system of individually dispersed thylakoids. Similar thylakoid arrangements are also found in the unicellular red alga *Porphyridium* (Brody and Vatter, 1959).

In all eukariotic algae the thylakoids are localized in the chloroplasts. In most of these, as for instance in red and brown algae and in diatoms and dinoflagellates, the subchloroplast grana are lacking. Manton (1966) used *Chlorella* as a distinction between 'lower' and 'higher' plants. According to her view 'higher' plants have grana and chlorophyll *b*, and 'lower' plants have grana-free chloroplast or free thylakoids and accessory pigments different from chlorophyll *b*.

Photosynthetic active pigments

Chlorophylls

All photosynthetic plants contain chlorophyll *a* which participates in the photochemical step of the photosynthetic reaction chain that converts energy contained in quanta into energy-rich chemical bonds. In photosynthetic bacteria other chlorophylls are active in this process (see Sybesma, Chapter 3 of this treatise). Extracted and purified chlorophyll *a* is identical in chemical and physical characteristics, irrespective of the plant source. Its chemical structure was elucidated through the work of several investigators (e.g. Willstätter, Stoll and Fischer). Fischer and coworkers made important contributions to the synthesis of components related to chlorophyll *a*, and its synthesis was completed in 1960 by Woodward and coworkers. Chlorophyll *b* is very similar to chlorophyll *a*. The only difference is that in the former an aldehyde replaces a methyl group. Chlorophyll *c* has been known for many years to be present in different algae, and like chlorophyll *b* its participation as an accessory pigment during photosynthesis has been demonstrated through quantum yield measurements and action spectra determination (Haxo, 1960; Haxo and Blinks, 1950; Tanda, 1951: Halldal, 1968, 1969). Two different chlorophyll *c* (c_1 and c_2) have been detected by Dougherty and coworkers (1966) in their chlorophyll *c* preparations from diatoms. These authors also suggested their structures based on comparative evidence; however, the different components were not isolated. Jeffrey (1968b) separated two spectrally different chlorophyll *c* components on polyethylene layers, which possibly correspond to the two components discovered by Dougherty and coworkers. Further analyses are still needed to arrive at the final confirmative evidence in this situation. Chlorophyll c_1 and c_2 are found in brown algae, diatoms and crysomonads, while dinoflagellates and cryptomonads contain only chlorophyll c_2. Chlorophyll *d* is found in small amounts in some red algae, but not in all (Manning and Strain, 1943; Smith and Benitez, 1955). Its structure has also been determined (Holt, 1965, 1966) and may be formed through an oxidation of chlorophyll *a*. *In vivo* absorption measurements never disclose the presence of chlorophyll *d*, and an increased photosynthetic rate in far red is never observed. As its red absorption maximum occurs at longer wavelengths than that of chlorophyll *a*, excitation energy transfer from chlorophyll *d* to chlorophyll *a* cannot take place. Its possible function in photosynthesis is entirely unknown. Chlorophyll *e* has been reported to be present in *Xanthophycea* and *Chrysophyceae* (Egle, 1960). Its structure and possible function in photosynthesis are unknown.

Since the middle part of the 1950's it was evident that different chlorophyll *a* forms exist in living plants. Data for such a conclusion were available both from absorption spectra measurements, and also from photophysiological analyses (for reviews see Butler, 1966; Smith and French, 1963). These different chlorophyll *a* forms may be observed directly in a spectrophotometer by the use of certain techniques (Halldal,

1958; Krasnovsky and Kosobutskaya, 1955; Thomas, 1962; Vorob'eva and Krasnovsky, 1956). These *in vivo* forms may be studied more clearly in instruments which record the first derivative of the absorption curve (Cramer and Butler, 1968; French, 1958; Halldal, 1969). Studies which directly led to the discovery of a particular chlorophyll *a* form were those by Kok and Hoch (1961), in which selective wavelength excitation was applied. In their measurements, reversible absorption changes occurred around 700 nm. They introduced the symbol P 700 for this particular chlorophyll *a* form and demonstrated further that it was an electron-carrier during photosynthesis. The number of different chlorophyll *a* forms has not been agreed on. Two forms are readily seen around 675 nm. They have been denoted differently by different authors. In this article they will be called chlorophyll *a* 670 (Chl 670) and chlorophyll *a* 682 (Chl 682). Another form with a red absorption peak at 695 was observed by French (1959), and by Brown and French (1961) from *Euglena* cells grown at low light levels. Further information of different chlorophyll forms in living cells has been reported by Allen, French and Brown (1960), and Govindjee (1963) in the blue-green algae *Anacystis*, by Govindjee, Cederstrand and Rabinowitch (1961) in different algae, and by Lippincott and coworkers (1962) in higher plants.

The investigations on *in vivo* forms of chlorophyll *a* are still in progress. A few recent observations may be mentioned. Halldal (1968) observed a chlorophyll *a* form which participated in the process of photosynthetic oxygen evolution from a green algae (Siphonales, *Ostreobium*), living inside massive corals in the Indo-Pacific. This chlorophyll had an absorption peak in the far red at 725 nm and was called Chl 725. Öquist (1969) further showed that it is possible to induce photosynthetic active chlorophyll *a* forms with absorption peaks around 700 nm in *Chlorella* cells under extreme far red light conditions.

It is generally assumed that all these different chlorophyll *a* forms which may be observed in living material have an identical chromophore, namely chlorophyll *a* absorbing at 662 nm in ether. The differences are supposed to be due to different chlorophyll *a*–macromolecular associations, or to different ways of aggregation in submicroscopical structures of the thylakoid membranes.

In order to analyse the function of these different chlorophyll forms it is necessary to perform high-precision photophysiological and spectrophotometric measurements on intact material and on isolated fractions of the photosynthetic apparatus. Below will be mentioned a few examples of such experiments.

In 1943 Emerson and Lewis discovered the low efficiency of long-wavelength red light in *Chlorella* photosynthesis. This led to the assumption that plants contained active and inactive chlorophyll *a* components. However, this discrepancy between quantum efficiency and algal absorption was explained by Emerson (1958) and by Emerson, Chalmers and Cederstrand (1957), when it was demonstrated that the efficiency of long-wavelength light was significantly raised by simultaneous irradiation with shorter wavelengths. This phenomenon has been designated the second Emerson effect. (The first Emerson effect is the carbon dioxide burst that occurs at the beginning of a light period during photosynthesis.)

A reaction type related to the second Emerson effect was discovered by Blinks (1957, 1959, 1960). This is called the Blinks chromatic transient effect and may be observed when the photosynthetic apparatus is exposed to sudden shifts between beams of different wavelengths, the intensity of which has been adjusted to give the same photosynthetic rate. Blinks observed that the amount of photosynthetically produced oxygen was dependent on combinations of wavelengths, and on which one was given first. These types of experiments support the idea that during photosynthesis involving oxygen evolution two different pigments, or pigment systems, cooperate. The same conclusion was arrived at by French, Myers and McLeod (1960) during action spectra determinations of the second Emerson effect. These experimental data fit a hypothesis formulated by Hill and Bendall (1960). They assumed that if cytochromes are directly involved in electron transfer in photosynthesis where oxygen is evolved and NADP is reduced, two light reactions would be required. This so-called Hill–Bendall model for electron transport during photosynthesis with oxygen evolution has been extended and modified during the last few years. In principle it involves two light reactions working in series. This model is accepted by the majority of workers in the field, though others, e.g. Arnon and coworkers (1965, 1968), favour a hypothesis which involves two light reactions operating in parallel.

By the irradiation of the photosynthetic apparatus with specific wavelengths, or by the supply or withdrawal of electrons through the additions of artificial electron donors or acceptors, electron carriers in the photosynthetic apparatus may be kept in reduced or oxidized form. In this way it is possible to map them according to a reduction–oxidation potential scale. Based upon data from such analyses, and assuming a two-light reaction system which operates in series, we might consider the mechanism behind photosynthesis as an electron pump with light as the driving power. This pump captures electrons from water, passes them over certain electron carriers to a final reduction of NADP, which then, together with photosynthetically formed ATP, are fed into a biochemical dark-reaction system

2

which reduces carbon dioxide to carbohydrates and to other energy-rich organic components. The model presented in Figure 2.3 involves the participation of five chlorophyll *a* forms, Chl 670, Chl 682, Chl 705, P 700 and Chl 695. It involves the electron carriers plastoquinone, cytochrome

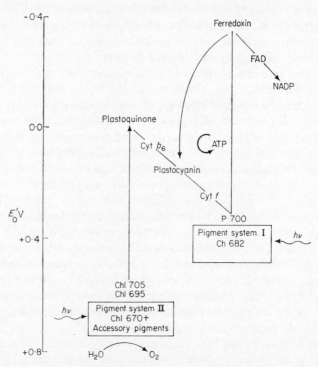

Figure 2.3 A simplified scheme for the electron transport chain during photosynthesis with oxygen evolution

b_6, plastocyanin, cytochrome *f*, P 700, ferredoxin, FAD and NADP. It also includes different methods of adenosine triphosphate (ATP)-formation. When the electrons are donated by water molecules, NADP is reduced to $NADPH_2$ according to the following scheme:

$$H_2O + NADP + P_i + ADP \xrightarrow{h\nu} NADPH_2 + \tfrac{1}{2}O_2 + ATP$$

This is called *non-cyclic photosynthetic phosphorylation*. During *cyclic photosynthetic phosphorylation* ATP is the only product. In this reaction chain only pigment system I is operating, and the electrons cycle from

Chl 682, P 700, over ferredoxin and back to Chl 682 via plastocyanin and cytochrome f:

$$ADP + P_i \xrightarrow{hv} ATP$$

During *pseudocyclic photosynthetic phosphorylation* FMN is first reduced with electrons donated by water. This FMN is then oxidized back again by the liberated oxygen:

$$H_2O + FMN + ADP + P_i \xrightarrow{hv} FMNH_2 + \tfrac{1}{2}O_2 + ATP$$
$$FMNH_2 + O_2 \longrightarrow FMN + H_2O_2$$
$$H_2O_2 \longrightarrow H_2O + \tfrac{1}{2}O_2$$

$$\overline{ADP + P_i \xrightarrow{hv} ATP}$$

Experiments which have led to the working model presented in Figure 2.3 have been performed by numerous investigators. References to these experiments are given in the survey books cited in the introduction to this chapter. A most interesting approach in the analysis of the electron transport chain during photosynthesis has been applied by Levine (1968). He used gene mutations of the unicellular alga *Chlamydomonas reinhardi* that cannot carry out normal photosynthesis because certain photosynthetic electron carriers are absent. By the use of selective inhibitors, excitation at specific wavelengths of light and artificial electron donors and acceptors, the electron pathway from water to NADP could be followed in great detail.

It is generally assumed that most, though possibly not all, of the accessory pigments are associated with Chl 670 in pigment system II. Several attempts have been made on a mechanical separation and concentration of thylakoid subunits, each of which contains pigments which participate in the two pigment systems. Different means of disintegration and separations are then followed by spectrophotometric analyses and tests for retained photochemical activity. Allen and coworkers (1963) and Allen and Murchio (1963) used a method which includes freezing, grinding and supersonic disintegration followed by density gradient centrifugation in chloroplasts from several algae. In another method used by Brown and Duranton (1964) tobacco chloroplasts were disintegrated with the detergent sodium dodecyl sulphate followed by mechanical treatment. In both these papers the conclusion is drawn that Chl 670 and chlorophyll *b* are physically associated in pigment system II, and Chl 682 in system I. This picture is also supported through experiments performed by Boardman and Anderson (1964) who removed Chl 682 from the combined pigment systems I and II, after treatment with the detergent digitonin. Thus, there is strong experimental evidence for the idea that one   the pigment systems (the one which involves Chl *a* 682) may be removed from the thylakoid photosynthetic system. This again suggests that the two pigment systems are morphologically separated, though conclusive support for this assumption has not been presented. Experimental results from the above and related experiments further indicate that Chl *a* 670 is morphologically associated with chlorophyll *b*, and that chlorophyll *b* functions as an accessory pigment in photosynthesis by transferring energy to Chl *a* 670. Analogous with this, one is inclined to assume that other accessory pigments (certain unknown carotenoids, fucoxanthin, peridinin, and the phycobilins) also function in this way during photosynthesis, though experimental data to support this idea do not exist.

In most models accessory pigments and Chl *a* 670 are placed in photosystem II, while Chl *a* 682, Chl *a* 705 and P 700 are supposed to be closely associated in system I. Based upon fluorescence analyses Brown (1967) favours another opinion. She assumes that Chl *a* 695 has similar properties to those of Chl *a* 705 observed by Butler (1960, 1961). This idea is also indicated in Figure 2.3, though the localization of the different chlorophyll *a* forms and their function during photosynthesis have by no means been settled.

If this model for pigment localization and energy transfer is accepted, five of the many *in vivo* chlorophyll *a* forms have been accounted for. However, one should be inclined to favour a high degree of flexibility for such a system. French (1959) and French, Myers and McLeod (1960) noted that the light conditions were of importance for the formation of Chl 695 in *Euglena* and *Ochromonas*, and Brown (1967) showed that this was also the case for the diatom *Phaeodactylum*. Halldal (1968) studied action spectra of photosynthesis and correlated the results with spectrophotometric analyses of *in vivo* samples of the green alga *Ostreobium*, which lives under most peculiar light conditions inside massive corals in the Indo-Pacific area (see page 36). In these experiments a chlorophyll *a* form with an absorption maximum at 725 nm (Chl 725) was recorded. If we accept the idea that the Kok pigment P 700 functions as an energy sink in pigment system I, and Chl 695 and Chl 705 operate in a similar way in pigment system II (see Figure 2.3), it is not possible to fit the photosynthetic active pigment Chl 725 into this scheme. As excitation of Chl 725 results in oxygen evolution, it should reasonably be placed in pigment system II. Because energy transfer from Chl 725 to Chl 695–705 or the Kok pigment P 700 is prohibited, the models presented in Figure 2.3 cannot be valid for photosynthesis in *Ostreobium*. One is therefore inclined to favour the assumption that the energy sink of system I, possibly also of system II, have undergone a general shift in absorption characteristics toward longer wavelengths. This hypothesis is supported by experimental data which show that both absorption characteristics and relative photosynthesis rates at different wavelengths around 700 nm may be altered through certain light treatment (Halldal, 1968; Öquist, 1969). Also, spectrophotometric analyses around 740 to 750 nm in different algae suggest the existence of certain chlorophyll forms around these wavelengths. The function of these forms has not been elucidated (Aghion, 1963; Gassner, 1962; Govindjee, 1963; Govindjee, Cederstrand and Rabinowitch, 1961; Lippincott and coworkers, 1962). Are they potential substitutes for P 700 or Chl 695–705?

Like chlorophyll *a*, chlorophyll *b* and *c* also undergo absorption changes when extracted. In chlorophyll *b* the *in vivo* absorption is around 650 nm (Yentsch and Guillard, 1969). When extracted and transferred to ether the red absorption maximum occurs at 642 nm. As the red absorption band of chlorophyll *b in vivo* overlaps with different chlorophyll *a* forms which also have several bands in this spectral region, it is difficult to analyse chlorophyll *b* for possible different forms in living algae. By direct optical measurements of the first derivative of the absorption from different plants in this spectral region, the shapes of the curves show great variations (French, 1958; Halldal, 1969). Due to interference with absorption bands of chlorophyll *a* a critical analysis of these curves has not been possible. The same situation exists for chlorophyll *c in vivo*.

Carotenoids

Action spectra analyses and quantum yield measurements clearly demonstrate that carotenoids participate in photosynthesis in most algal groups (Haxo, 1960). Important exceptions are found within the bluegreen and red algae where the carotenoids seem to be completely inactive as accessory pigments. The function of the often abundant carotenoid pigments of these algal groups has not been elucidated. As a general rule the carotenoid content in blue-green and red algae grown at high light levels increases simultaneously as the photosynthetic rate in blue light decreases and the photosynthetic capacity in general diminishes. Growth at low light level induces the opposite situation. The mechanism behind this effect is unknown.

The carotenoids might possibly serve as simple variable blue light filters. Some may function in the epoxide cycle in the deactivation of excited chlorophyll–oxygen complexes to prevent harmful photosensitized oxidation of chlorophyll. Several other possibilities also exist. This subject has been dealt with extensively by Krinsky (1968). Two identified carotenoids participate in photosynthesis; fucoxanthin in diatoms and brown algae and peridinin in the dinoflagellates. When extracted from living cells a distinct change in absorption characteristics to shorter wavelengths is observed. The absorption shifts are assumed to be caused through a release of carotenoids from macromolecules, presumably proteins or lipoproteins. This indicates that the carotenoids which participate as accessory pigments during photosynthesis must have the ability to form a carotenoid–macromolecular association. However, though this theory seems reasonable it does not have sufficient experimental support to be generally accepted.

The chemical constitution of fucoxanthin has been known for several years (Jensen, 1964). That of peridinin is unknown.

The principle for resonance energy transfer from one pigment to another has as one of the requirements that the emission spectrum of the donor molecule overlaps with the absorption spectrum of the acceptor. The efficiency of this transfer depends on the magnitude of overlap between the two spectra. It is extremely difficult to demonstrate any fluorescence from carotenoids in organic solvents. The situation in living algae seems to be somewhat different. Excitation spectra of chlorophyll *a* fluorescence in living plants show pronounced effects around 450 to 550 nm. This effect is particularly noticeable in measurements from dinoflagellates, where Shibata and Haxo (1970) demonstrated a maximum around 540 nm corresponding to a peak in the action spectrum of phtotosynthesis in this spectral region (Halldal, 1968).

Comparative analyses in living plant fractions and light petroleum-treated chloroplasts by Goedheer (1969) show that under these conditions energy transfer occurs from β-karotene to chlorophyll *a* in photosystem I in blue-green and red algae, while in green algae and greening leaves the transfer between β-karotene and chlorophyll *a* takes place in both photosystems I and II. No energy transfer between xanthophylls and chlorophyll *a* was observed.

In living algae the carotenoids are partly physically associated with thylakoids and function as accessory pigments exemplified by fucoxanthin and peridinin (Haxo, 1960). They may also be present in large amounts outside the chloroplast in the resting state in oil drops of the brine flagellate *Dunaliella salina* (Fox and Sargent, 1938). Blinks (1954) showed that blue light was completely inactive in the photosynthesis of such algae, which suggests that these dissolved carotenoids function as a yellow filter which prevents the blue light from reaching the chloroplast.

Table 2.1 Distribution of accessory pigments in different algal groups

Group	Chlorophylls	Carotenoids	Biliproteins	Reference[b]
Cyanophyta (blue-green algae)	a	—	phycocyanin, phycoerythrin, allophycocyanin?	(1), (2)
Rhodophyta (red algae)	a d?	—	phycocyanin, phycoerythrin, allophycocyanin?	(1), (2)
Cryptophyta	a c_2	yes	phycocyanin, phycoerythrin, allophycocyanin?	(1), (2)
Pyrrophyta (dinoflagellates)	a c_2	peridinin	—	(1), (3)
Bacillariophyta (diatoms)	a c_1, c_2	fucoxanthin	—	(1)
Phaeophyta (brown algae)	a c_1, c_2	fucoxanthin	—	(1)
Chrysophyta (silicoflagellates coccolithophorids)	a c_2	?	—	—
Xanthophyta	a	?	—	—
Euglenophyta	$a\,b$	—	—	—
Chlorophyta (green algae)	$a\,b$	yes[a]	—	(1)

[a] The carotenoid has not been identified.
[b] These literature references give review articles or more recent specific analyses: (1) Haxo, 1960; (2) Ó hEocha, 1966; (3) Halldal, 1966.

Biliproteins or phycobilins

Biliproteins are accessory pigments in blue-green algae, red algae and cryptophyta (see Table 2.1). These photosynthetic pigments have been dealt with extensively in several recent reviews (for references see Ó hEocha, 1966). The biliproteins are separated into two different groups, the red phycoerythrins which absorb maximally between 457 and 575 nm, with one, two or three absorption bands, and the blue phycocyanin with maximum absorption between 530 and 650 nm, with one or two bands. The different types of biliproteins are usually grouped according to their absorption characteristics and are denoted R-phycoerythrin, C-phycoery-thrin, B-phycoerythrin, R-phycocyanin, C-phycocyanin and allophy-cocyanin according to the number and po tion of peaks in the green to yellow portion of the spectrum (Haxo and Ó hEocha, 1960).

The biliproteins all have strong fluorescence. The red fluorescence of phycoerythrin has an emission maximum at 578 nm and the orange fluores-cence of phycocyanin and allophycocyanin in the spectral region between 630 and 690 nm.

The chromophore groups of biliproteins are tightly bound to their protein part. In fact this binding is so strong that it has been one of the main obstacles in the clarification of the structure of the pigment. This strong association also implies that no spectral shift, or only an insigni-ficant one, occurs when these water-soluble accessory pigments are removed from living algae.

Contradictory results are reported on the efficiency of excitation energy transfer between the protein and the chromophore group in biliproteins. Bannister (1954) reported constant quantum yield of C-phycocyanin fluorescence between 250 and 400 nm. This indicates around 100 per cent efficiency, while Eriksson and Halldal (1965) observed no energy transfer, or only an insignificant one, in phycoerythrin and phycocyanin isolated from the red algae *Porphyra*.

Action spectra measurements of blue-green algae, red algae and the chryptomonads clearly demonstrate that phycoerythrin and phycocyanin are accessory pigments with an energy transfer efficiency close to 100 per cent (Haxo, 1960).

The function of allophycocyanins is unknown, though analyses with automatic recordings of photosynthetic action spectra show details around 650 nm which point to their function as accessory pigments (Haxo and Halldal, and Halldal, unpublished results).

Are accessory pigments an absolute requirement for photosynthesis?

Normally chlorophyll *a* cooperates with accessory pigments during photosyntheses in all algae and higher plants. In green algae chlorophyll *b* is practically always present, and the blue-green and red algae with very few exceptions contain one or more biliproteins. The 'blue-green *Chlorella*', *Cyanidium caldarium*, isolated by Allen (1954) has an uncertain systematic position (see page 34). Volk and Bishop (1968) induced by ultraviolet radiation a mutant of *Cyanidium* which lacks phycocyanin. This mutant retained photosynthesis. The action spectra for quantum requirement were about the same for the mutant and for the wild type. Volk and Bishop further concluded that phycocyanin acts solely as an energy-gathering mechanism which increases the photosynthetic efficiency in the green to yellow portion of the spectrum. This seems to be experimental evidence indicating that the photosynthetic mechanism functions with high efficiency without the participation of accessory pigments. Unfortunately the action spectra measurements of Volk and Bishop for the quanta requirement are restricted to the spectral region between 560 and 750 nm. We therefore lack information about the response in blue light where the carotenoids absorb and are consequently not able to eliminate completely the possibility that carotenoids may substitute for phycocyanin as accessory pigments in the mutant, though the probability for the existence of such a mechanism admittedly is very low.

Several years earlier Allen (1958) had also studied photosynthesis in ultraviolet-induced mutants, but of the green algae *Chlorella* and *Nanochloris* which both lacked chlorophyll *b*. She concluded that certain mutants of both these genera could photosynthesize, though the efficiency was very low. These experimental analyses are important contributions toward the understanding of accessory pigments in photosynthesis. The analyses of Volk and Bishop are perhaps the most valuable, as they could demonstrate comparable photosynthetic rates in both the wild type and the mutant. *Cyanidium* is also a somewhat more interesting material than the green algae, as the latter has both chlorophyll *b* and carotenoid(s) as accessory pigments. After the elimination of chlorophyll *b*, carotenoid(s) might still function. In *Cyanidium* phycocyanin seems to be the only accessory pigment (Haxo, 1960). Allen concluded that chlorophyll *b* is involved in a mechanism which serves as a protection for high light level damage of the photosynthetic apparatus.

Adaptations to environmental conditions

The pigment composition of algae within different systematic groups has been subjected to extensive analyses. In most cases it is now possible, with a high degree of certainty, to predict which pigment will be found in a specimen that has been taxonomically identified according to morphological features. In some cases it is also possible to work in the opposite direction. A good example of this is the different attempts that have been made to place the so-called 'blue-green *Chlorella*' in the proper systematic group. Its morphology resembles that of a unicellular green alga. Like *Chlorella* it possesses a chloroplast however its pigment composition resembles that of a blue-green alga. It has been listed by different investigators both as a green and a blue-green alga (Allen, 1959). At present it is placed as a genus of uncertain position under the Cryptophyta by Silva (1962) and by Christensen (1962). In a similar way analyses of pigment composition may be useful tools for a check of taxonomical identifications (Jeffrey, 1968a; Jeffrey and Haxo, 1968).

When an action spectrum curve of photosynthesis reflects the absorption spectrum of the material and matches this exactly, we conclude that all the pigments present in the material participate in the process with equal effect. Deviations betweeen the action spectrum and the absorption curve are then assumed to be caused by inactive or screening pigments, or by insufficient excitation or cooperation of the pigments now known to be physically separated in the two different pigment systems which function during electron transport in photosynthesis (see page 26 of this chapter). A fair to good correspondence between the action spectrum of photosynthesis and the absorption spectrum of living material is found in several algal groups, e.g. green and brown algae, diatoms and dinoflagellates (Halldal, 1964, 1966, 1968, 1969; Haxo, 1960; Haxo and Blinks, 1950). A most striking difference between photosynthetic action spectra and the corresponding absorption curves is found among all red and blue-green algae, but this difference is significantly smaller in the cryptomonads which also contain biliproteins (Haxo, 1960; Haxo and Fork, 1959).

In studies on the flexibility of the photosynthetic apparatus it is often useful to manipulate the environmentally induced differences between the action spectra of photosynthesis and the corresponding absorption spectra. In addition analyses of light intensity and light quality effects on pigment content and photosynthetic rates and capacities have given important contributions to our knowledge on the mechanism behind environmental adjustments. In this respect nature itself is perhaps the best growth chamber. If we look and search we will discover natural growth conditions that are difficult to arrange under laboratory conditions. Too few have discovered nature as a source for experimental material, or as a place where ideas can be obtained for laboratory experimental designs.

Recently, extensive studies (including reviews) on the adaptive effect of light levels on the photosynthetic apparatus have been performed by

Brown (1968), Brown and Richardson (1968), Brown, Richardson and Vaughn (1967), Jørgensen (1964a, 1964b, 1968, 1969), Steemann Nielsen (1961, 1962), Steemann Nielsen and Hansen (1959), Steemann Nielsen and Jørgensen (1962, 1968a, 1968b) and Steemann Nielsen and Park (1964). Steemann Nielsen and coworkers distinguish two different main types of adaptation. One which they call the *Chlorella type* is found in a number of species. This type is characterized by light intensity-induced changes in chlorophyll content per cell with intensity. More chlorophyll per cell is formed at low light levels than at high. The second is the *Cyclotella type*. In algae following the principle of adaptation of this diatom, the chlorophyll content is about the same in cells grown at high and low light levels, while the actual photosynthetic rate is considerably higher in cells developed at high intensity. Steemann Nielsen and coworkers assumed that in the latter case increased enzymatic content in the dark reaction steps of photosynthesis raised the photosynthetic rate. Direct enzymatic studies to test this hypothesis were, however, not performed. More specific results from the adaptation experiments performed by Brown and Richardson will be dealt with below.

The chlorophyll a–*chlorophyll* b–*carotenoid system*

The chlorophyll b/a ratio in higher plants is roughly 0.3. Analyses on leaves exposed to bright sunlight and shaded environments in general show that relatively more chlorophyll b is formed at low light levels. A relatively high chlorophyll b content thus seems to be a characteristic feature of shade plants. The variation in the chlorophyll b/a ratio can be great; values between 0.25 and 0.5 have been reported (Egle, 1960). The green algae usually have rather high chlorophyll b content, and thus have the characteristic pigment ratio of a shade plant. In some other algal groups the opposite situation exists. The Xanthophyta contain only chlorophyll a.

A more extensive study of the chlorophyll b/a ratio in algae grown at different light levels was performed by Brown and Richardson (1968). For the three species analysed (*Chlorococcum wimmeri*, *Chlorella pyrenoidosa* and *Euglena gracilis*) the chlorophyll b/a ratio was significantly increased when algae were grown in dim light; for *Chlorella* a chlorophyll b/a ratio variation from 0.65 at about 500 lux to about 0.58 at 15 000 lux.

The light-induced changes in the chlorophyll b/a ratio thus seem to be a means of adjusting the photosynthetic process to light levels, though very few analyses in this direction have been performed. Particularly sparse are correlated measurements on action spectra. However, since photosynthetic action spectra of green algae as a rule closely follow the absorption spectra of the material in the spectral region from around 550 to 675 nm, we are inclined to assume that changes in pigment ratio will be reflected in the photosynthetic spectral response curve. A direct demonstration of such correlation was observed for the green algae *Ostreobium* (see page 37). The chlorophyll *b/a* ratio was 0·79 (Jeffrey, 1968a). This high clorophyll *b/a* ratio was also distinctly noticeable in the action spectrum of photosynthesis (Halldal, 1968). Öquist (1969) payed particular attention to comparative pigment and photosynthesis studies in the red and far red spectral region of *Chlorella*. His analyses were focused on different chlorophyll *a* forms, and not so much on induced changes in the chlorophyll *b/a* ratio. His curves, however, do not indicate any great variations in this ratio from algae grown in low or high light intensity or in far red light. If the variations occurred in the same order of magnitude as reported by Brown and Richardson (1968), they would be difficult to measure in absorption curves from living material and in point by point measurements for action spectra of photosynthesis.

Much attention has lately been paid to the different *in vivo* chlorophyll *a* forms, their function in the light-capturing process during photosynthesis, and the electron transport chain.

For some years now it has been evident that the ratio Chl 670/Chl 682 may undergo changes during development or different environmental conditions (Halldal, 1958; Krasnovsky and Kosobutskaya, 1955), though French and Huang (1957) did not find evidence for any effect of temperature and light intensity in *Chlorella*. Newer analyses, however, clearly demonstrate that the *in vivo* chlorophyll *a* system is very flexible and that the ratios of these chlorophyll forms may be significantly to drastically altered during extreme light conditions, e.g. *Ostreobium*. Its environmental situation within the coral *Favia* will be used to illustrate this.

The light conditions within a 12 cm wide and 8·5 cm high specimen of *Favia pallida* (Figure 2.4) were carefully analysed by Shibata and Haxo (1969). *Favia*, like all other reef-building corals, possesses unicellular symbiotic dinoflagellates (zooxanthellae) in their gastrodermal cells. In *Favia* these symbiotic dinoflagellates constitute a dense cover, which gives the impression of a surface-layer chocolate smear that captures nearly all the light that reaches the coral. Shibata and Haxo calculated that only 0·10 to 0·15 per cent of daylight penetrated to the top of the green algal layer between 400 and 678 nm. This corresponded to 500 to 750 erg/cm² second at full tropical sunlight. In the middle of the green layer they calculated that only 0·5 to 5 erg/cm² second were available for photosynthesis in the same spectral region. At the bottom of the layer only a fraction of an erg/cm² second was left. Above 680 nm the light situa-

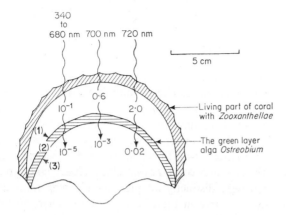

Figure 2.4 Schematic drawing of light conditions within the massive coral *Favia*, according to measurements by Shibata and Haxo (1970). **(1)**, **(2)** and **(3)** mark the positions of sampling for the light intensity/ response curves of Figure 2.5

tion was greatly improved. At 700 nm 0·6 per cent of daylight was transmitted to the top of the layer, and at 720 nm this was further improved to 2 per cent. At first it seemed impossible to visualize how light at this level could drive the photosynthetic apparatus in the spectral region where we expect to find photosynthesis at a significant rate, that is, between 350 and 700 nm. However, action spectra measurements correlated with absorption spectra analyses disclosed that the photosynthetic apparatus in *Ostreobium* was adjusted to the peculiar light conditions inside the coral in such a way that particularly light above 690 nm was captured. In *Ostreobium* the photosynthetic rate, measured as oxygen evolution, at 700 nm was equal to that at 675 nm, and at 725 nm it was up to 80 per cent of the maximum activity in red light at between 675 and 700 nm. Usually in green algae the photosynthetic oxygen evolution at 700 nm is only around one tenth of that at 670 nm, and at 725 nm it is zero or close to zero. In *Ostreobium* significant oxygen evolution was measured at 760 nm. Spectrophotometric analyses disclosed unusually high absorption above 675 nm, and a distinct shoulder occurred around 725 nm. It was further demonstrated that variations could be induced both in the photosynthetic rates and the relative absorption in this spectral region. Exposure to low intensity dim daylight reduced the relative photosynthetic activity above 675 nm, and the relative absorption decreased. The pigment system was evidently very flexible, and difference absorption spectra of paired samples where one piece was exposed to dim daylight for a few days, disclosed a chlorophyll *a* form with an absorption maximum at 725 nm (Chl 725, see page 28).

It was evident that the pigment system of *Ostreobium* was indeed unusual for a green alga. It was also very flexible. Therefore Öquist (1969) raised the question of whether a similar pigment and photosynthetic flexibility also occurred in other green algae. The commonly used experimental material *Chlorella* was selected for such analyses. Comparative experiments were performed on *Chlorella* cultures grown in 'white' light from incandescent lamps and cells developed under far red light conditions at wavelengths above 690 nm, the latter set-up thus simulating the spectral energy distribution inside the massive coral *Favia*. Under these conditions an alga grown in far red light distinctly adapted its pigment ratio and photosynthetic spectral response in the direction of the characteristic pattern of *Ostreobium*, though the features were far less pronounced. In algae grown in 'white' light oxygen evolution was recorded to 730 nm, while in those developed in far red light an increased photosynthetic rate was recorded above 675 nm, and these algae responded to 740 nm. The light absorption of living cells showed a corresponding increase. Difference absorption spectra between algae grown under these light conditions showed that the greatest increase in absorption in the far red grown algae occurred at 700 nm. Öquist (1969) further assumed that this increase was mainly a conversion of Chl 682 to possibly Chl 700. In comparative studies of algae grown at different intensities of 'white' light, samples from such cultures

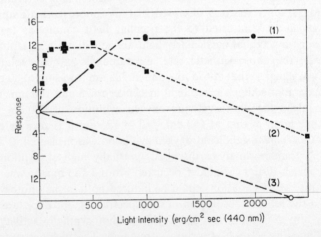

Figure 2.5 Light intensity/response curves for the green layer alga *Ostreobium* living inside the massive coral *Favia*. For sampling positions of the curves **(1)**, **(2)** and **(3)** see Figure 2.4. After Halldal, 1968

had the same pigment ratio and photosynthetic action spectra between 500 and 730 nm. A general conclusion was then drawn that the increased absorption and the correlated increase in photosynthetic spectral response above 675 nm were adaptations to the peculiar spectral energy distribution and not to light of the different levels.

In another way the green algal layer *Ostreobium* inside the massive coral gave an illustrating example of adaptation of the photosynthetic apparatus to the very low light levels where it grows (Halldal, 1968). Figure 2.5 shows three light intensity/photosynthetic response curves in blue light from a sample taken at the top of the algal layer (which was exposed to around 500 to 750 erg/cm² second in nature), another sample taken from the middle of the layer (in nature exposed to around 0·5 to 5 erg/cm² second), and a third from the bottom (naturally exposed to only a fraction of an erg/cm² second in the visible region of the spectrum). The sample taken from the top of the layer obtained light saturation at around 1000 erg/cm² second, and responded linearly to light intensity between 0 and this light level. For the sample taken from the middle of the green layer, light saturation was recorded at about one tenth of the intensity of that of the former, that is at 100 erg/cm² second. The algae from the bottom showed photooxidation at all light levels. The light intensity/ response curves for the top and the middle layer algae fit extraordinarily well the light conditions that naturally prevailed on the actual spot within the coral. The light saturation at these very low intensities made it necessary to check frequently whether action spectra measurements were actually performed on the linear part of the light intensity/response curve. In algal samples from the top of the green layer the photosynthetic capacity was significantly higher than in algae developed further down. No critical analyses were performed to examine whether the amount of pigment per unit material or other factors were responsible for these great variations in light saturation levels and photosynthetic capacities over these very short distances which amounted to a few mm. Visual analyses on the colour of the green layer and on samples taken from different parts of the layer and examined in the microscope indicated that the pigment content was about the same.

The effect of environmental conditions on the relative amount of carotenoids and their function during photosynthesis are best studied through action spectra or quantum yield measurements. It is generally found that high levels increase the relative amount of carotenoids to chlorophyll (Halldal and French, 1958; Öquist, 1969), and Öquist made comparative determinations of action spectra from such samples. The general conclusion was drawn that high light levels increased the amount of photosynthetically inactive carotenoids and vice versa. Relatively higher carotenoid content and lower photosynthetic activity were observed in *Chlorella* grown in 'white' light than in those developed in far red, where conditions for photosynthesis were poor. It is therefore assumed that green algae may use their carotenoid content to adjust their photosynthetic capacity. Relatively high contents of photosynthetically active carotenoid(s) are produced when the algae are exposed to poor light conditions for photosynthesis, and more inactive carotenoids are formed under surplus light conditions.

Under extreme environmental conditions the colour of many green algae changes to deep red, due to the accumulation of large amounts of carotenoids in oil drops. As an example, the red-coloured flagellate *Dunaliella salina* may be mentioned (Fox and Sargent, 1938), which may give Californian salt ponds their red colour. Blinks (1954) showed that blue light was completely inactive in the photosynthesis of such red-coloured algae, which suggests that these dissolved carotenoids function as yellow filters which prevent the blue light from reaching the chloroplast. When cultivated under normal salt conditions, the colour of these algae is pure green. Other examples of the accumulation of red pigments in green algae are the 'red snow' in alpine and arctic regions which are caused by *Chlamydomonas nivalis* and others (Marrè, 1962). Red *Haematococcus pluvialis* often occurs in almost pure stand in hollow and rocky ledges temporarily filled with rain water. Brown, Richardson and Vaughn (1967) carried out experimental analyses on the development of red pigments in the green algae *Chlorococcum wimmeri*. Their measurements included comparative Hill reaction studies and photosynthetic rate measurements for *Chlorococcum* versus *Chlorella*. They identified the red pigments as astacin, esterified astaxanthin and other carotenoids. The intensity of the light was the factor of greatest importance for the accumulation of red pigments in the liquid cultures. Even the carbon dioxide concentration affected the coloration. Although the capacities for photosynthesis and Hill reaction per volume packed cells were greater for *Chlorella* than for *Chlorococcum*, based upon chlorophyll *a* content the *Chlorococcum* cells were more effective. In studies of red pigment accumulation in the green alga *Haematococcus pluvialis*, Droop (1954) found that carbon dioxide concentration was most important. In studies with *Haematococcus lacustris* in crossed gradients of light intensity and temperature, Halldal and French (1958) reported that red coloration occurred at high light levels, while temperature appeared to be of little importance.

It thus seems as if a significant increase in carotenoid content takes place whenever certain green algae are exposed to extreme conditions, such as high salt concentrations (e.g. *Dunaliella salina*); exposure to low temperature (e.g. *Chlamydomonas nivalis*); high light levels (e.g. *Chlorococcum wimmeri* and *Chlamydomonas lacustris*); and low carbon dioxide concentration (e.g. *Haematococcus pluvialis*).

The chlorophyll a–*chlorophyll* c–*carotenoid system*

There is no doubt that carotenoids participate in photosynthesis of diatoms and brown algae and of dinoflagellates (Halldal, 1968, 1969; Haxo, 1960; Haxo and Blinks, 1950; Tanda, 1951). These algal groups also represent the only examples where specific known (or identified) carotenoids function as accessory pigments, e.g. fucoxanthin in diatoms and brown algae, and peridinin in the dinoflagellates. In action spectra measurements of these forms, great deviations may be observed in the violet to green portion of the spectrum. This indicates that the ratio of photosynthetically active/inactive carotenoids shows great variations and that the situation in this photosynthetic system is related to that which was described for the relative amount of carotenoids in the chlorophyll *a*–chlorophyll *b*–carotenoid system.

The variations in the chlorophyll *c/a* ratio with intensity have been analysed by Brown and Richardson (1968) for four species, the cryptophyt *Cryptomonas ovata*, the diatom *Nitzschia closterium*, the dinoflagellate *Amphidinium* sp., and the brown alga *Sphacelaria* sp. In *Cryptomonas*, *Nitzschia* and *Amphidinium*, the ratio chlorophyll *c/a* decreased significantly with increasing intensity from around 500 to 15 000 lux. In *Cryptomonas* it changed from 0·45 to 0·38, in *Amphidinium* from 0·45 to 0·30, and in *Nitzschia* from 0·30 to 0·10. In *Sphacelaria* the pattern was less regular. The ratio first dropped from 0·20 to 0·10 between 500 and 3000 lux. This drop was followed by an increase to 0·20 at around 15 000 lux; thus with this last exception the chlorophyll *c/a* ratio is greatest in dim light. It seems as if shade plants of this group produce a high content of chlorophyll *c*. Holm-Hansen and coworkers (1965) report extraordinarily lower chlorophyll *c/a* ratios (around 0·15) than reported by most others, e.g. Jeffrey (1963) at 0·76, Madgwick (1966) at 0·25–0·4, and Mann and Myers (1968). These data indicate that the chlorophyll *c/a* ratio undergo great variations both under natural conditions as well as in laboratory cultures. However, as the methods used during these determinations are rather different it is difficult to directly compare the data.

Only a few action spectra measurements have been performed for the dinoflagellates (Halldal, 1968; Haxo, 1960). Results from two of these measurements are rather similar, namely those of Haxo for *Goniaulax polyedra* and of Halldal for the symbiotic dinoflagellate (zooxanthellae). However, in one particular spectral region a distinct deviation is observed for the final drawn curves. For the zooxanthellae a distinct shoulder, sometimes a peak, was observed around 525 to 550 nm with automatic recordings of the photosynthetic action spectrum. This detail is absent in the action spectrum curve drawn for *Goniaulax*. However, from the individual values of the point-by-point measurements by Haxo it is not unrealistic to trace an average curve indicating a shoulder or a small maximum around 540 nm.

A study of the effect of environmental conditions on the pigment composition and photosynthetic spectral response in algae where the pigment system, chlorophyll *a*–chlorophyll *c*–carotenoids, is involved has just started, and it is impossible at this time to give a general picture. However, it is evident that the relative amounts of photosynthetically effective and ineffective carotenoids undergo significant variations. Under laboratory conditions *Amphidinium* sp. produced large amounts of inactive carotenoids both at high and low light levels, which had the effect that the action spectrum peak in the blue part occurred at 470 nm, evidently efficiently screening the chlorophyll *a* peak at 435 nm (Halldal, unpublished results). This peak was distinctly resolved both for *Goniaulax* and the zooxanthellae.

Brown and Richardson (1968) measured growth, photosynthesis, respiration and pigmentation at intensities up to about 12 000 lux in *Amphidinium* sp., *Cryptomonas ovata*, *Nitzschia closterium*, *Ochromonas danica* and *Sphacelaria* sp. As a general rule most photosynthetic pigment per volume cells was formed at low light levels, but their data do not indicate any drastic variations in the ratio chlorophyll $a + c$/accessory carotenoids. All these algae had a rather broad optical range of growth with an optimum around 9 000 lux. Optimum photosynthetic rates were also measured around the same intensity.

Changes in the pigment content of the diatom *Navicula pelliculosa* as a function of silicon starvation and growth was studied by Healey, Coombs and Volcani (1967). The synthetic rates of chlorophyll *a* and the accessory carotenoid fucoxanthin were similar. The formation of these pigments ceased after 5 to 7 hours of silicon starvation, but was resumed if silicon was added in the light but not in darkness. Diadinoxanthin synthesis continued in the light at all times, but at a somewhat lower rate during silicon starvation; thus the ratio photosynthetic/non-photosynthetic pigments decreased. The authors point to the fact that excess carotenoid production often takes place in stationary cultures, and that at least some of this excess, presumably non-photosynthetic pigments, serves in a photoprotective mechanism. It is then suggested that the decreases in the synthesis of the photosynthetic interrelated pigments, chlorophyll *a* and fucoxanthin are consistent with a general decrease in metabolic activity, and that the changes in diadinoxanthin synthesis suggest a function connected with photoprotection.

The chlorophyll a–biliprotein system

These algae are also very rich in carotenoids which show great variations in amounts due to environmental conditions (Brody and Emerson, 1959a, 1959b; Halldal, 1958; Yocum and Blinks, 1958). However, so far there is no experimental evidence that they are accessory pigments in blue-green or red algae (Emerson and Lewis, 1942; Haxo and Blinks, 1950; Yokum and Blinks, 1958).

The chlorophyll *a*–biliprotein system shows much greater flexibility in pigment ratio changes than the chlorophyll *a*–carotenoid combination. The adjustment for natural light conditions reflected through the vertical distribution of red algae, in particular those which have a pure red colour, has been selected as an example of this adaptation for a great many years. In coastal water the spectral energy distribution (Jerlov, 1968) is practically identical to the spectral response of such algae (Halldal, 1964, 1969; Haxo, 1960; Haxo and Blinks, 1950; Levring, 1947). The literature concerning the phylogenetic and ontogenetic adaptation of the chlorophyll *a*–biliprotein system was extensively reviewed by Rabinowitch (1951). In later years such studies have been extended more systematically to also include photosynthetic spectral response studies, fluorescence measurements and excitation energy transfer analyses.

Yocum and Blinks (1958) showed that the photosynthetic efficiency of chlorophyll *a* in the unicellular red alga *Porphyridium cruentum* was greatly increased when the cells were exposed to blue and red lights at low light levels (blue at 15 500 erg/cm^2 second, red at 20 000 erg/cm^2 second), while the algae grown in green light at 17 500 erg/cm^2 second and in darkness maintained their original low chlorophyll efficiency, the high effect of the green light being absorbed by the biliproteins. Yocum and Blinks further showed that these induced changes were rapidly reversed. In their analyses pigment ratio changes were also observed, but not in the direction of complementary chromatic adaptation. No indication of photosynthetic effective carotenoids was reported. Similar pigment ratio changes and alterations in the photosynthetic spectral response following different light treatments could also be observed with the thallus red algae *Porphyra*, which as *Porphyridium* is systematically placed in the algal class Bangiophyceae. Members of the other algal class, the Florideophyceae, did not have the ability to make such adjustments.

The effect of wavelengths and the intensity of light on the pigment ratio in *Poryphyridium cruentum* has also been analysed by Brody and Emerson (1959a). They draw the general conclusion that the intensity of the light will be decisive for the way in which chromatic adaptation will be directed, and that the tendency toward a complementary chromatic adaptation will be more apparent at low light levels than at high. *Porphyridium* thus seems to possess an adjustable pigment ratio system which makes it possible to arrive at a situation where the light absorbed by photosynthetically active pigments will keep the energy utilized within the limitation of the enzymatic steps following the photochemical reaction of the process.

In later experiments Brody and Emerson (1959b) also measured the quantum yield of photosynthesis in *Porphyridium* cultured to give different proportions of chlorophyll and biliproteins. In order to have an exact figure of absorbed light, totally absorbing suspensions were used. During such analyses it was demonstrated that at wavelengths shorter than 650 nm, chlorophyll *a* had a photosynthetic efficiency as high as in other algal groups and higher than that of the biliproteins. They concluded that the yield of photosynthesis for light absorbed by chlorophyll *a* is about the same in all algal groups and that there is no need for theories of the photosynthetic process which include inactive chlorophyll *a* in red algae. However, they point to the remarkable steep decline in photosynthetic efficiency in red algae from around 650 nm, which today, through the works of Emerson and others, is explained by the function of accessory pigments in the theory for excitation energy transfer and electron transport during photosynthesis (see page 25).

Later Brody and Brody (1962) continued analyses on adaptation in *Porphyridium*. They reported a rather high flexibility of the quantum yield in the green region of the spectrum, while that in the red was relatively constant. Thus the efficiency of chlorophyll-sensitized photosynthesis is less subject to change than that of phycoerythrin-sensitized photosynthesis. The photosynthetic quantum yield increase or decrease depended upon the pigment absorbing the 'adapting' light. If blue light was used for the adaptation (chlorophyll excitation) the quantum yield in green decreased, while adaptation in green light (phycoerythrin excitation) raised the quantum yield in the same spectral region. In their analyses they used as short a time of adaptation as possible. Exposure for 12 hours at 20 000–30 000 erg/cm² second was sufficient to give significant results. The experimental approaches of Yocum and Blinks (1958) and Brody and Emerson (1959a) are in one respect different from those of Brody and Brody (1962). Yocum and Blinks, and Brody and Emerson anlaysed the material after prolonged adaptations (several days), after extensive pigment ratio changes were induced, while Brody and Brody restricted the adaptation time to 12 hours. In this way they demonstrated that pigments could be 'activated' for photosynthesis. In the discussion they also make an attempt to explain the photosynthetic spectral response of red-coloured red algae. They write:

> 'It might be well to note that thalli of red algae, as they are found in their natural habit, can be considered comparable to cells grown in or adapted to green light, for sunlight filtering down through 2 or 3 m water . . . has had much of its blue and red components removed. It is in these cells that the yield of phycoerythrin-sensitized photosynthesis is at its highest. It is not surprising, therefore, that with freshly harvested red algae, when the yield in the maximum of the phycoerythrin band is compared to the yield in the region of the red decline, phycoerythrin seems more efficient than chlorophyll.'

Jones and Myers (1965) studied pigment variations in *Anacystis nidulans* induced by light of selected wavelengths. The variations in pigment ratios were estimated from *in vivo* absorption spectra by means of readings at 678 nm for chlorophyll *a*, 625 nm for phycocyanin and 490 nm for carotenoids. They concluded that light predominantly absorbed by chlorophyll *a* caused a dramatic lowering of the chlorophyll content, in some cases to one quarter of that present in algae grown in 'white' light. In phycocyanin and carotenoid content only minor changes occurred.

Growth and excitation energy transfer studies of the chlorophyll–biliprotein system have been performed for the blue-green alga *Anacystis nidulans* by Gosh and Govindjee (1966). The pigmentation of *Anacystis* depended strongly on the intensity of coloured light. As a general conclusion it was stated that the pigment which best absorbed the light supplied during the growth period is reduced when very strong light is used, i.e. high intensity red light gave a higher phycocyanin/chlorophyll *a* ratio than low intensity red light. High intensity orange light gave a low phycocyanin/chlorophyll *a* ratio, and algae exposed to low intensity 'white' light had a higher phycocyanin/chlorophyll *a* ratio than those grown at high light levels.

Gosh and Govindjee (1966) made detailed excitation energy transfer studies of adapted *Anacystis* populations. They focused their studies on the relative magnitude of the fluorescence peaks at 660 nm (phycocyanin fluorescence) and at 680 nm (chlorophyll *a* fluorescence), when excited at 580 nm. Algae grown in red light at high light levels (high phycocyanin/chlorophyll *a* ratio) had a distinct emission peak at 660 nm and no indication of chlorophyll *a* fluorescence, while a pronounced chlorophyll *a*

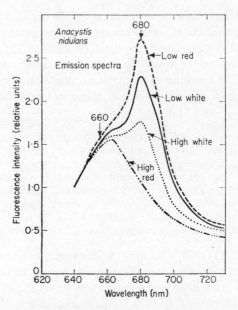

Figure 2.6 Fluorescence emission spectra of the blue-green alga *Anacystis nidulans* grown in red and white light. Exciting light at 580 nm. After Gosh and Govindjee, 1966

fluorescence occurred in algae exposed to red light at low light level (low phycocyanin/chlorophyll *a* ratio), (Figure 2.6). Algae grown in orange light of high intensity (low phycocyanin/chlorophyll *a* ratio) showed poor energy transfer from phycocyanin to chlorophyll *a*. Through comparative studies with algae exposed to different intensities of 'white' light, Gosh and Govindjee concluded that

> 'a decrease in efficiency of energy transfer from phycocyanin to chlorophyll seems to occur whenever the ratio chlorophyll *a* to phycocyanin deviates from the normal'.

By the exposures of blue-green algae to crossed gradients of light intensity and temperature, the changes in pigment ratios, and also the total number of pigments formed, could be studied under a great variety of conditions over a limited area ca. 30 × 30 cm (Halldal, 1958; Halldal and French, 1958). For *Anacystis* the ratio phycocyanin/chlorophyll *a* was rather complex. Great differences were observed in samples taken from different light intensity–temperature combinations, and the changes with time were very dynamic. For an *Anabaena* sp. the growth pattern was simpler, and the development of pigment ratios was also considerably more constant with time. The phycocyanin/chlorophyll *a* ratio showed a steady increase with decreasing intensity, at all temperatures where healthy growth occurred. In the algae grown at medium to high light levels no phycoerythrin was apparent in the absorption spectra of living algae, but a pronounced peak occurred at 550 nm in algae grown at low light intensity, indicating the presence of this accessory pigment in addition to phycocyanin. No action spectra or quantum yield measurements for photosynthesis were performed during these experiments, but induced changes in the biliprotein/chlorophyll *a* ratio is as a rule correlated with corresponding changes in the spectral response of photosynthesis (Brody and Brody, 1962; Brody and Emerson, 1959a; Gosh and Govindjee, 1966; Yocum and Blinks, 1958). One is therefore inclined to assume that even the relative rate of photosynthesis increased in the green part of the spectrum at low light levels in *Anabaena*.

Extensive studies of the effect of coloured light on the formation of biliproteins in blue-green algae, and analyses of the preconditions for the synthesis of these pigments have been performed by Fujita and Hattori (Fujita and Hattori, 1960, 1962a, 1962b; Hattori and Fujita, 1959a, 1959b). The algae were exposed to rather broad spectral bands isolated by means of dye filters. The energy at different spectral regions was adjusted in energy to give 1.6×10^5 erg/cm^2 second. From the experiments performed on *Tolypothrix tenuis* it was concluded that for the formation of phycoerythrin the effectiveness of the different spectral regions was as follows:

blue = green > yellow > orange-yellow > orange > red = pink

For the formation of phycocyanin and allophycocyanin:

purple = pink = red > orange > orange-yellow > yellow > blue = green

Thus the formation of phycoerythrin was stimulated by light that is not absorbed to any significant degree by the pigment itself. The pigment pattern is complicated, and the experimental approach has the weakness that only one light level is employed. Our knowledge today says that both spectral region and light levels combined will determine the biliprotein/chlorophyll *a* pigment ratio, and light conditions are also decisive factors for the spectral and absolute photosynthetic characteristics.

Leclerc (1969) measured pigment ratios in *Phorphyridium* grown at different salt levels. The chlorophyll *a* content was maximum at normal sea water salt concentration, and the cells showed a general increase in the chlorophyll *a*/phycoerythrin ratio with increasing salinity up to three times that of natural sea water.

The chlorophyll a–*chlorophyll* c–*biliprotein–carotenoid system*

Blue-green and red algae contain only one photosynthetically active chlorophyll, namely *a*. In these algae there is no indication of photosynthetic active carotenoids. The response in the blue to green region of the spectrum is so low, even in algae adapted to a variety of light levels and spectral regions, that a different conclusion is unrealistic. The cryptomonads belong to an algal group with unusual pigment composition. They contain chlorophyll *a* and *c*, and carotenoids and phycobilins. Measurements of the action spectra of photosynthesis from representatives of this group show that the photosynthetic rates in blue-green and red lights are about the same. For *Rhodomonas lens*, Haxo and Fork (1959) resolved a double peak in the blue region of the spectrum, indicating photosynthetic participation of chlorophyll *a* and either chlorophyll *c* or carotenoids, or both. Although the details in the measurements are not sufficient to draw any definite conclusion about this, one is inclined to include both chlorophyll *c*, phycobilins and carotenoid(s) as accessory pigments for these algae.

REFERENCES

Aghion, J. (1963). Properties des extraits aqueax de chloroplastes ayant un maximum d'absorption à 740 mμ. *Biochim. Biophys. Acta*, **66**, 212.

Allen, M. B. (1954). Studies on a blue-green *Chlorella*. *Rapp. Commun. 8 Intern. Bot. Congress*, 1954, Sec. **17**, 41.

Allen, M. B. (1958). Possible function of chlorophyll *b*. Studies with green algae that lack chlorophyll *b*. In *The Photochemical Apparatus. Its Structure and Function*, Brookhaven Symposia in Biology number 11, Brookhaven National Library, Upton, New York. p. 339–342.

Allen, M. B. (1959). Studies with *Cyanidium caldarium*, an anomalously pigmented chlorophyte. *Arch. Mikrobiol.*, **32**, 270.

Allen, M. B., C. S. French, and J. S. Brown (1960). Native and extractable forms of chlorophyll in various algal groups. In M. B. Allen (Ed.), *Comparative Biochemistry of Photoreactive Systems*, Academic Press, London, New York. pp. 33–52.

Allen, M. B., and J. C. Murchio (1963). Studies of the constitution and photochemical activity of an isolated chlorophyll complex. In *Photosynthetic Mechanisms in Green Plants*, Publ. 1145 Natl. Acad. Sci.–Natl. Research Coun. USA. pp. 486–495.

Allen, M. B., J. C. Murchio, S. W. Jeffrey, and S. A. Bendix (1963). Fractionation of the photosynthetic apparatus of *Chlorella pyrenoidosa*. In *Studies on Microalgae and Photosynthetic Bacteria*, University of Tokyo Press, Tokyo. pp. 407–412.

Arnon, D. I., H. Y. Tsujimoto, and B. D. McSwain (1965). Photosynthetic phosphorylation and electron transport. *Nature*, **207**, 1367.

Arnon, D. I., H. Y. Tsujimoto, B. D. McSwain, and R. K. Chain (1968). Separation of two photochemical systems of photosynthesis by fractionation of chloroplasts. In K. Shibata and coworkers (Eds.), *Comparative Biochemistry and Biophysics of Photosynthesis*, University of Tokyo Press, Tokyo and University Park Press, Pennsylvania, pp. 113–132.

Bannister, T. T. (1954). Energy transfer between chromophore and protein in phycocyanin. *Arch. Biochem. Biophys.*, **49**, 222.

Blinks, L. R. (1954). The role of accessory pigments in photosynthesis. In B. A. Fry and J. L. Peel (Eds.), *Autotrophic Microorganisms*, Cambridge Univ. Press, London and New York, pp. 224–246.

Blinks, L. R. (1957). Chromatic transients in photosynthesis of red algae. In H. Gaffron (Ed.), *Research in Photosynthesis*, Interscience, New York. pp. 444–449.

Blinks, L. R. (1959). Chromatic transients in photosynthesis of a green alga. *Plant Physiol.*, **34**, 200.

Blinks, L. R. (1960). Chromatic transients in the photosynthesis of green, brown and red algae. In M. B. Allen (Ed.), *Comparative Biochemistry of Photoreactive Systems*, Academic Press, London, New York. pp. 367–376.

Boardman, N. K., and J. M. Anderson (1964). Isolation from spinach chloroplasts of particles containing different proportions of chlorophyll *a* and chlorophyll *b* and the possible role in the light reaction of photosynthesis. *Nature*, **203**, 166.

Branton, D. (1968). Structure of the photosynthetic apparatus. In A. C. Giese (Ed.), *Photophysiology, Current Topics*, Vol. III. Academic Press, London, New York. pp. 197–224.

Brody, M., and S. S. Brody. (1962). Induced changes in the photosynthetic efficiency of *Porphyridium cruentum*. II. *Arch. Biochem. Biophys.*, **96**, 354.

Brody, M., and R. Emerson (1959a). The effect of wavelength and intensity of light on the proportion of pigments in *Porphyridium cruentum*. *Am. J. Botany*, **46**, 433.

Brody, M., and R. Emerson (1959b). The quantum yield of photosynthesis in *Porphyridium cruentum*, and the role of chlorophyll *a* in the photosynthesis of red algae. *J. Gen. Physiol.*, **43**, 251.

Brody, M., and A. Vatter (1959). Observation on cellular structures of *Porphyridium cruentum*. *J. Biophys. Biochem. Cytol.*, **5**, 289.

Brown, J. S. (1967). Fluoremetric evidence for the participation of chlorophyll *a*–695 in system 2 of photosynthesis. *Biochim. Biophys. Acta*, **143**, 391.

Brown, J. S., and J. Duranton (1964). Partial separation of the forms of chlorophyll *a* by sodium doedecyl sulfate. *Biochim. Biophys. Acta*, **79**, 209.

Brown, J. S., and C. S. French (1961). The long-wavelength forms of chlorophyll *a*. *Biophys. J.*, **1**, 539.

Brown, T. E. (1968). Spectral light requirement of algae. *Food laboratory, U.S. Army Natick Laboratories, Natick, Mass.* Technical report 96–45–FL.

Brown, T. E., and F. T. Richardson (1968). The effect of growth environment on the physiology of algae: light intensity. *Phycologia*, **4**, 38.

Brown, T. E., F. T. Richardson, and M. L. Vaughn (1967) .Development of red pigmentation *in Chlorococcum wimmeri*. *Phycologia*, **6**, 167.

Butler, W. L. (1960). A far red absorbing form of chlorophyll. *Biochem. Biophys. Res. Commun.*, **3**, 685.

Butler, W. L. (1961). A far-red absorbing form of chlorophyll, *in vivo. Arch. Biochem. Biophys.*, **93**, 413.

Butler, W. L. (1966). Spectral characteristics of chlorophyll in green plants. In L. P. Vernon and G. R. Seely (Eds.), *The Chlorophylls*, Academic Press, London, New York, pp. 343–379.

Christensen, T. (1962). *Systematisk Botanik Nr.* 2, Alger. Munksgaard–Copenhagen.

Cramer, W. A., and W. L. Butler (1968). Further resolution of chlorophyll pigments in photosystems 1 and 2 of spinach chloroplasts by low temperature derivative spectrophotometer. *Biochim. Biophys. Acta*, **153**, 889.

Dougherty, R. C., H. H. Strain, W. A. Svec, R. A. Uphaus, and J. J. Katz (1966). Structure of chlorophyll *c. J. American Chem. Soc.*, **88**, 5037.

Droop, M. R. (1954). Conditions governing haematochrome formation and loss in the alga *Haematococcus pluvialis* Flotow. *Arch. Mikrobiol.*, **20**, 391.

Duysens, L. N. M. (1952). *Transfer of Excitation Energy in Photosynthesis.* Doctoral thesis, University of Utrecht, The Netherlands.

Egle, K. (1960). Menge und Verhältnis der Pigmente. In W. Ruhland (Ed.), *Encyclopedia of Plant Physiology*, Vol. V, Part 1. Springer, Berlin. pp. 444–496.

Emerson, R. (1958). Yield of photosynthesis from simultaneous illumination with pairs of wavelengths. *Science*, **127**, 1059.

Emerson, R., and W. Arnold (1932). The photochemical reaction in photosynthesis. *J. Gen. Physiol.* **16**, 191.

Emerson, R., R. Chalmers, and C. Cederstrand (1957). Some factors influencing the long-wave limit of photosynthesis. *Proc. Natl Acad. Sc. USA*, **43**, 133.

Emerson, R., and C. M. Lewis (1942). The photosynthetic efficiency of phycocyanin in Chroococcus and the problem of carotenoid participation in photosynthesis. *J. Gen. Physiol.*, **25**, 579.

Emerson, R., and C. M. Lewis (1943). The dependence of quantum yield of *Chlorella* photosynthesis on wavelength of light. *Am. J. Botany*, **30**, 165.

Engelmann, Th. W. (1881). Neue Methode zur Untersuchung der Sauerstoffausscheidung pflanzlicher und thierischer Organismen. *Botan. Z.*, **39**, 441.

Engelmann, Th. W. (1882). Ueber Sauerstoffausscheidung von Pflanzenzellen im Mikrospectrum. *Botan. Z.*, **40**, 419.

Eriksson, C. E. A., and P. Halldal (1965). Purification of Phycobilins from red algae and their fluorescence excitation spectra in visible and ultraviolet. *Physiol. Plant.*, **18**, 146.

Fox, D. L., and M. C. Sargent (1938). Variations in chlorophyll and carotenoid pigments of the brine flagellate *Dunaliella salina*, induced by environmental concentrations of sodium chloride. *Chem. and Ind.*, **57**, 1111.

French, C. S. (1958). The variability of chlorophyll in plants. In *Photobiology*, Proceedings of the Nineteenth Annual Biological Colloquium Oregon State College, April 1958, pp. 52–64.

French, C. S. (1959). Various forms of chlorophyll *a* in plants. In *The Photochemical Apparatus. Its Structure and Function*, Brookhaven Symposia in Biology number 11, Brookhaven National Laboratory, Upton, New York. pp. 65–73.

French, C. S., and H. S. Huang (1957). The shape of the red absorption band of chlorophyll in live cells. *Carnegie Inst. of Wash. Year Book* 1956–57. p. 266.

French, C. S., J. Myers, and G. C. McLeod (1960). Automatic recording of photosynthesis action spectra used to measure the Emerson enhancement effect. In M. B. Allen (Ed.), *Comparative Biochemistry of Photoreactive Systems*, Academic Press, London, New York. pp. 361–365.

Fujita, Y., and A. Hattori (1960). Effect of chromatic light on phycobilin formation in a blue–green alga, *Tolypothrix tenuis. Plant and Cell Physiol.*, 1, 293.

Fujita, Y., and A. Hattori (1962). Photochemical interconversion between precursors of phycobilin chromoproteids in *Tolypothrix tenius. Plant and Cell Physiol.*, 3, 209.

Fujita, Y., and A. Hattori (1963). Effects of second illumination on phycobilin chromoprotein formation in chromatically preilluminated cells of *Tolypothrix tenuis. Microalgae and Photosynthetic Bacteria*, 431.

Gassner, E. B. (1962). On the pigment absorbing at 750 mμ occurring in some blue–green algae. *Plant Physiol.*, 37, 637.

Goedheer, J. C. (1969). Energy transfer from carotenoids to chlorophyll in blue–green, red and green algae and greening bean leaves. *Biochim. Biophys. Acta*, 172, 252.

Goodwin, T. W. (Ed.) (1965). *Chemistry and Biochemistry of Plant Pigments*, Academic Press, London, New York.

Goodwin, T. W. (Ed.) (1966). *Biochemistry of Chloroplasts*, Vol. I. Academic Press, London, New York.

Goodwin, T. W. (Ed.) (1967). *Biochemistry of Chloroplasts*, Vol. II. Academic Press, London, New York.

Gosh, A. K., and Govindjee (1966). Transfer of the excitation energy in *Anacystic nidulans* grown to obtain different pigment ratios. *Biophys. J.*, 6, 611.

Govindjee (1963). Observation on P 750 from *Anacystis nidulans. Naturviss.*, 23, 720.

Govindjee, C. Cederstrand, and E. Rabinowitch (1961). Existence of absorption bands at 730–740 and 750–760 millimicrons in algae of different divisions. *Science*, 134, 391.

Halldal, P. (1958). Pigment formation and growth in blue-green algae in crossed gradients of light intensity and temperature. *Physiol. Plant.*, 11, 401.

Halldal, P. (1964). Ultraviolet action spectra of photosynthesis and photosynthetic inhibition in a green and a red alga. *Physiol. Plant.*, 17, 414.

Halldal, P. (1966). Induction phenomena and action spectra analyses of photosynthesis in ultraviolet and visible light studied in green and blue-green algae, and in isolated chloroplast fragments. *Z. Pflanzenphysiol.*, 54, 28.

Halldal, P. (1968). Photosynthetic capacities and photosynthetic action spectra of endozoic algae of the massive coral *Favia. Biol. Bull.*, 134, 411.

Halldal, P. (1969). Automatic recording of action spectra of photobiological processes, spectrophotometric analyses, fluorescence measurements and recording of the first derivative of the absorption curve in one simple unit. *Photochem. Photobiol.*, 10, 23.

Halldal, P., and C. S. French (1958). Algal growth in crossed gradients of light intensity and temperature. *Plant. Physiol.*, 33, 249.

Hattori, A., and Y. Fujita (1959a). Formation of phycocyanin pigments in a blue-green alga, *Tolypothrix tenuis*, as induced by illumination with coloured lights. *J. Biochem.*, 46, 521.

Hattori, A., and Y. Fujita (1959b). Effect of preillumination on the formation of phycobilin pigments in a blue-green alga, *Tolypothrix tenuis. J. Biochem.*, **46**, 1259.

Haxo, F. T. (1960). The wavelength dependence of photosynthesis and the role of accessory pigments. In M. B. Allen (Ed.), *Comparative Biochemistry of Photoreactive Pigments*, Academic Press, New York, London, pp. 339–360.

Haxo, F. T., and L. R. Blinks (1950). Photosynthetic action spectra of marine algae. *J. Gen. Physiol.*, **33**, 389.

Haxo, F. T., and D. C. Fork (1959). Photosynthetically active accessory pigments of cryptomonads. *Nature*, **184**, 1051.

Haxo, F. T., and C. Ó hEocha (1960). Chromoproteins of algae. In W. Ruhland (Ed.), *Encyclopedia of Plant Physiology*, Vol. V/1, Springer, Berlin. pp. 497–510.

Healey, F. P., J, Coombs, and B. E. Volcani (1967). Changes in the pigment content of the diatom *Navicula pelliculosa* (Bréb.) Hilse in silicon-starvation synchrony. *Arch. Microbiol.*, **59**, 131.

Heslop-Harrison, J. (1962). Evanescent and persistent modifications of chloroplast ultrastructure induced by unnatural pyrimidine. *Planta*, **58**, 237.

Heslop-Harrison, J. (1963). Structure and morphogenesis of lamellar systems in grana-containing chloroplasts. *Plants*, **60**, 243.

Hill, R., and F. Bendall (1960). Function of the two cytochrome components in chloroplasts: A working hypothesis. *Nature*, **186**, 136.

Holm-Hansen, O., C. J. Lorenzen, R. W. Holmes, and J. D. H. Strickland (1965). Fluorometric determination of chlorophyll. *J. Cons. Int. Explor. Mer.*, **30**, 3.

Holt, A. S. (1965). Nature, properties and distribution of chlorophylls. In T. W. Goodwin (Ed.), *Chemistry and Biochemistry of Plant Pigments*, Academic Press, London, New York, pp. 3–28.

Holt, A. S. (1966). Recently characterized chlorophylls. In L. P. Vernon and G. R. Seely (Eds.), *The Chlorophylls*, Academic Press, London, New York. pp. 111–118.

Jeffrey, S. W. (1963). Purification and properties of chlorophyll *c* from *Sargassum flavicans. Biochem. J.*, **86**, 313.

Jeffrey, S. W. (1968a). Pigment composition of siphonales algae in the brain coral *Favia. Biol. Bull.*, **135**, 141.

Jeffrey, S. W. (1968b). Two spectrally distinct components in preparation of chlorophyll *c. Nature*, **220**, 1032.

Jeffrey, S. W. and F. T. Haxo (1968). Photosynthetic pigments of symbiotic dinoflagellates (*Zooxanthellae*) from corals and clams. *Biol. Bull.*, **135**, 149.

Jensen, A. (1964). Algal carotenoids IV. On the structure of fucoxanthin. *Acta Chem. Scand.*, **18**, 2005.

Jerlov, N. G. (1968). *Optical Oceanography*, Elsevier Publ. Co., Amsterdam, London, New York.

Ji, T. H., J. L. Hess, and A. A. Benson (1968). The nature of β-carotene association in chloroplast lamellae. In K. Shibaya and coworkers (Eds.), *Comparative Biochemistry and Biophysics of Photosynthesis*, University of Tokyo Press, Tokyo and University Park Press, Pennsylvania. pp. 36–49.

Jones, L. W., and J. Myers (1965). Pigment variations in *Anacystis nidulans* induced by light of selective wavelengths. *J. Phycol.* **1**, 7.

Jørgensen, E. G. (1964a). Adaptation to different light intensities in the diatom *Cyclotella Meneghiniana* Kütz. *Physiol. Plant.*, **17**, 136.

Jørgensen, E. G. (1964b). Chlorophyll content and rate of photosynthesis in relation to cell size of the diatom *Cyclotella Meneghiniana*. *Physiol. Plant.*, 17, 407.

Jørgensen, E. G. (1968). The adaptation of plankton algae. II. Aspects of the temperature adaptation of *Skeletonema costatum*. *Physiol. Plant.*, 21, 423.

Jørgensen, E. G. (1969). The adaptation of plankton algae. V. Variation in the photosynthetic charactersitics of *Skeletonema costatum* cells grown at low light intensity. *Physiol. Plant.* (in the press).

Kok, B., and G. Hoch (1961). Spectral changes in photosynthesis. In W. D. McElroy and B. Glass (Eds.), *Light and Life*, The Johns Hopkins Press, Baltimore, pp. 397–416.

Krasnovsky, A. A., and L. M. Kosobutskaya (1955). *Dokl. Akad. Nauk. SSSR*, 104, 440.

Krinsky, N. I. (1968). The protective function of carotenoid pigments. In A. C. Giese (Ed.), *Photophysiology*, Vol. III. Academic Press, London, New York. pp. 123–195.

Leclerc, J. C. (1969). Pigments et salinité chez *Porphyridium*. *Physiol. Plant.*, 22, 1013.

Levine, R. P. (1968). Genetic dissection of photosynthesis. *Science* 162, 768.

Levring, T. (1947). Submarine daylight and the photosynthesis of marine algae. *Göteborgs Kgl. Vetenskaps–Vitterhets–Samhäll. Handl. Ser. B*5 : 1.

Lippincott, J. A., J. Aghion, E. Porcile, and W. Bertsch (1962). The preparation and characterization of chloroplast fragments having an absorption maximum at 7400 Å. *Arch. Biochem. Biophys.*, 98, 17.

Lwowski, W. (1966). The synthesis of chlorophyll *a*. In L. P. Vernon and G. R. Seely (Eds.), *The Chlorophylls*, Academic Press, London, New York. pp. 119–143.

Madgwick, J. C. (1966). Chromatographic determination of chlorophylls in algal cultures and phytoplankton. *Deep-Sea Res.*, 13, 459.

Mann, J. E., and J. Myers (1968). Photosynthetic enhancement in the diatom *Phaeodactylum tricornatum*. *Plant Physiol.*, 43, 1991.

Manning, W. M., and H. H. Strain (1943). Chlorophyll *d*, a green pigment of red algae. *J. Biol. Chem.*, 151, 1.

Manton, I. (1966). Some possibly significant structural relations between chloroplasts and other cell components. In T. W. Goodwin (Ed.), *Biochemistry of Chloroplast*, Vol. I. Academic Press, London, New York. pp. 23–47.

Marrè, E. (1962). Temperature. In R. A. Lewin (Ed.), *Physiology and Biochemistry of Algae*, Academic Press, London, New York. pp. 541–550.

Menke, W. (1962). Structure and chemistry of plastids. *Ann. Rev. Plant. Physiol.*, 13, 27.

Moudrianakis, E. N., S. H. Howell, and A. E. Karu (1968). Characterization of the 'quantasome' and its role in photosynthesis. In K. Shibata and coworkers (Eds.), *Comparative Biochemistry and Biophysics of Photosynthesis*, University of Tokyo Press, Tokyo and University Park Press, Pennsylvania. pp. 67–81.

Mühlethaler, K. (1966). The ultrastructure of the plastid lamellae. In T. W. Goodwin (Ed.), *Biochemistry of Chloroplasts*, Vol. I. Academic Press, London, New York. pp. 49–64.

Mühlethaler, K. (1970). *J. Exp. Cell Res.* (in the press).

Ó hEocha, C. (1966). Biliproteins. In T. W. Goodwin (Ed.), *Biochemistry of Chloroplasts*, Vol. I. Academic Press, London, New York. pp. 407–421.

Öquist, G. (1969). Adaptations in pigment composition and photosynthesis by far red radiation in *Chlorella pyrenoidosa*. *Physiol. Plant.*, **22**, 516.

Park, R. B., and J. Biggins (1964). Quantasome: size and composition. *Science*, **144**, 1009.

Park, R. B., and D. Branton (1966). *Freeze-etening of Chloroplasts from Glutaraldehyde-fixed Leaves*, Brookhaven Symposia in Biology number 19, Brookhaven National Library, Upton, New York. p. 341.

Park, R. B., and N. G. Pon (1961). Correlation of structure with function in *Spinacea oleracea* chloroplasts. *J. Mol. Biol.*, **3**, 1.

Park, R. B., and L. K. Shumway (1968). The ultrastructure of fracture and deep etch faces of spinach thylakoids. In K. Shibata and coworkers (Eds.), *Comparative Biochemistry and Biophysics of Photosynthesis*, University of Tokyo Press, Tokyo and University Park Press, Pennsylvania. pp. 57–66.

Rabinowitch, E. I. (1951). *Photosynthesis and Related Processes*, Vol. II. Intersc. Publ., New York.

Shibata, K., and F. T. Haxo (1969). Light transmission and spectral distribution through epi- and endozoic algal layer in the brain coral, *Favia. Biol. Bull.*, **136**, 461.

Shibata, K., A. Takamiya, A. T. Jagendorf, and R. C. Fuller (Eds.) (1968). *Comparative Biochemistry and Biophysics of Photosynthesis*. University of Tokyo Press, Tokyo and University Park Press, Pennsylvania.

Silva, P. C. (1962). Classification of algae. In R. A. Lewin (Ed.), *Physiology and Biochemistry of Algae*, Academic Press, London, New York, pp. 827–837.

Smith, J. H. C., and A. Benitz (1955). Chlorophylls: analysis in plant materials. In K. Paech and M. V. Tracey (Eds.), *Modern Methods of Plant Analyses*, Vol. IV. Springer, Berlin, pp. 142–196.

Smith, J. H. C., and C. S. French (1963). The major and accessory pigments in photosynthesis. In L. Machlis and W. R. Briggs (Eds.), *Annual Review of Plant Physiology*, **14**, 181.

Steemann Nielsen, E. (1961). Chlorophyll concentration and the rate of photosynthesis in *Chlorella vulgaris*. *Physiol. Plant.*, **14**, 868.

Steemann Nielsen, E. (1962). Inactivation of the photochemical mechanism in photosynthesis as a means to protect the cells against too high light intensities. *Physiol. Plant.*, **15**, 161.

Steemann Nielsen, E., and V. K. Hansen (1959). Light adaptation in marine phytoplankton populations and its interrelation with temperature. *Physiol. Plant.*, **12**, 353.

Steemann Nielsen, E., and E. G. Jørgensen (1962). The adaptation to different light intensities in *Chlorella vulgaris* and the time dependence on transfer to a new light intensity. *Physiol. Plant.*, **15**, 505.

Steemann Nielsen, E., and E. G. Jørgensen (1968a). The adaptation of plankton algae I. General part. *Physiol. Plant.*, **21**, 401.

Steemann Nielsen, E., and E. G. Jørgensen (1968b). The adaptation of plankton algae. III. With special consideration of the importance in nature. *Physiol. Plant.*, **21**, 647.

Steemann Nielsen, E., and T. S. Park (1964). On the time course in adapting to low light intensity in marine phytoplankton. *J. Cons. Int. Explor. Mer.*, **29**, 19.

Tanda, T. (1951). The photosynthetic efficiency of carotenoid pigments in *Navicula minima. Am. J. Botany*, **38**, 276.

Thomas, J. B. (1962). Structure of the red absorption band of chlorophyll *a* in *Aspidistra elator. Biochim. Biophys. Acta*, **59**, 202.

Vernon, L. P. and G. R. Seely (Eds.) (1966). *The Chlorophylls*, Academic Press, London, New York.

Volk, S. L., and N. I. Bishop (1968). Photosynthetic efficiency of a phycocyaninless mutant of *Cyanidium. Photochem. Photobiol.*, **8**, 213.

Vorob'eva, V. M., and A. A. Krasnovsky (1956). The photochemical active form of chlorophyll in leaves and its transformation. *Biochimika*, **21**, 126.

Wehrmeyer, W. (1963). Über Membranbildungsprozesse im Chloroplasten. I. *Planta*, **59**, 280.

Wehrmeyer, W. (1964a). Zur Klärung der Strukturellen Variabilität der Chloroplastengrana des Spinats in Profil und Aufsicht. *Planta*, **62**, 272.

Wehrmeyer, W. (1964b). Über Membranbildungsprozesse im Chloroplasten. II. *Planta*, **63**, 13.

Weier, T. E. (1961). The ultramicrostructure of starch-free chloroplasts of fully expended leaves of *Nicotiana rustica. Amer. J. Bot.*, **48**, 615.

Weier, T. E., and A. A. Benson (1966). The molecular nature of chloroplast membranes. In T. W. Goodwin (Ed.), *Biochemistry of Chloroplasts*, Academic Press, London, New York. pp. 91–113.

Woodward, R. B., W. A. Ayer, J. M. Beaton, F. Bickelhaupt, R. Bonnett, P. Buchschacher, G. L. Closs, H. Dutler, J. Hannah, F. P. Hauck, S. Itô, A. Langemann, E. LeGoff, W. Leimgruber, W. Lwowski, J. Sauer, Z. Valenta, and H. Volz (1960). The total synthesis of chlorophyll. *J. Am. Chem. Soc.*, **82**, 3800.

Yentsch, C. S., and R. R. L. Guillard (1969). The absorption of chlorophyll *b* in vivo. *Photochem. Photobiol.* **9**, 385.

Yocum, C. S., and L. R. Blinks (1958). Light induced efficiency and pigment alterations in red algae. *J. Gen. Physiol.*, **41**, 1113.

CHAPTER 3

Photosynthetic bacteria

CHR. SYBESMA

Department of Physiology and Biophysics, University of Illinois, Urbana, Illinois 61801, U.S.A.

Introduction

The earliest recognized photoresponse in the group of organisms presently known as photosynthetic bacteria was phototaxis. In fact, Engelmann (1883) inferred photosynthetic activity from the phototactic behaviour of certain red-coloured bacteria and the correlation of phototaxis and photosynthesis of motile green algae. At that time, Winogradski (1887,

1888) was engaged in an investigation of the sulphur metabolism of both coloured and colourless bacteria. He found a definite hydrogen sulphide requirement for the growth of these organisms. Although a relation between the sulphur metabolism and the supposed photosynthetic activity was soon suspected, the two decades following these fundamental discoveries saw much confusion. This was mainly due to the failure of the investigators to detect oxygen production. The discovery of red bacteria which could grow in the dark using organic substrates (Molisch, 1907) did not contribute very much to the solution of the problem. Some 10 years later, Buder (1919) developed a theory which explained the failure to detect oxygen evolution by assuming that the photosynthetically evolved oxygen was used instantaneously for the oxidation of hydrogen sulphide and sulphur. However, this theory did not provide an answer to the problem of why many of these bacteria are unable to grow in the dark in the presence of hydrogen sulphide and exogeneous oxygen (Kluyver and Donker, 1926). The break-through came when Van Niel (1931, 1941, 1944) proposed a unified concept of photosynthesis in general. This concept is illustrated by the following reaction equation:

$$2H_2A + CO_2 \xrightarrow{h\nu} (HCOH) + H_2O + 2A$$

which indicates that the process essentially amounts to the transport of hydrogen (or electrons) from a donor H_2A to an acceptor, which is used for the carbon dioxide fixation by a reductive process. It is now generally accepted that in higher plants and algae the donor from which the oxygen is evolved is water and that in the autotrophic purple sulphur bacteria the donor is hydrogen sulphide. This explains the H_2S requirement for growth and the appearance of sulphur globules (Winogradski, 1887). Other photosynthetic bacteria can use simple organic compounds as hydrogen donors.

Three different groups of photosynthetic bacteria are presently recognized. These are the purple sulphur bacteria (Thiorhodaceae), the purple (and brown) non-sulphur bacteria (Athiorhodaceae) and the green sulphur bacteria (Chlorobacteriaceae). Photosynthesis in all these organisms occurs without oxygen evolution and only under anaerobic conditions. Most purple sulphur bacteria are photoautotrophs which assimilate carbon dioxide by the oxidation of hydrogen sulphide to sulphur (and sulphate); some species are also capable of photoassimilating organic compounds (see Chapter 4, p. 98). The purple non-sulphur bacteria use simple organic compounds as hydrogen donors and are inhibited by hydrogen sulphide (Pfennig, 1967). Most species in this group are capable of growing heterotrophically in the dark under aerobic conditions. The green sulphur bacteria are predominantly obligate photoautotrophs; only one species is

known, *Chloropseudomonas ethylicum*, which is able to grow photoorgano-
trophically (Balitskaya and Erokhin, 1963).

Phototaxis occurs in all flagellated purple sulphur and non-sulphur bacteria (Pfennig, 1967), and probably also in the motile green bacteria of the
Chloropseudomonas genus (Van Niel, 1963). In the purple (sulphur and
non-sulphur) bacteria, the tactic response is of the phobotaxic type (see,
however, Chapter 8, p. 237 of this treatise); there is no directional movement to or from the light but merely a reversal of direction following a
stimulation by light (Engelmann's *Schreckbewegung*). The action spectra
of the photophobotaxis in *Chromatium*, *Rhodospirillum rubrum*, and
Rhodospirillum molischianum (Duysens, 1952; Clayton, 1953) support
Manten's conclusion (1948) that the abrupt change in the rate of photosynthesis induces the phototatic response. This conclusion was confirmed and
extended by Clayton (1953, 1955, 1957). The subject is treated in more
detail in Chapter 8 of this treatise.

Since the early studies of Van Niel we know that bacterial photosynthesis
involves a light-induced transport of hydrogen (or electrons) from substrate
molecules to acceptor molecules; the purple sulphur bacteria (such as
Chromatium) demonstrate this clearly by the oxidation of inorganic sulphur
compounds and the reduction of carbon dioxide by light, thus showing a net
terminal electron flow. Frenkel (1956), however, discovered an additional
cyclic electron flow in which the light-induced formation of adenosine triphosphate (ATP) was demonstrated. Light, absorbed by the bacterial
photosynthetic apparatus, apparently can cause the electrons to flow in a
cyclic manner (in a closed system) and, in addition, to promote a net
transport of electrons from appropriate donors to an acceptor. The
products of this light-induced electron transport seem to be adenosine triphosphate (Frenkel, 1956) and reduced nicotinamide adenine dinucleotide
(Duysens and Sweep, 1957).

The fixation of carbon dioxide by Calvin's pentose phosphate reduction
cycle, mediated by adenosine triphosphate and reduced nicotinamide
adenine dinucleotide, is shown to occur in all strains of photosynthetic
bacteria tested (Pfennig, 1967). A key enzyme for this cycle is carboxydismutase. Synthesis of this enzyme is repressed under aerobic conditions, in
the light and in the dark; return to anaerobic conditions in the light results
in a resumption of the synthesis (Lascelles, 1960). In strict anaerobic organisms, as in the purple sulphur bacterium *Chromatium*, the concentration of
carboxydismutase depends on the phototrophic metabolism; cells grown
autotrophically with thiosulphate contained a high concentration of the
enzyme and carbon dioxide fixation occurred mostly via the Calvin cycle.
A change to photoorganotrophical conditions was followed by a rapid

decline of the carboxydismutase activity (Hurlbert and Lascelles, 1963). The Calvin cycle thus seems to be the predominant process for the fixation of carbon dioxide under photoautotrophic conditions.

Additional carbon dioxide fixation reactions can also be significant. Particulate fractions of *Chromatium* and *Chlorobium thiosulphatophilum* catalyse the synthesis of pyruvate from acetyl CoA and carbon dioxide in a reaction which depends on the presence of the non-haem iron protein ferredoxin (Buchanan, Bachofen and Arnon, 1964; Evans and Buchanan, 1965). Another ferredoxin-dependent reaction is the synthesis of α-keto-glutarate from succinyl CoA and carbon dioxide (Evans, Buchanan and Arnon, 1966). Based on the discovery of more enzymes essential for the Krebs cycle, these authors suggested the possibility of a reversal of the Krebs cycle, including the two ferredoxin-dependent carbon dioxide fixation reactions. Such a reversal had already been suggested by Van Niel (1962). However, no conclusive evidence for this is as yet available.

The photopigments

The chlorophylls

In the last three decades the isolation and purification of cultures of a wide variety of photosynthetic bacteria has substantially increased our knowledge of the photopigments, chlorophylls and carotenoids of these organisms. The chlorophyllous pigments in most purple bacteria show a more or less complicated spectrum *in vivo*, with one to three peaks in the near infrared spectral region. The early experiments of Katz and Wassink (1939) and of Wassink, Katz and Dorrestein (1939) have shown that extracts of the bacteria in organic solvents yield only one chlorophyll, bacterio-chlorophyll *a*. *In vitro* (ether solution) bacteriochlorophyll *a* has a red absorption band at 773 nm, an orange absorption band at 575 nm and Soret absorption bands at 390 and 360 nm. *In vivo*, the red band is shifted markedly to the near infrared, and shows, in most cases, a complicated spectral structure. This is illustrated by the absorption spectra of both a suspension of whole cells of *Chromatium* and of the bacteriochlorophyll *a* extracted by ether from these cells (Figure 3.1).

Another bacteriochlorophyll, bacteriochlorophyll *b* (Eimhjellen, Aas-mundrud and Jensen, 1963), appeared to exist as the only chlorophyllous pigment in the purple *Rhodopseudomonas* sp. strain NTHC 133 (*Rhodo-pseudomonas viridis*), and in the sulphur bacterium *Thiococcus* sp., strain Nidelven (Eimhjellen and coworkers, 1967). The major chlorophyll of the green sulphur bacteria is chlorobium chlorophyll (Larsen, 1953) earlier called bacterioviridin (Metzner, 1922). Stanier and Smith (1960) discovered two types of chlorobium chlorophyll, designated chlorobium chlorophyll 650 and chlorobium chlorophyll 660 according to their absorption maximum in ether solution. Any given strain of green bacteria contains one or the other chlorobium chlorophyll type, but not both. In addition to chlorobium chlorophyll, all green bacteria investigated contain bacteriochlorophyll *a* in an amount of about 5 to 10 per cent of the total chlorophyll content (Holt and coworkers, 1963; Jensen, Aasmundrud and Eimhjellen, 1964; Olson and Romano, 1962). The bacteriochlorophyll *a* bound to its

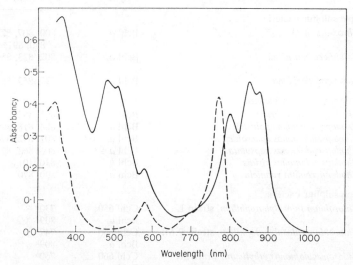

Figure 3.1. Absorption spectra of *Chromatium*, strain D. Solid curve: a suspension of whole cells in growth medium containing sulphide and thiosulphate as electron donors and bicarbonate as the carbon source. Broken curve: an ether extract of the same cells

native protein can be extracted from green bacteria as a soluble bacteriochlorophyll *a*–protein complex (Olson and Romano, 1962). The complex is probably made up of four identical subunits (with a molecular weight of about 38 000), each containing five bacteriochlorophyll *a* molecules (Thornber and Olson, 1968). The complex may be approximated by a prolate ellipsoid (Olson, Koenig and Ledbetter, 1969) in which the bacteriochlorophyll molecules are radially oriented (Olson, Jennings and Olson, 1969). The complex *in vitro* is not photochemically active. Sybesma and Olson (1963) have demonstrated that *in vivo* the chlorobium chlorophyll serves as a light-harvesting pigment in the green bacteria, transferring some 80 per cent of the absorbed light energy to the bacteriochlorophyll *a*.

Table 3.1 gives the distribution of the chlorophylls among some organisms, together with their spectral characteristics.

Table 3.1. Near infrared absorption bands of the chlorophylls of some photosynthetic bacteria

Organism	Chlorophyll type	Near infrared absorption maxima *in vivo* (nm)
Purple sulphur bacteria		
Chromatium D	Bchl *a*	800, 807, 823, 850, 891[a]
Rodothece conspicua	Bchl *a*	802, 823, 850, 891[a]
Thiocapsa floridana	Bchl *a*	797, 853, 889[a]
Purple non-sulphur bacteria		
Rhodospirillum rubrum	Bchl *a*	805, 883[a]
Rhodopseudomonas palustris	Bchl *a*	804, 818, 880[a]
Rhodopseudomonas spheroides	Bchl *a*	801, 852, 870[a]
Rhodopseudomonas capsulata	Bchl *a*	806, 863, 875[a]
Rhodopseudomonas viridis	Bchl *b*	830, 1014[b]
Rhodomicrobium vannielii	Bchl *a*	800, 870[b]
Green sulphur bacteria		
Chlorobium thiosulphatophilum, strain L	Cchl 650	730[b]
	Bchl *a*	809[b]
Chlorobium thiosulphatophilum, strain P.M.	Cchl 660	747[b]
	Bchl *a*	809[b]
Chloropseudomonas ethylicum, strain 2K	Cchl 660	750[b]
	Bchl *a*	809[b]

[a] Spectral data from Vredenberg and Amesz (1966).
[b] Spectral data from Olson and Stanton (1966).

The carotenoids

Carotenoids most commonly found in both purple sulphur and purple non-sulphur bacteria are members of the so-called normal Spirilloxanthin series, and are aliphatic intermediates in the transformation of lycopene to spirilloxanthin (Jensen, 1965). Species in which the end-product spirilloxanthin predominates are the Rhodoteceae, the Thiopedieae, *Rhodospirillum rubrum* and a few strains of *Chromatium*. In most *Chromatium* strains, *Thiocystis*, *Rhodopseudomonas palustris* and the hypomicrobace *Rhodomicrobium vannielii*, earlier members of the Spirilloxanthin series are also predominantly present. In *Rhodospirillum fulvum*, *Rhodospirillum molischianum* and in some purple sulphur bacteria, lycopene and rhodopene are the predominant carotenoids. Species in which intermediates from another transformation series leading to spirilloxanthin synthesis (such as spheroidene and hydroxyspheroidene) predominate are *Rhodopseudomonas spheroides*, *Rhodopseudomonas gelatinosa* and *Rhodopseudomonas capsulata*. In addition, some keto-carotenoids are found in several *Chromatium* species.

The carotenoid characteristics of the green sulphur bacteria are quite different. The major carotenoid in both *Chlorobium* and *Chloropseudomonas* is the alicyclic γ-carotene named chlorobactene (Jensen, Hegge and Jackman, 1964).

Carotenoid–chlorophyll interaction has been inferred (Bergeron and Fuller, 1959; Clayton and Arnold, 1961; Izawa and coworkers, 1963) to explain the wide variety of the structure of the near infrared bacteriochlorophyll *a* absorption spectra *in vivo*. However, evidence against a specific carotenoid effect is accumulating (Bril, 1963; Crounse, Sistrom and Nemser, 1963; Wassink and Kronenberg, 1962). It seems more likely that chlorophyll–protein interactions, chlorophyll–chlorophyll interactions or a combination of both are responsible for the *in vivo* appearance of the near infrared absorption spectra of bacteriochlorophyll. One could consider the carotenoids along with the lipids as non-polar elements of the *in vivo* environment of the chlorophylls. Whereas chlorophyll–chlorophyll interactions seem to offer a satisfactory explanation for the near infrared absorption spectra of the green bacteria (Krasnovskii, 1965; Krasnovskii Erokhin and Fedorovich, 1961; Krasnovskii, Erokhin and Yü-Ch'ün, 1962), a more specific theory seems to be required for the explanation of the near infrared absorption bands in purple bacteria (Olson and Stanton, 1966).

Carotenoids transfer about 30 to 50 per cent of the absorbed light energy to bacteriochlorophyll (Duysens, 1952), and are to such an extent functional as 'light-harvesting' pigments. A more fundamental function of carotenoids may be their protective action against harmful photooxidation of the bacteriochlorophylls (Clayton, 1966d). Several carotenoid-lacking mutant strains of photosynthetic bacteria are known to be extremely sensitive to such photooxidations (Sistrom, Griffiths and Stanier, 1956).

The specialized bacteriochlorophylls in the reaction centra

Duysens (1952) discovered in the sulphur bacterium *Chromatium* and in the non-sulphur bacterium *Rhodospirillum rubrum* light-induced changes in the absorption spectrum of bacteriochlorophyll *a in vivo*, which resemble those induced by chemical oxidation (Goedheer, 1960). The most conspicuous changes are a slight bleaching of a near infrared band at about 880 nm and a marked blue shift of an absorption band at about 800 nm. Similar light-induced absorbancy changes were found in the non-sulphur bacterium *Rhodopseudomonas spheroides* (Clayton, 1962a). The quantum requirement for the changes was found to be from 2 to 4, even in dried chromatophore preparations at $1°\text{K}$ (Clayton, 1962b). They are due, not to a light-induced alteration of part of the bacteriochlorophyll *a*, but to a light-induced alteration of specialized molecules present in an amount of about 2 to 5 per cent of the total bacteriochlorophyll *a*. For *Rps. spheroides* this has been demonstrated convincingly by Clayton (1966a) who was able to retain the full extent of the absorbancy changes in a particulate fraction of the organisms in which the major complement of the bacteriochlorophyll was destroyed. Extraction of this preparation with organic solvents yielded solutions of bacteriochlorophyll *a*, thus demonstrating that the specialized molecules indeed are bacteriochlorophyll *a*. The absorbancy changes in the preparation could be induced either by light or by chemical oxidation.

The specialized bacteriochlorophyll *a* molecules have been designated P 890 (Vredenberg and Duysens, 1963) for *Chromatium* and *Rhodospirillum rubrum* and P 870 (Clayton, 1963a) for *Rhodopseudomonas spheroides*. They are apparently components of so-called reaction centra, sites at which excitation energy is converted into chemical energy; light absorbed by the major pigments is transferred to these reaction centra, causing the oxidation of P 890 or P 870. The importance of the reaction centra is emphasized in mutants of the facultative aerobe *Rps. spheroides* which contain the full amount of photosynthetic pigments, but lack the specialized bacteriochlorophyll *a*, P 870. These mutants are unable to grow photosynthetically and do not exhibit any of the characteristic light-induced absorbancy changes (Sistrom and Clayton, 1964). The bacterial photosynthetic apparatus thus appears to be organized, like in higher plants and algae, in photosynthetic units.

The experiments of Clayton (1966a) also indicated that in *Rps. spheroides* the bleaching of the near infrared absorption band (at 865 nm) and the blue shift of the absorption band at 800 nm are alterations of two different molecular entities. Sauer, Dratz and Coyne (1968) investigated this more closely in a subcellular preparation of *Rps. spheroides* in which the light-harvesting bulk of bacteriochlorophyll was destroyed but the reaction centra were left intact (Reed and Clayton, 1968). By comparing the absorption and circular dichroism in reduced, oxidized and illuminated preparations, these authors were able to establish that the reaction centra contain complexes of at least three bacteriochlorophyll molecules. Excitation interaction results in a three-fold splitting of the energy levels in the complex, one level giving rise to the absorption band at 865 nm and the other two, almost degenerate levels, giving rise to the absorption band at about 800 nm. When the central molecule in the complex is oxidized (chemically or by illumination), causing a major shift in its absorption spectrum (Reed and Clayton, 1968), the interaction between the remaining bacteriochlorophyll molecules becomes much weaker; the absorption then is at only one (higher energy) level, thus explaining the blue shift.

Light-induced absorbancy changes in the purple bacterium *Rhodopseudomonas viridis* (NTHC 133) are analogous to those observed in other purple bacteria, but in a spectral region farther out to the red (Holt and Clayton, 1965). Bleaching of the reaction centre component P 985 is accompanied by a blue shift of a band at 830 nm (Olson and Clayton, 1966).

In the green photosynthetic bacteria, a light-induced bleaching of an absorption band around 840 nm (where there is very little absorption of the bulk pigments) can be considered as indicative of the oxidation of a reaction centre component P 840 (Sybesma and Vredenberg, 1963). Additional evidence for this is the relation of this bleaching with the oxidation of a cytochrome and its persistence at liquid nitrogen temperatures (Sybesma and Vredenberg, 1964). In contrast with the purple bacteria, this bleaching is not accompanied by another spectral shift.

The structure of the photosynthetic apparatus

The absence of chloroplasts from the cells of photosynthetic bacteria (and of blue-green algae) is one of the distinctive cellular properties which set these so-called procaryotic organisms apart from all other so-called eucaryotic organisms. This distinction raises a host of important problems in respect to the relations between structure and function, such as the problem of the existence of a structural unit for the photosynthetic apparatus.

In 1952 Pardee, Schachman and Stanier analysed extracts of *Rhodospirillum rubrum* in an analytical centrifuge and isolated a fraction having an absorption spectrum similar to that of the whole cells (Pardee, Schachman and Stanier, 1952; Schachman, Pardee and Stanier, 1952). The sedimentation constant of the fraction was 190 Svedberg units. Electron micrographs of such fractions (Thomas, 1952) showed disk-like structures with a diameter of about 1100 Å. Later, Frenkel (1954, 1956) demonstrated that these particles were able to carry out photo-induced phosphorylation. Since then these particles, called 'chromatophores', have been considered as the structural units of the photosynthetic apparatus. This view received impetus by results of thin section electron microscopy (Vatter and Wolfe, 1958) which showed cytoplasmic structures in light-grown cells of *R. rubrum* very similar to Pardee's chromatophores. Membrane-bounded vesicles, smaller in size but similar in appearance, were seen in thin sections of *Rhodopseudomonas spheroides* (Vatter and Wolfe, 1958) and *Chromatium D* (Bergeron, 1958). It soon became apparent, however, that these structural units may be artifacts of the preparation. Tuttle and Gest (1959) failed to get the chromatophores out of *Rhodospirillum rubrum* cells by mild osmotic lysis. Moreover, the observed vesicular structures did not seem to be a universal feature of the purple bacteria. Fine structure studies on *R. molischianum* (Drews, 1960), *Rhodomicrobium vannielii* (Boatman and Douglas, 1961), *R. fulvum, Rps. palustris* and *Rps. viridis* (Cohen-Bazire and Sistrom, 1966) showed streaks of regularly stacked lamellae situated near the periphery of the cell. From microscopic studies on *R. molischianum* Gibbs, Sistrom and Worden, 1965; Giesbrecht and Drews, 1962), *R. rubrum* (Cohen-Bazire and Kunisawa, 1963) and *Rps. spheroides* (Drews and Giesbrecht, 1963) evidence was obtained for a continuity of the lamellae and the vesicular structures with the cytoplasmic membrane (see Figure 3.2, Plate I). A mixture of vesicular and lamellar organization of intracytoplasmic membranes has been seen in *Chromatium D*, grown in high light intensity, and in *Thiocapsa* (Cohen-Bazire, 1963). These results seem to support a structural interpretation, first given by Tuttle and Gest (1959), in which the vesicular structure is seen as resulting from invaginations of the cytoplasmic membrane (Hickman and Frenkel, 1965).

Various external conditions, such as light intensity, temperature and (in facultative aerobes) oxygen tension, determine the pigment content of a number of purple bacteria (Cohen-Bazire and Kunisawa, 1960; Gibbs, Sistrom and Worden, 1965). In aerobically grown *R. rubrum*, thin section electron microscopy showed only very few membrane-limited vesicles. When pigment synthesis was resumed in the light under conditions of oxygen limitation, more vesicular invaginations appeared in the cytoplasm (Cohen-Bazire and Kunisawa, 1963). Cells grown in high intensity light with a low pigment content showed far less membrane-limited vesicles than cells grown in lower intensity light with a higher pigment content (Cohen-Bazire, 1963; Cohen-Bazire and Kunisawa, 1963). In the latter case, several vesicles could be seen to be formed out of one invagination of the cytoplasmic membrane and to be connected to each other. In cells with a very high bacteriochlorophyll content, the stacked lamellar structure seemed to be predominant (Gibbs, Sistrom and Worden, 1965). It seems reasonable to assume, therefore, that pigment synthesis goes together with infoldings of the cytoplasmic membrane which can spread out to form the vesicular structure seen in the thin sections. At high pigment concentrations the membrane infoldings become appressed together, thus forming the stacks of lamella. The fact that bacteriochlorophyll synthesis is closely linked with protein synthesis, which is necessary for the formation of membrane material (Bull and Lascelles, 1963; Gray, 1967; Lascelles, 1965), is in agreement with such an assumption. Drastic disruption of the cells by sonication or in a French pressure cell may result in a preferential shearing at the contact sites of the vesicular membranous invagination, thus showing the flat disks in the chromatophore preparations.

Disruption of the cells followed by differential centrifugation through a sucrose density gradient often yielded two fractions (a heavy fraction and a light fraction) which were different, not only in their sedimentation rates but also in properties such as the ratio between the different spectral forms of bacteriochlorophyll and the bacteriochlorophyll–carotenoid ratio. This has been demonstrated in *R. rubrum* (Cohen-Bazire and Kunizawa, 1960), in *Rps. spheroides* (Worden and Sistrom, 1964) and in *Chromatium* (Cusanovich and Kamen, 1968a). In the latter case, the light fraction appeared to be the most stable one, containing about 41 per cent of the bacteriochlorophyll and 26 per cent of the carotenoid found in the whole cells. It consisted for about 75 per cent of lipid and for about 25 per cent of protein, one-third of which could be ascribed to cytochromes. The particle was photosynthetically active. The sedimentation constant was 145 S which compares favourably to a value of 143 S found by Bergeron (1958) for *Chromatium* chromatophores. Electron microscopy showed disks (or spheres) with a diameter of about 650 Å, which is somewhat smaller than the diameter of Bergeron's chromatophores. Although this functional preparation is the most reproducable one obtained thus far, its relation to the *in vivo* structure of the photosynthetic apparatus is still a matter of speculation.

Data on the relation between structure and function in the green bacteria are scarce. The vesicular structure, a characteristic cytoplasmic feature of most purple bacteria, seemed to be absent in the green bacteria *Chlorobium limicola* (Vatter and Wolfe, 1958), *Chlorobium thiosulphatophilum* (Bergeron and Fuller, 1961; Fuller, Conti and Mellin, 1963) and *Chloropseudomonas ethylicum* (Cohen-Bazire, Pfennig and Kunisawa, 1964). A functional particle, approximately 150 Å in diameter, which contained almost all of the pigment of the cell was isolated from *Chlorobium thiosulphatophilum* (Bergeron and Fuller, 1961). It appeared later that these particles are probably macromolecular complexes derived by comminution of larger oblong vesicles which are 300 to 400 Å wide and 1000 to 1500 Å long (see Figure 3.3, Plate II) and are located at the periphery of the cell immediately under the cytoplasmic membrane (Cohen-Bazire, 1963; Holt, Conti and Fuller, 1966).

Photosynthetic electron transport

Light-induced cytochrome reactions

In the years following Van Niel's pioneering work, the concept of photosynthesis as a process which provides for a photochemical separation of oxidizing and reducing power received substantial support. A large part of the evidence resulted from the development of techniques for ultrasensitive absorption spectrophotometry (Chance, 1951; Duysens, 1952). Reversible changes in the absorption spectrum induced by light have been measured by such techniques in a variety of bacterial cells and chromatophore preparations. One class of such changes are the already mentioned reversible changes in the near infrared absorption, which reflect the oxidation of components of the reaction centra (Clayton, 1962a,, 1962b, 1962c; Duysens, 1952; Duysens and coworkers, 1956). Another class of light-induced absorbancy changes can be attributed to the oxidation of cytochromes (Chance, 1954; Duysens, 1954). By kinetic analysis of light-induced absorbancy changes in whole cells of *Chromatium*, grown autotrophically with sulphide and thiosulphate, Olson and Chance (1960) unscrambled the light-induced oxidation of three cytochromes. One of these cytochromes, designated C 552, is oxidized by dim light. The quantum requirement for this reaction was found to be close to one (Olson, 1962). The suggestion made by Olson and Chance (1960) that the rereduction of this cytochrome was mediated by the electron-donating substrates was confirmed by Morita, Edwards and Gibson (1965) who showed that the cytochrome was in the oxidized state in starved cells and could be reduced subsequently by the addition of sulphide, thiosulphate or molecular hydrogen. In the presence of oxygen, the cytochrome was oxidized and remained oxidized in the dark. Bright light oxidized the other two cytochromes, designated C 555 and *cc′*, in a reaction which was independent of the redox state of C 552[†]. The reactions of these two cytochromes were unaffected by oxygen. The kinetics of the reaction suggested that C 555 was rereduced, either through more intermediates or directly by cytochrome *cc′*. Vredenberg and Duysens (1964) reported a quantum requirement for the oxidation of C 555 which was about the same as that for the oxidation of C 552, close to one photon per electron transported. The rapid rereduction of the cytochromes C 555 and *cc′* when the light was turned off and the relatively high intensity necessary to change the redox state of the cytochromes led to the postulation (Olson and Chance,

---

[†]The light-induced absorbancy change, centered at 430 nm, is often ascribed to a reaction of a *b*-type cytochrome. Hind and Olson (1968) pointed out that the spectral characteristics of cytochrome *cc′* fit the data just as well. A function of *b*-type cytochromes in bacterial photosynthetic electron transport is doubted by these authors.

1960) that both cytochromes are involved in a cyclic electron transport chain coupled to the formation of adenosine triphosphate (Frenkel, 1956). Support for this suggestion was given by Morita, Edwards and Gibson (1965), who demonstrated that 2-n-nonyl-4-hydroxyquinoline-*N*-oxide (NQNO) and antimycin A retarded the rereduction of C 555 after illumination, but did not affect the rereduction of C 552.

These results seem to indicate that there are two independent electron transport chains in *Chromatium*, a non-cyclic one which is coupled to substrate oxidation and a cyclic one which is coupled to adenosine triphosphate formation. A recent confirmation for this came from Cusanovich, Bartsch and Kamen (1968). They made a careful study of light-induced absorbancy changes in the light chromatophore fraction obtained by sucrose density gradient centrifugation of ruptured cells of *Chromatium*, grown heterotrophically with succinate (Cusanovich and Kamen, 1968a). This fraction contained the bound cytochromes C 552, C 555 and *cc'*. As in the whole cells, low intensity light oxidized C 552 and high intensity light oxidized C 555 and *cc'* independently of the redox state of C 552. By titration of the absorbancy changes with various redox buffers at a pH of 7·5, an estimate of the midpoint potentials for the bound cytochromes was made. According to this estimate, the midpoint potentials of C 555 and *cc'* were + 0·32 V and + 0·18 V respectively, and the midpoint potential of the reaction centre component P 890 was found to be + 0·49 V. The midpoint potential of C 552 could not be established, but was certainly below 0 mV. The authors reported evidence for a two electron acceptor for this cytochrome, with a midpoint potential of − 0·13 V. From a subsequent study with the same system Cusanovich and Kamen (1968b) concluded that ATP-formation was coupled to electron transport in which only the cytochromes C 555 and *cc'* are involved.

A similar situation appeared to exist in other photosynthetic bacteria. In *Rhodospirillum rubrum* two cytochromes were found to be oxidized by light (Duysens, 1957). Sybesma and Fowler (1968) reported that only one of these, C 428, was oxidized in dim light, but that bright light also oxidized the other one, C 422. The kinetics of the reactions indicated that C 422 is involved in a cyclic electron transport. In starved cells the rereduction of C 428 is extremely slow (with a half-time of several seconds); a speeding-up can be accomplished by the addition of malate (Sybesma, unpublished), indicating that this cytochrome is coupled to substrate dehydrogenation. Cytochrome C 422 has been identified as cytochrome c_2, with a redox potential of + 0·31 V (Horio and Yamashita, 1963). It can be extracted from the cells as a soluble protein (Vernon, 1953). Chance and coworkers (1966) labelled C 428 as cytochrome *o*. Cytochrome *cc'* (also called RHP or cytochromoid C) has also been extracted from this organism (Bartsch, 1963; Vernon, 1953). Its exact function *in vivo*, however, is not certain.

In the bacteriochlorophyll *b* containing bacterium *Rhodopseudomonas viridis*, two of the three *c*-type cytochromes present function in photosynthesis. Cytochrome C 553 seemed to have the same function as C 552 in *Chromatium* and C 428 in *R. rubrum*, while cytochrome C 558 appeared to function in a cycle like C 555 in *Chromatium* and C 422 in *R. rubrum* (Olson and Nadler, 1965). In the green bacterium *Chloropseudomonas ethylicum*, a cytochrome, C 422, was found to be coupled to substrate dehydrogenation and another cytochrome, C 419, was shown to be involved in a cyclic electron transport chain (Sybesma, 1967). *Rhodomicrobium vannielii* contains a high potential cytochrome, C 550, and a lower potential particle bound cytochrome, C 553 (Morita and Conti, 1963; Morita, Olson and Conti, 1964). Two *c*-type cytochromes and one *b*-type cytochrome were isolated from *Rhodopseudomonas spheroides* (Orlando, 1962, 1967; Orlando and Horio, 1961). One of the *c*-type cytochromes (C 553) was thought to be functional in respiration (Orlando, 1962). The function *in vivo* of the other two cytochromes could not be asserted.

Light-induced reactions of the specialized bacteriochlorophyll

Chance and Nishimura (1960) reported the light-induced oxidation of cytochrome C 552 in *Chromatium* at 77°K, which occurred, albeit irreversibly, with a quantum efficiency roughly equal to that at room temperature. This observation has led these authors to suggest that the light-induced oxidation of cytochromes in photosynthetic bacteria is a temperature-independent reaction, and, indeed, the primary photochemical reaction (Nishimura and coworkers, 1964). This idea has been challenged, predominantly by Duysens and his coworkers; they maintained that the oxidation of the specialized bacteriochlorophyll components in the reaction centra, P 890 or P 870, is the primary photochemical act in bacterial photosynthesis (Beugeling and Duysens, 1966; Duysens, 1966). Several experimental results were in their favour. The light-induced oxidation of the cytochromes in a number of purple and green bacteria did not appear to be a temperature-independent reaction (Sybesma and Vredenberg, 1964; Vredenberg and Duysens, 1964). In contrast, the light-induced oxidation of P 890 and P 870 in chromatophore preparations of *Chromatium*, *Rhodospirillum rubrum* and *Rhodopseudomonas spheroides* occurred with a high efficiency at temperatures as low as 1°K (Arnold and Clayton, 1960; Clayton, 1962b), and the light-induced oxidation of P 840 in *Chloropseudomonas ethylicum* was demonstrated at 77°K, a temperature at which no oxidation of cyto-chrome occurs (Sybesma and Vredenberg, 1964). Poisons like phenyl mercuric acetate, which inhibit the cytochrome reaction, promote the light-induced oxidation of P 890, P 870 or P 840, but inhibit its rereduction in the dark (Clayton, 1963a; Sybesma and Vredenberg, 1964).

The matter was finally settled by Parson (1968), who measured the kinetics of P 890 and cytochrome C 555 oxidation induced by 30 nsec flashes from a Q-switched ruby laser in a *Chromatium* chromatophore preparation. He not only found that the light-induced oxidation of P 890 was about four times faster than the light-induced oxidation of the cyto-chrome C 555, but also that the half-time of the rereduction of P 890 after the flash corresponds to the oxidation half-time of C 555. The primary photoact must therefore be the oxidation of P 890 in the reaction centre followed by the oxidation of a cytochrome.

Based on the results obtained by varying the environmental redox potential of their *Chromatium* chromatophore preparation, Cusanovich, Bartsch and Kamen (1968), have suggested that the two independent electron transport chains are driven by two different types of reaction centra. This was confirmed by Morita (1968) who found that in whole cells of *Chromatium* the action spectra for the oxidation of the cytochromes C 555 and C 552 were different. In whole cells of *R. rubrum* also, the action spectra for the oxidation of the cytochromes C 422 and C 428 were found to be dif-ferent (Sybesma and Fowler, 1968). A closer study of the relation between the reaction centra and the cytochromes in *R. rubrum* (Sybesma, 1969b) indicated that, whereas the cyclic reaction was driven by the light-induced oxidation of P 890 molecules which are part of a complex of at least three specialized bacteriochlorophyll molecules (Sauer, Dratz and Coyne, 1968), the non-cyclic reaction was driven by the light-induced oxidation of 'iso-lated' P 890 molecules. Both types of reaction centre could operate quite independently from each other.

In *Chromatium* the cyclic reaction is also driven by a light-induced reac-tion of a specialized bacteriochlorophyll complex (P 800 · P 890 · P 800), as shown by Cusanovich, Bartsch and Kamen (1968). The non-cyclic reaction, according to these authors, is driven by the light-induced reaction of other specialized bacteriochlorophylls. The appearance of an absorption band at 905 nm upon illumination of the chromatophore preparation in a low potential environment was taken as an indication of these specialized bacteriochlorophyll molecules. This light-induced increase of absorbance was seen also by Clayton (1963b) in *Rps. spheroides* and explained by him as a reduction of bacteriochlorophyll in an independent light reaction. Vredenberg and Amesz (1966) however explained this phenomenon as a

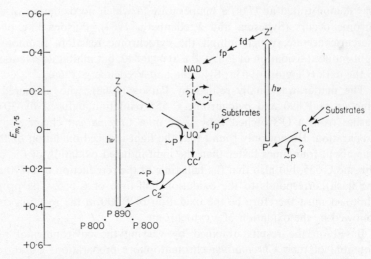

Figure 3.4. Electron transport schema for *Chromatium* and *Rhodospirillum rubrum*. C_1 and C_2 designate the cytochromes C 552 and C 555 in *Chromatium* and the cytochromes C 428 and c_2 in *R. rubrum*. Z and Z' are the (unknown) electron acceptors for the cyclic and non-cyclic reactions respectively. P' is the primary donor for the non-cyclic reaction. Ferredoxin, flavoprotein, ubiquinone and nicotiamide adenine dinucleotide are symbolized by fd, fp, UQ and NAD respectively. $\sim$ I and $\sim$ P designate high-energy intermediates and high-energy phosphates (adenosine triphosphate)

light-induced band shift in the absorption spectrum of the light-harvesting chlorophyll.

The electron transport schema presented in Figure 3.4 summarizes the interpretation of the recent experimental results on *Chromatium* and *R. rubrum*. It has no dogmatic character and ought to be regarded as a working hypothesis only, or at most as a snapshot of the present state of affairs in research on bacterial photosynthetic electron transport.

Light-induced reduction of pyridine nucleotide

An efficient light-induced reduction of nicotinamide adenine dinucleotide (NAD^+) in intact photosynthetic bacteria has been observed by Duysens and Sweep (1957), Olson (1958) and Amesz (1963). Photoreduction of NAD^+ in cell-free preparations of *Rhodospirillum rubrum* can occur in the presence of substrate molecules, such as succinate or reduced dichlorophenol indophenol (Nozaki, Tagawa and Arnon, 1961; Vernon and Ash,

1960). Similar reactions were found in a small particle fraction from *Chromatium* (Hood, 1964). The light reaction was stimulated by the addition of flavoprotein together with the non-haem iron protein ferredoxin.

There is no general agreement about the mechanism of the light-induced reduction of NAD^+ *in vivo*. Bose and Gest (1962) have suggested that NAD^+ is reduced not by a direct photochemically induced electron transport but by an energy-linked reversal of electron flow, such as the succinate coupled reduction of pyridine nucleotides in a reaction linked to energy supplied by ATP which has been reported for mitochondria (Chance and Hollunger, 1961). The main function of the light, then, would be the generation of ATP (Losada and coworkers, 1960; Stanier and coworkers, 1959). Amesz (1964), however, reported that in whole cells of *R. rubrum* the light-induced reduction of NAD^+ is not inhibited by high concentrations of 2-n-heptyl-4-hydroxyquinoline-*N*-oxide (HOQNO), a poison which is known to inhibit (cyclic) photophosphorylation at lower concentrations in whole cells (Nishimura, 1963; Smith and Baltscheffsky, 1959) and in bacterial extracts (Baltscheffsky and Baltscheffsky, 1960). This would rule out a direct involvement of ATP in the photoreduction of NAD^+. The possibility (Newton, 1964) of a competition for a high-energy intermediate between the final phosphorylation and the reduction of NAD^+ cannot be ruled out, however. Suggestive in this respect is the observation that the addition of a phosphorylating system in chromatophore fragments of *R. rubrum* inhibited the reduction of NAD^+ (Horio, Yamashita and Nishikawa, 1963). Hinkson (1965) studied the effects of several inhibitors on the photoreduction of NAD^+ and ATP formation in chromatophore preparations of *Chromatium* and *R. rubrum*. He concluded that the existence of a common high energy intermediate for both reactions still cannot be proven or disproven. However, as Hind and Olson (1968) pointed out, a direct photochemical reduction of NAD^+ does not exclude an energy-linked reversed electron flow also leading to NAD^+ reduction. It is quite possible that the bacteria utilize both types of non-cyclic electron flow.

The primary electron acceptor; the role of ubiquinone

The reaction centre can be visualized as a complex in the form $Cyt \cdot P\,890 \cdot Z$ in which Z is an electron acceptor. The primary photoact then can be represented by

$$Cyt \cdot P\,890 \cdot Z \xrightarrow{h\nu} Cyt \cdot (P\,890)^* \cdot Z \xrightarrow{k_1} Cyt \cdot (P\,890)^+ \cdot Z^-$$

followed by

$$Cyt \cdot (P\,890)^+ \cdot Z^- \xrightarrow{k_2} Cyt^+ \cdot P\,890 \cdot Z^-$$

in which $k_1 > 1\cdot4 \times 10^6$ and $k_2 \approx 0\cdot3 \times 10^6$ in *Chromatium* (Parson, 1968). The identity of the primary electron acceptor Z is not known. Several investigations have suggested ubiquinone (coenzyme Q 10) for such a function. Zaugg, Vernon and Tirpack (1964) have demonstrated a photoreduction of ubiquinone by reduced phenasine methosulphate (PMS), catalysed by either chromatophore preparations or bacteriochlorophyll in solution. Quinones are detected in all purple and green photosynthetic bacteria (Pennock, 1966; Redfearn, 1966). Purple bacteria contain ubiquinone (Takamiya, Nishimura and Takamiya, 1967), and in the facultative photoheterotrophes the level in light-grown cells exceeds the level in dark-grown cells (Carr, 1964; Fuller and coworkers, 1961). Ubiquinone 10 appeared to be an essential factor for cyclic photophosphorylation (Okayama and coworkers, 1968). Green bacteria contain menaquinone and vinyl-1-4-naphthaquinone (Frydman and Rapoport, 1963; Takamiya, Nishimura and Takamiya, 1967).

Clayton (1962d) observed in dried chromatophore preparations of several purple bacteria light-induced changes in the ultraviolet spectral region with a difference spectrum similar to an oxidized–minus–reduced spectrum of ubiquinone. Based on such types of measurements, he suggested that ubiquinone may be the primary electron acceptor under conditions in which the cytochrome remains oxidized in the dark (Clayton, 1963b). Recent experiments of Parson (1967) could indicate that this may be the case. If the spectral characteristics of the light-induced ultraviolet absorbancy changes are considered as evidence for the reduction of ubiquinone in the primary reaction, it must involve the transport of two electrons (Beugeling, 1968). Evidence for a $-0\cdot02$ V two electron acceptor for P 890 was found by Loach (1966) in *Rhodospirillum rubrum* chromatophores. Parson (1967) reported that his results would indicate the involvement of two different pools of ubiquinone which do not communicate rapidly. Some evidence for two ubiquinone pools can be found in recent work of Ke and coworkers (1968). The spectroscopic and kinetic characteristics of the light-induced absorbancy changes, however, are of such a complex nature that it seems wise to withhold any conclusions about their origin. They may, for example, also reflect changes of the redox state of bacteriochlorophyll in the reaction centre.

The relative abundancy of quinones in the cells (Takamiya, Nishimura and Takamiya, 1967) and a midpoint potential of ubiquinone of $+0\cdot10$ V at pH 7·4 at 25°C (Moret, Pinamonti and Fornasari, 1961) favours a function as a redox pool in the centre of the cyclic reaction (see Figure 3.4). Ubiquinone may also interact with substrates, such as succinate, and NAD^+ through specific transhydrogenases (flavoproteins). There is indirect evidence (Sybesma, 1969a) that a common pool of ubiquinone may react in both the cyclic and non-cyclic electron transport chains.

Non-haem iron proteins

Low potential ($-$ 0·5 V) non-haem iron proteins known as ferredoxins are found in *Chromatium* (Bachofen and Arnon, 1966), *Rhodospirillum rubrum* (Tagawa and Arnon, 1962), *Chlorobium thiosulphatophilum* (Evans and Buchanan, 1965). Arnon (1965) has proposed a central role of ferredoxin in $NADP^+$ reduction and cyclic phosphorylation in higher plants and, by implication, in photosynthetic bacteria. There is no evidence, however, for the participation of ferredoxin in the cyclic electron transport chain. In fact, ferredoxin appeared to be absent in a washed chromatophore preparation of *R. rubrum* which was capable of cyclic photophosphorylation (Horio and coworkers, 1968). Some evidence may be quoted for a function of ferredoxin in the non-cyclic electron transport chain. Flavoprotein together with ferredoxin stimulated the photoreduction of NAD^+ in a small-particle fraction from *Chromatium*; ferredoxin alone, however, inhibited the reaction (Hood, 1964). A ferredoxin linked reduction of NAD^+ by hydrogen in the dark has been found in extracts of *Chromatium* (Weaver, Tinker and Valentine, 1965). Trebst, Pistorius and Baltscheffsky (1967) concluded, however that ferredoxin was not required for the NAD^+ photoreduction in *R. rubrum*. They found that the light-induced reduction of NAD^+ by *p*-phenylenediamines was not inhibited by disalycylidene propanediamine disulphonic acid, an inhibitor presumably specific for ferredoxin.

A role for ferredoxin in the light-dependent nitrogen fixation is supported by the work of Tagawa and Arnon (1962). A general function of ferredoxin in the hydrogen and nitrogen metabolism in photosynthetic bacteria is proposed by Bennett and Fuller (1964) and Buchanan, Bachofen and Arnon (1964).

Photophosphorylation

Cyclic photophosphorylation

One of the major functions of the bacterial photosynthetic process is the formation of adenosine triphosphate (ATP) in light-induced phosphorylation reactions coupled to electron transport. Whole cells, chromatophore fragments and cell-free preparations catalyse this reaction without involving the net oxidation or reduction of added substrates. The electron transport reactions responsible for the phosphorylation thus appear to be cyclic in nature. In the original experiments of Frenkel (1954), phosphorylation in a washed chromatophore preparation from *Rhodospirillum rubrum* appeared to be dependent on the addition of substrates, such as α-ketoglutarate or succinate. No stoichiometric relationship could be detected, however. It was later found (Bose and Gest, 1963; Frenkel, 1956) that phosphorylation is sensitive to the redox state of the electron transport components; the substrate effects thus could be explained by their action on the redox state of these components, the electron transport process itself being a cyclic one. Reducing agents, such as ascorbate, reduced phenazine methosulphate (PMS), and hydrogen affected the redox state of the components and inhibited the photophosphorylation; the deleterious effect of oxygen upon photophosphorylation, on the other hand, could be overcome by the addition of reducing agents (Geller and Lipmann, 1960; Newton and Kamen, 1957; Vernon and Ash, 1960). This type of cyclic photophosphorylation has been observed in the purple sulphur bacterium *Chromatium* (Newton and Kamen, 1957), the purple non-sulphur bacteria *Rhodospirillum rubrum* (Frenkel, 1956; Geller and Lipmann, 1960; Vernon and Ash, 1960) and *Rhodospirillum molischianum* (Williams, 1956), and the green sulphur bacteria *Chlorobium limicola* (Williams, 1956) and *Chlorobium thiosulphatophilum* (Hughes, Conti and Fuller, 1963).

Smith and Baltscheffsky (1959) have demonstrated that in *R. rubrum* low concentrations of 2-n-heptyl-4-hydroxiquinoline-*N*-oxide (HOQNO), a specific inhibitor of mitochondrial electron transport, inhibited photophosphorylation and accelerated the light-induced oxidation of cytochrome c_2. Antimycin A had the same effect on photophosphorylation and the light-induced cytochrome c_2 oxidation, but was not effective in oxidative phosphorylation in this facultative photoheterotroph (Geller, 1962). The addition of PMS abolished the effect of HOQNO and antimycin A (Baltscheffsky and Baltscheffsky, 1958; Geller and Lipmann, 1960) and greatly accelerated the reduction rate of the cytochromes after illumination (Morita, Edwards and Gibson, 1965). Furthermore, Nishimura (1962) who examined the kinetics of photophosphorylation in *R. rubrum* fragments using a flashing light technique, concluded that the rate-limiting step in photophosphorylation was the electron transport process at the site which was blocked by HOQNO and antimycin A. An analysis of his experiments indicates that HOQNO inhibited the 'dark' electron transport, possibly from cytochrome *cc'* to cytochrome c_2 after a light flash, and that this HOQNO-sensitive site was bypassed by PMS. Thus, it appears that in *R. rubrum* photophosphorylation is closely associated with the oxidation–reduction reactions involving cytochrome c_2, and that PMS provides for a non-physiological bypass around the electron transport site inhibited by HOQNO and antimycin A. The restoration of photophosphorylation as a result of this bypassing implies a phosphorylation site outside that part of the electron transport which is bypassed by PMS; no phosphorylation step seems to be involved in the reaction mediated by PMS (Zaugg, Vernon and Helmer, 1967). Evidence for two phosphorylating sites in the light-induced cyclic electron transport chain in *R. rubrum* chromatophore preparations is given by Baltscheffsky and Arwidsson (1962) who observed that the un-coupler valinomycin inhibited photophosphorylation to a maximum of about 50 per cent in the absence of PMS, but did not inhibit photophos-phorylation in the presence of PMS. These observations are consistent with two phosphorylation sites in the physiological cycle. One of these sites is at the level of cytochrome c_2 and is sensitive to the electron transport inhibi-tors HOQNO and antimycin A and to the uncoupler valinomycin. The other site is possibly at the level of ubiquinone (Zaugg, Vernon and Helmer, 1967) and is insensitive to these reagents. Ubiquinone 10 seems to be essential for photophosphorylation in 'intact' *R. rubrum* chromatophores (Okayama and coworkers, 1968).

The situation in the purple sulphur bacterium *Chromatium* seems to be similar. In whole cells, light of a saturating intensity increased the endo-genous ATP/ADP ratio from 1 to 10, independently of the absence or presence of a substrate (Gibson & Morita, 1967). Antimycin A and HOQNO blocked the transport of electrons from cytochrome cc' to cytochrome C 555 and PMS bypassed this pathway (Morita, Edwards and Gibson, 1965). There are some indications that the non-cyclic cytochrome C 552 is a part of the PMS-mediated artificial cyclic pathway.

Non-cyclic photophosphorylation

The possibility of a phosphorylation associated with the non-cyclic electron pathway is indicated by Nozaki, Tagawa and Arnon (1961) in a *Rhodospirillum rubrum* chromatophore preparation. They measured a formation of ATP coupled stoichiometrically with the photoreduction of NAD+ by reduced dichlorophenol–indophenol (DPIP) in a system in which cyclic electron flow was inhibited by antimycin A or HOQNO. The inter-pretation of their results, however, was challenged by Bose and Gest (1962) who pointed out that DPIP could bypass the antimycin A and HOQNO sensitive site in a way similar to PMS. They contended that the ATP formation was associated with the cyclic electron flow, and that the observed stoichiometry was fortuitous. The flashing light experiments of Geller (1967) in which the yield per flash of the production of ATP was compared with the kinetics of the flash-induced absorbancy changes at 420, 428 and 435 nm also indicated that phosphorylation is associated with the cyclic electron flow in *R. rubrum*.

In *Chromatium*, the addition of sodium sulphide to whole cells accelerated the rate of ATP formation in low intensity light, thus suggesting a phosphorylation site in the non-cyclic electron flow (Gibson and Morita, 1967). Cusanovich and Kamen (1968b), however, found that in their purified chromatophore preparation of *Chromatium* ATP formation was coupled mainly, if not exclusively, to the oxidation–reduction reactions of the cytochromes *cc'* and C 555 in the cyclic electron transport chain. Apparently there is a need for additional data to resolve this question.

Light absorption and excitation energy transfer

Excitation energy transfer in the photosynthetic unit

Light energy absorbed by a pigment molecule in photosynthetic organisms is transferred through an aggregate of dissimilar and/or similar pigment molecules until it is trapped at a specialized site, a reaction centre, where it can be converted into chemical energy in the form of separated oxidizing and reducing power. This general concept of the physical mechanism of the photosynthetic process has developed in the last three decades from the conclusions of Emerson and Arnold (1932) and Gaffron and Wohl (1936) about the photosynthetic unit, and of Van Niel (1931, 1941, 1944) about the primary photochemistry of photosynthesis.

A photosynthetic unit is defined as a single reaction centre plus the amount of 'light-harvesting' pigments associated with it. Such a unit could be visualized as a distinct morphological unit or as an extended aggregate with a random distribution of many reaction centra. There is as yet little evidence to indicate which of these possibilities occurs *in vivo*. It may be that the actual situation in some or all photosynthetic organisms is a combination of both.

The size of a photosynthetic unit in photosynthetic bacteria has been estimated from experiments with a flashing light (Nishimura, 1962) to be about 50 to 100 molecules of bacteriochlorophyll per reaction centre. Clayton (1963c) estimated the molecular ratio of bacteriochlorophyll to photochemically reactive components (P 870) in the reaction centre to be about 50 in *Rhodopseudomonas spheroides*. The apparent existence of more than one type of reaction centre in purple bacteria (Morita, 1968; Sybesma, 1969; Sybesma and Fowler, 1968) obviously confuses such estimates. The evidence is abundant, however, that excitation energy is transferred from many light-harvesting pigment molecules to each reaction centre.

In the purple sulphur and non-sulphur bacteria light energy absorbed by the carotenoids is transferred to the bacteriochlorophyll with an efficiency ranging from about 30 to about 50 per cent (Duysens, 1952). Eventually, this energy reaches the reaction centra where it causes the oxidation of P 890 or P 870 (Clayton, 1962a, 1962b, 1962c; Vredenberg and Duysens, 1963). In the green sulphur bacteria light energy absorbed by the major pigment chlorobium chlorophyll is transferred to the minor pigment bacteriochlorophyll *a* with an efficiency ranging from about 35 to 80 per cent (Olson and Sybesma, 1963; Sybesma and Olson, 1963) and from there to the reaction centra, causing the oxidation of P 840 (Sybesma and Vredenberg, 1963). A resonant interaction between electric dipole oscillations in a donor–acceptor couple has been proposed as the physical mechanism of such a transfer of excitation energy (Duysens, 1952, 1964). An extended formulation of such a mechanism in a system of dissimilar molecules has been given by Förster (1948). The efficiency of the transfer is proportional to the square of the dipole–dipole interaction energy and hence to the inverse sixth power of the distance between the donor and the acceptor. The weakness of the interaction allows a relaxation of the vibrational states in the excited donor molecules before the transfer and the efficiency of the transfer is determined by the amount of overlap of the emission spectrum of the donor and the absorption spectrum of the acceptor. The excitation can be localized at any moment and carries out a random walk through the system before it is trapped or lost by emission or radiationless transitions. Highly efficient transfer of energy between pairs of dissimilar molecules can be explained satisfactorily by such a mechanism, when there is a large overlap between a fluorescence band of the donor and an absorption band of the acceptor. This is the case for the chlorophyll *b*–chlorophyll *a* system in higher plants and green algae, and the phycobilin–chlorophyll *a* system in red and blue-green algae. Transfer of energy among similar molecules can be explained by the Förster mechanism under weak dipole–dipole coupling conditions, such as in dilute solutions. In the chlorophyll *a* and bacteriochlorophyll aggregates of photosynthetic units, however, dipole–dipole coupling is strong enough to warrant the requirement of a delocalized treatment of excitation energy transfer (Bay and Pearlstein, 1963a, 1963b; Robinson, 1966). The excitation then, is 'shared' by the aggregate as a whole, and localization occurs only as a result of trapping. The efficiency in this case is linearly proportional to the interaction energy and hence to the inverse third power of the intermolecular distance.

Among other mechanisms proposed for the transfer of energy to the reaction centre sites is a diffusion of electrons and holes after a local photoionization in the pigment aggregate (Arnold and Maclay, 1959). This is unlikely, however, since the efficiency of such a process is probably no more than about 0·1 per cent (Clayton, 1966d); quantum efficiencies for the light-

induced oxidation of P 890, P 870, and the cytochromes in bacteria range
between 50 and 100 per cent (Clayton, 1962b; Olson, 1962; Vredenberg and
Duysens, 1964).

Delocalized excitation energy transfer in the bacteriochlorophyll aggregate can be from a singlet or from a triplet state. In spite of the weaker
interaction energy, triplet excitation migration can be highly efficient due to
the much longer lifetime of the triplet state (Robinson, 1963). The triplet
state has been implied as an intermediate in photochemical reactions of
chlorophyll *in vitro* (Livingston, 1960). There is indirect evidence, however,
that triplet states are not involved in the actual process *in vivo*. Variations in
the fluorescence yield of the light-harvesting bacteriochlorophyll (see later)
can be correlated with the photochemical oxidation of the component P 890
(P 870) of the reaction centra (Clayton, 1966c; Vredenberg and Duysens,
1963). One would not expect such variations if the fluorescent singlet state
is converted to the triplet state or to any other metastable state before transfer to the reaction centre.

Clayton (1966d) proposed that the triplet state is involved in the (harmful)
photooxidation of the light-harvesting bacteriochlorophyll. The bacteriochlorophyll in carotenoidless mutants of several purple bacteria is destroyed
when the organisms are exposed to light in the presence of oxygen (Sistrom,
Griffiths and Stanier, 1956). Carotenoids with nine or more conjugated
double bonds may protect the cells against such harmful photooxidations
by quenching the triplet state. This may be the main function of the
carotenoids in photosynthetic organisms.

If the singlet excited state of the trapping centre is at a lower energy level
than the singlet excited state of the light-harvesting pigment molecules,
trapping of energy in the form of a singlet excitation of the reaction centre
can be visualized. This may be the case in higher plants and algae, where the
singlet excitation energy of the reaction centre component, P 700, is about
1·3 kcal/mole lower than that of the light-harvesting chlorophyll *a*. In most
photosynthetic bacteria, however, there is no difference in energy between
the singlet excited state of the trap and that of the light-harvesting bacteriochlorophyll. In the bacteriochlorophyll *b* containing the bacterium *Rhodopseudomonas viridis*, the energy level of the singlet excited state of the
trapping molecule, P 985, is even higher than that of the light-harvesting
bacteriochlorophyll *b*. Trapping of energy in the form of a singlet excitation
of the trapping molecule is therefore unlikely. Based on these arguments
and on the observation that in photochemically active preparations no
fluorescence from the reaction centra could be detected, Clayton (1966d)
concluded that the trapping of singlet excitation energy must lead to a rapid
transition (in a time much smaller than the intrinsic fluorescence lifetime) to

some metastable state of the reaction centre. Such a metastable state may be a charge transfer complex involving the specialized bacteriochlorophyll (P 890) and an electron acceptor.

Variations of the bacteriochlorophyll fluorescence yield

The quantum yield of the fluorescence emitted by all photosynthetic organisms varies and is dependent on the rate of the photochemical process. If it is assumed that the fluorescence is emitted by the light-harvesting chlorophyll and not by the specialized chlorophylls in the reaction centre, the emission of radiation can be seen as a transition which competes with excitation energy transfer and ultimately with trapping of this energy in the reaction centre. That fluorescence is indeed emitted by the light-harvesting bacteriochlorophyll in *Rhodopseudomonas spheroides* has been demonstrated by the lack of detectable fluorescence in a photochemically active preparation in which the light-harvesting pigments are destroyed (Clayton, 1966b). If emission, radiationless transitions and excitation energy transfer to the reaction centre are first-order processes, and the rate-constant for transfer is proportional to the concentration of unbleached P 890, the fluorescence yield ϕ is given by

$$\phi = \frac{k_e}{k_e + k_i + k[\text{P 890}]} \tag{3.1}$$

in which k_e and k_i are first-order rate-constants for emission and radiationless transitions respectively. This relation requires the assumption that the photosynthetic unit is an extended bacteriochlorophyll aggregate, in which many reaction centra have equal and independent probabilities to be excited.

According to equation (3.1), the inverse of the fluorescence yield at a given intensity of the actinic light should be proportional to the amount of unbleached P 890 which in turn is proportional to the amount of bleaching at 890 nm, ΔA, induced by the actinic light at the given intensity:

$$1/\phi = \alpha + \beta(\Delta A) \tag{3.2}$$

in which α and β are constants derived from the rate constants and the extinction coefficient of P 890. Vredenberg and Duysens (1963) have shown that in the steady state reached at different intensities of the actinic light in whole cells of *Rhodospirillum rubrum* equation (3.2) is satisfied.

Since

$$\frac{d(1/\phi)}{dt} = \beta \frac{d(\Delta A)}{dt} \tag{3.3}$$

a unique validity of the relation between the fluorescence yield and the amount of bleaching, given in equation (3.2), should result in identical time courses for $1/\phi$ and ΔA during the transition from the dark steady state to the light steady state. Clayton (1966c) tested this with whole cells and extracts of a number of purple bacteria and obtained inconclusive results. In some cases equation (3.3) seems to be satisfied; in many others deviations were found. A number of reasons, none of them proven or disproven, can be given for these deviations. The existence of two different types of reaction centra, one for the cyclic and the other for the non-cyclic reaction, could be one of them. Suggestive in this respect is the fact that Clayton found an adherence to equation (3.3) for aerobic suspensions of whole cells of *R. rubrum* and a deviation when the suspension became anaerobic. Oxygen is known to keep the non-cyclic cytochrome in the oxidized state, and thus the non-cyclic reaction inoperative, in *Chromatium* (Olson and Chance, 1960) and *R. rubrum* (Sybesma, unpublished). The relations of the light-harvesting bacteriochlorophyll with the two types of reaction centra, however, is still a matter of pure speculation.

Summary and conclusions

Photosynthesis in photosynthetic bacteria differs from the process in other phototrophs (higher plants and algae) in that it occurs without the evolution of oxygen and only under anaerobic conditions. The process depends on the presence of external substrates, such as reduced sulphur compounds or organic compounds. However, Van Niel's concept of photosynthesis as a photoinduced separation of oxidizing and reducing power seems to be generally valid.

The photosynthetic apparatus seems to be embedded in membranous structures which are related to the cytoplasmic membrane. It is organized in photosynthetic units, aggregates of light-harvesting pigments which, when excited, transfer the excitation energy to reaction centra, sites at which the photoinduced separation of oxidizing and reducing power occurs. The transfer probably is in the form of a singlet excitation, and trapping of the singlet excitation in the reaction centra is the result of a rapid conversion of these centra into some metastable state. This conversion ultimately results in an oxidation of a specialized bacteriochlorophyll component with the concomitant reduction of an acceptor, the identity of which is still unknown.

There is little doubt that the light-induced electron flow in photosynthetic bacteria occurs in two distinct pathways. One of these pathways is cyclic and able to generate adenosine triphosphate (ATP), probably at two sites in the cycle. The other pathway is a non-cyclic electron flow from the external substrates to nicotinamide adenine dinucleotide (NAD^+). Both ATP and NADH are required for the fixation of carbon dioxide in the Calvin cycle, a process which seems to predominate in photosynthetic bacteria under photoautotrophic conditions. The photoreduction of NAD^+ can be conceived as a direct light-induced electron flow from the substrates to NAD^+ or as an energy-linked reversal of electron flow in the dark, mediated by high energy intermediates produced in the light-induced cyclic electron transport. Although the available evidence seems to favour the former of the two mechanisms in whole cells, a definite decision cannot be made yet. It may be possible that conditions for both mechanisms are available to the cell.

For two species of photosynthetic bacteria, the purple sulphur bacterium *Chromatium* and the purple non-sulphur bacteria *Rhodospirillum rubrum*, there is evidence that the two electron transport pathways are driven by two separate light reactions. In *R. rubrum* the reaction centre for the cyclic reactions seems to consist of a complex of at least three bacteriochlorophyll molecules associated with a donor–acceptor couple. The non-cyclic electron flow is driven by a reaction centre which may have only one bacteriochlorophyll molecule associated with a donor–acceptor couple.

REFERENCES

Amesz, J. (1963). Kinetics, quantum requirement and action spectrum of light-induced phosphopyridine nucleotide reduction in *Rhodospirillum rubrum* and *Rhodopseudomonas spheroides*. *Biochim. et Biophys. Acta*, **66**, 22.

Amesz, J. (1964). *Intracellular Reactions of Nicotinamide adenine dinucleotide in Photosynthetic Organisms*. *Ph.D. Thesis*, University of Leiden.

Arnold, W., and R. K. Clayton (1960). The first step in photosynthesis: evidence for its electronic nature. *Proc. Natl. Acad. Sci. US*, **46**, 769.

Arnold, W., and H. K. Maclay (1959). Chloroplast and chloroplast pigments as semiconductors. Brookhaven Symposia in Biology Number 11, Brookhaven National Laboratory, Upton, New York, p. 1.

Arnon, D. I. (1965). Ferredoxin and photosynthesis. *Science*, **149**, 1460.

Bachofen, R., and D. I. Arnon (1966). Crystalline ferredoxin from the photosynthetic bacterium *Chromatium*. *Biochim. Biophys. Acta*, **120**, 259.

Balitskaya, R. M., and Yu. E. Erokhin (1963). Development of green sulphur bacteria and their formation of pigments in various regions of the spectrum. *Dokl. Akad. Nauk. SSSR*, **153**, 460.

Baltscheffsky, H., and B. Arwidsson (1962). Evidence for two phosphorylation sites in bacterial cyclic photophosphorylation. *Biochim. Biophys. Acta*, **65**, 425.

Baltscheffsky, H., and M. Baltscheffsky (1958). On light-induced phosphorylation in *Rhodospirillum rubrum*. *Acta Chem. Scand.*, **12**, 1333.

Baltscheffsky, H., and M. Baltscheffsky (1960). Inhibitor studies on light-induced phosphorylation in extracts of *Rhodospirillum rubrum*. *Acta Chem. Scand.*, **14**, 257.

Bartsch, R. G. (1963). Spectroscopic properties of purified cytochromes of photosynthetic bacteria. In H. Gest, A. San Pietro and L. P. Vernon (Eds.), *Bacterial Photosynthesis*, The Antioch Press, Yellow Springs, Ohio. p. 475.

Bay, Z., and R. M. Pearlstein (1963a). Delocalized versus localized pictures in resonance energy transfer. *Proc. Natl Acad. Sci. US*, **50**, 962.

Bay, Z., and R. M. Pearlstein (1963b). A theory of energy transfer in the photosynthetic unit. *Proc. Natl Acad. Sci. US*, **50**, 1071.

Bennett, R., and R. C. Fuller (1964). The pyruvate phosphoroclastic reaction in *Chromatium*; a possible role for ferredoxin in a photosynthetic bacterium. *Biochem. Biophys. Res. Comm.*, **16**, 300.

Bergeron, J. A. (1958). The bacterial chromatophore. Brookhaven Symposia in Biology Number 11, Brookhaven National Laboratory, Upton, New York. p. 118.

Bergeron, J. A., and R. C. Fuller (1959). Influence of carotenoids on the infrared spectrum of bacteriochlorophyll in *Chromatium*. *Nature*, **184**, 1340.

Bergeron, J. A., and R. C. Fuller (1961). The photosynthetic macromolecules of *Chlorobium thiosulphatophilum*. In T. W. Goodwin and O. Lindberg (Eds.), *Biological Structure and Function*, Vol. II. Academic Press, London and New York. p. 307.

Beugeling, T. (1968). Photochemical activities of $K_3Fe(CN)_6$-treated chromatophores from *Rhodospirillum rubrum*. *Biochim. Biophys. Acta*, **153**, 143.

Beugeling, T., and L. N. M. Duysens (1966). P 890 and cytochrome C 422 in *Chromatium*. In J. B. Thomas and J. C. Goedheer (Eds.), *Currents in Photosynthesis*, Ad. Donker, Rotterdam. p. 49.

Boatman, E. S., and H. C. Douglas (1961). Fine structure of the photosynthetic bacterium *Rhodomicrobium vannielii*. *J. Biophys. Biochem. Cytol.*, **11**, 469.

Bose, S. K., and H. Gest (1962). Electron transport systems in purple bacteria. Hydrogenase and light-stimulated electron transfer reactions in photosynthetic bacteria. *Nature*, **195**, 1168.

Bose, S. K., and H. Gest (1963). Bacterial photophosphorylation: regulation by redox balance. *Proc. Natl Acad. Sci. US*, **49**, 337.

Bril, C. (1963). Studies on bacterial chromatophores. II. Energy transfer and photo-oxidative bleaching of bacteriochlorophyll in relation to structure in normal and carotenoid-depleted *Chromatium*. *Biochim. Biophys. Acta*, **66**, 50

Buchanan, B. B., R. Bachofen, and D. I. Arnon (1964). Role of ferredoxin in the reductive assimilation of CO_2 and acetate by extracts of the photosynthetic bacterium *Chromatium*. *Proc. Natl Acad. Sci.*, **52**, 839.

Buder, J. (1919). Zur Biologie des Bakteriopurpurins und der purpur Bakterien *Jahrb. Wissensch. Botan.*, **58**, 525.

Bull, M. J., and J. Lascelles (1963). The association of protein synthesis with the formation of pigments in some photosynthetic bacteria. *Biochem. J.*, **87**, 15.

Carr, N. G. (1964). Ubiquinone concentrations in Athiorhodaceae. *Biochem. J.* **91**, 28P.

Chance, B. (1951). Rapid and sensitive spectrophotometry. III. A double beam apparatus. *Rev. Sci. Inst.*, **22**, 634.

Chance, B. (1954). Spectrophotometry of intracellular respiratory pigments *Science*, **120**, 767.

Chance, B., and G. Hollunger (1961). The interaction of energy and electron transfer reactions in mitochondria. *J. Biol. Chem.*, **236**, 1577.

Chance, B., T. Horio, M. D. Kamen, and L. S. Taniguchi (1966). Kinetic studies on the oxidase systems of photosynthetic bacteria. *Biochim. Biophys. Acta*, **112**, 1.

Chance, B., and M. Nishimura (1960). On the mechanism of chlorophyll–cytochrome interaction: the temperature insensitivity of light-induced cytochrome oxidation in *Chromatium*. *Proc. Natl Acad. Sci.*, **46**, 19.

Clayton, R. K. (1953). Studies in the phototaxis of *Rhodospirillum rubrum*. *Arch. Mikrobiol.*, **19**, 107.

Clayton, R. K. (1955). Tactic responses and metabolic activities in *Rhodospirillum rubrum*. *Arch. Mikrobiol.*, **22**, 213.

Clayton, R. K. (1957). Phototaxis of purple bacteria. *Encyclopedia of Plant Physiol.*, **17**, 371.

Clayton, R. K. (1962a). Primary reactions in bacterial photosynthesis. I. The nature of light-induced absorbancy changes in chromatophores; evidence for a special bacteriochlorophyll component. *Photochem. Photobiol.*, **1**, 201.

Clayton, R. K. (1962b). Primary reactions in bacterial photosynthesis. II. The quantum requirement for bacteriochlorophyll conversion in the chromatophore. *Photochem. Photobiol.*, **1**, 305.

Clayton, R. K. (1962c). Primary reactions in bacterial photosynthesis. III. Reactions of carotenoids and cytochromes in illuminated bacterial chromatophores. *Photochem. Photobiol.*, **1**, 313.

Clayton, R. K. (1962d). Evidence for the photochemical reduction of coenzyme Q in chromatophores of photosynthetic bacteria. *Biochem. Biophys. Res. Comm.*, **9**, 49.

Clayton, R. K. (1963a). Photosynthesis: primary physical and chemical processes. *Ann. Rev. Plant Physiol.*, **14**, 159.

Clayton, R. K. (1963b). Two light reactions of bacteriochlorophyll *in vivo*. *Proc. Natl Acad. Sci. US*, **50**, 583.

Clayton, R. K. (1963c). Toward the isolation of a photochemical reaction center in *Rhodopseudomonas spheroides*. *Biochim. Biophys. Acta*, **75**, 312.

Clayton, R. K. (1966a). Spectroscopic analysis of bacteriochlorophylls *in vitro* and *in vivo*. *Photochem. Photobiol.*, **5**, 669.

Clayton, R. K. (1966b). Fluorescence from major and minor bacteriochlorophyll components *in vivo*. *Photochem. Photobiol.*, **5**, 679.

Clayton, R. K. (1966c). Relations between photochemistry and fluorescence in cells and extracts of photosynthetic bacteria. *Photochem. Photobiol.*, **5**, 807.

Clayton, R. K. (1966d). Physical processes involving chlorophylls *in vivo*. In L. P. Vernon and G. R. Seely (Eds.), *The Chlorophylls*, Academic Press, London and New York. p. 609.

Clayton, R. K., and W. Arnold (1961). Absorption spectra of bacterial chromatophores at temperatures from 300°K to 1°K. *Biochim. Biophys. Acta*, **48**, 319.

Cohen-Bazire, G. (1963). Some observations on the organization of the photosynthetic apparatus in purple and green bacteria. In H. Gest, A. San Pietro and L. P. Vernon (Eds.), *Bacterial Photosynthesis*, Antioch Press, Yellow Springs, Ohio. p. 89.

Cohen-Bazire, G., and R. Kunisawa (1960). Some observations on the synthesis and function of the photosynthetic apparatus in *Rhodospirillum rubrum*. *Proc. Natl Acad. Sci. US*, **46**, 1543.

Cohen-Bazire, G., and R. Kunisawa (1963). The fine structure of *Rhodospirillum rubrum. J. Cell Biol.*, **16**, 401.

Cohen-Bazire, G., N. Pfennig, and R. Kunisawa (1964). The fine structure of green bacteria. *J. Cell Biol.*, **22**, 207.

Cohen-Bazire, G., and W. R. Sistrom (1966). The procaryotic photosynthetic apparatus. In L. P. Vernon and G. R. Seely (Eds.), *The Chlorophylls*, Academic Press, London and New York. p. 313.

Crounse, J., W. R. Sistrom, and S. Nemser (1963). Carotenoid pigments and the *in vivo* spectrum of bacteriochlorophyll. *Photochem. Photobiol.*, **2**, 361.

Cusanovich, M. A., R. G. Bartsch, and M. D. Kamen (1968). Light-induced electron transport in *Chromatium* strain D. II. Light-induced absorbancy changes in *Chromatium* chromatophores. *Biochim. Biophys. Acta*, **153**, 397.

Cusanovich, M. A., and M. D. Kamen (1968a). Light-induced electron transport in *Chromatium* strain D. I. Isolation and characterization of *Chromatium* chromatophores. *Biochim. Biophys. Acta*, **153**, 376.

Cusanovich, M. A., and M. D. Kamen (1968b). Light-induced electron transfer in *Chromatium* strain D. III. Photophosphorylation by *Chromatium* chromatophores. *Biochim. Biophys. Acta*, **153**, 418.

Drews, G. (1960). Uentersuchungen zur Substruktur der 'Chromatophoren' von *Rhodospirillum rubrum* und *Rhodospirillum molischianum. Arch. Mikrobiol.*, **36**, 99.

Drews, G., and P. Giesbrecht (1963). Zur Morphogenese der Bakterien-'Chromatophoren' (-Thylakoide) und zur Synthese des Bacteriochlorophylls bei *Rhodopseudomonas spheroides* und *Rhodospirillum rubrum. Zentr. Bakteriol. Parasitenk.*, Abt. I. Orig., **190**, 508.

Duysens, L. N. M. (1952). *Transfer of Excitation Energy in Photosynthesis. Ph.D. Thesis*, University of Utrecht, The Netherlands.

Duysens, L. N. M. (1954). Reversible photo-oxidation of a cytochrome pigment in photosynthesizing *Rhodospirillum rubrum. Nature*, **173**, 692.

Duysens, L. N. M. (1957). Investigations in the photosynthetic mechanism of purple bacteria by means of sensitive absorption spectrophotometry. In H. Gaffron (Ed.), *Research in Photosynthesis*, Intersci. Publ. Inc., New York. p. 164.

Duysens, L. N. M., W. J. Huiskamp, J. J. Vos, and J. M. van der Hart (1956). Reversible changes in bacteriochlorophyll in purple bacteria upon illumination. *Biochim. Biophys. Acta*, **19**, 188.

Duysens, L. N. M. (1964). Photosynthesis. *Progress in Biophysics*, **14**, 1.

Duysens, L. N. M. (1966). Kinetics of components of the photochemical reaction center in purple bacteria. In J. B. Thomas and J. C. Goedheer (Eds.), *Currents in Photosynthesis*, Ad. Donker, Rotterdam. p. 263.

Duysens, L. N. M., and G. Sweep (1957). Fluorescence spectrophotometry of pyridine nucleotide in photosynthesizing cells. *Biochim. Biophys. Acta*, **25**, 13.

Eimhjellen, K. E., O. Aasmundrud, and A. Jensen (1963). A new bacterial chlorophyll. *Biochem. Biophys. Res. Comm.*, **10**, 232.

Eimhjellen, K. E., H. Steensland, and J. Traetteberg (1967). A *Thiococcus* sp. nov. gen., its pigments and internal membrane system. *Arch. Mikrobiol.*, **59**, 82.

Emerson, R., and W. Arnold (1932). The photochemical reaction in photosynthesis. *J. Gen. Physiol.*, **16**, 191.

Engelmann, T. W. (1883). *Bacterium photometricum*. Ein Beitrag zur vergleichenden Physiologie des Licht- und Farbensinnes. *Pflügers Arch. Ges. Physiol.*, **30**, 95.

Evans, M. C. W., and B. B. Buchanan (1965). Photoreduction of ferredoxin and its use in CO_2 fixation by a subcellular system from a photosynthetic bacterium. *Proc. Natl Acad. Sci. US*, **53**, 1420.

Evans, M. C. W., B. B. Buchanan, and E. I. Arnon (1966). A new ferredoxin-dependent carbon reduction cycle in a photosynthetic bacterium. *Proc. Natl Acad. Sci. US*, **55**, 928.

Förster, T. (1948). Zwischenmolekuläre Energiewanderung und Fluoreszens. *Ann. der Physik*, **2**, 55

Frenkel, A. W. (1954). Light-induced phosphorylation by cell-free preparations of photosynthetic bacteria. *J. Am. Chem. Soc.*, **76**, 5568

Frenkel, A. W. (1956). Photophosphorylation of adenine nucleotides by cell-free preparations of purple bacteria. *J. Biol. Chem.*, **222**, 823.

Frydman, B., and H. Rapoport (1963). Non-chlorophyllous pigments of *Chlorobium thiosulphatophilum*. Chlorobiumquinone. *J. Am. Chem. Soc.*, **85**, 823.

Fuller, R. C., S. F. Conti, and D. B. Mellin (1963). The structure of the photosynthetic apparatus in the green and purple sulphur bacteria. In H. Gest, A. San Pietro and L. P. Vernon (Eds.), *Bacterial Photosynthesis*, The Antioch Press, Yellow Springs, Ohio. p. 71.

Fuller, R. C., R. M. Smillie, N. Rigopoulos, and V. Yount (1961). Comparative studies of some quinones in photosynthetic systems. *Arch. Biochem. Biophys.*, **95**, 197.

Gaffron, H., and K. Wohl (1936). Zur Theorie der Assimilation. *Naturwissensch.*, **24**, 81.

Geller, D. M. (1962). Oxidative phosphorylation in extracts of *Rhodospirillum rubrum*. *J. Biol. Chem.*, **237**, 2947.

Geller, D. M. (1967). Correlation of light-induced absorbance changes with photophosphorylation in *Rhodospirillum rubrum* extracts. *J. Biol. Chem.*, **242**, 40.

Geller, D. M., and F. Lipmann (1960). Photophosphorylation in extracts of *Rhodospirillum rubrum*. *J. Biol. Chem.*, **235**, 2478.

Gibbs, S. P., W. R. Sistrom, and P. B. Worden (1965). The photosynthetic apparatus of *Rhodospirillum molischianum*. *J. Cell Biol.*, **26**, 395.

Gibson, J., and S. Morita (1967). Changes in adenine nucleotides of intact *Chromatium D* produced by illumination. *J. Bacteriol.*, **93**, 1544.

Giesbrecht, P., and G. Drews (1962). Electronenmikroskopische Uentersuchungen über die Entwickelung der 'Chromatophoren' von *Rhodospirillum molischianum*. *Arch. Mikrobiol.*, **43**, 152.

Goedheer, J. C. (1960). Spectral and redox properties of bacteriochlorophyll in its natural state. *Biochim. Biophys. Acta*, **38**, 389.

Gray, E. D. (1967). Studies on the adaptive formation of photosynthetic structures in *Rhodopseudomonas spheroides*. I. Synthesis of macromolecules. *Biochim. Biophys. Acta*, **138**, 550.

Hickman, D. D., and A. W. Frenkel (1965). Observations on the structure of *Rhodospirillum rubrum*. *J. Cell Biol.*, **25**, 279.

Hind, G., and J. M. Olson (1968). Electron transport pathways in photosynthesis. *Ann. Rev. Plant Physiol.*, **19**, 249.

Hinkson, J. W. (1965). Nicotinamide adenine dinucleotide photoreduction with *Chromatium* and *Rhodospirillum rubrum* chromatophores. *Arch. Biochem. Biophys.*, **112**, 478.

Holt, A. S., and R. K. Clayton (1965). Light-induced absorbancy changes in Eimhjellen's *Rhodopseudomonas*. *Photochem. Photobiol.*, **4**, 829.

Holt, A. S., D. W. Hughes, H. J. Kende, and J. W. Purdie (1963). Chlorophylls of green photosynthetic bacteria. *Plant and Cell Physiol.*, **4**, 49.

Holt, S. C., S. F. Conti, and R. C. Fuller (1966). Photosynthetic apparatus in the green bacterium *Chloropseudomonas ethylicum. J. Bacteriol.*, **91**, 311.

Hood, S. L. (1964). Photoreduction of nicotinamide adenine dinucleotide by a cell-free system from *Chromatium. Biochim. Biophys. Acta*, **88**, 461.

Horio, T., K. Nishikawa, Y. Horiuchi, and T. Kakuno (1968). Mode of coupling of the phosphorylating system to the electron transport system in *Rhodospirillum rubrum* chromatophores. In K. Shibata, A. Takamiya, A. T. Jagendorf and R. C. Fuller (Eds.), *Comparative Biochemistry and Biophysics of Photosynthesis*, University of Tokyo Press, Tokyo and University Park Press, Pennsylvania. p. 408.

Horio, T., and J. Yamashita (1963). Electron transport system in facultative photoheterotroph: *Rhodospirillum rubrum*. In H. Gest, A. San Pietro and L. P. Vernon (Eds.), *Bacterial Photosynthesis*, The Antioch Press, Yellow Springs, Ohio. p. 275.

Horio, T., J. Yamashita, and K. Nishikawa (1963). Photosynthetic adenosine triphosphate formation and photoreduction of diphosphopyridine nucleotide with chromatophores of *Rhodospirillum rubrum. Biochim. Biophys. Acta*, **66**, 37.

Hughes, D. E., S. F. Conti, and R. C. Fuller (1963). Inorganic polyphosphate metabolism in *Chlorobium thiosulphatophilum. J. Bacteriol.*, **85**, 577.

Hurlbert, R. E., and J. Lascelles (1963). Ribulose diphosphate carboxylase in Thiorhodaceae. *J. Gen. Microbiol.*, **33**, 445.

Izawa, S., M. Itoh, T. Ogawa, and K. Shibata (1963). Absorption spectra of solubilized chromatophores and their changes upon oxidation. In *Studies on Microalgae and Photosynthetic Bacteria*. Special issue of *Plant and Cell Physiology*, Tokyo. p. 413.

Jensen, L. S. (1965). Biosynthesis and function of carotenoid pigments in microorganisms. *Ann. Rev. Microbiol.*, **19**, 163.

Jensen, A., O. Aasmundrud, and K. E. Eimhjellen (1964). Chlorophylls of photosynthetic bacteria. *Biochim. Biophys. Acta*, **88**, 466.

Jensen, S. L., E. Hegge, and L. M. Jackman (1964). Bacterial carotenoids. XVII. The carotenoids of photosynthetic green bacteria. *Acta Chem. Scand.*, **18**, 1703.

Katz, E., and E. C. Wassink (1939). Infrared absorption spectra of chlorophyllous pigments in living cells and in extra-cellular states. *Enzymol.*, **7**, 97.

Ke, B., L. P. Vernon, A. Garcia, and E. Ngo (1968). Coupled photo-oxidation of bacteriochlorophyll P 890 and photoreduction of ubiquinone in a photochemically active subchromatophore particle derived from *Chromatium. Biochem.*, **9**, 31.

Kluyver, A. J., and H. J. L. Donker (1926). Die Einheit in der Biochemie. *Chem. Zelle Gewebe*, **13**, 134.

Krasnovskii, A. A. (1965). Photochemistry and spectroscopy of chlorophyll, bacteriochlorophyll and bacterioviridin in model systems and photosynthesizing organisms. *Photochem. Photobiol.*, **4**, 641.

PLATE I

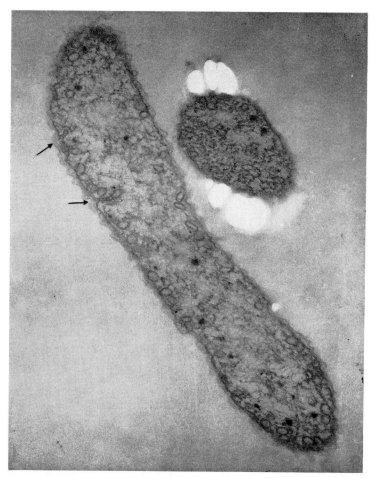

Figure 3.2. Section of *Rhodospirillum rubrum*, treated with ribonuclease in order to show more clearly the continuity between the cytoplasmic membrane and the membranes forming the vesicles (arrows)— × 112 000. Courtesy of Dr. G. Cohen-Bazire

PLATE II

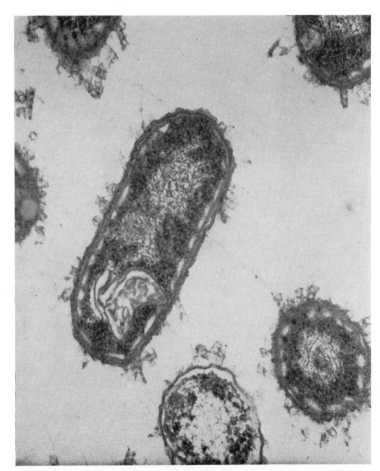

Figure 3.3. Sections of *Chlorobium thiosulphatophilum* embedded in Vestopal, showing the electron-transparent *Chlorobium* vesicles lining the periphery of the cell. The vesicles seem to be contiguous to the cytoplasmic membrane. A large mesosomal element is present in the lower part of the cell—× 120 000. Courtesy of Dr. G. Cohen-Bazire

Krasnovskii, A. A., Yu. E. Erokhin, and I. B. Fedorovich (1961). The fluorescence of green photosynthesizing bacteria and the state of the bacterioviridin in them. *Doklady Akad. Nauk. SSSR*, **134**, 1232.

Kranovskii, A. A., Yu. E. Erokhin, and H. Yü-Ch'ün (1962).Fluorescence of aggregated forms of bacterioviridin and chlorophyll in relation to the state of pigments in photosynthesizing organisms. *Doklady Akad. Nauk. SSSR*, **143**, 456.

Larsen, H. (1953). On the microbiology and biochemistry of the photosynthetic green sulphur bacteria. *Kgl. Norske Videnskab. Selskabs Skrifter*, **1**, 1.

Lascelles, J. (1960). The formation of ribulose 1 : 5-diphosphate carboxylase by growing cultures of Athiorhodaceae. *J. Gen. Microbiol.*, **23**, 499.

Lascelles, J. (1965). Comparative aspects of structures associated with electron transport. *Symp. Soc. Gen. Microbiol.*, **15**, 32.

Livingston, R. (1960). Photochemistry of chlorophyll related to the primary act of photosynthesis. *Radiation Res. Suppl.*, **2**, 196.

Loach, P. A. (1966). Primary oxidation-reduction changes during photosynthesis in *Rhodospirillum rubrum*. *Biochem.*, **5**, 592.

Losada, M., A. V. Trebst, S. Ogata, and D. I. Arnon (1960). Equivalence of light and adenosine triphosphate in bacterial photosynthesis. *Nature*, **186**, 753.

Manten, A. (1948). *Phototaxis, Phototropism and Photosynthesis in Purple Bacteria and Blue-green Algae. Ph.D. Thesis*, University of Utrecht, The Netherlands.

Metzner, P. (1922). Ueber den Farbstoff der grünen Bakterien. *Ber. Deut. Botan. Ges.*, **40**, 125.

Molisch, H. (1907). *Die Purper Bakterien nach neuen Untersuchungen*, Jena.

Moret, V., S. Pinamonti, and E. Fornasari (1961). Polarographic study on the redox potential of ubiquinones. *Biochim. Biophys. Acta*, **54**, 381.

Morita, S. (1968). Evidence for three photochemical systems in *Chromatium D*. *Biochim. Biophys. Acta*, **153**, 241.

Morita, S., and S. F. Conti (1963). Localization and nature of the cytochromes of *Rhodomicrobium vannielii*. *Arch. Biochem. Biophys.*, **100**, 302.

Morita, S., M. Edwards, and J. Gibson (1965). Influence of metabolic conditions on light-induced absorbancy changes in *Chromatium D*. *Biochim. Biophys. Acta*, **109**, 45.

Morita, S., J. M. Olson, and S. F. Conti (1964). Light-induced reactions of cytochromes and carotenoids in *Rhodomicrobium vannielli*. *Arch. Biochem. Biophys.*, **104**, 346.

Newton, J. W. (1964). Interaction of photophosphorylation and electron transport systems in bacterial chromatophores. *J. Biol. Chem.*, **239**, 3038.

Newton, J. W., and M. D. Kamen (1957). Photophosphorylation by subcellular particles from *Chromatium*. *Biochim. Biophys. Acta*, **25**, 462.

Nishimura, M. (1962). Studies on bacterial photophosphorylation. IV. On the maximum amount of delayed photophosphorylation induced by a single flash. *Biochim. Biophys. Acta*, **59**, 183.

Nishimura, M. (1963). Studies on the electron-transfer systems in photosynthetic bacteria. II. The effect of heptylhydroxyquinoline-*N*-oxide and antimycin A on the photosynthetic and respiratory electron transfer systems. *Biochim. Biophys. Acta*, **66**, 17.

4

Nishimura, M., S. B. Roy, H. Schleyer, and B. Chance (1964). Studies on the electron-transfer systems in photosynthetic bacteria. IV. Kinetics of light-induced cytochrome reactions and analysis of electron transfer paths. *Biochim. Biophys. Acta*, **88**, 251.

Nozaki, M., K. Tagawa, and D. I. Arnon (1961). Noncyclic photophosphorylation in photosynthetic bacteria. *Proc. Natl Acad. Sci. US*, **47**, 1334.

Okayama, S., N. Yamamoto, K. Nishikawa, and T. Horio (1968). Roles of ubiquinone-10 and rhodoquinone in photosynthetic formation of ATP by chromatophores from *Rhodospirillum rubrum. J. Biol. Chem.*, **243**, 2445.

Olson, J. M. (1958). Fluorometric identification of pyridine nucleotide changes in photosynthetic bacteria and algae. Brookhaven Symposia in Biology Number 11, Brookhaven National Laboratory, Upton, New York, p. 316.

Olson, J. M. (1962). Quantum efficiency of cytochrome oxidation in a photosynthetic bacterium. *Science*, **135**, 101.

Olson, J. M., and B. Chance (1960). Oxidation-reduction reactions in the photosynthetic bacterium *Chromatium*. I. Absorption spectrum changes in whole cells. II. Dependence of light reactions on intensity of irradiation and quantum efficiency of cytochrome oxidation. *Arch. Biochem. Biophys.*, **88**, 26.

Olson, J. M., and R. K. Clayton (1966). Sensitization of photoreactions in Eimhjellen's *Rhodopseudomonas* by a pigment absorbing at 830 mμ. *Photochem. Photobiol.*, **5**, 655.

Olson, J. M., D. F. Koenig, and M. C. Ledbetter (1969). A model of the bacteriochlorophyll protein from green photosynthetic bacteria. *Arch. Biochem. Biophys.*, **129**, 42.

Olson, J. M., and K. D. Nadler (1965). Energy transfer and cytochrome function in a new type of photosynthetic bacterium. *Photochem. Photobiol.*, **4**, 783.

Olson, J. M., and C. A. Romano (1962). A new chlorophyll from green bacteria. *Biochim. Biophys. Acta*, **59**, 726.

Olson, J. M., and E. K. Stanton (1966). Absorption and fluorescence spectra of bacterial chlorophylls *in situ*. In L. P. Vernon and G. R. Seely (Eds.), *The Chlorophylls*, Academic Press, London and New York. p. 381.

Olson, J. M., and C. Sybesma (1963). Energy transfer and cytochrome oxidation in green bacteria. In H. Gest, A. San Pietro and L. P. Vernon (Eds.), *Bacterial Photosynthesis*, The Antioch Press, Yellow Springs, Ohio, p. 413.

Olson, R. A., W. H. Jennings, and J. M. Olson (1969). Chlorophyll orientation in crystals of bacteriochlorophyll protein from green photosynthetic bacteria. *Arch. Biochem. Biophys.*, **129**, 30.

Orlando, J. A. (1962). *Rhodopseudomonas spheroides* cytochrome 553. *Biochim. Biophys. Acta*, **57**, 373.

Orlando, J. A. (1967). *Rhodopseudomonas spheroides* cytochrome c_2. *Biochim. Biophys. Acta*, **143**, 634.

Orlando, J. A., and T. Horio (1961). Observations on a *b*-type cytochrome from *Rhodopseudomonas spheroides. Biochim. Biophys. Acta*, **50**, 367.

Pardee, A. B., H. K. Schachman, and R. Y. Stanier (1952). Chromatophores of *Rhodospirillum rubrum. Nature*, **169**, 282.

Parson, W. W. (1967). Observations on the change in ultraviolet absorbance caused by succinate and light in *Rhodospirillum rubrum. Biochim. Biophys. Acta*, **143**, 263.

Parson, W. W. (1968). Role of P 870 in bacterial photosynthesis. *Biochim. Biophys. Acta*, **153**, 248.

Pennock, J. F. (1966). Occurrence of vitamins K and related quinones. *Vitamins, Hormones*, **24**, 307.

Pfennig, N. (1967). Photosynthetic bacteria. *Ann. Rev. Microbiol.*, **21**, 285.

Redfearn, E. R. (1966). Mode of action of ubiquinones (coenzymes Q) in electron transport systems. *Vitamins, Hormones*, **24**, 465.

Reed, D. W., and R. K. Clayton (1968). Isolation of a reaction center fraction from *Rhodopseudomonas spheroides*. *Biochem. Biophys. Res. Comm.*, **30**, 471.

Robinson, G. W. (1963). Dynamic role of triplet states in photosynthesis. *Proc. Natl Acad. Sci. US*, **49**, 521.

Robinson, G. W. (1966). Excitation transfer and trapping in photosynthesis. In *Energy Conversion by the Photosynthetic Apparatus*, Brookhaven Symposia in Biology Number 19, Brookhaven National Laboratory, New York. p. 16.

Sauer, K., E. A. Dratz, and L. Coyne (1968). Circular dichroism spectra and the molecular arrangement of bacteriochlorophylls in the reaction centers of photosynthetic bacteria. *Proc. Natl Acad. Sci. US*, **61**, 17.

Schachman, H. K., A. B. Pardee, and R. Y. Stanier (1952). Studies on the macromolecular organization of microbial cells. *Arch. Biochem. Biophys.*, **38**, 245.

Sistrom, W. R., and R. K. Clayton (1964). Studies on a mutant of *Rhodopseudomonas spheroides* unable to grow photosynthetically. *Biochim. Biophys. Acta*, **88**, 61.

Sistrom, W. R., M. Griffiths, and R. Y. Stanier (1956). The biology of a photosynthetic bacterium which lacks colored carotenoids. *J. Cell. Comp. Physiol.*, **48**, 473.

Smith, L., and M. Baltscheffsky (1959). Respiration and light-induced phosphorylation in extracts of *Rhodospirillum rubrum*. *J. Biol. Chem.*, **234**, 1575.

Stanier, R. Y., M. Doudoroff, R. Kunisawa, and R. Contopoulou (1959). The role of organic substrates in bacterial photosynthesis. *Proc. Natl Acad. Sci. US*, **45**, 1246.

Stanier, R. Y., and J. H. C. Smith (1960). The chlorophylls of green bacteria. *Biochim. Biophys. Acta*, **41**, 478.

Sybesma, C. (1967). Light-induced cytochrome reactions in the green photosynthetic bacterium *Chloropseudomonas ethylicum*. *Photochem. Photobiol.*, **6**, 261.

Sybesma. C. (1969a). Two light-induced reactions in photosynthetic bacteria. In P. Metzner (Ed.), *Progress in Photosynthesis Research*, Tübingen. p. 1091.

Sybesma, C. (1969b). Light-induced reactions of P 890 and P 800 in the purple photosynthetic bacterium *Rhodospirillum rubrum*. *Biochim. Biophys. Acta*, **172**, 177.

Sybesma, C., and C. F. Fowler (1968). Evidence for two light-driven reactions in the purple photosynthetic bacterium *Rhodospirillum rubrum*. *Proc. Natl Acad. Sci. US*, **61**, 1343.

Sybesma, C., and J. M. Olson (1963). Transfer of chlorophyll excitation energy in green photosynthetic bacteria. *Proc. Natl Acad. Sci. US*, **49**, 248.

Sybesma, C., and W. J. Vredenberg (1963). Evidence for a reaction center P 840 in the green photosynthetic bacterium *Chloropseudomonas ethylicum*. *Biochim. Biophys. Acta*, **75**, 439.

Sybesma, C., and W. J. Vredenberg (1964). Kinetics of light-induced cytochrome oxidation and P 840 bleaching in green photosynthetic bacteria under various conditions. *Biochim. Biophys. Acta*, **88**, 205.

Tagawa, K., and D. I. Arnon (1962). Ferredoxins as electron carriers in photosynthesis and in the biological production and consumption of hydrogen gas. *Nature*, **195**, 537.

Takamiya, K., M. Nishimura, and A. Takamiya (1967). Distribution of quinones in some photosynthetic bacteria and algae. *Plant and Cell Physiol.*, **8**, 79.

Thomas, J. B. (1952). A note on the occurrence of grana in algae and in photosynthesizing bacteria. *Kon. Ned. Acad. Wetensch. Proc.*, **C55**, 207.

Thornber, J. P., and J. M. Olson (1968). The chemical composition of a crystalline bacteriochlorophyll–protein complex isolated from the green bacterium *Chloropseudomonas ethylicum*. *Biochemistry*, **7**, 2242.

Trebst, A., E. Pistorius, and H. Baltscheffsky (1967). *p*-Phenylenediamines as electron donors for photosynthetic pyridine nucleotide reduction in chromatophores from *Rhodospirillum rubrum*. *Biochim. Biophys. Acta*, **143**, 257.

Tuttle, A. L., and H. Gest (1959). Subcellular particulate systems and the photochemical apparatus of *Rhodospirillum rubrum*. *Proc. Natl Acad. Sci. US*, **45**, 1261.

Van Niel, C. B. (1931). On the morphology and physiology of the purple and green sulphur bacteria. *Arch. Mikrobiol.*, **3**, 1.

Van Niel, C. B. (1941). The bacterial photosyntheses and their importance for the general problem of photosynthesis. *Advan. Enzymol.*, **1**, 263.

Van Niel, C. B. (1944). The culture, general physiology, morphology and classification of the non-sulphur purple and brown bacteria. *Bacteriol. Rev.*, **8**, 1.

Van Niel, C. B. (1962). The present status of the comparative study of photosynthesis. *Ann. Rev. Plant. Physiol.*, **13**, 1.

Van Niel, C. B. (1963). A brief survey of the photosynthetic bacteria. In H. Gest. A. San Pietro and L. P. Vernon (Eds.), *Bacterial Photosynthesis*, The Antioch Press, Yellow Springs, Ohio. p. 459.

Vatter, A. E., and R. S. Wolfe (1958). The structure of photosynthetic bacteria. *J. Bacteriol.*, **75**, 480.

Vernon, P. P. (1953). Cytochrome *c* content of *Rhodospirillum rubrum*. *Arch. Biochem. Biophys.*, **43**, 492.

Vernon, L. P., and O. K. Ash (1960). Coupled photo-oxidation and photoreduction reactions and associated phosphorylation by chromatophores of *Rhodospirillum rubrum*. *J. Biol. Chem.*, **235**, 2721.

Vredenberg, W. J., and J. Amesz (1966). Absorption bands of bacteriochlorophyll in purple bacteria and their response to illumination. *Biochim. Biophys. Acta*, **126**, 244.

Vredenberg, W. J., and L. N. M. Duysens (1963). Transfer of energy from bacteriochlorophyll to a reaction centre during bacterial photosynthesis. *Nature*, **197**, 355.

Vredenberg, W. J., and L. N. M. Duysens (1964). Light-induced oxidation of cytochromes in photosynthetic bacteria between 20 and −170°. *Biochim. Biophys. Acta*, **79**, 456.

Wassink, E. C., E. Katz, and K. Dorrestein (1939). Infrared absorption spectra of various strains of purple bacteria. *Enzymologia*, **7**, 113.

Wassink, E. C., and Kronenberg (1962). Strongly carotenoid deficient *Chromatium* strain D cells with 'normal' bacteriochlorophyll peaks in the 800–850 mμ region. *Nature*, **194**, 553.

Weaver, P., K. Tinker, and R. C. Valentine (1965). Ferredoxin linked DPN reduction by the photosynthetic bacteria *Chromatium* and *Chlorobium*. *Biochem. Biophys. Res. Comm.*, **21**, 195.

Williams, A. M. (1956). Light-induced uptake of inorganic phosphate in cell-free extracts of obligately anaerobic photosynthetic bacteria. *Biochim. Biophys. Acta*, **19**, 570.

Winogradski, S. (1887). Ueber Schwefelbakterien. *Botan. Ztg.*, **45**, 489.

Winogradski, S. (1888). *Zur Morphologie und Physiologie der Schwefelbakterien*, A. Felix, Leipzig.

Worden, P. B., and W. R. Sistrom (1964). The preparation and properties of bacterial chromatophore fractions. *J. Cell Biol.*, **23**, 135.

Zaugg, W. S., L. P. Vernon, and G. Helmer (1967). Light-induced electron transfer reactions and adenosine triphosphate formation by *Rhodospirillum rubrum* chromatophores. *Arch. Biochem. Biophys.*, **119**, 560.

Zaugg, W. S., L. P. Vernon, and A. Tirpack (1964). Photoreduction of ubiquinone and photo-oxidation of phenazine methosulfate by chromatophores of photosynthetic bacteria and bacteriochlorophyll. *Proc. Natl Acad. Sci. US*, **51**, 232.

CHAPTER 4

Photometabolism of organic substrates

WOLFGANG WIESSNER

Department of Plant Physiology, University of Göttingen, Germany

Introduction

Photometabolism of organic substrates is generally understood as the light-dependent conversion of extracellular organic compounds into cell material by which the organic substrates are used directly as sources of organic carbon. Only organisms containing bacteriochlorophyll or

95

chlorophyll can transform the energy of visible light into the chemical energy used in photometabolism.

The ability to photometabolize organic compounds is very common among microorganisms, especially in the following systematic groups:

(1) Procaryota
  (a) Photosynthetic bacteria
    (i) Athiorhodaceae (non-sulphur purple bacteria)
    (ii) Thiorhodaceae (sulphur purple bacteria)
    (iii) Chlorobacteriaceae (green photosynthetic bacteria)
    (iv) Rhodomicrobium. (This organism is metabolically similar to the Athiorhodaceae; in its morphology, however, it is entirely different and belongs to the budding bacteria.)
  (b) Cyanophyta (blue-green algae)
(2) Eucaryota
  (a) Euglenophyta
      Euglenales
  (b) Chlorophyta
    (i) Volvocales
    (ii) Chlorococcales

As will be seen later on, the metabolic pathways involved in photometabolism may differ significantly according to the systematic position of the organisms. However, in spite of variations between the different groups of microorganisms, several general aspects of photometabolism are common to all of them.

In the first aspect light energy is converted by photophosphorylation into chemical energy in the form of adenosine triphosphate (ATP) in the course of a light-dependent electron transport (see Halldal, Chapter 2, and Sybesma, Chapter 3, in this treatise). This photoprocess can be the only way to generate ATP, or it may increase the cellular ATP content above the level achieved by oxidative phosphorylation. In any case, ATP generated in the light promotes energy-requiring reactions necessary to metabolize organic substrates. These can be of two kinds:

(1) ATP is required to activate the carbon source metabolized, a reaction necessary to introduce the compound into cell metabolism, e.g. the formation of acetyl-CoA from acetate and coenzyme A.
(2) ATP activates products formed during metabolism of organic compounds already participating in cell metabolism. Without such reactions the metabolism of the organic substrates cannot proceed.

A reaction of this kind is the formation of phosphoenolpyruvate during the light promoted conversion of malate into carbohydrates.

ATP can be produced by cyclic or, at least in green plants and algae, non-cyclic electron transport (see Halldal, Chapter 2 in this treatise). In cyclic electron transport ATP is the only product of the light reaction. In non-cyclic electron transport pyridine nucleotides can be reduced simultaneously with ATP-formation. They have to be reoxidized to facilitate continuous photophosphorylation. This can be done by using the intermediates of the organic substrate metabolism as hydrogen acceptors, which become reduced. Thus, the type of photophosphorylation can determine the nature of the products finally formed during photometabolism. For example, if cyclic photophosphorylation prevails, organic acids (or amino acids) and polysaccharides are produced from acetate by green algae. Non-cyclic electron transport, however, facilitates the occurrence of lipids as the main products of acetate metabolism.

In the second aspect of photometabolism the chemical configuration of the assimilated substrates determines the nature of the products formed in conjunction with the different metabolic pathways present in the organisms. This is the reason why we have to discuss later on in detail the photometabolism of various organic compounds according to their chemical constitution. However, as more or less general rules, the following conditions play a role in determining the nature of the products of photometabolism:

(1) the presence of nitrogen sources. Amino acid synthesis generally minimizes the transformation of organic substrates into carbohydrates and other storage products;

(2) a significant supply of reduced pyridine nucleotides. This favours the formation of reduced compounds, such as lipids;

(3) the presence of carbon dioxide. The influence of carbon dioxide on photometabolism can be three-fold:

(a) carboxylation of intermediates of the photometabolism alters the products appearing during photometabolism. Concurrent photoautotrophic carbon dioxide fixation, on the other hand, can change the nature of the assimilatory products formed, either because of

(b) the competition between photoassimilation and this process for the limited amount of ATP (and reduced pyridine nucleotides available), or because

(c) products of photosynthesis and of the metabolism of organic compounds can interact.

Most of the work on photometabolism has been done with photosynthetic bacteria. Therefore we will start this discussion of several specific aspects of photometabolism with the photosynthetic bacteria.

Photosynthetic bacteria

Introduction and historical remarks

The growth of many photosynthetic bacteria is favoured by light. This has been known since the early experiments of Winogradsky (1888), who added calcium butyrate, calcium formate or sodium acetate to crude cultures of Thiorhodaceae. Molisch (1907) pointed out that the organic substrates may be used by the photosynthetic bacteria as readily assimilable carbon sources in the light. He regarded the photoassimilation of organic compounds as a special type of photosynthesis, different from the familiar photosynthetic carbon dioxide fixation. His point of view was shared by Muller (1933), who worked with sulphur purple bacteria. and by Gaffron (1933, 1934, 1935a, 1935b, 1936), who worked on metabolic processes in non-sulphur purple bacteria. Both authors interpreted the photometabolism of organic substrates as a light-dependent direct conversion of organic carbon compounds into cell constituents.

This concept, however, was dropped during the late thirties in favour of the general hypothesis of Van Niel (Gaffron, 1936; Van Niel, 1935, 1936, 1941), applicable to all photosynthetic organisms. According to this concept organic compounds were viewed with emphasis on their role as hydrogen donors for the photofixation of carbon dioxide, just as inorganic reduced sulphur compounds can be the sources of reducing power for carbon dioxide fixation in Thiorhodaceae or Chlorobacteriaceae. The function of organic compounds serving as carbon sources was thought to be secondary. However, it is worth noting that Van Niel never disregarded completely the possible role of organic substrates as sources of carbon. He stated, for example, that the partly dehydrogenated organic substrate might be assimilated and converted to cell material. He further suggested that organic substrates can function as carbon dioxide acceptors in the course of the light-dependent carbon dioxide assimilation in reactions similar to the Wood–Werkman reaction (Van Niel, 1936, 1941). As will be seen later, this progressive idea of Van Niel's has shown its merits more recently.

Although nowadays Van Niel's hypothesis is no longer used as a general concept it still can be applied to certain metabolic events. This has been shown by Foster (1940), whose investigations have been confirmed by Siegel and Kamen (1950). *Rhodopseudomonas gelatinosa* can oxidize isopropanol to acetone with the simultaneous reduction of carbon dioxide.

Thus isopropanol can function as a hydrogen donor in the photosynthetic reduction of carbon dioxide. Similar results were obtained with several other strains of non-sulphur purple bacteria, all being able to utilize various alcohols as hydrogen donors in photosynthesis (Foster, 1944).

The final abolishment of Van Niel's hypothesis as a general concept is based on three main categories of experimental results:

(1) *Rhodospirillum rubrum* and other Athiorhodaceae photoassimilate acetate and fatty acids even if carbon dioxide is excluded (Gaffron, 1935a, 1935b; Stanier and coworkers, 1959).

(2) The activity of several enzymes of the Calvin–Benson cycle, which might be a measure for photosynthetic carbon dioxide fixation, is lowered in purple bacteria grown in the presence of organic compounds. This was demonstrated in *Chromatium* (Fuller and Gibbs, 1959; Fuller and coworkers, 1961; Hurlbert and Lascelles, 1963), *Thiopedia* (Hurlbert and Lascelles, 1963), and *Rhodospirillum rubrum* (Anderson and Fuller, 1967a, 1967b; Hoare, 1963). Ribulose-1,5-diphosphate carboxylase especially is drastically suppressed if acetate, pyruvate or, in some, malate have been added to the growth media. Experiments from Elsden's laboratory (Elsden, 1962b; Elsden and Ormerod, 1956) lead to the same conclusion. It was shown that photometabolism of acetate is not affected by 10^{-3} M cyanide, the inhibitor of the carboxydismutase reaction (Trebst, Losada and Arnon, 1960).

(3) Light-induced electron transport has been evaluated to be the fundamental event in photosynthesis. This process can be used in bacteria as in other photosynthetic organisms via photophosphorylation to convert light energy into chemical energy in the form of adenosine triphosphate (ATP) (Arnon, 1959; Arnon and coworkers, 1960; Frenkel, 1954, 1956, 1959a, 1959b; Geller and Lipmann, 1960; Losada and coworkers, 1960; Newton and Kamen, 1957; Nishimura, 1961; Smith and Baltscheffsky, 1959; Vernon and Ash, 1960; Wiessner, 1965, 1966b). Because ATP is necessary (as outlined earlier) for the participation of organic carbon sources in metabolism, this light-dependent ATP synthesis may be the basic reaction in the photoassimilation of many organic substrates (Stanier, 1961).

There is, therefore, no question nowadays that organic compounds can indeed serve as directly assimilable carbon sources for different synthetic processes; the main point of interest is focused on the biochemical pathways involved in their photodirected assimilation.

A number of organic compounds have been tested for their ability to serve as carbon sources in photometabolism. Not all of them can be used equally for growth (Pfennig, 1967; Thiele, 1966). Therefore, this survey is restricted to only a few organic compounds, the photometabolism of which has been investigated in detail.

Photometabolism of acetate

The first step in the metabolism of acetate is its activation. This is mediated in all photosynthetic bacteria so far tested by acetyl-CoA synthetase: *Rhodospirillum rubrum* (Benedict and Rinne, 1964; Eisenberg, 1955, 1957), *Chromatium* (Fuller and coworkers, 1961), and *Chlorobium thiosulphatophilum* (Hoare and Gibson, 1964a).

Acetyl-CoA is channelled into various metabolic pathways depending on experimental conditions. Therefore, the metabolic pathways will be divided into those occurring in resting cells and those predominant in growing organisms. This, however, does not imply that there are metabolic processes which take place only in the former or only in the latter.

According to Glover and Kamen (1951) and Glover, Kamen and Van Genderen (1952), 30 to 50 per cent of acetate carbon fixed by *resting* cells of *Rhodospirillum rubrum* appear in water-insoluble cell material, both carbon atoms being incorporated approximately equally (Elsden, 1962a). The water-insoluble compound formed from acetate is mainly poly-β-hydroxybutyrate (Doudoroff and Stanier, 1959; Fuller and Dykstra, 1962; Stanier and coworkers, 1959). The polymer has also been identified in *Chromatium* (Arnon, Das and Anderson, 1963; Schlegel, 1962; Schlegel and Gottschalk, 1962) and *Rhodomicrobium vannielii* (Hoare and Gibson, 1964b). The water-insoluble product synthesized from acetate in Rhodobacillus (Gaffron, 1933, 1935a) was most likely also poly-β-hydroxybutyrate. This compound can be formed from acetate if both ATP and reduced pyridine nucleotides are available. The latter can be generated by a degradation of part of the acetate via the tricarboxylic acid cycle which is operating in *Rhodospirillum rubrum* (Cutinelli and coworkers, 1951b; Eisenberg, 1953; Elsden and Ormerod, 1956; Ormerod, 1956; Ormerod and Gest, 1962; Vernon and Kamen, 1953; Woody and Lindstrom, 1955) and in *Rhodopseudomonas palustris* (Morita, 1961; Suzuki, 1959). This requirement of tricarboxylic acid cycle activity for acetate photoassimilation in resting cells, as also in growing cells of certain purple bacteria, explains the sensitivity to monofluoroacetate, as has been shown in *Rhodospirillum rubrum* (Elsden and Ormerod, 1956; Gest, Ormerod and Ormerod, 1962). Acetate metabolism is much less sensitive to fluoroacetate if the TCA cycle is not involved; this was shown in *Rhodopseudomonas spheroides*

(Tsuiki, Muto and Kikuchi, 1963). Tuboi and Kikuchi (1966) obtained evidence that illumination suppresses the TCA cycle activity.

The operation of the tricarboxylic acid cycle to provide reduction equivalents for NAD reduction is unnecessary in *Rhodospirillum rubrum* if hydrogen gas is available (Stanier and coworkers, 1959). Under these conditions pyridine nucleotides are reduced in a ferredoxin-dependent reaction. Similarly in *Rhodopseudomonas capsulata* NAD can be reduced with molecular hydrogen (Klemme and Schlegel, 1967).

Chlorobium thiosulphatophilum and *Chlorobium limicola* (Hoare and Gibson, 1964a, 1964b; Sadler and Stanier, 1960), as well as *Chloropseudomonas ethylicum* (Callely, Rigopoulos and Fuller, 1968) do not possess the complete tricarboxylic acid cycle. *Chromatium* also lacks the complete cycle, malate dehydrogenase (Fuller, 1959; Sisler and Fuller, 1959) and α-ketoglutarate dehydrogenase (Fuller and coworkers, 1961) being absent. *Chromatium* can therefore synthesize poly-β-hydroxybutyrate only when the ferredoxin linked reduction of pyridine nucleotides occurs with external hydrogen donors (Buchanan and Bachofen, 1968; Buchanan, Bachofen and Arnon, 1964; Salder and Stanier, 1960; Weaver, Tinker and Valentine, 1965). In extracts of *Chlorobium* the ferredoxin–NAD system could be demonstrated only if hydrogenase from *Clostridium pasteurianum* was added to supply reduced ferredoxin (Weaver, Tinker and Valentine, 1965). Recently also Buchanan and Evans (1969) described a subcellular system from the same organism in which NAD and NADP were reduced in a ferredoxin-dependent light reaction. The discovery of the ferredoxin-dependent reduction of pyridine nucleotides in *Chromatium* explains observations from Arnon's laboratory (Arnon, Das and Anderson, 1963). They found in growing cells of *Chromatium* that under a hydrogen atmosphere 74 per cent of the acetate carboxyl appears in poly-β-hydroxybutyrate and only 19 per cent in protein. Under an argon atmosphere, however, only 7 per cent is incorporated into the polymer, but 64 per cent into protein. Apparently the ferredoxin-dependent reduction of pyridine nucleotides maintains these in the reduced state, favouring the synthesis of highly reduced compounds such as poly-β-hydroxybutyric acid.

It has been shown with *Rhodospirillum rubrum* by Stanier and coworkers (1959) and with *Chromatium okenii* by Schlegel (1962) and by Schlegel and Gottschalk (1962) that poly-β-hydroxybutyric acid is an intracellular reserve of reducing power, which can be used for a subsequent fixation of carbon dioxide for the synthesis of carbohydrates and nitrogenous compounds.

In *growing* cells in the presence of nitrogen sources, amino acids, especially glutamic acid, are the major products of acetate assimilation

(Glover, Kamen and Van Genderen, 1952). In those purple bacteria possessing a complete TCA-cycle the withdrawal of intermediates from the latter for amino acid synthesis during growth on acetate would stop further operation of the cycle, if new intermediates were not supplied by special reactions. Such a special mechanism for the synthesis of four-carbon acids is the glyoxylate cycle (Kornberg and Krebs, 1957), the enzymes of which have been shown to be present in *Rhodopseudomonas palustris*, *Rhodopseudomonas capsulata* and *Chromatium*, if the bacteria are cultivated in acetate-containing media (Fuller and coworkers, 1961; Kornberg and Lascelles, 1960; Losada and coworkers, 1960).

In *Chlorobium*, *Rhodospirillum rubrum* and *Rhodopseudomonas spheroides* no, or only insignificant, activity of the isocitrate lyase has been found, and the normal glyoxylate cycle cannot account for the net synthesis of cell constituents from acetate (Fuller, 1959; Kornberg and Lascelles, 1960). In *Rhodopseudomonas spheroides*, however, glyoxylate has been detected (Kikuchi and coworkers, 1963; Tsuiki, Muto and Kikuchi, 1963), which besides glycollate and malate is one of the earliest products formed from acetate. Its appearance is insensitive to fluoroacetate. This suggests a rather direct conversion of acetate to glyoxylate by a so-far unknown mechanism. From glyoxylate and acetyl-CoA malate is synthetized; the malate synthase has been shown to be present in this non-sulphur purple bacterium (Kikuchi and coworkers, 1963; Kornberg, 1959). In addition, it must be mentioned that the following enzymatic reactions exist in *Rhodo-pseudomonas spheroides*, leading to complete degradation of glyoxylate formed via cleavage of malate (Okuyama, Tsuiki and Kikuchi, 1965):

(1) condensation of glyoxylate and α-ketoglutarate to form α-keto-β-hydroxyadipate and carbon dioxide; (2) decarboxylation of α-keto-β-hydroxyadipate to form α-hydroxy-glutarate; and (3) dehydrogenation of α-hydroxyglutarate to regenerate α-ketoglutarate.

Rhodospirillum rubrum, *Chlorobium* and *Chromatium* possess other synthetic mechanisms to provide the carbon skeletons for amino acid synthesis from acetate. Interest in these metabolic pathways started with the observations that in *Rhodospirillum rubrum* (Cutinelli and coworkers, 1951a, 1951b; Glover, Kamen and van Genderen, 1952; Hoare, 1962b, 1963) and *Chromatium* (Losada and coworkers, 1960; Nozaki, Tagawa and Arnon, 1961) glutamate is one of the first products found labelled with carbon-14 from radioactive acetate, and that succinate, malate and α-ketoglutarate also become labelled much earlier than citrate or isocitrate. Furthermore, experiments with *Rhodospirillum rubrum* (Cutinelli and coworkers, 1951a, 1951b; Hoare, 1962b, 1963) and *Chlorobium thio-sulphatophilum* (Hoare and Gibson, 1964a, 1964b) on the labelling pattern in several amino acids formed from acetate in the light in the presence of carbon dioxide, revealed that the carboxyl carbon of acetate enters into C_2 of alanine, serine and aspartic acid, and the methyl carbon into C_3 of these amino acids, their carboxyl groups being derived from carbon dioxide. This suggests a photodirected condensation of acetate with carbon dioxide, forming a three-carbon acid (Cutinelli and coworkers, 1951a) from which alanine and serine are formed. Further carboxylation in the C_3 position yields oxalacetic acid. This is the precursor of malate, succinate and aspartate (Benedict, 1962; Cutinelli and coworkers, 1951a, 1951b; Glover, Kamen and van Genderen, 1952). A carboxylation of succinyl-CoA could probably account for the early appearance of α-ketoglutarate in the course of acetate assimilation (Benedict and Rinne, 1964; Glover, Kamen and van Genderen, 1952). The enzymes necessary to introduce acetate into, and to operate, such a reductive carboxylic acid pathway (pyruvate synthase, phosphoenolpyruvate carboxylase, succinyl-CoA synthetase, α-ketoglutarate synthase) have been shown to be present in *Rhodospirillum rubrum* (Buchanan, Evans and Arnon, 1967), in *Chlorobium thiosulphatophilum* (Buchanan and Evans, 1965; Buchanan, Evans and Arnon, 1965; Evans and Buchanan, 1965; Evans, Buchanan and Arnon, 1966) and in *Chlorobium ethylicum* (Callely, Rigoupoulos and Fuller, 1968; Evans, 1968). In *Chromatium* the pyruvate synthase has been found (Buchanan and Arnon, 1966; Buchanan, Bachofen and Arnon, 1964; Buchanan, Evans and Arnon, 1965), but not yet the α-ketoglutarate synthase. The reduction equivalents required for the pyruvate synthase and the α-ketoglutarate synthase reactions are provided as reduced ferredoxin (Arnon, 1969; Buchanan, Bachofen and Arnon, 1964; Buchanan and Evans, 1965; Buchanan, Evans and Arnon, 1965; Evans and Buchanan, 1965). The energy requirements for the operation of the cycle are accounted for by the ability of the organisms to generate reduced ferredoxin and ATP at the expense of radiant energy. The isotope distribution after the application of radioactive acetate and carbon dioxide, as observed in alanine, serine and aspartic acid (Cutinelli and coworkers, 1951a, 1951b; Hoare, 1962b, 1963; Hoare and Gibson, 1964a, 1964b), is in accordance with the function of a part of this reductive carboxylic acid cycle for the synthesis of three- and four-carbon compounds from acetate and carbon dioxide.

However, a few cautionary remarks should be made at this point. Although Shigesada and coworkers (1966) were able to confirm that glutamate can be synthetized from acetate and carbon dioxide as observed in the light, the operation of the reductive carboxylic acid cycle cannot explain the isotope distribution in glutamate as observed in *Rhodospirillum rubrum* by Cutinelli and coworkers (1951a) and by Hoare (1962b, 1963). According to Hoare, only the C_1-carboxyl is derived from carbon dioxide, while acetate-2-C^{14} is incorporated in the C_3- and C_4-positions and acetate-1-C^{14} into the C_2-carbon and into the C_5-carboxyl. Similar, but not identical, isotope distribution has been observed in *Chlorobium thiosulphatophilum*, where the acetate carboxyl enters into the carbon atoms 3 to 5 of glutamate (Hoare and Gibson, 1964a, 1964b). Incorporation of carbon dioxide together with acetate by the reductive carboxylic acid cycle, however, should lead to a fixation of carbon dioxide in the carboxyl carbons as well as in the C_2-carbon of glutamate. It is therefore most probable that this cycle is of significance *in vivo* only for the biosynthesis of three- and four-carbon acids from acetate.

The citramalate pathway, although reported in *Rhodospirillum rubrum* (Benedict, 1962; Benedict and Rinne, 1964) and *Chromatium* (Losada and coworkers, 1960), cannot account either for the labelling of glutamate as observed. Benedict and Rinne (1964), Ehrensvärd and Gatnebeck (1965) and Rinne, Buckman and Benedict (1965) suggested that glutamate is formed in *Rhodospirillum* from α-ketoglutarate which has been synthetized by the following reaction sequence: $\text{acetate} \xrightarrow{+CO_2} C_3 \xrightarrow{+CO_2} C_4 \xrightarrow{+\text{acetate}} C_6 \xrightarrow{-CO_2} C_5$.

Acetate can be converted to carbohydrates via phosphoenolpyruvate in growing as well as in resting cells of purple bacteria by decarboxylation of four-carbon acids formed from acetate. The reduced pyridine nucleotides required for the conversion of phosphoglyceric acid to triosephosphate would be supplied either by the anaerobic tricarboxylic acid cycle or in some organisms by a ferredoxin-dependent reduction of pyridine nucleotides in those cases where additional external hydrogen donors are supplied. However, in general other synthetic pathways predominate during photometabolism of acetate.

Photometabolism of propionate

Photoassimilation of propionate starts with the transformation to propionyl-CoA by propionyl-CoA synthetase and depends on a continuous supply of carbon dioxide as demonstrated with *Rhodospirillum rubrum*, *Rhodopseudomonas capsulata* and *Chlorobium thiosulphatophilum* (Clayton, 1957; Knight, 1962; Larsen, 1951, 1953). Tracer experiments showed that carbon dioxide enters almost exclusively into the carboxyl groups of succinate, fumarate, malate and aspartate, suggesting the following reaction sequence (Knight, 1962):

$$\text{propionate} \rightarrow \text{propionyl-CoA} \xrightarrow{+CO_2} \text{methylmalonyl-CoA}$$
$$\rightarrow \text{succinate} \rightarrow \text{fumarate} \rightarrow \text{malate}$$

The early appearance of labelled glutamate during propionate photo-assimilation is most probably due to the amination of α-ketoglutarate formed from succinate (Buchanan and Evans, 1965; Shigesada and co-workers, 1966). Recently Buchanan (1969) discovered a ferredoxin-dependent α-ketobutyrate synthesis in *Chromatium*, by which propionate via propionyl-CoA can be channelled into isoleucine and α-aminobutyrate.

Photometabolism of butyrate

Butyrate is converted to poly-β-hydroxybutyrate during photometa-bolism by resting cells of Athiorhodaceae or Thiorhodaceae (Doudoroff and Stanier, 1959; Gaffron, 1933, 1935; Stanier and coworkers, 1959). Manometric data from Elsden's laboratory showed that approximately 0·4 mole of carbon dioxide is fixed per mole of butyrate metabolized (Elsden and Ormerod, 1956; Ormerod, 1956). The concomitant reduction of carbon dioxide during butyrate photometabolism has also been observed by Muller (1933). Poly-β-hydroxybutyrate can serve as a carbon source for various syntheses under appropriate conditions and in the presence of carbon dioxide (Schlegel, 1962; Schlegel and Gottschalk, 1962; Stanier and coworkers, 1959).

Photometabolism of pyruvate

The photometabolism of pyruvate in photosynthetic bacteria can follow different pathways, depending on growth and experimental conditions.

First we will consider the synthesis of carbohydrates, of organic acids, and of amino acids from pyruvate. In *Rhodospirillum rubrum*, *Chromatium* and *Chlorobium thiosulphatophilum*, pyruvate, the light-dependent uptake of which in *Chromatium* is decreased by oxygen (Hurlbert, 1967), can be transformed into phosphoenolpyruvate (PEP) by the ATP-requiring PEP synthetase (Buchanan and Evans, 1966) reaction. The latter can be channelled into the synthesis of carbohydrates, as has been shown with *Rhodospirillum rubrum* (Schlegel and Gottschalk, 1962; Stanier and coworkers, 1959), or into the anaerobic reductive carboxylic acid cycle (Evans, Buchanan and Arnon, 1966), from which organic acids as carbon skeletons for amino acid synthesis are derived. Carbohydrate synthesis takes place under conditions especially favouring the accumulation of storage compounds, such as high light intensity or high substrate concentration (Bosshard-Heer and Bachofen, 1969; Schlegel and Gottschalk, 1962).

Poly-β-hydroxybutyrate has been shown to be synthesized from pyruvate in *Rhodopseudomonas spheroides* (Ohashi, Ishihara and Kikuchi, 1967) and *Rhodospirillum rubrum* (Bosshard-Heer and Bachofen, 1969). Its formation requires the decarboxylation of pyruvate, which can proceed via the ferredoxin-dependent clastic cleavage. This reaction has been shown to occur in cell-free extracts of *Chromatium* (Bennett and Fuller, 1964; Bennett, Rigopoulos and Fuller, 1964; Buchanan, Bachofen and Arnon, 1964), *Chlorobium thiosulphatophilum* (Evans and Buchanan, 1965) and *Rhodospirillum rubrum* (Bosshard-Heer and Bachofen, 1969; Buchanan, Evans and Arnon, 1967). The NADH required for the synthesis of poly-β-hydroxybutyrate can be provided by NAD reduction with reduced ferredoxin obtained during the clastic pyruvate cleavage.

The formation of poly-β-hydroxybutyrate (PHBA) under participation of clastic pyruvate cleavage is a dark reaction. However, according to Bosshard-Heer and Bachofen (1969), PHBA synthesis from pyruvate in *Rhodospirillum rubrum* is much higher in the light than in the dark. The reasons for this light requirement are not known.

Photoevolution of hydrogen from pyruvate—and other organic substrates—has been observed in *Chromatium* (Bennett and Fuller, 1964), *Rhodospirillum rubrum* (Kohlmiller and Gest, 1951; Stiffler and Gest, 1954), *Rhodopseudomonas gelatinosa* (Siegel and Kamen, 1951) and *Rhodobacillus palustris* (Morita, Suzuki and Takashima, 1951). Hydrogen can be produced during the clastic cleavage of pyruvate. A complete light-induced decomposition of pyruvate resulting in the liberation of equimolar quantities of carbon dioxide and hydrogen (Kohlmiller and Gest, 1951; Stiffler and Gest, 1954) can take place under participation of the anaerobic tricarboxylic acid cycle, if the latter is present. Hydrogen evolution is diminished if the organisms have been grown in the presence of ammonium salts (Gest, Kamen and Bregoff, 1950; Gest, Ormerod and Ormerod, 1962; Kohlmiller and Gest, 1951; Morita, Suzuki and Takashima, 1951; Ormerod, Ormerod and Gest, 1961; Stiffler and Gest, 1954). It has been suggested that ammonium ions suppress the biosynthesis of the hydrogen evolving enzyme system (Gest, Ormerod and Ormerod, 1962). However, Bose, Gest and Ormerod (1961) obtained high hydrogenase activity in *Rhodospirillum rubrum* cultivated in ammonium sulphate-containing media. Photoevolution of hydrogen is also lowered under nitrogen atmosphere (Bennett, Rigopoulos and Fuller, 1964; Kohlmiller and Gest, 1951; Stiffler and Gest, 1954). It is possible that this results from a coupling of the clastic reaction to nitrogen fixation similar to the mechanisms proposed by Mortenson (1964) and D'Eustachio and Hardy (1964) (see also Bennett and Fuller, 1964; Bennett, Rigopoulos and Fuller, 1964). Light-dependent nitrogen fixation proceeds in *Chromatium* at the expense of hydrogen evolution, and is even stimulated by pyruvate.

In *Rhodopseudomonas palustris*, *Rhodopseudomonas capsulata*, *Rhodopseudomonas gelatinosa* and *Rhodospirillum molischianum* the presence of a thiamine pyrophosphate activated pyruvate decarboxylase has been established. The reaction leads to the formation of carbon dioxide, acetaldehyde and acetylmethylcarbinol as products of pyruvate degradation (Quadri and Hoare, 1967). This enzyme was not detectable in *Rhodopseudomonas spheroides*, *Rhodospirillum rubrum* or *Rhodomicrobium vannielii*.

Photometabolism of lactate

Little information is available concerning lactate photometabolism. Experiments with *Rhodopseudomonas palustris* (Morita, 1955) have shown that all carbon atoms are assimilated directly into cell material. This suggests a primary oxidation of lactate to pyruvate which is channelled into various metabolic pathways as outlined in the foregoing section, probably including its degradation with evolution of hydrogen.

Photometabolism of malate

Photometabolism of malate by resting cells of *Rhodospirillum rubrum* mainly yields polysaccharides (Ormerod, Ormerod and Gest, 1961; Tuboi and Kikuchi, 1966). The reaction is not sensitive to monofluoroacetate (Elsden and Ormerod, 1956) and approximately 1 mole of carbon dioxide is evolved per mole of malate assimilated (Bregoff and Kamen, 1952; Elsden and Ormerod, 1956; Ormerod, 1956). This can be explained by the following sequence of reactions:

malate → oxalacetate → phosphoenolpyruvate → 3-phosphoglycerate
    → dihydroxyacetonephosphate/3-phosphoglyceraldehyde
    → fructose-1,6-diphosphate → polysaccharides.

The reaction proceeds without the external supply of reduction equivalents because the reduced pyridine nucleotides required for the reduction of 1,3-diphosphoglycerate to triosephosphate arise from the oxidation of malate to oxalacetate (Elsden and Ormerod, 1956; Ormerod and Gest, 1962).

However, this formation of polysaccharides from malate predominantly takes place only if the hydrogen-producing enzyme system is absent or its action inhibited. Otherwise, photometabolism of malate in *Rhodospirillum rubrum* can lead to its partial or complete degradation to carbon dioxide and hydrogen (Bregoff and Kamen, 1952; Gest, 1951; Gest and Kamen, 1949a, 1949b; Kohlmiller and Gest, 1951; Ormerod and Gest, 1962; Siegel and Kamen, 1951) under participation of the anaerobic tricarboxylic acid cycle (Anderson, Brenneman and Gest, 1963; Gest, Ormerod and Ormerod, 1962). Because more hydrogen is produced from L-malate than from D-malate it is possible that the two isomers are metabolized through different routes (Anderson, Brenneman and Gest, 1963).

In *Chromatium* malate dehydrogenase has not been detected (Fuller and coworkers, 1961; Sisler and Fuller, 1959). In order to serve as carbon skeletons for various syntheses, malate must first be oxidized by the malic enzyme to pyruvate. This acid is then carboxylated to oxalacetic acid (Fuller and coworkers, 1961), either directly by the pyruvate carboxylase (Utter and Keech, 1960) or after transformation into phosphoenolpyruvate (Buchanan and Evans, 1966) by the phosphoenolpyruvate carboxylase.

In all the metabolic processes outlined above, the ATP-requiring reactions necessary to metabolize malate always follow a preceding transformation of malate that does not require ATP. However, malate metabolism can also start with ATP-requiring reactions, such as enzymatic cleavage to acetyl-CoA and glyoxylate and the activation of malate to malonyl-CoA, which is also split to acetyl-CoA and glyoxylate as outlined by Stern (1963) for *Rhodospirillum rubrum* and by Mue, Tuboi and Kikuchi (1964) and Tuboi and Kikuchi (1965) for *Rhodopseudomonas spheroides*. These reactions can lead to an excessive formation of acetyl-CoA, which itself can be the substrate of many synthetic reactions, explaining, for example, the formation of poly-β-hydroxybutyrate (PHBA) from malate as observed by Anderson, Brenneman and Gest (1963). The accumulation of PHBA is higher from D-malate than from L-malate. It is not formed if the concentration of inorganic phosphate is high (e.g. 0·01 M). However, if the nutritional conditions are unbalanced by decreasing the inorganic phosphate concentration to 0·0001 M, accumulation of PHBA is observed.

Photometabolism of succinate

Photometabolism of succinate under nitrogen-gasphase in *Rhodospirillum rubrum* and *Rhodopseudomonas spheroides* is not inhibited by monofluoroacetate (Elsden and Ormerod, 1956; Kikuchi, Abe and Muto, 1961). This, as well as the liberation of 1 mole of carbon dioxide per mole of succinate—the carbon dioxide deriving from the carboxyl groups (Knight, 1962)—and the predominant accumulation of polyglucose (Stanier and coworkers, 1959), suggests that in these organisms the complete tricarboxylic acid cycle is not involved in succinate metabolism by resting cells. The main succinate pathway follows, after the conversion of this compound to malate, the sequence of reactions outlined under malate (Elsden and Ormerod, 1956; Evans, 1965; Kikuchi, Abe and Muto, 1961; Knight, 1962; Stanier and coworkers, 1959). Kikuchi and coworkers (1963) suggested that in *Rhodopseudomonas spheroides* the oxidation of succinate to fumarate, required for malate formation, is mediated by a photo-oxidant. According to Keister and Yike (1967), succinate can be the direct hydrogen donor for photoreduction of NAD (Frenkel, 1959a, 1959b) by chromatophores of *Rhodospirillum rubrum* if ATP or pyrophosphate are supplied.

Poly-β-hydroxybutyrate can be formed from succinate in significant amounts in addition to other products (Stanier and coworkers, 1959). This results from the formation of excessive acetyl-CoA, either by an enzymatic cleavage of malate or malonyl-CoA to glyoxylate and acetyl-CoA, or more probably by the decarboxylation of pyruvate produced during succinate metabolism.

Carbon dioxide can stimulate the light-induced metabolism of succinate as has been shown with *Rhodospirillum rubrum* by Elsden and Ormerod (1956). The reductive carboxylation of succinate to α-ketoglutarate (Shigesada and coworkers, 1966) may explain these findings. However, the significance of this reaction for the direct synthesis of the C_5-skeleton for glutamic acid synthesis remains obscure. It has been demonstrated by Ehrensvärd and Gatnebeck (1965) that the activity of 1,4-^{14}C-succinate enters almost exclusively into C_1 of glutamic acid. This indicates that no carboxylation of succinyl-CoA took place, but that succinate has been converted via the normal tricarboxylic acid cycle (see also the section on photometabolism of acetate).

Hydrogen can be evolved during succinate metabolism as well as from fumarate (Gest and Kamen, 1949a, 1949b; Gest, Ormerod and Ormerod, 1962; Kamen, 1950; Ormerod and Gest, 1962), the experimental conditions and the reaction mechanisms being the same as described for the hydrogen photoevolution from pyruvate or malate.

Photometabolism of amino acids

Although many purple bacteria can utilize amino acids for growth, only a few systematic studies have been published concerning their role as carbon sources for light-stimulated growth. Light does not only increase the amount of acids metabolized, as has been shown with *Rhodospirillum rubrum*, but also changes the products formed (Coleman, 1956, 1958; Fuller and Dykstra, 1962; Gibson and Wang, 1968). In the dark, glutamate as well as other amino acids are converted into carbon dioxide and ammonia as principal extracellular products; in addition, hydrogen can be liberated from glutamate. In the case of light-dependent glutamate assimilation 80 per cent of the carbon and 55 to 70 per cent of the nitrogen of the glutamic acid, metabolized under participation of the TCA cycle (Fuller and Dykstra, 1962; Gibson and Wang, 1968), are converted into cellular protein. External carbon dioxide increases the glutamate uptake, indicating that carboxylation reactions might play a role during metabolism of compounds formed from glutamate. In *Rhodopseudomonas spheroides* glycine metabolism is insensitive to monofluoroacetate (Tsuiki and Kikuchi, 1962). It starts with glycine conversion to glyoxylate, from which the glycine carbon can be introduced into several biochemical pathways, for example, by way of the malate synthase reaction (Tsuiki and Kikuchi, 1962).

Photometabolism of carbohydrates

According to Van Niel's (1944) experiments carried out anaerobically in the light, *Rhodopseudomonas capsulata* can grow in media containing glucose or fructose; *Rhodopseudomonas gelatinosa* can be cultivated with glucose, fructose or mannose; and *Rhodopseudomonas spheroides* with glucose or mannose. *Rhodospirillum fulvum* can use glucose as a suitable substrate for growth. *Rhodopseudomonas palustris* and *Rhodospirillum rubrum* did not grow in solutions with sugars. Gibson and Wang (1968) obtained growing cultures of *Rhodospirillum rubrum* with fructose, but not with glucose as a carbon source; light increased the utilization only slightly above the aerobic dark level. Fructose is catabolized exclusively via glycolysis.

An extensive study was carried out by Szymona and Doudoroff (1960) with *Rhodopseudomonas spheroides*. In a wild-type strain more than half of the carbon of newly formed cells grown anaerobically in the light were found to be derived from glucose. Mannose and ribose were oxidized but did not contribute appreciably to growth. Aldonic acids and 2-keto-3-deoxygluconic acids accumulated in the media. Mutants have been selected which are capable of growing on glucose without the accumulation of acids. In the hexose metabolism the constitutive Embden–Meyerhof pathway appears to be of very limited importance because of a low degree of aldolase activity. In addition, no good evidence for the occurrence of the pentosephosphate oxidative pathway was obtained. Carbohydrates apparently are metabolized via dehydrative pathways, according to which glucose or fructose/mannose are oxidized to the corresponding aldonic acid lactones and thereafter converted to 2-keto-3-deoxygluconic acid. The action of 2-keto-3-deoxygluconic acid kinase leads to the corresponding phosphate, which by an aldolase reaction is split into glyceraldehyde-phosphate and pyruvate.

Requirement of carbon dioxide for photoassimilation of organic substrates

The requirement of carbon dioxide for photoassimilation of organic substrates as well as the interaction between photoassimilation of carbon dioxide (photosynthesis) and of organic compounds depend on the organism, the type of organic substrate assimilated and the cellular products synthesized.

Chlorobium limicola (Sadler and Stanier, 1960) and *Chloropseudomonas ethylicum* (Callely, Rigopoulos and Fuller, 1968; Nesterov, Gogotov and Kondratieva, 1966) assimilate acetate only if carbon dioxide is present. These organisms lack the metabolic mechanisms liberating carbon dioxide from acetate, which itself is indispensable for several synthetic processes, as described in the foregoing sections. *Rhodopseudomonas gelatinosa* assimilates acetone anaerobically and converts it to cell material via acetoacetate and acetate in the light only if carbon dioxide is supplied (Siegel, 1950, 1954, 1956). In *Rhodospirillum rubrum*, as in other photo-synthetic bacteria, poly-β-hydroxybutyrate is stored which can be converted into carbohydrates and into nitrogenous cell components only if carbon dioxide is furnished (Fuller and Dykstra, 1962; Schlegel, 1962; Schlegel and Gottschalk, 1962; Stanier and coworkers, 1959).

The presence of organic compounds can change the products formed from carbon dioxide (Hoare, 1962a; Stoppani, Fuller and Calvin, 1955). Normally it enters mainly into phosphate esters; if acetate is present, or pyruvate, or succinate, it is incorporated preferentially into malate and glutamate (Hurlbert and Lascelles, 1963). For the mechanisms involved see the foregoing sections.

In all the cases mentioned, the interaction between assimilation of organic compounds and carbon dioxide is due to their joint participation in various biochemical reactions; the simultaneous assimilation of carbon dioxide and organic compounds usually results in an increased fixation of the former, a fact already observed by Roelofsen (1934) and by Gaffron (1935b). It is obvious, however, that this is not the case if organic substrates are assimilated, the metabolism of which itself leads to the liberation of carbon dioxide.

On the other hand, the simultaneous assimilation of carbon dioxide and organic compounds such as acetate can also decrease the $^{14}CO_2$-incorporation rate (Ormerod, 1956). The competition between the two assimilatory processes for the ATP and the limited reducing power available is thought to be responsible for this interaction.

Blue-green algae

Blue-green algae have usually been considered to be obligately autotrophic organisms (Allen, 1952; Holm-Hansen, 1968; Kratz and Myers, 1955; Pringsheim, 1914). The reason for the obligate autotrophy of several Cyanophyta appears to be the incomplete tricarboxylic acid cycle. Several enzymes, such as α-ketoglutarate dehydrogenase, malate dehydrogenase, succinyl-CoA synthetase, succinate dehydrogenase or NADH oxidase, can be absent or have only very low activity. It has to be noted in addition that according to Leach and Carr (1968) the NADPH oxidase, is markedly more active than the NADH oxidase in *Anabaena variabilis*, *Anabaena cylindrina*, *Fremyella diplosiphon*, *Mastigocladus laminosus* and *Nostoc muscorum*. The tricarboxylic acid cycle cannot then function as a catabolic cycle. Its enzymes play a role only in a few synthetic processes (Hoare, Hoare and Moore, 1967; Pearce and Carr, 1967b; Pearce, Leach and Carr, 1969; Smith, London and Stanier, 1967). Nevertheless a number of Cyanophyta (*Anacystis nidulans*, syn. *Lauterbornia nidulans* (Pringsheim, 1968); *Anabaena variabilis*; *Anabaena flos aquae*; *Chlorogloea fritschii*; *Coccochloris piniocystis*; *Gloeocapsa alpicola*; *Nostoc muscorum*; *Synechococcus cedrorum*; *Tolypothrix tenuis*) have been tested, and several are able to photoassimilate organic substrates. Out of a great number of substrates the influence of light on assimilation and metabolism of acetate and of glucose or sucrose is most pronounced.

Photometabolism of acetate

Acetate is the most readily assimilated organic acid. Its activation is mediated either by the acetate kinase, as in *Anabaena variabilis* (Pearce and Carr, 1966, 1967a), or by the acetyl-CoA synthetase, as in *Anacystis nidulans* or *Chlorogloea fritschii* and in *Anabaena flos aquae* (Hoare, Hoare and Moore, 1967; Hoare and Moore, 1965; Pearce and Carr, 1966, 1967a). Although a light-dependent acetate assimilation has been reported for several organisms, acetate does not stimulate growth in the light (Hoare, Hoare and Moore, 1967; Hoare and Moore, 1965). This might be due to the fact that acetate is mainly converted into lipids and not cell material generally. Poly-β-hydroxybutyrate has been found in *Chlorogloea fritschii* (Carr, 1966). The predominant lipid formation as well as the extreme sensitivity of acetate photoassimilation to 3(3,4-dichlorophenyl)1-1-dimethyl urea (DCMU) (Hoare, Hoare and Moore, 1967), a potent inhibitor of non-cyclic photophosphorylation, demonstrate that a non-cyclic electron transport is favoured in the photophosphorylation by Cyanophyta (Petrack and Lipmann, 1961), leading to light-dependent acetate assimilation. Intermediates of the fatty acid synthesis serve as hydrogen acceptors in the course of non-cyclic photophosphorylation.

Acetate carbon enters into amino acids, especially into glutamic acid, arginine, proline and leucine, and also into organic acids, but to a much lesser extent (Allison and coworkers, 1953; Hoare, Hoare and Moore, 1967). They are synthesized by conventional biochemical pathways, including the participation of parts of the incomplete tricarboxylic acid cycle (Hoare, Hoare and Moore, 1967; Pearce and Carr, 1966, 1967a; Smith, London and Stanier, 1967) and probably the glyoxylate cycle (Kornberg and Krebs, 1957). Experiments in Hoare's laboratory with *Anacystis nidulans* on incorporation of acetate gave no evidence for a glyoxylate cycle operation (Hoare, Hoare and Moore, 1967), even though its enzymes are constitutive and were detected at low activities in extracts of *Anacystis nidulans* and *Anabaena variabilis*. They do not show alterations in activity if acetate is absent from the growth medium (Pearce and Carr, 1967a). The function of the glyoxylate cycle for regeneration of malate and succinate is, however, not excluded in other Cyanophyta (Pearce, Leach and Carr, 1969). Additionally, it might be worthwhile to mention that in *Beggiatoa*, which might be a colourless Cyanophyta, the enzymes of the glyoxylate cycle are constitutive too, and that poly-β-hydroxy-butyrate is also formed from acetate (Pringsheim and Wiessner, 1963).

Regarding the influence of carbon dioxide on acetate assimilation and vice versa in blue-green algae, several observations have been reported. A complete absence of carbon dioxide decreases the fixation rate of acetate in *Nostoc muscorum* (Allison and coworkers, 1953). Its requirement for fatty acid synthesis might be the reason. In *Anacystis nidulans* and in *Anabaena variabilis* (Carr and Pearce, 1966) the presence of acetate reduces the carbon dioxide-fixation rate. Likewise the incorporation of glucose is decreased if acetate is assimilated simultaneously. The competition for the light-generated ATP explains these observations.

Photoassimilation of other organic acids

Other organic acids are assimilated to a much smaller extent. In the case of fatty acids, the low capacity of the acyl-CoA synthetase to activate fatty acids besides acetate or propionate might be responsible (Hoare, Hoare and Moore, 1967). Permeability barriers, at least in some blue-green algae, might also be a reason for the non-utilization of most organic acids (Kratz and Myers, 1955).

Photoassimilation of carbohydrates

Certain blue-green algae such as *Tolypothrix tenuis* or *Chlorogloea fritschii* also photoassimilate carbohydrates, e.g. glucose or sucrose. In *Anabaena variabilis* growth takes place on glucose only when carbon dioxide is present in the gasphase (Pearce and Carr, 1968). Carbohydrates are either preferentially stored as polysaccharides in resting cells, or are used for synthetic purposes in growing organisms. The oxidative pentose-phosphate cycle is probably the major pathway of carbohydrate break-down; however, at least in the light, the Embden–Meyerhof pathway seems to participate (Cheung, Busse and Gibbs, 1964; Cheung and Gibbs, 1966; Pearce and Carr, 1968; Wildon and Rees, 1965). Kiyohara and coworkers (1962) observed a stimulating effect of light on glucose assimilation in *Tolypothrix tenuis* only under anaerobic conditions. Sucrose photoassimilation is decreased by simultaneous photosynthesis (Fay, 1965).

Green algae and flagellatae

Photometabolism of acetate

A light-promoted growth on acetate or of acetate uptake has been reported for several green flagellatae and algae: *Chlamydobotrys stellata* (Pringsheim and Pringsheim 1959: Pringsheim and Wiessner, 1960), *Chlamydomonas globosa* (Pimenova and Kondratieva, 1965), *Chlamydomonas mundana* (Eppley and Maciasr, 1962a, 1962b, 1963; Gee, Saltman and Eppley, 1962), *Chlamydomonas pseudagloe* (Luksch, 1933), *Chlamydomonas pseudococcum* (Luksch, 1933), *Chlamydomonas reinhardii* (Stross, 1960), *Chlorogonium euchlorum* (Pringsheim, 1937a, 1937b, 1937d; Wiessner, 1968), *Euglena gracilis* (Cook, 1965, 1967; Cook and Heinrich, 1965; Lynch and Calvin, 1953; Ohmann, 1964a; Pringsheim, 1937c, 1937d), *Chlorella ellipsoidea* (Oaks, 1962), and *Chlorella pyrenoidosa* (Goulding and Merrett, 1966; Merrett and Goulding, 1967a, 1967b; Schlegel, 1956, 1959; Syrett, Bocks and Merrett, 1964; Syrett, Merrett and Bocks, 1963). Two of these organisms are restricted to light for growth on acetate: *Chlamydobotrys stellata* and *Chlamydomonas mundana*.

Occasionally nutritional experiments have led to the classification of algae and flagellatae as nutritional physiological types. It seems desirable to say a few words about the importance of such schemes which describe the observed different nutritional requirements (Lwoff, 1932, 1935, 1943; Pringsheim, 1937c, 1959). I would like to suggest that not too much attention be paid to such schemes, at least as long as they are applied to algae or flagellatae. Our present knowledge about the real nutritional requirements of most flagellatae and algae is still very limited, and the results reported vary significantly depending on methods of cultivation. As long as we do not know the exact biochemical reasons for the limited growth requirements of a certain organism, its description in terms of nutritional schemes based only on nutritional experiments can lead to confusion.

Acetate assimilation starts either with its activation and transformation to acetyl-CoA mediated by the acetyl-CoA synthetase, as has been shown for *Euglena gracilis* (Abraham and Bachhawat, 1962; Ohmann, 1964a) and *Chlamydobotrys stellata* or by the acetate kinase as demonstrated for several Chlorococcales: *Stichococcus bacillaris, Chlorella pyrenoidosa* and *Scenedesmus* sp. (Ohmann, 1964b).

The fate of the activated acetate depends on the following three conditions: the type of photophosphorylation responsible for the provision of ATP, the presence or absence of the glyoxylate cycle, and the presence or absence of carbon dioxide.

Cyclic photophosphorylation dominates in *Chlamydobotrys stellata* and *Chlamydomonas mundana* (Eppley, 1962; Eppley, Gee and Saltman, 1963; Gaffron, Wiessner and Homann, 1963; Wiessner, 1963, 1965, 1966b; Wiessner and Gaffron, 1964b). In *Chlamydobotrys stellata* cyclic photophosphorylation proceeds with a quantum requirement of 2 Einstein/Mol ATP at wavelengths longer than 700 nm (Wiessner, 1965, 1966b). In these two Volvocales the light-dependent uptake of acetate into the cells is resistant to DCMU under aerobic conditions. The further metabolism of acetate to organic acids, amino acids and carbohydrates proceeds with participation of the tricarboxylic acid cycle and the glyoxylate pathway (Kornberg and Krebs, 1957). Proof for the participation of the tricarboxylic acid cycle comes from experiments with carbon-14 labelled acetate. Most of the cycle intermediates are labelled 10 to 60 seconds after the application of radioactive acetate to suspensions of *Chlamydobotrys stellata* (Goulding and Merrett, 1967a). The process can be inhibited with $4 \cdot 10^{-3}$ M monofluoroacetate. Proof for the operation of the glyoxylate cycle is derived from work on the fate of the methyl- and the carboxyl-carbon of acetate (Wiessner, 1962). From 90 to 94 per cent of the carbon dioxide evolved in the course of the light-dependent acetate assimilation is derived from the acetate carboxyl. The methyl-carbon is mostly incorporated into cell material. Subsequent to these observations, the presence of the leading enzymes of the glyoxylate cycle, isocitrate lyase and malate synthase, was shown in *Chlamydobotrys stellata* (Goulding and Merrett, 1967a; Wiessner and Kuhl, 1962) and *Chlamydomonas mundana* (Eppley, Gee and Saltman, 1963).

The main difference between *Chlamydobotrys stellata* or *Chlamydomonas mundana* and *Euglena gracilis* or several Chlorococcales is the dominating role of non-cyclic electron transport responsible for photophosphorylation in the latter. Here, reducing power is generated in addition to ATP, and consequently acetate is converted mainly into fats, fat synthesis being limited by the supply of reduced NADP (Oaks, 1962). Carbohydrates can be synthesized by these organisms too. However, their synthesis is a special problem which we shall deal with later on. Fats have been shown to be a main product of acetate photoassimilation in *Scenedesmus* sp. (Calvin and coworkers, 1951), *Chlorella pyrenoidosa* (Schlegel, 1956, 1959) and *Euglena gracilis* strain L (Cook, 1967).

The enzymes of the glyoxylate cycle have only a very low activity in autotrophically grown Chlorococcales or *Euglena gracilis*. In one strain of *Chlorella vulgaris* isocitrate lyase activity is high even in autotrophically grown cells (Harrop and Kornberg, 1966), whereas malate synthase activity is low. Glycollate is excreted by these algae as the product of acetate metabolism. The activities of the glyoxylate enzymes increase significantly in most algae if acetate is present during growth (Cook and Carver, 1966; Goulding and Merrett, 1966; Haigh and Beevers, 1964; Harrop and Kornberg, 1963; Reeves, Kadis and Ajl, 1962; Syrett, 1966; Syrett, Bocks and Merrett, 1964; Syrett, Merrett and Bocks, 1963; Wiessner and Kuhl, 1962). According to Harrop and Kornberg (1966) the isocitrate lyase participates in the glyoxylate cycle only when incorporated into particulate structures. As a consequence of high glyoxylate cycle activity, fat synthesis decreases a little in acetate-grown *Euglena gracilis* and *Chlorella* or *Scenedesmus*. Carbohydrates and protein synthesis are favoured (Syrett, Bocks and Merrett, 1964; Fischer, 1969). However, it must be pointed out that the observed synthesis of carbohydrates and protein is not necessarily the consequence of acetate *photo*-metabolism. It was shown by Gibbs and his coworkers (see Marsh, Galmiche and Gibbs, 1963, 1965) that the turnover rate of the tricarboxylic acid cycle is not influenced by light, and in several organisms, such as *Chlorella pyrenoidosa* (Syrett, Merrett and Bocks, 1963) or *Euglena gracilis* (Cook and Carver, 1966; Wiessner and Kuhl, 1962), light leads to a decrease in the activity of the glyoxylate cycle enzymes.

It should also be mentioned that synthesis of isocitrate lyase is promoted by light-dependent cyclic photophosphorylation (Syrett, 1966; see also Ramirez, del Campo and Arnon, 1968).

Recently glycollic acid and succinic acid have been found to become labelled first during assimilation of ^{14}C-acetate by *Chlorella pyrenoidosa*, whereas a labelling of glyoxylic acid could not be detected and the labelling of citrate was very low (Goulding and Merrett, 1967b; Merrett and Goulding, 1967a, 1967b). From this it has been suggested that glycollate might be a possible precursor of succinic acid (Merrett and Goulding, 1967a). However, it is more likely that the formation of glycollate from acetate results from the reduction of glyoxylate by a NADPH: or NADH: glyoxylate reductase reaction (Harrop and Kornberg, 1966; Hess and Tolbert, 1967; Hess and coworkers, 1965; Wiessner, 1967). The results on glyoxylate reductase activity in *Chlorella pyrenoidosa* were confirmed by Goulding, Lord and Merrett (1969). Merrett and Goulding (1968), however, pointed out that the low isocitrate lyase activity in autotrophic *Chlorella pyrenoidosa* (Syrett, 1966; Syrett, Merrett and Bocks, 1963) could suggest another, unknown, pathway leading from acetate to glycollate than via the formation of glyoxylate.

It had been reported from Pringsheim's laboratory (Pringsheim and Pringsheim, 1959; Pringsheim and Wiessner, 1960, 1961) for *Chlamydobotrys stellata* and from Eppley's group (Eppley, Gee and Saltman, 1963; Eppley and Maciasr, 1963) for *Chlamydomonas mundana* that these organisms are able to grow on acetate in the light under completely anaerobic conditions. This problem was taken up again by Wiessner and Gaffron (1964a, 1964b). They could show that normal photosynthetic carbon dioxide fixation is essential for acetate assimilation if external oxygen is excluded from the system. Photosynthetically released oxygen is required for the reoxidation of intracellular hydrogen acceptors, which became reduced during acetate metabolism. As could also be demonstrated for *Chlorella pyrenoidosa* (Goulding and Merrett, 1966) and for *Chlorella vulgaris* (Nührenberg, Lesemann and Pirson, 1968), green algae do not grow on any organic substrate in the light under completely anaerobic conditions if photosynthetic oxygen production is inhibited.

Goulding and Merrett (1966) reported that in *Chlorella pyrenoidosa* inhibition of acetate assimilation by DCMU induces a pattern of acetate photoassimilation similar to that in the dark, accompanied by carbon dioxide evolution. Particularly fats and polysaccharides are formed to a much lesser extent. The smaller incorporation of acetate carbon into lipids obviously results from the eliminated non-cyclic electron transport. However, it is surprising that polysaccharide syntheses is also decreased.

This is unexpected if polysaccharide is synthesized mainly via gluconeogenesis in combination with the glyoxylate and tricarboxylic acid pathway,
the ATP being derived from cyclic photophosphorylation. It has been
suggested therefore, that in the light, acetate carbon enters into polysaccharides preferentially by photosynthetic fixation of carbon dioxide released
during acetate oxidation in the Chlorococcales, and to a lesser extent also
in *Euglena gracilis* (Wiessner, 1968). Support for this type of acetate
carbon assimilation, which has been called indirect assimilation of acetate
carbon, came from experiments in which polysaccharide synthesis from
acetate could be lowered significantly by the presence of a carbon dioxide-
absorbing potassium hydroxide solution during photoassimilation of
acetate (Wiessner, 1969a, 1969b). In *Euglena gracilis* gluconeogenesis from
acetate proceeds via phosphoenolpyruvate carboxykinase (Fischer, 1969).

The influence of photoassimilation of acetate on that of carbon dioxide
and vice versa also depends on the physiological type of organism in
question. In Chlorococcales and *Euglena gracilis*, particularly under anaerobic conditions, competition for ATP and reduced pyridine nucleotides
can decrease the rates of carbon dioxide and of acetate photoassimilation.
Generally, carbon dioxide fixation is less affected, and the simultaneous
presence of acetate and carbon dioxide during growth of these organisms does not lead to drastic changes in the enzymatic activity of the
Calvin–Benson cycle (Wiessner, 1968). In *Euglena gracilis* the additional
presence of carbon dioxide increases acetate incorporation into proteins
and lipids and decreases carbohydrate synthesis from acetate (Fischer and
Wiessner, 1968). In the two Volvocales, *Chlamydobotrys stellata* and
Chlamydomonas mundana, on the other hand, growth on acetate decreases
distinctively the activity of several enzymes required for photosynthetic
carbon dioxide fixation—ribulose-1,5-diphosphate carboxylase, ribulose-
5P-kinase, aldolase, NADP-glyceraldehyde-3P-dehydrogenase (Russel and
Gibbs, 1964, 1966; Wiessner, 1962, 1968). Consequently, photosynthetic
carbon dioxide fixation by these Volvocales is diminished. Furthermore,
chloroplasts from autotrophically grown *Chlamydobotrys stellata* differ
significantly from those of photoheterotrophically cultivated organisms.
In the latter the thylakoids are separated from each other by stroma and
only a few thylacoid packages are present. In photoautotrophic organisms,
however, a characteristic folding structure of thylakoid membranes results
in the formation of grana (Wiessner and Amelunxen, 1969a). The submicroscopic chloroplast structure in photoheterotrophically cultivated *Chlamydobotrys stellata* changes into one of the autotrophic organisms as a
consequence of the transfer of the algae to autotrophic growth conditions
(Wiessner and Amelunxen, 1969b).

Photometabolism of other organic acids

Extensive investigations on photometabolism of organic acids other than acetate by green algae have not been performed so far. The conversion of malic, citric, succinic or pyruvic acid into polysaccharides is only slightly increased by light in *Chlorella pyrenoidosa* (Schlegel, 1956). The light-favoured pyruvate uptake by *Chlorella ellipsoidea* or *Scenedesmus* leads mainly to its appearance in cellular lipids (Milhaud, Benson and Calvin, 1956; Oaks, 1962), because of its primary conversion to acetyl-CoA (Marsh, Galmiche and Gibbs, 1965). In addition, it can enter the tricarboxylic acid cycle as acetyl-CoA and by a pathway involving its carboxylation to malate (Marsh, Galmiche and Gibbs, 1965).

Photometabolism of carbohydrates

Several hexoses have been tested as substrates for growth in the light (compare Luksch, 1933, and Dvoráková-Hladká, 1966). Their utilization as organic carbon sources differs, and also depends on the organism in question. For certain strains of *Euglena gracilis* the absence of hexokinase and of fructose-6-P-kinase has been postulated as an explanation for their inability to grow on sugars (Ohmann, 1963). Belsky (1957), however, reported for the same organism, *Euglena gracilis*, that hexokinase is quite labile in the absence of glucose, and that utilization of glucose, fructose and sucrose needs adaptation.

A promoting effect of visible radiation on glucose uptake has been observed in *Ankistrodesmus braunii* (Bishop, 1961; Simonis, 1956), *Chlamydomonas humicula* (Luksch, 1933), *Chlorella pyrenoidosa* (Dvoráková-Hladká, 1966; Moses and coworkers, 1959; Myers, 1947; Wiessner, 1966a; Whittingham, Bermingham and Hiller, 1963), *Chlorella vulgaris* (Bergmann, 1955; Kandler, 1954, 1955), *Chlorella* sp. (Butt and Peel, 1963), *Nitella translucens* (Smith, 1967), *Scenedesmus obliquus* (Dvoráková-Hladká, 1966) and *Scenedesmus quadricauda* (Taylor, 1960).

In the presence of oxygen, oxidative phosphorylation nearly suffices to supply ATP for such reactions as the hexokinase or the phosphorylase to proceed at full rate. Therefore, glucose assimilation is only slightly increased by additional photophosphorylation.

More than 85 per cent of the total glucose assimilated anaerobically enters into oligo- and polysaccharides (Kandler and Tanner, 1966). Glycolysis can be limited because of a deficiency of inorganic phosphate resulting from the competition between photophosphorylation and glycolysis (Kandler and Haberer-Liesenkötter, 1963). Aerobic conditions lead to a slightly increased formation of glycolic acid from glucose (Marker and Whittingham, 1966; Whittingham, Bermingham and Hiller, 1963).

Cyclic photophosphorylation dominates in supplying ATP for glucose photoassimilation. This follows from experiments concerning the effect of antimycin A or 3(3,4-dichlorophenyl)-1-1-dimethyl urea (DCMU) on glucose assimilation carried out by Butt and Peel (1963), in Kandler's laboratory (Kandler and Tanner, 1966; Tanner, Dächsel and Kandler, 1965; Tanner, Loos and Kandler, 1966) and by Wiessner (1966a). Antimycin A inhibits glucose uptake by up to 70 per cent. DCMU in concentrations completely inhibiting non-cyclic photophosphorylation diminishes glucose photoassimilation to only 45 per cent. Salicylaldoxime and the uncouplers desaspidine or carbonyl-p-trifluoromethoxyphenylhydrazone (CFP) also strongly inhibit anaerobic glucose assimilation (Kandler and Tanner, 1966; Tanner, Loffler and Kandler, 1969). Strongest support for the role of cyclic photophosphorylation comes from results according to which glucose assimilation proceeds at wavelengths which only facilitate cyclic photophosphorylation (Kandler and Tanner, 1966; Tanner, Loos and Kandler, 1966, 1968; Wiessner, 1966a), that is at wavelengths longer than 700 nm. The quantum requirement per one mole of glucose under anaerobic conditions was found to be 6·0 Einstein at 658 nm, and 4·1 Einstein at 712 nm (Tanner and coworkers, 1968). Glucose assimilation is saturated at much lower light intensities of white light (1200 lux) than photosynthesis (Tanner, Dächsel and Kandler, 1965). Bishop's *Scenedesmus* mutant 11, which is defective in lightreaction I (Bishop, 1962), does not show any significant glucose assimilation. Mutant 8, however, which has an incomplete lightreaction II (Weaver and Bishop, 1963), exhibits very active glucose assimilation (Tanner, Zinecker and Kandler, 1967).

The fact that DCMU does not leave glucose assimilation by *Chlorella* unaffected can also be explained differently. First, DCMU at higher concentrations poisons not only non-cyclic but also cyclic photophosphorylation (Asahi and Jagendorf, 1963). Such an inhibition of cyclic photophosphorylation not only decreases glucose assimilation because of an ATP deficiency, but for the same reason also decreases the enzymatic adaptation to glucose uptake (Tanner, Dächsel and Kandler, 1965; Tanner and Kandler, 1967; and compare Syrett, 1966). Second, DCMU prevents pseudo-cyclic electron transport involving oxygen as the final acceptor of reduction equivalents. Butt and Peel (1963) observed that α-hydroxysulphonates also decrease glucose assimilation significantly. Because this inhibitor acts on the glycollate oxidase (Zelitsch, 1957), they postulated the participation of glyoxylate reduction and glycollate oxidation in photophosphorylation during glucose photometabolism.

Photosynthesis and glucose photoassimilation both require ATP and compete with each other. However, the decrease of photosynthesis during concomitant glucose uptake can be observed only if the concentration of anorganic phosphate is low (Simonis, 1956; Moses and coworkers, 1959). Anaerobic photosynthesis, e.g. photoreduction, is drastically lowered during glucose assimilation (Bishop, 1961).

Photometabolism of other organic compounds

It has been reported by Enebo (1967) that a certain strain of *Chlorella* can use methane in the light as a carbon source.

REFERENCES

Abraham, A., and B. K. Bachhawat (1962). Studies on aceto-coenzyme A kinase from *Euglena gracilis. Biochim. Biophys. Acta* (*Amst.*), **62**, 376.

Allen, M. B. (1952). The cultivation of Myxophyceae. *Arch. Mikrobiol.*, **17**, 34.

Allison, R. K., H. E. Skipper, M. R. Reid, W. A. Short, and G. L. Hogan (1953). Studies on the photosynthetic reaction. I. The assimilation of acetate by *Nostoc muscorum. J. Biol. Chem.*, **204**, 197.

Anderson, L., C. Brenneman, and H. Gest (1963). Alternative metabolic pathways in bacterial photosynthesis. *Bact. Proc.* (*Baltimore*), p. 123.

Anderson, L., and R. C. Fuller (1967a). Photosynthesis in *Rhodospirillum rubrum* II. Photoheterotrophic carbon dioxide fixation. *Plant Physiol.*, **42**, 491.

Anderson, L., and R. C. Fuller (1967b). Photosynthesis in *Rhodospirillum rubrum* III. Metabolic control of reductive pentose phosphate and tricarboxylic acid cycle enzymes. *Plant Physiol.*, **42**, 497.

Arnon, D. I. (1959). Conversion of light into chemical energy in photosynthesis. *Nature*, **184**, 10.

Arnon, D. I. (1969). Role of ferredoxin in photosynthesis. *Naturwiss.*, **56**, 295.

Arnon, D. I., V. S. R. Das, and J. D. Anderson (1963). Metabolism of photosynthetic bacteria I. Effect of carbon source and hydrogen gas on biosynthetic patterns in *Chromatium*. In *Studies on Microalgae and Photosynthetic Bacteria*, Special issue of *Plant and Cell Physiol.*, Tokyo. pp. 529–545.

Asahi, T., and A. T. Jagendorf (1963). A spinach enzyme functioning to reverse the inhibition of cyclic electron flow by *p*-chlorophenyl-1,1-dimethylurea at high concentrations. *Arch. Biochem. Biophys.*, **100**, 531.

Belsky, M. M. (1957). The metabolism of glucose and other sugars by the algal flagellate *Euglena gracilis. Bact. Proc.* (*Baltimore*), p. 123.

Benedict, C. R. (1962). Early products of (^{14}C) acetate incorporation in resting cells of *Rhodospirillum rubrum. Biochim. Biophys. Acta* (*Amst.*), **56**, 620.

Benedict, C. R., and R. W. Rinne (1964). Glutamic acid synthesis from acetate units and bicarbonate in extracts of photosynthetic bacteria. *Biochem. Biophys. Res. Comm.*, **14**, 474.

Bennett, R., and R. C. Fuller (1964). The pyruvate phosphoroclastic reaction in *Chromatium*. A probable role for ferredoxin in a photosynthetic bacterium. *Biochem. Biophys. Res. Comm.*, **16**, 300.

122 Photobiology of Microorganisms

Bennett, R., N. Rigopoulos, and R. C. Fuller (1964). The pyruvate phosphoroclastic reaction and light-dependent nitrogen fixation in bacterial photosynthesis. *Proc. Nat. Acad. Sci.* (*Wash.*), **52**, 762.

Bergmann, L. (1955). Stoffwechsel und Mineralsalzernährung einzelliger Grünalgen. II. Vergleichende Untersuchungen über den Einfluss mineralischer Faktoren bei heterotropher und mixotropher Ernährung. *Flora*, **142**, 493.

Bishop, N. I. (1961). The photometabolism of glucose by an hydrogen-adapted alga. *Biochim. Biophys. Acta* (*Amst.*), **51**, 323.

Bishop, N. I. (1962). Separation of the oxygen evolving system of photosynthesis from the photochemistry in a mutant of *Scenedesmus. Nature*, **195**, 55.

Bosshard-Heer, E. und R. Bachofen (1969). Synthese von Speicherstoffen aus Pyruvat durch *Rhodospirillum rubrum. Arch. Mikrobiol.*, **65**, 61.

Bregoff, H. M., and M. D. Kamen (1952). Studies on the metabolism of photosynthetic bacteria. XIV. Quantitative relations between malate dissimilation, photoproduction of hydrogen, and nitrogen metabolism in *Rhodospirillum rubrum. Arch. Biochem. Biophys.*, **36**, 202.

Buchanan, B. B. (1969) Role of ferredoxin in the synthesis of α-ketobutyrate from propionyl coenzyme A and carbon dioxide by enzymes from photosynthetic and nonphotosynthetic bacteria. *J. Biol. Chem.*, **244**, 4218.

Buchanan, B. B., and D. I. Arnon (1966). Ferredoxin in plant and bacterial photosynthesis. In H. Peeters (Ed.), *Protides of the Biological Fluids*, Vol. 14. Elsevier Publishing Company, Amsterdam. pp. 143–158.

Buchanan, B. B., and R. Bachofen (1969). Ferredoxin-dependent reduction of nicotinamide-adenine dinucleotides with hydrogen gas by subcellular preparations from the photosynthetic bacterium *Chromatium. Biochim. Biophys. Acta.* (*Amst.*), **162**, 607.

Buchanan, B. B., R. Bachofen, and D. I. Arnon (1964). Role of ferrodoxin in the reductive assimilation of CO_2 and acetate by extracts of the photosynthetic bacterium, *Chromatium. Proc. Nat. Acad. Sci.* (*Wash.*), **52**, 839.

Buchanan, B. B., and M. C. W. Evans (1965). The synthesis of α-ketoglutarate from succinate and carbon dioxide by a subcellular preparation of a photosynthetic bacterium. *Proc. Nat. Acad. Sci.* (*Wash.*), **54**, 1212.

Buchanan, B. B., and M. C. W. Evans (1966). The synthesis of phosphoenolpyruvate from pyruvate and ATP by extracts of photosynthetic bacteria. *Biochem. Biophys. Res. Comm.*, **22**, 484.

Buchanan, B. B., and M. C. W. Evans (1969). Photoreduction of ferredoxin and its use in NAD(P)$^+$ reduction by a subcellular preparation from the photosynthetic bacterium, *Chlorobium thiosulphatophilum. Biochim. Biophys. Acta* (*Amst.*), **180**, 123.

Buchanan, B. B., M. C. W. Evans, and D. I. Arnon (1965). Ferredoxin-dependent pyruvate synthesis by enzymes of photosynthetic bacteria. In A. San Pietro (Ed.), *Non-Heme Iron Proteins: Role in Energy Conversion*, Antioch Press, Yellow Springs, Ohio. pp. 175–188.

Buchanan, B. B., M. C. W. Evans, and D. I. Arnon (1967). Ferredoxin-dependent carbon assimilation in *Rhodospirillum rubrum. Arch. Mikrobiol.*, **59**, 32.

Butt, V. S., and M. Peel (1963). The participation of glycollate oxidase in glucose uptake by illuminated *Chlorella* suspensions. *Biochem. J.*, **88**, 31p.

Cally, A. G., N. Rigopoulos, and R. C. Fuller (1968). The assimilation of carbon by *Chloropseudomonas ethylicum. Biochem. J.*, **106**, 615.

Calvin, M., J. A. Bassham, A. A. Benson, V. H. Lynch, C. Quellet, L. Schou, W. Stepka, and N. E. Tolbert (1951). Carbon dioxide assimilation in plants. *Symposia Soc. exp. Biology V*, 284.

Carr, N. G. (1966). The occurrence of poly-β-hydroxybutyrate in the blue-green alga, *Chlorogloea fritschii*. *Biochim. Biophys. Acta (Amst.)*, **120**, 308.

Carr, N. G., and J. Pearce (1966). Photoheterotrophism in blue-green algae. *Biochem. J.*, **99**, 28 p.

Cheung, W. Y., M. Busse, and M. Gibbs (1964). The dark and photometabolism of glucose by a blue-green alga. *Fed. Proc.*, **23**, 226.

Cheung, W. Y., and M. Gibbs (1966). Dark and photometabolism of sugars by a blue-green alga: *Tolypothrix tenuis*. *Plant Physiol.*, **41**, 731.

Clayton, R. K. (1957). Photosynthetic metabolism of propionate in *Rhodospirillum rubrum*. *Arch. Mikrobiol.*, **26**, 29.

Coleman, G. S. (1956). The dissimilation of amino acids by *Rhodospirillum rubrum*. *J. Gen. Microbiol.*, **15**, 248.

Coleman, G. S. (1958). The incorporation of amino acid carbon by *Rhodospirillum rubrum*. *Biochim. Biophys. Acta (Amst.)*, **30**, 549.

Cook, J. R. (1965). Influence of light on acetate utilization in green *Euglena*. *Plant Cell Physiol.*, **6**, 301.

Cook, J. R. (1967). Photo-assimilation of acetate by an obligate phototrophic strain of *Euglena gracilis*. *J. Protozool.*, **14**, 382.

Cook, J. R., and M. Carver (1966). Partial photo-repression of the glyoxylate by-pass in *Euglena gracilis*. *Plant Cell Physiol.*, **7**, 377.

Cook, J. R., and B. Heinrich (1965). Glucose vs. acetate metabolism in *Euglena*. *J. Protozool.*, **12**, 581.

Cutinelli, C., G. Ehrensvärd, G. Högström, L. Reio, E. Saluste, and R. Stjernholm (1951b). Acetic acid metabolism in *Rhodospirillum rubrum* under anaerobic conditions. III. *Arkiv Kemi*, **3**, 501.

Cutinelli, C., G. Ehrensvärd, L. Reio, E. Saluste, and R. Stjernholm (1951a). Acetic acid metabolism in *Rhodospirillum rubrum* under anaerobic conditions. II. *Arkiv Kemi*, **3**, 315.

D'Eustachio, A. J., and R. W. E. Hardy (1964). Reductants and electron transport in nitrogen fixation. *Biochem. Biophys. Res. Comm.*, **15**, 319.

Doudoroff, M., and R. Y. Stanier (1959). Role of poly-β-hydroxybutyric acid in the assimilation of organic carbon by bacteria. *Nature*, **183**, 1440.

Dvoráková-Hladká, J. (1966). Utilization of organic substrates during mixotrophic and heterotrophic cultivation of algae. *Biol. Plantarum (Praha)*, **8**, 354.

Ehrensvärd, G., and S. Gatnebeck (1965). Metabolic connection between oxalacetate and glutamate in *Rhodospirillum rubrum*. *Acta chem. scand.*, **19**, 2006.

Eisenberg, M. A. (1953). The tricarboxylic acid cycle in *Rhodospirillum rubrum*. *J. Biol. Chem.*, **203**, 815.

Eisenberg, M. A. (1955). The acetate-activating enzyme of *Rhodospirillum rubrum*. *Biochim. Biophys. Acta (Amst.)*, **16**, 58.

Eisenberg, M. A. (1957). The acetate-activating mechanism in *Rhodospirillum rubrum*. *Biochim. Biophys. Acta (Amst.)*, **23**, 327.

Elsden, S. R. (1962a). Assimilation of organic compounds by photosynthetic bacteria. *Fed. Proc.*, **21**, 1047.

Elsden, S. R. (1962b). Photosynthesis and lithotrophic carbon dioxide fixation. In I. C. Gunsalus and R. Y. Stanier (Eds.), *The Bacteria*, Vol. 3. Academic Press, London, New York. pp. 1–40.

Elsden, S. R., and J. G. Ormerod (1956). The effect of monofluoroacetate on the metabolism of *Rhodospirillum rubrum*. *Biochem. J.*, **63**, 691.

Enebo, L. (1967). A methane-consuming green alga. *Acta chem. scand.*, **21**, 625.

Eppley, R. W. (1962). Light-induced acetate incorporation in *Chlamydomonas mundana*. *Plant Physiol.*, **37**, lix.

Eppley, R. W., R. Gee, and P. Saltman (1963). Photometabolism of acetate by *Chlamydomonas mundana*. *Physiol. Plantarum*, **16**, 777.

Eppley, R. W., and F. M. Maciasr (1962a). Rapid growth of sewage lagoon *Chlamydomonas* with acetate. *Physiol. Plantarum*, **15**, 72.

Eppley, R. W., and F. M. Maciasr (1962b). Metabolism of *Chlamydomonas* in sewage lagoons. *Amer. J. Bot.*, **49**, 671.

Eppley, R. W., and F. M. Maciasr (1963). Role of the alga *Chlamydomonas mundana* in anaerobic waste stabilization lagoons. *Limnology and Oceanography*, **8**, 411.

Evans, M. C. W. (1965). The photoassimilation of succinate to hexose by *Rhodospirillum rubrum*. *Biochem. J.*, **95**, 669.

Evans, M. C. W. (1968). Ferredoxin dependent synthesis of α-ketoglutarate and pyruvate by extracts of the green photosynthetic bacterium *Chloropseudomonas ethylicum*. *Biochim. Biophys. Res. Comm.*, **33**, 146.

Evans, M. C. W., and B. B. Buchanan (1965). Photoreduction of ferredoxin and its use in carbon dioxide fixation by a subcellular system from a photosynthetic bacterium. *Proc. Nat. Acad. Sci.* (*Wash.*), **53**, 1420.

Evans, M. C. W., B. B. Buchanan, and D. I. Arnon (1966). A new ferredoxin-dependent carbon reduction cycle in a photosynthetic bacterium. *Proc. Nat. Acad. Sci.* (*Wash.*), **55**, 928.

Fay, P. (1965). Heterotrophy and nitrogen fixation in *Chlorogloea fritschii*. *J. Gen. Microbiol.*, **39**, 11.

Fischer, E. (1969). *Acetatstoffwechsel in* Euglena gracilis. *Doctoral Thesis*, University of Göttingen, Germany.

Fischer, E., and W. Wiessner (1968). Acetate utilization in *Euglena gracilis*. *Plant Physiol.*, **43**, S-31.

Foster, J. W. (1940). The role of organic substrates in the photosynthesis of purple bacteria. *J. Gen. Physiol.*, **24**, 123.

Foster, J. W. (1944). Oxidation of alcohols by non-sulfur photosynthetic bacteria. *J. Bact.*, **47**, 355.

Frenkel, A. W. (1954). Light-induced phosphorylation by cell-free preparations of photosynthetic bacteria. *J. Amer. Chem. Soc.*, **76**, 5568.

Frenkel, A. W. (1956). Photophosphorylation of adenosine nucleotides by cell-free preparations of purple bacteria. *J. Biol. Chem.*, **222**, 823.

Frenkel, A. W. (1959a). Light-induced reactions of chromatophores of *Rhodospirillum rubrum*. *Brookhaven Symposia in Biol.*, **11**, 276.

Frenkel, A. W. (1959b). Light-induced reactions of bacterial chromatophores and their relation to photosynthesis. *Ann. Rev. Pl. Physiol.*, **10**, 53.

Fuller, R. C. (1959). Bacterial photosynthesis subsequent to the photochemical act. *Abstracts Proc. IX International Botanical Congress, Montreal*, Vol. **II**. 125.

Fuller, R. C., and A. P. Dykstra (1962). The photometabolism of glutamic acid in *Rhodospirillum rubrum*. *Plant Physiol.*, **37**, iv.

Fuller, R. C., and M. Gibbs (1959). Intracellular and phylogenetic distribution of ribulose-1,5-diphosphate carboxylase and *d*-glyceraldehyde-3-phosphate dehydrogenases. *Plant Physiol.*, **34**, 324.

Fuller, R. C., R. M. Smillie, E. C. Sisler, and H. L. Kornberg (1961). Carbon metabolism in *Chromatium. J. Biol. Chem.*, **236**, 2140.

Gaffron, H. (1933). Uber den Stoffwechsel der schwefelfreien Purpurbacterien I. *Biochem. Z.*, **260**, 1.

Gaffron, H. (1934). Uber die Kohlensäure-Assimilation der roten Schwefelbakterien I. *Biochem. Z.*, **269**, 447.

Gaffron, H. (1935a). Uber den Stoffwechsel der schwefelfreien Purpurbacterien II. *Biochem. Z.*, **275**, 301.

Gaffron, H. (1935b). Uber die Kohlensäureassimilation der roten Schwefelbakterien II. *Biochem. Z.*, **279**, 1.

Gaffron, H. (1936). Zur Theorie der Assimilation. *Naturwiss.*, **7**, 103.

Gaffron, H., W. Wiessner, and P. Homann (1963). Utilization of far-red light by green algae and the problem of oxygen evolution. In *Photosynthetic Mechanisms of Green Plants*, Publ. 1145 Nat. Acad. Sciences—Nat. Res. Council, Wash., D.C. pp. 436–440.

Gee, R. W., P. Saltman, and R. Eppley (1962). Pathways of acetate metabolism in *Chlamydomonas mundana. Plant Physiol.*, **37**, lix.

Geller, D. M., and F. Lipmann (1960). Photophosphorylation in extracts of *Rhodospirillum rubrum. J. Biol. Chem.*, **235**, 2478.

Gest, H. (1951). Metabolic patterns in photosynthetic bacteria. *Bact. Rev.*, **15**, 183.

Gest, H., and M. D. Kamen (1949a). Studies on the metabolism of photosynthetic bacteria. IV. Photochemical production of molecular hydrogen by growing cultures of photosynthetic bacteria. *J. Bact.*, **48**, 215.

Gest, H., and M. D. Kamen (1949b). Photoproduction of molecular hydrogen by *Rhodospirillum rubrum. Science*, **109**, 558.

Gest, H., M. D. Kamen, and H. M. Bregoff (1950). Studies on the metabolism of photosynthetic bacteria V. Photoproduction of hydrogen and nitrogen fixation by *Rhodospirillum rubrum. J. Biol. Chem.*, **182**, 153.

Gest, H., J. G. Ormerod, and K. S. Ormerod (1962). Photometabolism of *Rhodospirillum rubrum*: light-dependent dissimilation of organic compounds to carbon dioxide and molecular hydrogen by an anaerobic citric acid cycle. *Arch. Biochem. Biophys.*, **97**, 21.

Gibson, M. S., and C. H. Wang (1968). Utilization of fructose and glutamate by *Rhodospirillum rubrum. Canad. J. Microbiol.*, **14**, 493.

Glover, J., and M. D. Kamen (1951). Observations of the simultaneous metabolism of acetate and carbon dioxide by resting cell suspensions of *Rhodospirillum rubrum. Fed. Proc.*, **10**, 190.

Glover, J., M. D. Kamen, and H. Van Genderen (1952). Studies on the metabolism of photosynthetic bacteria. XII. Comparative light and dark metabolism of acetate and carbonate by *Rhodospirillum rubrum. Arch. Biochem. Biophys.*, **35**, 384.

Goulding, K. H., and M. J. Merrett (1966). The photometabolism of acetate by *Chlorella pyrenoidosa. J. exp. Bot.*, **17**, 678.

Goulding, K. H., and M. J. Merrett (1967a). The photoassimilation of acetate by *Pyrobotrys stellata. J. gen. Microbiol.*, **48**, 127.

Goulding, K. H., and M. J. Merrett (1967b). The role of glycollic acid in the photoassimilation of acetate by *Chlorella pyrenoidosa. J. exp. Bot.*, **18**, 620.

Goulding, K. H., M. J. Lord, and M. J. Merrett (1969). Glycollate formation during the photorespiration of acetate by *Chlorella. J. exp. Bot.*, **20**, 34.

Haigh, W. G., and H. Beevers (1964). The occurrence and assay of isocitrate lyase in algae. *Arch. Biochem. Biophys.*, **107**, 147.

Harrop, L. C., and H. L. Kornberg (1963). Enzymes of the glyoxylate cycle in *Chlorella vulgaris. Biochem. J.*, **88**, 42p.

Harrop, L. C., and H. L. Kornberg (1966). The role of isocitrate lyase in the metabolism of algae. *Proc. Roy. Soc.* B, **166**, 11.

Hess, J. L., M. G. Huck, F. H. Liao, and N. E. Tolbert (1965). Glycollate metabolism in algae. *Plant Physiol.*, **40**, xlii.

Hess, J. L., and N. E. Tolbert (1967). Glycollate pathway in algae. *Plant Physiol.*, **42**, 371.

Hoare, D. S. (1962a). The photometabolism of $(1-^{14}C)$acetate and $(2-^{14}C)$acetate by washed-cell suspensions of *Rhodospirillum rubrum. Biochim. Biophys. Acta (Amst.)*, **59**, 723.

Hoare, D. S. (1962b). The photoassimilation of acetate to glutamate in washed cell suspensions of *Rhodospirillum rubrum. Biochem. J.*, **84**, 94p.

Hoare, D. S. (1963). The photoassimilation of acetate by *Rhodospirillum rubrum. Biochem. J.*, **87**, 284.

Hoare, D. S., and J. Gibson (1964a). Acetate assimilation in *Chlorobium thiosulphatophilum. Bact. Proc. (Baltimore)*, 101.

Hoare, D. S., and J. Gibson (1964b). Photoassimilation of acetate and the biosynthesis of amino acids by *Chlorobium thiosulphatophilum. Biochem. J.*, **91**, 546.

Hoare, D. S., S. L. Hoare, and R. B. Moore (1967). The photoassimilation of organic compounds by autotrophic blue-green algae. *J. gen. Microbiol.*, **49**, 351.

Hoare, D. S., and R. B. Moore (1965). Photoassimilation of organic compounds by autotrophic blue-green algae. *Biochim. Biophys. Acta (Amst.)*, **109**, 622.

Holm-Hansen, O. (1968). Ecology, physiology and biochemistry of blue-green algae. *Ann. Rev. Microbiol.*, **22**, 47.

Hurlbert, R. E. (1967). Effect of oxygen on viability and substrate utilization in *Chromatium. J. Bact.*, **93**, 1346.

Hurlbert, R. E., and J. Lascelles (1963). Ribulose diphosphate carboxylase in Thiorhodaceae. *J. Gen. Microbiol.*, **33**, 445.

Kamen, M. D. (1950). Hydrogenase activity and photoassimilation. *Fed. Proc.*, **9**, 543.

Kandler, O. (1954). Uber die Beziehungen zwischen Phosphathaushalt und Photosynthese II. Gesteigerter Glucoseeinbau im Licht als Indikator einer lichtabhängigen Phosphorylierung. *Z. Naturforschg.*, **9b**, 625.

Kandler, O. (1955). Uber die Beziehungen zwischen Phosphathaushalt und Photosynthese. III. Hemmungsanalyse der lichtabhängigen Phosphorylierung. *Z. Naturforschg.*, **10b**, 38.

Kandler, O. und I. Haberer-Liesenkötter (1963). Uber den Zusammenhang zwischen Phosphathaushalt und Photosynthese. V. Regulation der Glykolyse durch die Lichtphosphorylierung bei *Chlorella. Z. Naturforschg.*, **18b**, 718.

Kandler, O. und W. Tanner (1966). Die Photoassimilation von Glucose als Indikator für die Lichtphosphorylierung *in vivo*. *Ber. deut. bot. Ges. General versammlungsheft*, **79**, 48.

Keister, D. L., and N. J. Yike (1967). Energy-linked reactions in photosynthetic bacteria. I. Succinate-linked ATP-driven NAD$^+$ reduction by *Rhodospirillum rubrum* chromatophores. *Arch. Biochem. Biophys.*, **121**, 415.

Kikuchi, G., S. Abe, and A. Muto (1961). Carboxylic acid metabolism and its relation to porphyrin biosynthesis in *Rhodopseudomonas spheroides* under light-anaerobic conditions. *J. Biochem. (Tokyo)*, **49**, 570.

Kikuchi, G., S. Tsuiki, A. Muto, and H. Yamada (1963). Metabolism of carboxylic acids in non-sulfur purple bacteria under light anaerobic conditions. In *Studies on Microalgae and Photosynthetic Bacteria*, Special issue of *Plant and Cell Physiology*, Tokyo. pp. 547–565.

Kiyohara, T., Y. Fujita, A. Hattori, and A. Watanabe (1962). Effect of light on glucose assimilation in *Tolypothrix tenuis*. *J. Gen. Appl. Microbiol.*, **8**, 165.

Klemme, J.-H. und H. Schlegel (1967). Photoreduktion von Pyridinnucleotid durch Chromatophoren aus *Rhodopseudomonas capsulata* mit molekularem Wasserstoff. *Arch. Mikrobiol.*, **59**, 185.

Knight, M. (1962). The photometabolism of propionate by *Rhodospirillum rubrum*. *Biochem. J.*, **84**, 170.

Kohlmiller, E. F., Jr., and H. Gest (1951). A comparative study of the light and dark fermentation of organic acids by *Rhodospirillum rubrum*. *J. Bact.*, **61**, 269.

Kornberg, H. L. (1959). Aspects of terminal respiration in microorganisms. *Ann. Rev. Microbiol.*, **13**, 49.

Kornberg, H. L., and H. A. Krebs (1957). Synthesis of cell constituents from C$_2$-units by a modified tricarboxylic acid cycle. *Nature*, **179**, 988.

Kornberg, H. L., and J. Lascelles (1960). The formation of isocitratase by the Athiorhodaceae. *J. Gen. Microbiol.*, **23**, 511.

Kratz, W. A., and J. Myers (1955). Photosynthesis and respiration of three blue-green algae. *Plant Physiol.*, **30**, 275.

Larsen, H. (1951). Photosynthesis of succinic acid by *Chlorobium thiosulphatophilum*. *J. Biol. Chem.*, **193**, 167.

Larsen, H. (1953). On the microbiology and biochemistry of the photosynthetic green sulfur bacteria. *Kgl. Norske Videnskabers Selskabs Skrifter*, Nr. 1, 1.

Leach, C. K., and N. G. Carr (1968). Reduced nicotinamide adenine dinucleotide phosphate oxidase in the autotrophic blue-green alga, *Anabaena variabilis*. *Biochem. J.*, **109**, 4p.

Losada, M., A. V. Trebst, S. Ogata, and D. I. Arnon (1960). Equivalence of light and adenosine triphosphate in bacterial photosynthesis. *Nature*, **186**, 753.

Luksch, I. (1933). Ernährungsphysiologische Untersuchungen an *Chlamydomonadeen*. *Beih. Bot. Centralbl.*, **Erste Abteilung, 50**, 64.

Lwoff, A. (1932). Recherches biochimiques sur la nutrition des protozoaires. *Monogr. Inst. Pasteur*, Paris.

Lwoff, A. (1935). L'oxytrophie et les organismes oxytrophes. *C. r. Soc. Biol. (Paris)*, **119**, 87.

Lwoff, A. (1943). L'évolution physiologique, étude des pertes de fonctions chez les microorganismes, Hermann and Cie., Paris.

Lynch, V. H., and M. Calvin (1953). CO$_2$ fixation by *Euglena*. *Ann. New York Acad. Sciences*, **56**, 890.

Marker, A. F. H., and C. P. Whittingham (1966). The photoassimilation of glucose in *Chlorella* with reference to the role of glycolic acid. *Proc. Roy. Soc.* B, **65**, 473.

Marsh, H. V., Jr., J. M. Galmiche, and M. Gibbs (1963). Effect of light on the citric acid cycle in *Scenedesmus*. *Plant Physiol.*, **38**, iv.

Marsh, H. V., Jr., J. M. Galmiche, and M. Gibbs (1965). Effect of light on the tricarboxylic acid cycle in *Scenedesmus*. *Plant Physiol.*, **40**, 1013.

Merrett, M. J., and K. H. Goulding (1967a). Short-term products of ^{14}C-acetate assimilation by *Chlorella pyrenoidosa* in the light. *J. exp. Bot.*, **18**, 128.

Merrett, M. J., and K. H. Goulding (1967b). Glycollate formation during the photoassimilation of acetate by *Chlorella*. *Planta* (*Berl.*), **75**, 275.

Merrett, M. J., and K. H. Goulding (1968). The glycollate pathway during the photoassimilation of acetate by *Chlorella*. *Planta* (*Berl.*), **80**, 321.

Milhaud, G., A. A. Benson, and M. Calvin (1956). Metabolism of pyruvic acid-2-C^{14} and hydroxypyruvic acid-2-C^{14} in algae. *J. biol. Chem.*, **218**, 599.

Molisch, H. (1907) *Die Purpurbakterian nach neuen Untersuchungen*, Gustav Fischer Verlag, Jena.

Morita, S. (1955). The effect of light on the metabolism of lactic acid by *Rhodopseudomonas palustris*. I. *J. Biochem.* (*Tokyo*), **42**, 533.

Morita, S. (1961). Metabolism of organic acids in *Rhodopseudomonas palustris* in light and dark. *J. Biochem.* (*Tokyo*), **50**, 190.

Morita, S., K. Suzuki, and S. Takashima (1951). On the case of the s-shaped rate–light intensity relationship in the photosynthesis of purple bacteria. *J. Biochem.* (*Tokyo*), **38**, 255.

Mortenson, L. E. (1964). Ferredoxin and ATP requirement for biological N_2 fixation. *Fed. Proc.*, **23**, 430.

Moses, V., O. Holm-Hansen, J. A. Bassham, and M. Calvin (1959). The relationship between the metabolic pools of photosynthetic and respiratory intermediates. *J. Molec. Biol.*, **1**, 21.

Mue, S., S. Tuboi, and G. Kikuchi (1964). On malyl-coenzyme A synthetase. *J. Biochem.* (*Tokyo*), **56**, 545.

Muller, F. M. (1933). On the metabolism of the purple sulphur bacteria in organic media. *Arch. Mikrobiol.*, **4**, 131.

Myers, J. (1947). Oxidative assimilation in relation to photosynthesis in *Chlorella*. *J. gen. Physiol.*, **30**, 217 .

Nesterov, A. I., I. N. Gogotov, and E. N. Kondratieva (1966). Effect of light intensity on utilization of various carbon sources by photosynthetic bacteria. *Mikrobiologija* (*Mosk.*), **35**, 193.

Newton, J. W., and M. D. Kamen (1957). Photophosphorylation by subcellular particles from *Chromatium*. *Biochim. Biophys. Acta* (*Amst.*), **25**, 462.

Nishimura, M. (1961). Photophosphorylation by intermittent light. *Fed. Proc.*, **20**, 374.

Nozaki, M., K. Tagawa, and D. I. Arnon (1961). Noncyclic photophosphorylation in photosynthetic bacteria. *Proc. Natl. Acad. Sci.* (*Wash.*), **47**, 1334.

Nührenberg, B., D. Lesemann und A. Pirson (1968). Zur Frage eines anaeroben Wachstums von einzelligen Grünalgen. *Planta* (*Berl.*), **79**, 162.

Oaks, A. (1962). Influence of glucose and light on pyruvate metabolism by starved cells of *Chlorella elipsoidea*. *Plant Physiol.*, **37**, 316.

Ohashi, A., N. Ishihara, and G. Kikuchi (1967). Pyruvate metabolism in *Rhodopseudomonas spheroides* under light-anaerobic conditions. *J. Biochem.* (*Tokyo*), 62, 497.

Ohmann, E. (1963). Uber den Ausfall glykolytischer Enzyme in *Euglena gracilis*. *Naturwiss.*, 50, 552.

Ohmann, E. (1964a). Acetataktivierung in Grünalgen I. Oxydation und Aktivierung des Acetats in *Euglena gracilis*. *Biochim. Biophys. Acta* (*Amst.*), 8, 325.

Ohmann, E. (1964b). Acetataktivierung in Grünalgen II. Die Bildung von Acetylphosphat in *Stichococcus bacillaris* und anderen Arten der Chlorococcales. *Biochim. Biophys. Acta* (*Amst.*), 90, 249.

Okuyama, M., S. Tsuiki, and G. Kikuchi (1965). α-Ketoglutarate-dependent oxidation of glyoxylic acid catalyzed by enzymes from *Rhodopseudomonas spheroides*. *Biochim. Biophys. Acta* (*Amst.*), 110, 66.

Ormerod, J. G. (1956). The use of radioactive carbon dioxide in the measurement of carbon dioxide fixation in *Rhodospirillum rubrum*. *Biochem. J.*, 64, 373.

Ormerod, J. G., and H. Gest (1962). Symposium on metabolism of inorganic compounds IV. Hydrogen photosynthesis and alternative metabolic pathways in photosynthetic bacteria. *Bact. Rev.*, 26, 51.

Ormerod, J. G., K. S. Ormerod, and H. Gest (1961). Light-dependent utilization of organic compounds and photoproduction of molecular hydrogen by photosynthetic bacteria; relationships with nitrogen metabolism. *Arch. Biochem. Biophys.*, 94, 449.

Pearce, J., and N. G. Carr (1966). Enzymes of acetate metabolism in blue-green algae. *J. gen. Microbiol.*, 45, i.

Pearce, J., and N. G. Carr (1967a). The metabolism of acetate by the blue-green algae, *Anabaena variabilis* and *Anacystis nidulans*. *J. gen. Microbiol.*, 49, 301.

Pearce, J., and N. G. Carr (1967b). An incomplete tricarboxylic acid cycle in the blue-green alga, *Anabaena variabilis*. *Biochem. J.*, 105, 45p.

Pearce, J., and N. G. Carr (1968). The incorporation and metabolism of glucose by *Anabaena variabilis*. *J. gen. Microbiol.*, 54, 451.

Pearce, J., C. K. Leach, and N. G. Carr (1969). The incomplete tricarboxylic acid cycle in the blue-green alga, *Anabena variabilis*. *J. gen. Microbiol.*, 55, 371.

Petrack, B., and F. Lipmann (1961). Photophosphorylation and photohydrolysis in cell-free preparations of blue-green algae. In W. D. McElroy and B. Glass (Eds.), *Light and Life*, John Hopkins Press, Baltimore. pp. 621–630.

Pfennig, N. (1967). Photosynthetic bacteria. *Ann. Rev. Microbiol.*, 21, 285.

Pimenova, M. N., and T. F. Kondratieva (1965). A contribution to the use of acetate by *Chlamydomonas globosa*. *Mikrobiologija* (*Mosk.*), 34, 230.

Pringsheim, E. G. (1914). Kulturversuche mit chlorophyllführenden Mikroorgnismen. III. Mitteilung. Zur Physiologie der Schizophyceen. *Beitr. Biol. Pflanzen*, 12, 49.

Pringsheim, E. G. (1937a). Algenreinkulturen. *Beih. bot. Zentralbl.*, 57, 105.

Pringsheim, E. G. (1937b). Beiträge zur Physiologie saprophytischer Algen und Flagellaten. 1. Mitteilung: *Chlorogonium* und *Hyalogonium*. *Planta* (*Berl.*), 26, 631.

Pringsheim, E. G. (1937c). Beiträge zur Physiologie saprotropher Algen und Flagellaten. 3. Mitteilung: Die Stellung der Azetatflagellaten in einem physiologischen Ernährungssystem. *Planta* (*Berl.*), 27, 61.

Pringsheim, E. G. (1937d). Assimilation of different organic substances by saprophytic flagellatae. *Nature,* **139,** 196.

Pringsheim, E. G. (1959). Heterotrophie bei Algen und Flagellaten. In W. Ruhland (Ed.), *Handbuch der Pflanzenphysiologie,* Vol. XI. Springer-Verlag, Berlin, Göttingen, Heidelberg, 1959. pp. 303–326.

Pringsheim, E. G. (1968). Kleine Mitteilungen über Flagellaten und Algen. XVI. *Lauterbornia (Anacystis) nidulans* (Richter) nov. gen., nov. comb. Cyanophyceae. *Arch. Mikrobiol.,* **63,** 1.

Pringsheim, E. G. und O. Pringsheim (1959). Die Ernährung koloniebildender Volvocales. *Biol. Zentralbl.,* **78,** 937.

Pringsheim, E. G., and W. Wiessner (1960). Photo-assimilation of acetate by green organisms. *Nature,* **188,** 919.

Pringsheim, E. G. und W. Wiessner (1961) Ernährung und Stoffwechsel von *Chlamydobotrys* (Volvocales). *Arch. Mikrobiol.,* **40,** 231.

Pringsheim, E. G., and W. Wiessner (1963). Minimum requirements for heterotrophic growth and reserve substance in *Beggiatoa. Nature,* **197,** 102.

Quadri, S. M., and D. S. Hoare (1967). Pyruvate decarboxylase in photosynthetic bacteria. *Biochim. Biophys. Acta (Amst.),* **148,** 304.

Ramirez, J. M., F. F. del Campo, and D. I. Arnon (1968). Photosynthetic phosphorylation as energy source for protein synthesis and carbon dioxide assimilation by chloroplasts. *Proc. Nat. Acad. Sciences (Wash.),* **59,** 606.

Reeves, H. C., S. Kadis, and S. Ajl (1962). Enzymes of the glyoxylate by-pass in *Euglena gracilis. Biochim. Biophys. Acta (Amst.),* **57,** 403.

Rinne, R. W., R. W. Buckman, and C. R. Benedict (1965) Acetate and bicarbonate metabolism in photosynthetic bacteria. *Plant Physiol.,* **40,** 1066.

Roelofsen, P. A. (1934). On the metabolism of the purple sulphur bacteria. *Proc. Koninklijke Akademie van Wetenschappen te Amsterdam,* **37,** 660.

Russel, G. K., and M. Gibbs (1964). Regulation by acetate of photosynthetic capacity in *Chlamydomonas mundana. Plant Physiol.,* **39,** xlv.

Russel, G. K., and M. Gibbs (1966). Regulation of photosynthetic capacity in *Chlamydomonas mundana. Plant Physiol.,* **41,** 885.

Sadler, W. R., and R. Y. Stanier (1960). The function of acetate in photosynthesis by green bacteria. *Proc. Nat. Acad. Sciences (Wash.),* **46,** 1328.

Schlegel, H. G. (1956) Die Verwertung organischer Säuren durch *Chlorella* im Licht. *Planta (Berl.),* **47,** 510.

Schlegel, H. G. (1959) Die Verwertung von Essigsäure durch *Chlorella* im Licht. *Z. Naturforschg.,* **14b,** 246.

Schlegel, H. G. (1962). Die Speicherstoffe von *Chromatium okenii. Arch. Mikrobiol.,* **42,** 110.

Schlegel, H. G. und G. Gottschalk (1962). Poly-β-hydroxybuttersäure, ihre Verbreitung, Funktion und Biosynthese. *Angew. Chemie,* **74,** 342.

Shigesada, K., K. Hidaka, H. Katsuki, and S. Tanaka (1966). Biosynthesis of glutamate in photosynthetic bacteria. *Biochim. Biophys. Acta (Amst.),* **112,** 182.

Siegel, J. M. (1950). The metabolism of acetone by the photosynthetic bacterium *Rhodopseudomonas gelatinosa. J. Bact.,* **60,** 595.

Siegel, J. M. (1954). The photosynthetic metabolism of acetone by *Rhodopseudomonas gelatinosa. J. Biol. Chem.,* **208,** 205.

Siegel, J. M. (1956). The application of carbon-14 to studies on bacterial photosynthesis. *Proc. Intern. Conf. on the Peaceful Uses of Atomic Energy, Geneva, XII*, United Nations, New York. pp. 334–339.

Siegel, J. M., and M. D. Kamen (1950). Studies on the metabolism of photosynthetic bacteria. VI. Metabolism of isopropanol by a new strain of *Rhodopseudomonas gelantinosa. J. Bact.*, **59**, 693.

Siegel, J. M., and M. D. Kamen (1951). Studies on the metabolism of photosynthetic bacteria. VII. Comparative studies on the photoproduction of H_2 by *Rhodopseudomonas gelatinosa* and *Rhodospirillum rubrum. J. Bact.*, **61**, 215.

Simonis, W. (1956). Untersuchungen zur lichtabhängigen Phosphorylierung II. *Z. Naturforschg.*, **11b**, 354.

Sisler, E. C., and R. C. Fuller (1959). The citric acid and glyoxylate cycles in *Chromatium. Fed. Proc.*, **18**, 325.

Smith, A. J., J. London and R. Y. Stanier (1967) Biochemical basis of obligate autotrophy in blue-green algae and thiobacilli. *J. Bact.*, **94**, 972.

Smith, F. A. (1967). Links between glucose uptake and metabolism in *Nitella translucens. J. exp. Bot.*, **18**, 348.

Smith, L., and M. Baltscheffsky (1959). Respiration and light-induced phosphorylation in extracts of *Rhodospirillum rubrum. J. Biol. Chem.*, **234**, 1575.

Stanier, R. Y. (1961). Photosynthetic mechanisms in bacteria and plants: development of a unitary concept. *Bact. Rev.*, **25**, 1.

Stanier, R. Y., M. Doudoroff, R. Kunisawa, and R. Contopoulou (1959). The role of organic substrates in bacterial photosynthetis. *Proc. Nat. Acad. Sci. (Wash.)*, **45**, 1246.

Stern, J. R. (1963). Enzymic activation and cleavage of D- and L-malate. *Biochim. Biophys. Acta (Amst.)*, **69**, 435.

Stiffler, H. J., and H. Gest (1954). Effects of light intensity and nitrogen growth source on hydrogen metabolism in *Rhodospirillum rubrum. Science*, **120**, 1024.

Stoppani, A. O. M., R. C. Fuller, and M. Calvin (1955). Carbon dioxide fixation by *Rhodopseudomonas capsulatus. J. Bact.*, **69**, 491.

Stross, R. G. (1960). Growth response of *Chlamydomonas* and *Haematococcus* to the volatile fatty acids. *Can. J. Microbiol.*, **6**, 611.

Suzuki, K. (1959). Studies on oxidation of TCA cycle members in the purple bacterium, *Rhodopseudomonas palustris. The Science Reports of the Saitama University*, **Series B, 3**, 161.

Syrett, P. J. (1966). The kinetics of isocitrate lyase formation in *Chlorella*: evidence for the promotion of enzyme synthesis by photophosphorylation. *J. exp. Bot.*, **17**, 641.

Syrett, P. J., S. M. Bocks, and M. J. Merrett (1964). The assimilation of acetate by *Chlorella vulgaris. J. exp. Bot.*, **15**, 35.

Syrett, P. J., M. J. Merrett, and S. M. Bocks (1963). Enzymes of the glyoxylate cycle in *Chlorella vulgaris. J. exp. Bot.*, **14**, 249.

Szymona, M., and M. Doudoroff (1960). Carbohydrate metabolism in *Rhodopseudomonas spheroides. J. gen. Microbiol.*, **22**, 167.

Tanner, W., L. Dächsel, and O. Kandler (1965). Effects of DCMU and antimycin A on photoassimilation of glucose in *Chlorella. Plant Physiol.*, **40**, 1151.

Tanner, W. und O. Kandler (1967). Die Abhängigkeit der Adaptation der Glucose-Aufnahme von der oxydativen und der photosynthetischen Phosphorylierung bei *Chlorella vulgaris. Z. Pflanzenphysiol.*, **58**, 24.

Tanner, W., M. Loffler, and O. Kandler (1969). Cyclic photophosphorylation *in vivo* and its relation to photosynthetic CO_2-fixation. *Plant Physiol.*, **44**, 422.

Tanner, W., E. Loos, and O. Kandler (1966). Glucose assimilation by *Chlorella* in monochromatic light of 658 and 711 mμ. In J. B. Thomas and J. C. Goedheer (Eds.), *Currents in Photosynthesis*, Ad. Donker Publisher, Rotterdam. pp. 243–251.

Tanner, W., E. Loos, W. Klob, and O. Kandler (1968). The quantum requirement for light dependent anaerobic glucose assimilation by *Chlorella vulgaris*. *Z. Pflanzenphysiol.*, **59**, 301.

Tanner, W., U. Zinecker und O. Kandler (1967). Die anaerobe Photoassimilation von Glucose bei Photosynthese-Mutanten von *Scenedesmus*. *Z. Naturforschg.*, **22b**, 358.

Taylor, F. J. (1960). The absorption of glucose by *Scenedesmus quadricauda*. II. The nature of the absorptive process. *Proc. Roy. Soc. London, B.*, **151**, 483.

Thiele, H. H. (1966). *Wachstumsphysiologische Untersuchungen an Thiorhodaceae; Wasserstoff–Donatoeren und Sulfatreduction.* Doctoral Thesis, University of Göttingen, Germany.

Trebst, A. V., M. Losada, and D. I. Arnon (1960). Photosynthesis by isolated chloroplasts. XII. Inhibitors of CO_2 assimilation in a reconstituted chloroplast system. *J. Biol. Chem.*, **235**, 840.

Tsuiki, S., and G. Kikuchi (1962). Catabolism of glycine in *Rhodopseudomonas spheroides*. *Biochim. Biophys. Acta (Amst.)*, **64**, 514.

Tsuiki, S., A. Muto, and G. Kikuchi (1963). A possible route of acetate oxidation in *Rhodopseudomonas spheroides*. *Biochim. Biophys. Acta (Amst.)*, **69**, 181.

Tuboi, S., and G. Kikuchi (1962). Enzymatic cleavage of malate to glyoxylate and acetyl-coenzyme A. *Biochim. Biophys. Acta (Amst.)*, **62**, 188.

Tuboi, S., and G. Kikuchi (1965). Enzymatic cleavage of malyl-coenzyme A into acetyl-coenzyme A and glyoxylic acid. *Biochim. Biophys. Acta*, **96**, 148.

Tuboi, S., and G. Kikuchi (1966). Regulation by illumination of the citric acid cycle activity in *Rhodopseudomonas spheroides*. *J. Biochem. (Tokyo)*, **59**, 456.

Utter, M. P., and D. B. Keech (1960). Formation of oxaloacetate from pyruvate and CO_2. *J. Biol. Chem.*, **235**, PC 17.

Van Niel, C. B. (1935). Photosynthesis of bacteria. *Cold Spring Harbor Symp. on Quant. Biol.*, **3**, 138.

Van Niel, C. B. (1936). On the metabolism of the Thiorhodaceae. *Arch. Mikrobiol.*, **7**, 323.

Van Niel, C. B. (1941). The bacterial photosyntheses and their importance for the general problem of photosynthesis. *Advances in Enzymol.*, **1**, 263.

Van Niel, C. B. (1944). The culture, general physiology, morphology, and classification of the non-sulfur purple and brown bacteria. *Bact. Rev.*, **8**, 1.

Vernon, L. P., and O. K. Ash (1960). Coupled photooxidation and photoreduction reactions and associated phosphorylation by chromatophores of *Rhodospirillum rubrum*. *J. Biol. Chem.*, **235**, 2721.

Vernon, L. P., and M. D. Kamen (1953). Studies on the metabolism of photosynthetic bacteria. XV. Photoautoxidation of ferrocytochrom *c* in extracts of *Rhodospirillum rubrum*. *Arch. Biochem. Biophys.*, **44**, 298.

Weaver, E. C., and N. I. Bishop (1963). Photosynthetic mutants separate electron paramagnetic resonance signals of *Scenedesmus*. *Science*, **140**, 1095.

Weaver, P., K. Tinker, and R. C. Valentine (1965). Ferredoxin linked DPN reduction by the photosynthetic bacteria *Chromatium* and *Chlorobium*. *Biochem. biophys. Res. Comm.*, **21**, 195.

Whittingham, C. P., M. Bermingham, and R. G. Hiller (1963). The photometabolism of glucose in *Chlorella*. *Z. Naturforschg.*, **18b**, 701.

Wiessner, W. (1962). Kohlenstoffassimilation von *Chlamydobotrys* (Volvocales). *Arch. Mikrobiol.*, **43**, 402.

Wiessner, W. (1963). The non-photosynthetic, light-dependent metabolism in *Chlamydobotrys* (Volvocales). *Plant Physiol.*, **38**, xxviii.

Wiessner, W. (1965). Quantum requirement for acetate assimilation and its significance for quantum measurements in photophosphorylation. *Nature*, **205**, 56.

Wiessner, W. (1966a). Relative quantum yields for anaerobic photoassimilation of glucose. *Nature*, **212**, 403.

Wiessner, W. (1966b). Vergleichende Studie zum Quantenbedarf der Photoassimilation von Essigsäure durch photoheterotrophe Purpurbakterien und Grünalgen. *Ber. deut. bot. Ges.*, **79**, **1. Generalversammlungsheft**, 58.

Wiessner, W. (1967). The problem of glycollate formation from acetate in green algae. *Arch. Mikrobiol.*, **58**, 366.

Wiessner, W. (1968). Enzymaktivität und Kohlenstoffassimilation bei Grünalgen unterschiedlichen ernährungsphysiologischen Typs. *Planta* (*Berl.*), **79**, 92.

Wiessner, W. (1969a). Photo-assimilation and photosynthesis in green algae. In H. Metzner (Ed.), *Progress in Photosynthesis Research*, Tübingen. Vol. I. pp. 442–449.

Wiessner, W. (1969b). Effect of autotrophic or photo-heterotrophic growth conditions on *in vivo* absorption of visible light by green algae. *Photosynthetica* (*Prag.*), 3, 225.

Wiessner, W. und F. Amelunxen (1969a). Beziehung zwischen submikroskopischer Chloroplastenstruktur und Art der Kohlenstoffquelle unter phototrophen Ernährungsbedingungen bei *Chlamydobotrys stellata*. *Arch. Mikrobiol.*, **66**, 14.

Wiessner, W. und F. Amelunxen (1969b). Umwandlung der submikroskopischen Chloroplastenstruktur parallel zur Veränderung der stoffwechselphysiologischen Leistung von *Chlamydobotrys stellata*. *Arch. Mikrobiol.* **67**, 357.

Wiessner, W., and H. Gaffron (1964a). The role of photosynthesis in the light induced assimilation of acetate by *Chlamydobotrys*. *Nature*, **201**, 725.

Wiessner, W., and H. Gaffron (1964b). Acetate assimilation at λ 723 in *Chlamydobotrys*. *Fed. Proc.*, **23**, 226.

Wiessner, W. und A. Kuhl (1962). Die Bedeutung des Glyoxylsäurezyklus für die Photoassimilation von Acetat bei phototrophen Algen. In *Beiträge zur Physiologie und Morphologie der Algen*, Gustav Fischer Verlag, Stuttgart. pp. 102–108.

Wildon, D. C., and T. ap Rees (1965). Metabolism of glucose-C^{14} by *Anabaena cylindrica*. *Plant Physiol.*, **40**, 332.

Winogradsky, S. (1888). Zur Morphologie und Physiologie der Schwefelbakterien. In *Beiträge zur Morphologie und Physiologie der Bakterien*, Heft 1, Leipzig.

Woody, B. R., and E. S. Lindstrom (1955). The succinic dehydrogenase from *Rhodospirillum rubrum*. *J. Bact.*, **69**, 353.

Zelitsch, J. (1957). α-hydroxysulfonates as inhibitors of the enzymatic oxidation of glycolic and lactic acid. *J. Biol. Chem.*, **224**, 251.

CHAPTER 5

Light effects on ion fluxes in microalgae

WOLFGANG WIESSNER

Department of Plant Physiology, University of Göttingen, Germany

Introductory remarks

The effect of light on ion fluxes is one of the unsolved problems in the physiology of algae. It is an outstanding example how difficulties arose by trying to solve physiological problems without well-founded knowledge of the ultra and molecular structure of cells and their organelles. It is therefore not surprising that the molecular mechanisms of the effects of light on ion fluxes are still obscure, hypothetical and controversial.

This paper is restricted to work done with microalgae. However the large coenocytic Charophyceae are the most studied class. Their cells are several thousand times larger than those of most microorganisms; Charophyceae cannot be regarded as microalgae. Nevertheless, it will become evident that the experimental results obtained with these organisms can, in general, be parallelled to the situation in other algae. One argument against this parallellization is the presence of the large vacuole in mature cells of Charophyceae, which occupies 80 to 90 per cent of the cell volume. The

135

vacuoles of unicellular algae—such as *Chlorella*—approximately occupy only 1 per cent (Schaedle and Jacobsen, 1965). Thus cells of Charophyceae might represent two distinct compartments, and cells of unicellular algae only one (for example, see Barber, 1968b). However, cells would behave as single compartments with regard to ion movement only, if the rate-limiting step for the ion flux is located solely in the cell surface membrane (plasmalemma). In respect to different ions each cell, with its many organ-elles, represents many compartments with different ion concentrations, the vacuole being only one among others. Moreover, the study of ion fluxes across the tonoplast in Charophyceae shows us that the cell surface mem-brane is not the only site where active transport mechanisms can be located. That the cell of unicellular algae sometimes behaves like one single com-partment is a simplification, due to present experimental limitations.

Furthermore, it is worthwhile to keep in mind that any active change in ion concentration in any compartment of the cells may inevitably lead to changes in other neighbouring compartments, either by active or by passive ion translocation. Therefore it is often difficult to define the actual site of the initiating ion movement. For example, it is possible that occasionally the most observed ion flux, the one across the cell boundary, is only one—the first or the last—step in a whole series of consecutive ion movements between adjacent cell compartments.

Even though this article is devoted to the influence of light on ion fluxes in microalgae, general aspects of ion flux, which also apply to other organisms, will be discussed occasionally. This is necessary in order to compare the results obtained for algae with those for other biological objects.

General aspects of ion fluxes

We will consider first some general aspects of ion movement into and inside the cells. This will be done very briefly and sometimes in a simplified version. For detailed information the reader is referred to Albers (1967), Bennett (1956), Danielli (1954), Dainty (1962), Epstein, (1956, 1960, 1966), Jennings (1963), Lüttge (1968), Lüttge and Bauer (1968), MacRobbie (1969), Miller (1960), Robertson (1956, 1968), Schnepf (1968), Stein (1967) or Sutcliffe (1959, 1962).

The algal cell in its natural environment or suspended in a nutrient medium as a whole and the several organelles inside the cell represent many phases, separated from each other by special membranes. Between these phases large differences of ion content and ion concentration can exist. For example, in *Nitella axillaris* only 1·6 per cent of the total K^+ content are in the cytoplasm and in the chloroplasts, while 98 per cent are in the vacuole (Diamond and Solomon, 1959). The contents of Na^+ and Cl^- in the vacuole of *Nitella translucens* are 4·5 times respectively, 2·5 times higher than in the cytoplasm (Spanswick, Stolarek and Williams, 1967; Spanswick and Williams, 1964). In the same organism the K^+ and Na^+ content of the chloroplasts exceeds that of the cytoplasm and the vacuole two to three times (MacRobbie, 1962). Such differences in ion content generally correspond to ion concentration differences. For example, the concentration of Na^+ is mostly lower in the cytoplasm than in the vacuole (MacRobbie and Dainty, 1958b; Saltman, Forte and Forte, 1963). In addition, the ion concentration in the cytoplasm itself is not homogeneous. This has been demonstrated for *Nitellopsis* (MacRobbie and Dainty, 1958b). Here, even though the concentrations for Na^+ and K^+ are in toto lower in the total cytoplasm than in the vacuole, they are as high in a 5 to 6 μ thick cytoplasma zone next to the vacuole. Large ion concentration differences also exist between the cytoplasm and the chloroplasts. Especially K^+ is accumulated and maintained in high concentrations in these organelles, whereas Na^+ is excluded (Larkum, 1968; Saltman, Forte and Forte, 1963). Mitochondria also maintain ions in high concentrations (Pullman and Schatz, 1967; Robertson, 1956).

As to concentration differences between the cells and the surrounding media, in *Nitella* (MacRobbie, 1962; Kishimoto, Nagai and Tazawa, 1965; Kishimoto and Tazawa, 1965a), in *Lamprothamnium succinctum* (Kishimoto and Tazawa, 1965b) and in *Hydrodictyon africanum* (Raven, 1967a), the concentrations of K^+ and Cl^- inside the vacuole exceed those in the external medium by factors of between 30 and 800. Under the experimental conditions applied (temperature, actual ion concentration, pH), these concentration differences correspond to electric potential differences between -90 and -160 mV. The internal K^+ and Cl^- concentrations in *Chlorella pyrenoidosa* also exceed those in the medium (Barber, 1968b). Vice versa, for some ions such as sodium, the ion concentration in the medium can be higher than inside the vacuole, for example, in *Nitella translucens* (Smith, 1967), *Lamprothamnium succinctum* (Kishimoto and Tazawa, 1965b) and *Valonia ventricosa* (Gutknecht, 1966).

It is obvious that ions can be transported from one compartment into another against such concentration and potential differences only with the expenditure of energy. Furthermore, additional work has to be performed to transfer the ions across the membrane against their internal resistance, the latter being necessary to diminish free osmotic movement in order to preserve the concentration differences on both sides of the membranes. This work is required even in the absence of any differences in concentration or electrical potential. Wherever the current resistance of membranes has been measured, it is high. For example, according to Walker (1957, 1960) the plasmalemma resistance is between 5 and 50 kΩ cm^2 and the tonoplast resistance is around 3 kΩ cm^2 (see also Findlay and Hope, 1964; Hope and Walker, 1961; Umrath, 1956; Walker, 1955). Skierczyńska (1968a, 1968b) determined the electrical resistance of the plasmalemma of *Chara australis* as 7 kΩ cm^2, while that of the tonoplast is much lower. To find out about the difficulties in measuring the electrical resistance of membranes, especially of the tonoplast, see Skierczyńska (1968b) and Umrath (1956). It should be kept in mind that no direct relation might exist between the ion permeability rate and the electrical resistance of membranes if the ion can penetrate with a carrier molecule in an uncharged form.

Active transport in living cells depends on a continuous supply of chemical energy, which may for instance be supplied by mechanisms *converting radiant into chemical energy*. If such mechanisms exist, light can influence ion fluxes significantly.

At this point it is necessary to briefly summarize our present knowledge about the ultrastructure and the chemical composition of membranes as sites of active transport mechanisms and as the fundamental organelles where light energy is used in order to accomplish active ion movement. All hypotheses about osmotic as well as active ion flux should consider the ultrastructure and the chemical composition of the biological membranes as the cytological basis for these processes.

Early investigations revealed that under the electronmicroscope all kinds of biological membranes exhibit similar features, even though they can hardly be expected to be completely alike. These features—the existence of a central bimolecular lipid leaflet and of two outer protein layers—have led to the term 'unit membrane', which describes all membrane types: the plasmalemma, the endoplasmatic reticulum, and the tonoplast, chloroplast and mitochondrial membranes (Danielli, 1954; Danielli and Davson, 1935; Danielli and Harvey, 1935; Davson, 1962; Nelson, 1965; Robertson, 1967). The validity of the Danielli–Davson–Robertson unit membrane concept as a general one has repeatedly been questioned (Green and Perdue, 1966; Korn, 1966, 1969a, 1969b; Lenard and Singer, 1966; Sjöstrand, 1967; Stoeckenius, 1967), especially since new techniques in electronmicroscopy (negative staining, freeze-etching) and the roentgenographic analysis have led to the recognition of repeating particulate subunits in cellular membranes, probably attached to or embedded into central lipids. Such subunits have been seen, for example, in thylakoid membranes (Kreutz and Menke, 1962; Menke, 1965; Mühlethaler, Moor and Szarkowski, 1965; Weier and coworkers, 1965), in cytoplasmic membranes of yeast cells or root cells

(Branton and Moor, 1964; Moor and Mühlethaler, 1963), as well as in membranes of zoological objects (Nielsson, 1964, 1965; Sjöstrand, 1963), to mention only a few earlier papers. For further references and detailed information see Branton (1969), Clowes and Juniper (1968) or Korn (1969a, 1969b). At present there is no strong basis for maintaining the paucimolecular unit membrane theory as a general concept. Most likely biological membranes differ from each other to such an extent that they cannot be described by one unifying concept.

Several answers are possible to the important question of the existence of pores in the membranes as the possible sites of activated ion transport.

(1) The tripled membrane structure can be perforated in its central lipid part by small pores, which are covered with polypeptid chains. In membranes composed of repeating subunital elements pores might also exist. The existence of pores has been presumed repeatedly, such as for *Beggiatsa mirabilis* by Ruhland and Hoffmann (1925) or Schönfelder (1931), and for *Dunaliella parva* by Ginzburg (1969). Fensom and Wanless (1967) calculated for *Nitella* 10^8 to 10^9 pores per cm^2.

(2) Biological membranes are dynamic structures transforming between several different configurations in the course of different membrane functions (Kavanau, 1965, 1966). Pores might be opened and closed temporarily during such transformation processes.

The molecular substructure of different membranes is still more or less the object of speculation, even though their lipid and protein composition has been the subject of numerous investigations. For references see, for example, Korn (1966, 1969b), O'Brien (1967), Rothfield and Finkelstein (1968) and Stein (1967).

Sites where special ion transport systems (carriers) might be located in the membranes have not been seen. The apparent uniformity of the gross membrane morphology leads to the inevitable view that they must be incorporated without perceptible local alterations of the general ultra-structure.

Carriers are thought to be proteins combining with ions in the course of the transport process. For the properties of carrier mechanisms, see for example, Weigl (1967) or Wilbrandt (1967). Proteins which might be engaged in ion transport mechanisms binding special ions have been isolated during the last few years. For example, Pardee's group (Pardee, 1967, 1968; Pardee and Prestidge, 1966; Pardee and coworkers, 1966) isolated from *Salmonella typhimurium* a protein which specifically binds sulphate. Wassermann and Taylor (Wassermann and Taylor, 1966; Taylor and Wassermann, 1967) isolated one from chick duodenal mucosa which binds calcium. The protein of the (Na^+–K^+)-activated ATPase (see p. 142) has also been enriched (Medzihradsky, Kline and Hokin, 1967). For further information see, for example, Pardee (1968). It is certain that all these proteins are involved in ion transport; their mode of action, however, is still obscure.

General aspects of light effects on ion movement

Light of the spectral region between 320 and 750 nm can be transformed by photosynthetic microalgae into chemical energy in the form of adenosine triphosphates (ATP), or other high-energy intermediates, states and reservoirs (Brierley and coworkers, 1963; Hind and Jagendorf, 1963a, 1963b; Hinkson and Boyer, 1965; Hodges and Hanson, 1965; Izawa, Winget and Good, 1966; Jagendorf and Hind, 1965; Kylin and Tillberg, 1967a; Mitchell, 1961, 1966; Skye, Shavit and Boyer, 1967; Vose and Spencer, 1967). For further references see Avron and Neumann (1968), Jagendorf and Uribe (1967) and Pullman and Schatz (1967). These may supply energy for several processes. Furthermore, the light-induced electron translocation in electron transporting enzyme systems of the chloroplast itself can be active in transporting ions in these organelles. Later on we will discuss the possibility of whether or not the ion transport across other cell membranes can be linked directly to such a light-induced electron translocation.

Influence of light on non-active ion movement

ATP generated photosynthetically can be used to maintain the energy-requiring membrane structures in a state which is highly favourable for the regulation of the free ion movement. This applies to any type of membrane, especially to dynamic concepts. Proof of such light-induced changes in ion permeability through the plasmalemma comes from several experimental results. Raven (1968a) observed with *Hydrodictyon africanum* that light brings about a 20 to 30 per cent higher passive influx of K^+ and Na^+, and a 40 per cent higher passive efflux of K^+. The efflux of Cl^- is also increased by 10 per cent. In *Nitella translucens* light decreases the efflux of Cl^- significantly, which is also attributed to changes in membrane permeability for free osmotic movement (Hope, Simpson and Walker, 1966). Blinks (1940) observed that the electrical resistance of the plasmalemma of *Nitella clavata* is reduced four- or five-fold, compared to its dark values, by exposing the cells to full sunlight for a day or two. Recently Hogg, Williams and Johnston (1969) reported a 30 per cent decrease in membrane resistance of *Nitella translucens* due to light, simultaneously increasing the permeability coefficient for Na^+, but not the one for K^+. Neither is the mechanism of this light effect fully understood, nor do we know how far the reduction of the electrical resistance can alter the permeability of all membrane types.

In connection with this problem we also have to remember Brauner and Brauner's experiments with artificial membranes (Brauner, 1956; Brauner and Brauner, 1937). In their experiments light, especially ultraviolet radiation, decreases the mobility of cations through membranes, simultaneously increasing the motility of certain anions. Changes in the negative membrane charge, caused by setting photoelectrons free, could account for this phenomenon, which may be parallelled to the so-called Hallwachs effect at metal surfaces.

Influence of light on active ion movement

(a) *Ion fluxes requiring the supply of ATP*

Light can influence the active ion flux in several ways (Danielli, 1954; Robertson, 1968; Stein, 1967; Sutcliffe, 1959, 1962). In the following discussion the possible functions of ATP, derived from photophosphorylation, in active transport processes will be summarized. Information about the use of high-energy intermediates or states prior to ATP formation, or of energy reservoirs as alternatives to ATP as storage devices, is scarce. This topic will be mentioned in the special chapters on activated cation and anion transport.

(1) ATP is required for the synthesis of carrier molecules or of energy-rich substances as parts of the transport systems. A parallel inhibition of protein synthesis and of the light-dependent ion transport by chloramphenicol or pyromycin has been observed (Brenner and Maynard, 1966; MacRobbie, 1966b; and for the action of chloramphenicol see Ellis, 1963). This points to the direct energy-dependent synthesis of carriers or of their parts required for the light-dependent ion flux.

(2) ATP is necessary to initiate either the ion movement itself, or for the formation and breakdown of a complex between the carrier and the ion to be transported. In the latter cases the actual transport does not require necessarily the direct expenditure of energy. If the complex is formed at one and is destroyed at the other side of the membrane, then it moves *along* a concentration gradient, even if the ion is carried *against* a concentration gradient existing between both sides of the membrane.

Not all the different model perceptions about activated transport can be summarized here. I would like to mention only a few representative ones (for the biochemical aspects of active transport see Albers, 1967). Many hypotheses about carrier-mediated active transport mechanisms are based on the active movement of the carrier molecules. It is necessary to remember in this respect Vidaver's (1966) considerations. He pointed out theoretically that alternate forms of carriers involving the transition of a combining site between two alternate shapes, or the transition of a combining site between two alternate hydrogen-bonded states, have different directional properties and are kinetically equivalent to mobile carriers.

The active transport of sodium and of potassium is one of the best studied. In Jardetzky's (1966) concept about the transport of these two cations through membranes, the alternative absorption of Na^+ and K^+ triggers phosphorylation and dephosphorylation of active sites in the membrane. This leads to allosteric rearrangements in the latter as the mechanisms for an imparting direction to a diffusion process not requiring mobile carrier molecules. Activated transport of other ions can be based on similar allosteric transitions of the membrane proteins. Several other concepts have evolved from the knowledge about a (Na^+-K^+)-activated transfer ATPase (Albers, 1967; Bader, Post and Bond, 1968; McIlwain, 1962; Opit and Charnock, 1965; Skou, 1960, 1964, 1965, 1967). Such an ATP-hydrolysing enzyme engaged in the transport of Na^+ and K^+, which can be rather specifically inhibited by cardiatonic steroids, has been demonstrated first in animal objects, such as in crab nerve cells (Skou, 1957). It is known to be associated with the endoplasmatic reticulum of guinea kidney cortex or with erythrocyte membranes, and has been found also in other zoological objects and tissues (Charnock and Post, 1963; Dunham and Glynn, 1961; Glynn, 1962; Post and coworkers, 1960; Richardson, 1968; Ruoho and coworkers, 1968; Whittam, 1962). Recently a plant ATPase has been extracted from the roots of wheat seedlings which shows properties similar to the animal membrane-transport ATPases (Salyaev, Kamenkova and Podolyakina, 1969). The enzyme is activated by the alkali ions in two sites. One has a preference for sodium, the other for potassium (or rubidium). However, the mechanisms by which the vectorial transport of the ions in opposite directions is mediated is largely unknown. Skou (1964, 1967) proposed that the specific affinities of the binding sites of the enzyme system for Na^+ and K^+ change under the influence of an activating effect of ATP when the latter is bound to the enzyme. As the consequence of this activation, K^+ replaces Na^+ at the binding sites— and vice versa—leading to opposite movement of the two cations. Post and Sen (1965) considered that the transport ATPase goes through several phosphorylated states of progressively lower energy with special combining preferences for Na^+ and K^+, so that the ions can be transported in opposite directions. In continuation of these hypotheses Stone (1968) developed a theory, according to which the carrier molecules of the transport system has successively three forms which take up K^+ or Na^+ with various affinities. In its first form, X, the carrier is confined to the inside of the membrane. It has roughly equal affinities for Na^+ and K^+. XNa is converted into a phosphorylated form, Y, which is confined to the outside of the membrane and has a higher affinity for K^+ than for Na^+. How this carrier translocation is achieved remains obscure. Y is converted to the third carrier form, Z, which is still phosphorylated. This conversion requires K^+ and is inhibited by Na^+. Finally Z diffuses (!) to the inside of the membrane, where it is hydrolysed to give X and phosphate.

There is good evidence that a phosphorylated intermediate really exists in the overall enzyme reaction (Bader, Post and Bond, 1968; Bader, Sen and Post, 1966; Hems and Rodnight, 1966; Hokin and coworkers, 1965; Kahlenberg, Galsworthy and Hokin, 1967; Nagano and coworkers, 1965). This phosphorylation of the protein very likely involves the formation of a high-energy acyl phosphate bond, probably L-glutamyl-γ-phosphate (Kahlenberg, Galsworthy and Hokin, 1967).

It is necessary to point out that an uptake of divalent cations (Ba^{2+}, Ca^{2+}, Sr^{2+}) into chloroplasts, and their deposition at the inner surfaces of the thylakoid membranes under conditions of an ATP hydrolysis, has been observed by Nobel's group (Nobel, 1967a, 1967b; Nobel and Murakami, 1967; Nobel, Murakami and Takamiya, 1966; Nobel and Packer, 1964a, 1964b).

(b) *Ion fluxes initiated by light-dependent electron translocation*

Light may act upon ion movement via the translocation of electrons in membranes; this is similar to Lundegårdh's electrochemical concept (Conway, 1953, 1954, 1955; Lundegårdh, 1945, 1954, 1955). The fundamental conditions for the function of this mechanism is the presence of an oxidation–reduction system in a vectorial arrangement within the membrane and the requirement of membrane-bound pigments able to absorb light and to use the quantum energy for electron translocation, or the presence of pigments which receive the energy absorbed by pigments elsewhere, and transport this energy to the membrane oxidation–reduction system. So far, it remains obscure in which membranes these conditions are fulfilled, except in grana disc membranes of chloroplasts. Here the electron translocation indirectly plays a role in ion fluxes. From the work of Deamer and Packer (1967), Dilley (1964, 1967), Dilley and Vernon

(1965), Jagendorf and Uribe (1966, 1967), Packer, Deamer and Crofts (1967), and the many experiments based on the chemiosmotic hypothesis of Mitchell (1961, 1966), the following concept evolved: the light-dependent electron flow in thylakoid membranes is accompanied by a vectorial inward transport of protons. This can lead to an efflux of cations such as K^+ or Mg^{2+}, and possibly also to an influx of anions such as Cl^-.

Finally, very likely more than one of the above-mentioned transport possibilities will be engaged in the overall active ion transport across membranes, a conclusion which can be drawn from many experiments, as will be seen in the following paragraphs.

Light effects on anion fluxes

The light-dependent uptake of halogen anions has been studied in several *Nitella species* (Hoagland and Davis, 1924; Hoagland, Hibbard and Davis, 1927; MacRobbie, 1962, 1964, 1966a, 1966b), *Chara australis* (Coster and Hope, 1968), *Chara corallina* (Smith and West, 1969), *Tolypella intricata* (Smith, 1968b), *Chlorella pyrenoidosa* (Barber, 1968a, 1968b; Nielsen, 1963), *Scenedesmus* (Kylin, 1967a) and *Hydrodictyon africanum* (Raven, 1967a, 1967b, 1968a, 1968b, 1968d, 1969). In *Chlorella pyrenoidosa* the influx of Cl^- into the cell in the light is about six to ten times higher than in the dark. In *Nitella translucens* light increases the flux to the cell vacuole to about sixteen times that of the dark value (MacRobbie, 1962).

According to some authors this light-stimulated Cl^- flux into the cell or the vacuole is directly linked to the light-driven electron transfer of photosynthesis (Hope, Simpson and Walker, 1966; MacRobbie, 1965, 1966a; Raven, 1967b, 1969; Smith, 1967, 1968b). This relationship is postulated because in *Nitella translucens, Tolypella intricata* or *Hyrodictyon africanum* the light-dependent Cl^- uptake shows an action spectrum similar to the one of photosynthesis (Raven, 1968d, 1969); it is not supported by far red light ($\lambda > 705$ nm), its quantum efficiency is lowered with increasing wavelengths of red light, and it is sensitive to poisoning with DCMU. In *Hydrodictyon africanum*, 1×10^{-7} M DCMU decreases Cl^- influx to 20 per cent of the control value. The same DCMU concentration lowers the photosynthetic carbon dioxide fixation to 15 per cent of the control. In this alga Cl^- influx is much less linked to photophosphorylation. Inhibitors of the latter such as imidazole (1×10^{-3} M) or carbonyl cyanide *m*-chlorophenyl hydrazone (CCCP, 5×10^{-6} M) leave Cl^- uptake nearly unaffected. *Tolypella intricata*, however, is much more sensitive against CCCP. It has been shown that 2×10^{-6} M inhibit photosynthesis completely and the Cl^- influx to about 30 per cent of the control (Smith,

1968b). That not all organisms respond alike to the metabolic inhibitors is also demonstrated with *Chlorella pyrenoidosa*. Here, 5×10^{-6} M DCMU inhibit photosynthetic oxygen evolution completely and the Cl⁻ uptake, however, only to about 40 per cent of the control value (Barber, 1968a; Nielsen, 1963). Furthermore, the cytochrome *b* inhibitor antimycin A (Chance, Bonner and Storey, 1968; Hind and Olson, 1967), which uncouples photophosphorylation and inhibits also the phosphorylation coupled electron transport (Drechsler, Nelson and Neumann, 1969; Izawa and coworkers, 1967), lowers the chloride influx to 65 per cent of the control (1×10^{-5} M) and to 20 per cent of the control (1×10^{-4} M) respectively. CCCP also proved to be an inhibitor.

It seems certain that a connexion exists between the Cl⁻ flux and the light-dependent flow of electrons in the grana thylakoid membranes. Here Cl⁻ ions can be transported into the thylakoid lumina to counterbalance the light-dependent proton uptake. It is difficult, however, to explain how electron flow across grana membranes can yield energy directly usable for Cl⁻ translocation across the plasmalemma or the tonoplast, without the interaction of a chemical high-energy compound. Another possibility that the Cl⁻ transport into the grana lumina leads to a concentration or activity gradient across the plasmalemma with a passive flow from the medium into the cells, is difficult to prove and has not yet been observed. Therefore the differing results obtained with the above-mentioned metabolic inhibitors can be explained differently. Firstly, an unknown electron carrier, being reduced during the photosynthetic electron transport, transfers energy from the chloroplast to the plasmalemma. Secondly, ATP derived from cyclic (DCMU-insensitive) as well as from non-cyclic or pseudo-cyclic (DCMU-sensitive) photophosphorylation can be utilized for chloride translocation across non-grana membranes. All types of photophosphorylation might be differently engaged in Cl⁻ movement—differently also with respect to the organisms in question.

The latter conclusion is supported by experiments with higher plants. For example, Jeschke (1967) observed that the light-dependent Cl⁻ uptake into *Elodea densa* in a nitrogen atmosphere and in the absence of carbon dioxide is resistant against DCMU concentrations which block photosynthetic oxygen evolution almost completely. In the presence of 0·5 vol. per cent carbon dioxide in nitrogen, low concentrations of DCMU, however, partly inhibit the light-dependent Cl⁻ uptake.

It is possible that ATP is not used as an energy source in anion transport, but some unknown high-energy intermediate or state prior to ATP formation, or energy reservoirs other than ATP. It has been shown by Kylin (1964b) that the uptake of sulphate into phosphorus-starved cells of *Scenedesmus*, as well as the uptake of chloride (Kylin, 1967a), is inhibited in the light or in the dark by the readdition of phosphate. ATP formation, which requires inorganic phosphate, and Cl^- or sulphate influx might be alternatives in using the energy available as a high-energy intermediate or state. Coster and Hope's (1968) data on the inhibition of Cl^- influx into *Chara australis* by phlorizin could also be taken as evidence that a high-energy intermediate or state of photophosphorylation is engaged in anion transport in addition to ATP. This inhibitor, which might poison photophosphorylation at reactions prior to the generation of ATP from $\sim$P (Izawa, Winget and Good, 1966; Izawa and coworkers, 1967; Winget, Izawa and Good, 1969), lowers the light-dependent Cl^- uptake only by about 50 per cent. The results of Raven, MacRobbie and Neumann (1969) that concentrations of Dio-9 which poison photosynthesis do not decrease the Cl^- influx into *Nitella translucens*, *Tolypella intricata* or *Hydrodictyon africanum* are evidence that intermediates of photophosphorylation prior to ATP-formation (Gromet-Elhanan, 1968; Karlish and Avron, 1968; MacCarty and coworkers, 1965) participate in active anion movement. In *Chara australis*, however, the Cl^- influx is more inhibited by Dio-9 than is photosynthesis (Smith and West, 1969). Jeschke (1968) reported similar results on the action of Dio-9 on the Cl^- uptake in *Elodea densa*.

The efflux of Cl^- from *Nitella translucens* also responds to inhibitors of photosynthesis (Hope, Simpson and Walker, 1966). DCMU increases the efflux and decreases the influx.

Light lowers the uptake of J^- by *Ulva lactuca* (Roche and Andrée, 1963).

As to light effects on fluxes of other anions, a light-favoured uptake has also been observed for NO_3^- in *Nitella clavata* (Hoagland and Davis, 1924) and *Nitzschia closterium* (Ketchum, 1939). It can be explained by the increase in nitrate reduction in the light (Kessler, 1959, 1964; Losada and coworkers, 1965). A light-dependent sulphate uptake is present in *Chlorella pyrenoidosa* (Wedding and Black, 1960) and *Scenedesmus* sp. (Kylin, 1964a, 1967c). In *Chara australis* (Robinson, 1969) concentrations of DCMU which poison photosynthetic CO_2-fixation stimulate the influx of sulphate in the light. The light influx is insensitive to CCCP (carbonyl cyanide-*m*-phenylhydrazone) when photosynthesis is totally inhibited.

Earlier studies on the light-influenced phosphate uptake into the cells and its accumulation are reviewed by Kamen and Spiegelmann (1948), Kuhl (1962, 1968) and Rowan (1966). Recently work has been done with *Nitella translucens* (Smith, 1966), *Scenedesmus* sp. and *Scenedesmus obliquus* (Dvořak, Dvořakova-Hladka and Fialova, 1966; Kylin, 1966a; Kylin and Tillberg, 1967a, 1967b), *Chlorella pyrenoidosa* and *Coccomyxa solorinae*

(Dvořak, Dvořakova-Hladka and Fialova, 1966). As Smith (1966) demonstrated with *Nitella translucens*, phosphate influx might be supported by cyclic but not by non-cyclic photophosphorylation, and proceeds very well at wavelengths above λ 705 nm. It has been found that 5×10^{-6} M CCCP inhibit the phosphate influx completely; DCMU in concentrations up to 1×10^{-6} M has no effect.

However, all results about the light-dependent influx or uptake of phosphate are difficult to judge. Light effects on phosphate uptake, taken as evidence for an active influx, can also result from the utilization of phosphate in photophosphorylation itself, or in other light-favoured cellular syntheses. Activated phosphate transport and such phosphate-requiring, light-dependent metabolic processes can respond differently to inhibitors, which complicates the interpretation of all observations concerning the overall light-dependent phosphate influx. For example, according to Kylin and Tillberg (1967b) DCMU decreases the light-induced phosphate uptake into phosphorus-starved cells of *Scenedesmus* (see also Urbach and Simonis, 1964), but increases the level of ATP in the algae. On the other hand, the uptake of phosphate is increased as well as its storage in the form of polyphosphates, if the formation of ATP is prevented by phlorizin or the inhibitor-β-complex from potato (Kylin and Tillberg, 1967a). One can argue that in these experiments on the action of DCMU the phosphate uptake is reduced either because non-cyclic electron transport is required for activated phosphate influx, or because phosphate cannot be used in synthetic processes which need non-cyclic electron transport. The increase of the ATP level of the cells evidently results from an increased cyclic photophosphorylation (Arnon, 1967; Arnon, Tsujimoto and McSwain, 1967). On the other hand, a reduced ATP formation does not necessarily lower the phosphate uptake. Apparently phosphate can be transferred from a high-energy intermediate to some other form of bound storage phosphate.

A connexion between the phosphate uptake and light-dependent cellular syntheses has been shown also in experiments reported from Simonis' laboratory (Simonis, 1960; Simonis and Urbach, 1963a, 1963b; Urbach and Simonis, 1962, 1964) on phosphorus uptake and incorporation into *Ankistrodesmus braunii*. Light increases preferentially the incorporation into organic compounds. It is sensitive to DCMU (5×10^{-5} M). The DCMU-resistant phosphate incorporation can be inhibited with antimycin A. It is strongly influenced by carbon dioxide and its action spectrum is similar to the one for photosynthetic oxygen evolution (Simonis and Mechler, 1963).

Furthermore, because of Mechsner's (1959) and Simonis and Urbach's

(1963a, 1963b) results on the action of K^+ and Na^+ on phosphate uptake, it is probable that phosphate influx can also be linked with an active cation transport.

Light effects on cation fluxes

First we have to consider light effects on cation fluxes against concentration or potential differences. Such light-dependent influxes of potassium have been observed in *Nitella flexilis* (Nagai and Tazawa, 1962), *Nitella translucens* (MacRobbie, 1962, 1964, 1965, 1966a; Raven, MacRobbie and Neumann, 1969), *Tolypella intricata* (Raven, MacRobbie and Neumann, 1969; Smith, 1968b), *Chlorella pyrenoidosa* (Barber, 1968a, 1968c; Dvořak, Dvořakova-Hladka and Fialova, 1966; Scott, 1945), *Scenedesmus obliquus* (Dvořak, Dvořakova-Hladka and Fialova, 1966), *Hydrodictyon africanum* (Raven, 1967a, 1967b, 1968a, 1968b, 1969; Raven, MacRobbie and Neumann, 1969), *Coccomyxa solorinae saccata* (Dvořak, Dvořakova-Hladka and Fialova, 1966), *Ulva lactuca* (Scott and Hayward, 1953a, 1953b, 1953c, 1954; West and Pitman, 1967) and *Rhodymenia palmata* (MacRobbie and Dainty, 1958a). The concentration and potential differences for K^+ and Na^+ on both sides of the plasmalemma are mostly arranged oppositely. Therefore, both ions are generally transported actively in opposite directions (Barber, 1968c; Barr and Broyer, 1964; MacRobbie and Dainty, 1958a; Smith, 1967). The light increased efflux of sodium has been studied in *Chlorella pyrenoidosa* (Barber, 1968c), *Scenedesmus obliquus* (Kylin, 1966b), *Hydrodictyon africanum* (Raven, 1967a, 1967b, 1968a, 1968b, 1968d, 1969; Raven, MacRobbie and Neumann, 1969) and *Ulva* (Cummins, Strand and Vaughan, 1969; Scott and Hayward, 1953a, 1954; West and Pitman, 1967).

It is generally assumed that both ion fluxes depend on the supply of ATP, supplemented mainly from cyclic photophosphorylation (Barber, 1968a, 1968c; MacRobbie, 1965, 1966a; Raven, 1967b, 1968b, 1969; Raven, MacRobbie and Neumann, 1969). Evidence comes from the following results:

(1) the relative quantum efficiency increases with increasing wavelength,
(2) uncouplers of photophosphorylation such as CCCP or imidazole are inhibitory, and
(3) concentrations of DCMU, which poison photosynthesis completely, do not affect the Na^+ efflux but sometimes affect the K^+ influx.

It is possible that the light-dependent potassium and sodium fluxes are mediated by the (K^+-Na^+)-activated transport ATPase (see page 142) known from zoological objects. Raven (1967a) could inhibit the fluxes by high concentrations of the cardiac glycoside ouabain. In Barber's (1968c) and Bonting's and Caravaggio's (1966) experiments, however, this poison did not lower the light induced Na^+ efflux. It is possible that—depending on the organism—ouabain did not reach the transport site. Doubts about the complete coupling of the Na^+/K^+ fluxes by a transport ATPase have also been raised by Barber (1968c) because the sizes of the two fluxes are not identical.

Indications that not only ATP but also a high-energy intermediate or state of photophosphorylation might be engaged in the movement of these cations comes from experiments with phlorizin. This inhibitor, which blocks the formation of ATP (Izawa, Winget and Good, 1966; Izawa and coworkers, 1967; Winget, Izawa and Good, 1969), blocks the active K^+ influx and Na^+ efflux to the same extent as can also be inhibited with CCCP, but not the fluxes left after poisoning with this inhibitor.

Non-cyclic electron transport facilitating non-cyclic and pseudo-cyclic photophosphorylation might be engaged in cation flux in *Chlorella pyrenoidosa* (Barber, 1968a) in addition to cyclic photophosphorylation. Here 5×10^{-6} M DCMU inhibit the K^+ influx to about 40 per cent of control. The rest flux is sensitive against antimycin A, the inhibitor of photophosphorylation and of electron transport (Drechsler, Nelson and Neumann, 1969; Izawa and coworkers, 1967). It has to be considered that part of the light-dependent K^+ flux is only passive, and is linked to an active inward transport of chloride such as, for example, in *Hydrodictyon africanum* (Raven, 1968a, 1969). The uptake of potassium into *Tolypella intricata* might also be linked to the active Cl^- influx (Raven, MacRobbie and Neumann, 1969; Smith, 1968b). Here, all inhibitors decreasing the Cl^- flux inhibit the K^+ influx to about the same extent.

Light can also enhance the Na^+ flux which takes place in the direction of the concentration and potential difference (Barr and Broyer, 1964; Schaedle and Jacobsen, 1966; Smith, 1967). This passive sodium flux, the size of which is also determined by the presence of other cations, might be linked to the inward anion pump, because it can be decreased considerably by removing Cl^- ions from the solution (Smith, 1967). A coupling of the passive sodium flux with the active phosphate uptake has to be considered too. In phosphorus-deficient cells of *Scenedesmus* practically no sodium is taken up without the addition of phosphate (Kylin, 1964b, 1966b). Furthermore, changes in membrane permeability, increasing the possibility for osmotic movement, have to be considered as reasons for the light-favoured sodium influx (Barr and Broyer, 1964). Evidence comes from the results of Scott and Hayward (1953c) on the light-dependent passive K^+ efflux in *Ulva lactuca*, of Raven (1967a, 1968a) on K^+ and Na^+ influx and efflux in *Hydrodictyon africanum*, and of Smith (1968b) on the K^+ and Cl^- flow in *Tolypella intricata*.

Light increases the uptake of Rb^+ into *Scenedesmus* sp. (Kylin, 1967b) and *Euglena gracilis* var. *bacillaris* (Brenner and Maynard, 1966), of Cs^+ into *Ulva lactuca* (Scott, 1954), or of Sr^{2+} and Ca^{2+} into *Scenedesmus* sp. (Kylin, 1967b) and *Nitella translucens* (Spanswick and Williams, 1964). The influx of these cations is highest when together with phosphorus (see also Frank, 1962), the optimal phosphate concentrations being different for the various cations (Kylin, 1967b). Their uptake might be—at least partly—a passive one, linked to the light-dependent active phosphate uptake for reasons of electrochemical neutrality. The transport ATPase might be involved in the translocation of Rb^+ and Cs^+. These two ions may partly replace potassium as counterions to compensate for extruded sodium.

A light-dependent uptake of zinc and cobalt has been reported for *Chlorella pyrenoidosa* from Broda's laboratory (Broda, 1968; Broda and Findenegg, 1967; Findenegg and Broda, 1966).

In several algae no effect of illumination on cation uptake could be observed. For example, light did not increase the uptake of potassium into *Nitella flexilis* (Jacques and Osterhout, 1935), or of potassium, sodium or calcium into *Chara australis* (Hope, 1963) or *Chaetomorpha darwinii* (Dodd, Pitman and West, 1966).

Experiments from Packer's laboratory (Crofts, Deamer and Packer, 1967; Nobel and Packer, 1965; Packer, 1967) on the light influx of cations into chloroplasts in cell-free preparations have to be mentioned here. They are related to the observations with intact algae. Their results indicate that such cations as K^+, Na^+ or Ca^{2+} are taken up into chloroplasts under conditions favouring ATP hydrolysis. Their uptake into illuminated chloroplasts might also compete for the energy required for photophosphorylation. According to Nobel (1969) also the efflux of K^+ from the chloroplasts depends on a high-energy state, or ATP.

Influence of carbon dioxide on light-affected ion movement

There are only a few observations concerning the influence of carbon dioxide or bicarbonate on light-dependent ion fluxes in algae. A relationship between carbon dioxide and ion movement has to be considered, however. As we will see, the presence of carbon dioxide can decrease or increase the light-induced ion fluxes considerably.

An increase of the alkali ion influx in the light by the presence of carbon dioxide has been observed by Barr and Broyer (1964), Hope (1965) and Scott (1954). Likewise, the uptake of sulphate (Kylin, 1967c) can be accelerated. An explanation of these observations is difficult. Unfortunately, the problem has not been studied intensively enough and we are left with several hypotheses. Firstly, influx of cations accompanies the movement of bicarbonate anions across cell membranes (Cummins, Strand and Vaughan, 1969; Hope, 1965; Osterlind, 1951). The characteristics of the observed bicarbonate flux are similar to those of the active chloride uptake. This has been shown by Raven (1968c) with *Hydrodictyon africanum* and by Smith (1968a) with several Charophyceae. It has been proposed, therefore, that the light-dependent active bicarbonate influx depends on the photosystem II and is independent of photophosphorylation. However, it has to be considered that the passive influx of bicarbonate in the light will also be lowered by all inhibitors blocking the photosynthetic use of bicarbonate. Next, the demand of photosynthetic reactions for particulate ions determines their influx in the light. Photosynthesis can render carbon skeletons for various syntheses requiring, for example, phosphate, nitrate or sulphate. Thirdly, optimal Hill reaction in chloroplasts requires a certain carbon dioxide concentration (the Warburg effect: Good, 1963, 1965; Heise and Gaffron, 1963; Warburg and Krippahl, 1958, 1960; West and Hill, 1967). Therefore, any movement of ions directly depending on the electron flow in photosystem II (or mediated by non-cyclic photophosphorylation) might be reduced under conditions of carbon dioxide deficiency. Finally, a stimulating influence of bicarbonate on photophosphorylation has been observed by Punnet and Iver (Punnet and Iver, 1964; Batra and Jagendorf, 1965). Therefore, ion fluxes which require the supply of ATP should be decreased if the carbon dioxide concentrations for photophosphorylation are suboptimal; and vice versa.

An inhibitory effect of carbon dioxide on the ion uptake has also been observed. According to Kylin (1966a) carbon dioxide lowers the phosphate uptake in the light into *Scenedesmus* cells. This effect can be explained by a hypothesis of Wintermanns (1955). He suggested that an active recycling of orthophosphate because of an intensive use of ATP by synthetic processes, such as photosynthesis, restricts the uptake of phosphate from the external medium.

REFERENCES

Albers, R. W. (1967). Biochemical aspects of active transport. *Ann. Rev. Biochem.*, **36**, 727.

Arnon, D. I. (1967). Photosynthetic phosphorylation: facts and concepts. In T. W. Goodwin (Ed.), *Biochemistry of Chloroplasts*, Vol. II. Academic Press, London and New York. pp. 461–503.

Arnon, D. I., H. Y. Tsujimoto, and B. D. McSwain (1967). Ferredoxin and photosynthetic phosphorylation. *Nature (Lond.)*, **214**, 562.

Avron, M., and J. Neumann (1968). Photophosphorylation in chloroplasts. *Ann. Rev. Pl. Physiol.*, **19**, 137.

Bader, H., R. L. Post, and G. H. Bond (1968). Comparison of sources of a phosphorylated intermediate in transport ATPase. *Biochim. Biophys. Acta*, **150**, 41.

Bader, H., A. K. Sen, and R. L. Post (1966). Isolation and characterization of a phosphorylated intermediate in the $(Na^+ + K^+)$ system-dependent ATPase. *Biochim. Biophys. Acta*, **118**, 106.

Barber, J. (1968a). Light-induced uptake of potassium and chloride by *Chlorella pyrenoidosa*. *Nature (Lond.)*, **217**, 876.

Barber, J. (1968b). Measurement of the membrane potential and evidence for active transport of ions in *Chlorella pyrenoidosa*. *Biochim. Biophys. Acta*, **150**, 618.

Barber, J. (1968c). Sodium efflux from *Chlorella pyrenoidosa*. *Biochim. Biophys. Acta*, **150**, 730.

Barr, C. E., and T. C. Broyer (1964). Effect of light on sodium influx, membrane potential, and protoplasma streaming in *Nitella*. *Plant Physiol.*, **39**, 48.

Batra, P. B., and A. T. Jagendorf (1965). Bicarbonate effects of the Hill reaction and photophosphorylation. *Plant Physiol.*, **40**, 1074.

Bennett, H. S. (1956). The concept of membrane flow and membrane vesiculation as mechanisms for active transport and ion pumping. *J. biophys. biochem. Cytol.*, **2**, 94.

Blinks, L. R. (1940). The relation of bioelectric phenomena to ionic permeability and to metabolism in large plant cells. *Cold Spring Harbor Symp. Quant. Biol.*, **8**, 204.

Bonting, S. L., and L. L. Caravaggio (1966). Studies on Na^+-K^+-activated adenosine triphosphatase. XVI. Its absence from the cation transport system of *Ulva lactuca*. *Biochim. Biophys. Acta*, **112**, 519.

Branton, D. (1969). Membrane structure. *Ann. Rev. Pl. Physiol.*, **20**, 209.

Branton, D., and H. Moor (1964). Fine structure in freeze-etched *Allium cepa* root tips. *J. Ultrastructure Res.*, **11**, 401.

Brauner, L. (1956). Die Beeinflussung des Stoffaustausches durch das Licht. In W. Ruhland (Ed.), *Encyclopedia of Plant Physiology*, Vol. 2. Springer-Verlag, Berlin, Göttingen, Heidelberg. pp. 381–397.

Brauner, L. und M. Brauner (1937). Untersuchungen über den photoelektrischen Effekt an Membranen. I. Weitere Beiträge zum Problem der Lichtpermeabilitäts-reaktionen. *Protoplasma*, **28**, 230.

Brenner, M. L., and D. N. Maynard (1966). A study of rubidium accumulation in *Euglena gracilis*. *Plant Physiol.*, **41**, 1285.

Brierley, G., E. Murer, E. Bachman, and D. Green (1963). Studies on ion transport. II. The accumulation of inorganic phosphate and magnesium ions by heart mitochondria. *J. Biol. Chem.*, **238**, 3482.

Broda, E. (1968). The uptake of some heavy trace elements by *Chlorella*. *Abhandl. Deutsch. Akad. Wissenschaften, Berlin, Klasse f. Medizin*, **4**, 109–116.

Broda, E. und G. R. Fendenegg (1967). Spezifität der Zinkaufnahme durch *Chlorella. Experientia, Basel,* **23**, 18.

Chance, B., W. D. Bonner, Jr., and B. T. Storey (1968). Electron transport in respiration. *Ann. Rev. Pl. Physiol.*, **19**, 295.

Charnock, J. S., and R. L. Post (1963). Studies on the mechanism of cation transport. I. The preparation and properties of a cation stimulated adenine triphosphatase from guinea-pig kidney cortex. *Aust. J. expt. Biol. med. Sci.*, **41**, 547.

Clowes, F. A. L., and B. E. Juniper (1968). *Plant Cells*, Blackwell Scientific Publ., Oxford and Edinburgh.

Conway, E. J. (1953). A redox pump for the biological performance of osmotic work and its relation to the kinetics of free-ion diffusion across membranes. *Intern. Rev. Cytol.*, **2**, 419.

Conway, E. J. (1954). Some aspects of ion transport through membranes. *Symp. Soc. Exp. Biol.*, **8**, 297.

Conway, E. J. (1955). Evidence for a redox pump in active transport of cations. *Intern. Rev. Cytol.*, **4**, 377.

Coster, H. G. L., and A. B. Hope (1968). Ionic relation of cells of *Chara australis.* XI. Chloride fluxes. *Aust. J. biol. Sci.*, **21**, 243.

Crofts, A. R., D. W. Deamer, and L. Packer (1967). Mechanisms of light-induced structural change in chloroplasts. II. The role of ion movements in volume changes. *Biochim. Biophys. Acta*, **131**, 97.

Cummins, J. T., J. A. Strand, and B. E. Vaughan (1969). The movement of H^+ and other ions at the onset of photosynthesis in *Ulva. Biochim. Biophys. Acta*, **173**, 198.

Dainty, J. (1962). Ion transport and electrical potentials in plant cells. *Ann. Rev. Pl. Physiol.*, **13**, 379.

Danielli, J. F. (1954). Morphological and molecular aspects of active transport. *Symp. Soc. Exp. Biol.*, **8**, 502.

Danielli, J. F., and E. N. Davson (1935). A contribution to the theory of permeability of thin films. *J. Cellular Comp. Physiol.*, **5**, 495.

Danielli, J. F., and E. N. Harvey (1935). The tension at the surface of mackerel egg oil, with remarks on the nature of the cell surface. *J. Cellular Comp. Physiol.*, **5**, 483.

Davson, H. (1962). Growth of the concept of paucimolecular membrane. *Circulation*, **26**, 1022.

Deamer, D. W., and L. Packer (1967). Correlation of ultrastructure with light-induced ion transport in chloroplasts. *Arch. Biochem. Biophys.*, **119**, 83.

Diamond, J. M., and A. K. Solomon (1959). Intracellular potassium compartments in *Nitella axillaris. J. Gen. Physiol.*, **42**, 1105.

Dilley, R. A. (1964). Light-induced potassium efflux from spinach chloroplasts. *Biochem. Biophys. Res. Commun.*, **17**, 716.

Dilley, R. A. (1967). Ion and water transport processes in spinach chloroplasts. *Brookhaven Symp. Biol.*, **19**, 258.

Dilley, R. A., and L. P. Vernon (1965). Ion and water transport process related to the light-dependent shrinkage of spinach chloroplasts. *Arch. Biochem. Biophys.*, **111**, 365.

Dodd, W. A., M. G. Pitman, and K. R. West (1966). Sodium and potassium transport in the marine alga *Chaetomorpha darwinii. Aust. J. biol. Sci.*, **19**, 341.

6

Drechsler, Z., N. Nelson, and J. Neumann (1969). Antimycin A as an uncoupler and electron transport inhibitor in photoreactions of chloroplasts. *Biochim. Biophys. Acta,* **189**, 65.

Dunham, E. T., and I. M. Glynn (1961). Adenosinetriphosphatase activity and the active movement of alkali metal ions. *J. Physiol., London,* **156**, 274.

Dvořak, M., J. Dvořakova-Hladka, and S. Fialova (1966). Sorption of some ions by algae related to their trophic conditions. *Biol. Plant., Praha,* **8**, 362.

Ellis, R. J. (1963). Chloramphenicol and uptake of salts in plants. *Nature (Lond.),* **200**, 596.

Epstein, E. (1956). Mineral nutrition of plants: mechanisms of uptake and transport. *Ann. Rev. Pl. Physiol.,* **7**, 1.

Epstein, E. (1960). Spaces, barriers, and ion carriers; ion absorption by plants. *Am. J. Bot.,* **47**, 393.

Epstein, E. (1966). Dual pattern of ion absorption by plant cells and by plants. *Nature (Lond.),* **212**, 1324.

Fensom, D. S., and J. R. Wanless (1967). Further studies of electro-osmosis in *Nitella* in relation to pores in membranes. *J. exp. Bot.,* **18**, 563.

Findenegg, G. R. und E. Broda (1966). Stoffwechselabhängige Aufnahme von Zink durch *Chlorella. Naturwiss.,* **53**, 358.

Findlay, G. P., and A. B. Hope (1964). Ionic relations of cells of *Chara australis.* VII. The separate electrical characteristics of the plasmalemma and tonoplast. *Aust. J. biol. Sci.,* **17**, 62.

Frank, E. (1962). Vergleichende Untersuchungen zum Calcium-, Kalium- und Phosphathaushalt von Grünalgen. I. Calcium, Phosphat und Kalium bei *Hydrodictyon* und *Sphaeroplea* in Abhängigkeit von der Belichtung. *Flora,* **152**, 139.

Ginzburg, M. (1969). The unusual membrane permeability of two halophilic unicellular organisms. *Biochim. Biophys. Acta,* **173**, 370.

Glynn, J. M. (1962). Activation of adenosine triphosphatase activity in a cell membrane by external potassium and internal sodium. *J. Physiol., London,* **160**, 18P.

Good, N. E. (1963). Carbon dioxide and the Hill reaction. *Plant Physiol.,* **38**, 298.

Good, N. E. (1965). Interpretations of the carbon dioxide dependence of the Hill reaction. *Canad. J. Bot.,* **43**, 119.

Green, D. E., and J. F. Perdue (1966). Membranes as expressions of repeating units. *Proc. Natl. Acad. Sci., U.S.,* **55**, 1295.

Gromet-Elhanan, Z. (1968). Energy transfer inhibitors and electron transport inhibitors in chloroplasts. *Arch. Biochem. Biophys.,* **123**, 447.

Gutknecht, J. (1966). Sodium, potassium and chloride transport and membrane potentials in *Valonia ventricosa. Biol. Bull. mar. biol. Lab. Woods Hole,* **130**, 331.

Heise, J. J., and H. Gaffron (1963). Catalytic effects of carbon dioxide in carbon dioxide assimilating cells. *Pl. Cell Physiol., Tokyo,* **4**, 1.

Hems, D. A., and R. Rodnight (1966). Properties of phosphate-bound cerebral microsomes during adenosinetriphosphatase activity. *Biochim. J.,* **101**, 516.

Hind, G., and A. T. Jagendorf (1963a). Separation of light and dark stages in photophosphorylation. *Proc. Nat. Acad. Sci., U.S.,* **49**, 715.

Hind, G., and A. T. Jagendorf (1963b). Operation of cofactors in two-stage photophosphorylation. *Z. Naturforschg.,* **18b**, 689.

Hind, G., and J. M. Olson (1967). Light-induced changes in cytochrome b_6 in spinach chloroplasts. *Brookhaven Symp. Biology*, **19**, 188.

Hinkson, J. W., and Boyer, P. D. (1965). The light-induced formation of rapidly phosphorylated compounds in chloroplasts. *Arch. Biochim. Biophys.*, **110**, 16.

Hoagland, D. R., and A. R. Davis (1924). Further experiment on the absorption of ions by plants, including observations on the effect of light. *J. Gen. Physiol.*, **6**, 47.

Hoagland, D. R., P. L. Hibbard, and A. R. Davis (1927). The influence of light, temperature and other conditions on the ability of *Nitella* cells to concentrate halogens in the cell sap. *J. Gen. Physiol.*, **10**, 121.

Hodges, T. K., and J. B. Hanson (1965). Calcium accumulation by maize mitochondria. *Plant Physiol.*, **40**, 101.

Hogg, J., E. J. Williams, and R. J. Johnston (1968). Light intensity and the membrane parameters of *Nitella translucens*. *Biochim. Biophys. Acta*, **173**, 564.

Hokin, L. E., P. S. Sastry, P. R. Galsworthy, and A. Yoda (1965). Evidence that a phosphorylated intermediate in a brain transport adenosine triphosphatase is an acyl phosphate. *Proc. Nat. Acad. Sci., U.S.*, **54**, 177.

Hope, A. B. (1963). Ionic relations of cells of *Chara australis*. VI. Fluxes of potassium. *Aust. J. biol. Sci.*, **16**, 429.

Hope, A. B. (1965). Ionic relations of cells of *Chara australis*. X. Effects of bicarbonate ions on electrical properties. *Aust. J. biol. Sci.*, **18**, 789.

Hope, A. B., A. Simpson, and N. A. Walker (1966). The efflux of chloride from cells of *Nitella* and *Chara*. *Aust. J. biol. Sci.*, **19**, 355.

Hope, A. B., and N. A. Walker (1961). Ionic relations of cells of *Chara australis* R. Br. IV. Membrane potential differences and resistances. *Aust. J. biol. Sci.*, **14**, 26.

Izawa, S., T. N. Connoly, G. D. Winget, and N. E. Good (1967). Inhibition and uncoupling of photophosphorylation in chloroplasts. *Brookhaven Symp. Biol.*, **19**, 169.

Izawa, S., G. D. Winget, and N. E. Good (1966). Phlorizin, a specific inhibitor of photophosphorylation and phosphorylation-coupled electron transport in chloroplasts. *Biochem. Biophys. Res. Commun.*, **22**, 223.

Jacques, A. G., and W. J. V. Osterhout (1935). The kinetics of penetration. XI. Entrance of potassium into *Nitella*. *J. Gen. Physiol.*, **18**, 967.

Jagendorf, A. T., and G. Hind (1965). Relation between electron flow, photophosphorylation and a high energy state of chloroplasts. *Biochem. Biophys. Res. Commun.*, **18**, 702.

Jagendorf, A. T., and E. Uribe (1966). ATP formation caused by acid base transition of spinach chloroplasts. *Proc. Nat. Acad. Sci., U.S.*, **55**, 170.

Jagendorf, A. T., and E. Uribe (1967). Photophosphorylation and the chemiosmotic hypothesis. *Brookhaven Symp. Biol.*, **19**, 215.

Jardetzky, O. (1966). Simple allosteric model for membrana pumps. *Nature (Lond.)*, **211**, 969.

Jennings, D. H. (1963). *The Absorption of Solutes by Plant Cells*, Oliver and Boyd, Edinburgh and London.

Jeschke, W. D. (1967). Die cyclische und die nichtcyclische Photophosphorylierung als Energiequellen der lichtabhängigen Chloridionenaufnahme bei *Elodea*. *Planta (Berl.)*, **73**, 161.

Jeschke, W. D. (1968). On the connexion between electron transport and ion transport. *Abhandl. Deut. Akad. Wissenschaften, Berlin, Klasse f. Medizin*, **4**, 127–143.

Kahlenberg, A., P. R. Galsworthy, and L. E. Hokin (1967). Sodium–potassium adenosine triphosphatase: acyl phosphate 'intermediate' shown to be L-glutamyl-γ-phosphate. *Science*, **157**, 434.

Kamen, M. D., and S. Spiegelmann (1948). Studies on the phosphate metabolism of some unicellular organisms. *Cold Spring Harbor Symp. Quant. Biol.*, **13**, 151.

Karlish, S. J. D., and M. Avron (1968). Analysis of light-induced proton uptake in isolated chloroplasts. *Biochim. Biophys. Acta*, **153**, 878.

Kavanau, J. (1965). *Structure and Function in Biological Membranes*, Holden-Day, San Francisco.

Kavanau, J. (1966). Membrane structure and functions. *Fed. Proc.*, **2**, 1096.

Kessler, E. (1959). Reduction of nitrate by green algae. *Symp. Soc. expt. Biol.*, **13**, 87.

Kessler, E. (1964). Nitrate assimilation by plants. *Ann. Rev. Pl. Physiol.*, **15**, 57.

Ketchum, B. H. (1939). The absorption of phosphate and nitrate by illuminated cultures of *Nitzschia closterium. Am. J. Bot.*, **26**, 399.

Kishimoto, U., R. Nagai, and M. Tazawa (1965). Plasmalemma potential in *Nitella. Pl. Cell Physiol., Tokyo*, **6**, 519.

Kishimoto, U., and M. Tazawa (1965a). Ionic composition of the cytoplasm of *Nitella flexilis. Pl. Cell Physiol., Tokyo*, **6**, 507.

Kishimoto, U., and M. Tazawa (1965b). Ionic composition and electric response of *Lamprothamnium succinctum. Pl. Cell Physiol., Tokyo*, **6**, 529.

Korn, E. D. (1966). Structure of biological membranes. *Science*, **153**, 1491.

Korn, E. D. (1969a). Current concepts of membrane structure and functions. *Fed. Proc.*, **28**, 6.

Korn, E. D. (1969b). Cell membranes: structure and synthesis. *Ann. Rev. Biochem.*, **38**, 263.

Kreutz, W. und W. Menke (1962). Strukturuntersuchungen an Plastiden. II. Röntgenographische Untersuchungen an isolierten Chloroplasten und Chloroplasten lebender Zellen. *Z. Naturforschg.*, **17b**, 675.

Kuhl, A. (1962). Inorganic phosphorus uptake and metabolism. In R. A. Lewin (Ed.), *Physiology and Biochemistry of Algae*, Academic Press, London, New York, pp. 211–229.

Kuhl, A. (1968). Phosphate metabolism of green algae. In D. F. Jackson (Ed.), *Algae, Man and the Environment*, Syracuse Univ. Press, Syracuse, U.S.A. pp. 37–52.

Kylin, A. (1964a). Sulphate uptake and metabolism in *Scenedesmus* as influenced by phosphate, carbon dioxide, and light. *Physiol. Plant.*, **17**, 422.

Kylin, A. (1964b). An outpump balancing phosphate-dependent sodium uptake in *Scenedesmus. Biochem. Biophys. Res. Commun.*, **16**, 497.

Kylin, A. (1966a). The influence of photosynthetic factors and metabolic inhibitors on the uptake of phosphate in P-deficient *Scenedesmus. Physiol. Plant.*, **19**, 644.

Kylin, A. (1966b). Uptake and loss of Na^+, Rb^+, and Cs^+ in relation to an active mechanism for extrusion of Na^+ in *Scenedesmus. Plant Physiol.*, **41**, 579.

Kylin, A. (1967a). Ion transport in P-deficient *Scenedesmus* upon readdition of phosphate in light and darkness. I. Uptake and loss of Cl⁻, and measurements of oxygen consumption. *Z. Pflanzenphysiol.*, **56**, 70.

Kylin, A. (1967b). Ion transport in P-deficient *Scenedesmus* upon readdition of phosphate in light and darkness. II. Uptake of Rb⁺, Cs⁺, Ca²⁺, and Sr²⁺. *Z. Pflanzenphysiol.*, **56**, 81.

Kylin, A. (1967c). The uptake and metabolism of sulphate in *Scenedesmus* as influenced by citrate, carbon dioxide and metabolic inhibitors. *Physiol. Plant.*, **20**, 139.

Kylin, A., and J. E. Tillberg (1967a). Action sites of the inhibitor-β complex from potato and of phlorizin in light-induced energy transfer in *Scenedesmus*. *Z. Pflanzenphysiol.*, **57**, 72.

Kylin, A., and J. E. Tillberg (1967b). The relation between total photophosphorylation, level of ATP, and oxygen evolution in *Scenedesmus* as studied with DCMU and antimycin A. *Z. Pflanzenphysiol.*, **58**, 165.

Larkum, A. W. D. (1968). Ionic relations of chloroplasts *in vivo*. *Nature (Lond.)*, **218**, 447.

Lenard, J., and S. J. Singer (1966). Protein conformation in cell membrane preparations as studied by optical rotatory dispersion and circular dichroism. *Proc. Natl Acad. Sci., U.S.*, **56**, 1828.

Losada, M., J. M. Ramirez, A. Paneque, and F. F. del Campo (1965). Light and dark reduction of nitrate in a reconstituted chloroplast system. *Biochim. Biophys. Acta*, **109**, 86.

Lundegårdh, H. (1945). Absorption, transport and exudation of inorganic ions by the roots. *Ark. Bot.*, **32A**, 1.

Lundegårdh, H. (1954). Anion respiration. The experimental basis of a theory of absorption, transport and exudation of electrolytes by living cells and tissues. *Symp. Soc. exp. Biol.*, **8**, 262.

Lundegårdh, H. (1955). Mechanism of absorption, transport, accumulation and secretion of ions. *Ann. Rev. Pl. Physiol.*, **6**, 1.

Lüttge, U. (1968). Betrachtung des Ionentransportes im Zusammenhang mit der Kompartmentierung pflanzlicher Zellen. *Abhandl. Deutsch. Akad. Wissenschaften, Berlin, Klasse f. Medizin*, **4**, 95–107.

Lüttge, U., and Bauer, K. (1968). Evaluation of ion uptake isotherms and analysis of individual fluxes of ions. *Planta (Berl.)*, **80**, 52.

McCarty, R. E., R. J. Guillory, and E. Racker (1965). Dio-9, an inhibitor of coupled electron transport and phosphorylation in chloroplasts. *J. biol. Chem.*, **240**, PC 4822.

MacRobbie, E. A. C. (1962). Ionic relations of *Nitella translucens*. *J. Gen. Physiol.*, **45**, 861.

MacRobbie, E. A. C. (1964). Factors affecting the fluxes of potassium and chloride in *Nitella translucens*. *J. Gen. Physiol.*, **47**, 859.

MacRobbie, E. A. C. (1965). The nature of the coupling between light energy and active ion transport in *Nitella translucens*. *Biochim. Biophys. Acta*, **94**, 64.

MacRobbie, E. A. C. (1966a). Metabolic effects of ion fluxes in *Nitella translucens*. I. Active fluxes. *Aust. J. biol. Sci.*, **19**, 363.

MacRobbie, E. A. C. (1966b). Metabolic effects on ion fluxes in *Nitella translucens*. II. Tonoplast fluxes. *Aust. J. biol. Sci.*, **19**, 371.

MacRobbie, E. A. C. (1969). Ion fluxes to the vacuole of *Nitella translucens*. *J. exp. Bot.*, **20**, 236.

MacRobbie, E. A. C., and J. Dainty (1958a). Sodium and potassium distribution and transport in the seaweed *Rhodymenia palmata* (L.) Grev. *Physiol. Plant.*, **11**, 782.

MacRobbie, E. A. C., and J. Dainty (1958b). Ion transport in *Nitellopsis obtusa*. *J. Gen. Physiol.*, **42**, 335.

McIlwain, H. (1962). Appraising enzymic actions of central depressants by examining cerebral tissues. In J. L. Mongar and A. V. S. Reucke (Eds.), *CIBA Found. Symp. Enzymes and Drug Action*, J. and A. Churchill Ltd., London. pp. 170–198.

Mechsner, K. (1959). Untersuchungen an *Chlorella vulgaris* über den Einfluss der Alkali-Ionen auf die Lichtphosphorylierung. *Biochim. Biophys. Acta*, **33**, 150.

Medzihradsky, F., M. H. Kline, and L. E. Hokin (1967). Studies on the characterization of the sodium–potassium transport adenosinetriphosphatase. I. Solubilization, stabilization, and estimation of apparent molecular weight. *Arch. Biochem. Biophys.*, **121**, 311.

Menke, W. (1965). Versuche zur Aufklärung der molekularen Struktur der Thylakoidmembranen. *Z. Naturforschg.*, **20b**, 802.

Miller, D. M. (1960). The osmotic pump theory of selective transport. *Biochim. Biophys. Acta*, **37**, 448.

Mitchell, P. (1961). Coupling of phosphorylation to electron and hydrogen transfer by a chemi-osmotic type of mechanism. *Nature* (*Lond.*), **191**, 144.

Mitchell, P. (1966). Chemi-osmotic coupling in oxidative and photosynthetic phosphorylation. *Biol. Rev.*, **41**, 445.

Moor, H., and K. Mühlethaler (1963). Fine structure in frozen-etched yeast cells. *J. Cell. Biol.*, **17**, 609.

Mühlethaler, K., H. Moor, and J. W. Szarkowski (1965). The ultrastructure of the chloroplast lamellae. *Planta* (*Berl.*), **67**, 305.

Nagai, R., and M. Tazawa (1962). Changes in resting potential and ion absorption induced by light in a single plant cell. *Pl. Cell Physiol.*, *Tokyo*, **3**, 323.

Nagano, K., T. Kanazawa, N. Mizuno, Y. Tashima, T. Nakao, and M. Nakao (1965). Some acyl phosphate-like properties of p^{32}-labelled sodium–potassium-activated adenosine triphosphatase. *Biochem. Biophys. Res. Commun.*, **19**, 759.

Nelson, C. D. (1965). Membrane transport of electrolytes and sugars in single cells of *Nitella*. *Trans. Roy. Soc. Can.*, iii, lv, 201.

Nielsen, P. T. (1963). Light-promoted uptake of chloride in *Chlorella*. *Plant Physiol.*, **38**, iv.

Nielsson, S. E. G. (1964). Receptor cell outer segment development and ultrastructure of the disk membranes in the retina of the tadpole (*Rana pipiens*). *J. Ultrastructure Res.*, **11**, 581.

Nielsson, S. E. G. (1965). The ultrastructure of the receptor outer segments in the retina of the leopard frog (*Rana pipiens*). *J. Ultrastructure Res.*, **12**, 207.

Nobel, P. S. (1967a). Calcium uptake, ATPase and photophosphorylation by chloroplasts *in vitro*. *Nature* (*Lond.*), **214**, 875.

Nobel, P. S. (1967b). Relation of swelling and photophosphorylation to light-induced ion uptake by chloroplasts *in vitro*. *Biochim. Biophys. Acta*, **131**, 127.

Nobel, P. S. (1969). Light-induced changes in the ionic content of chloroplasts in *Pisum sativum*. *Biochim. Biophys. Acta*, **172**, 134.

Nobel, P. S., and S. Murakami (1967). Electron microscopic evidence for the location and amount of ion accumulation by spinach chloroplasts. *J. Cell. Biol.*, **32**, 209.

Nobel, P. S., S. Murakami, and A. Takamiya (1966). Localization of light-induced strontium accumulation in spinach chloroplasts. *Pl. Cell Physiol., Tokyo*, **7**, 263.

Nobel, P. S., and L. Packer (1964a). Energy-dependent ion uptake in spinach chloroplasts. *Biochim. Biophys. Acta*, **88**, 453.

Nobel, P. S., and L. Packer (1964b). Studies on ion translocation by spinach chloroplasts. *J. Cell. Biol.*, **23**, 67A.

Nobel, P. S., and L. Packer (1965). Light-dependent ion translocation in spinach chloroplasts. *Plant Physiol.*, **40**, 633.

O'Brien, J. S. (1967). Cell membranes—composition: structure: function. *J. Theoret. Biol.*, **15**, 307.

Opit, L. J., and J. S. Charnock (1965). A molecular model for a sodium pump. *Nature (Lond.)*, **208**, 471.

Osterlind, S. (1951). Inorganic carbon sources of green algae. IV. Photo-activation of some factors necessary for bicarbonate assimilation. *Physiol. Plant.*, **4**, 514.

Packer, L. (1967). Effect of nigericin upon light-dependent monovalent cation transport in chloroplasts. *Biochem. Biophys. Res. Commun.*, **28**, 1022.

Packer, L., D. W. Deamer, and A. R. Crofts (1967). Conformational changes in chloroplasts. *Brookhaven Symp. Biol.*, **19**, 281.

Paneque, A., F. F. del Campo, J. M. Rámirez, and M. Losada (1965). Flavin nucleotide nitrate reductase from spinach. *Biochim. Biophys. Acta*, **109**, 79.

Pardee, A. B. (1967). Crystallization of a sulfate-binding protein (permease) from *Salmonella typhimurium. Science*, **156**, 1627.

Pardee, A. B. (1968). Membrane transport proteins. *Science*, **162**, 632.

Pardee, A. B., and L. S. Prestige (1966). Cell-free activity of a sulfate binding site involved in active transport. *Proc. Nat. Acad. Sci. U.S.*, **55**, 189.

Pardee, A. B., L. S. Prestige, M. B. Whipple, and J. Dreyfuss (1966). A binding site for sulfate and its relation to sulfate transport into *Salmonella typhimurium. J. Biol. Chem.*, **241**, 3962.

Post, R. L., C. R. Merritt, C. R. Kinsolving, and C. D. Albright (1960). Membrane adenosine triphosphatase as a participant in the active transport of sodium and potassium in human erythrocyte. *J. Biol. Chem.*, **235**, 1796.

Post, R. L., and K. Sen (1965). An enzymatic mechanism of active sodium and potassium transport. *J. Histochem. Cytochem.*, **13**, 105.

Pullman, M. E., and G. Schatz (1967). Mitochondrial oxidations and energy coupling. *Ann. Rev. Biochemistry*, **36**, II, 539.

Punnet, T., and J. V. Iver (1964). The enhancement of photophosphorylation and the Hill reaction by carbon dioxide. *J. Biol. Chem.*, **239**, 2335.

Raven, J. A. (1967a). Ion transport in *Hydrodictyon africanum. J. Gen. Physiol.*, **50**, 1607.

Raven, J. A. (1967b). Light stimulation of active transport in *Hydrodictyon africanum. J. Gen. Physiol.*, **50**, 1627.

Raven, J. A. (1968a). The linkage of light-stimulated Cl influx to K and Na influxes in *Hydrodictyon africanum. J. exp. Bot.*, **19**, 233.

Raven, J. A. (1968b). The action of phlorizin on photosynthesis and light-stimulated ion transport in *Hydrodictyon africanum*. *J. exp. Bot.*, **19**, 712.

Raven, J. A. (1968c). The mechanism of photosynthetic use of bicarbonate by *Hydrodictyon africanum*. *J. exp. Bot.*, **19**, 193.

Raven, J. A. (1968d). Photosynthesis and light-stimulated ion transport in *Hydrodictyon africanum*. *Abhandl. Deutsch. Akad. Wissenschaften, Berlin, Klasse f. Medizin*, **4**, 145–151.

Raven, J. A. (1969). Action spectra for photosynthesis and light-stimulated ion transport process in *Hydrodictyon africanum*. *New Phytol.*, **68**, 45.

Raven, J. A., E. A. C. MacRobbie, and J. Neumann (1969). The effect of Dio-9 on photosynthesis, and ion transport in *Nitella*, *Tolypella* and *Hydrodictyon*. *J. exp. Bot.*, **20**, 221.

Richardson, S. H. (1968). An ion translocase system from rabbit intestinal mucosa. Preparation and properties of the (Na^+-K^+)-activated ATPase. *Biochim. Biophys. Acta*, **150**, 572.

Robertson, J. D. (1967). Origin of the unit membrane concept. *Protoplasma*, **63**, 218.

Robertson, R. N. (1956). The mechanism of absorption. In W. Ruhland (Ed.), *Encyclopedia of Plant Physiology*, Vol. 2. Springer-Verlag, Berlin, Göttingen, Heidelberg. pp. 449–467.

Robertson, R. N. (1968). *Protons, Electrons, Phosphorylation and Active-Transport*, University Press, Cambridge.

Robinson, J. B. (1969). Sulphate influx in Characean cells. II. Links with light and metabolism in *Chara australis*. *J. exp. Bot.*, **20**, 212.

Roche, J. et S. Andrée (1963). Sur la concentration des iodures (^{131}J) par une algue marine (*Ulva lactuca* L.): action de la lumiére et rythme nycthéméral. *Compt. Rend. Soc. Biol.*, **157**, 1412.

Rothfield, L., and A. Finkelstein (1968). Membrane biochemistry. *Ann. Rev. Biochem.*, **37**, 463.

Rowan, K. S. (1966). Phosphorus metabolism in plants. *Intern. Rev. Cytol.*, **19**, 301.

Ruhland, W. und C. Hoffmann (1925). Die Permeabilität von *Beggiatoa mirabilis*. Ein Beitrag zur Ultrafiltertheorie des Plasmas. *Planta (Berl.)*, **1**, 1.

Ruoho, A. E., L. E. Hokin, R. J. Hemingway, S. M. Kupchan (1968). Hellebrigenin 3-haloacetates: Potent site-directed alkylators of transport adenosine-triphosphatase. *Science*, **159**, 1354.

Saltman, P., J. G. Forte, and G. M. Forte (1963). Permeability studies on chloroplasts from *Nitella*. *Exp. Cell Res.*, **29**, 504.

Salyaev, R. K., L. D. Kamenkova, and L. A. Podolyakina (1969). Variation of adsorptional properties of ATP-ase proteins as a result of conformational reconstruction, and the relation of these phenomena to membrane transport. *Abstracts XI. Intern. Bot. Congr., Seattle, Wash. USA*, 188.

Schaedle, M., and L. Jacobsen (1965). Ion absorption and retention by *Chlorella pyrenoidosa*. I. Absorption of potassium. *Plant Physiol.*, **40**, 214.

Schaedle, M., and L. Jacobsen (1966). Ion absorption and retention by *Chlorella pyrenoidosa*. II. Permeability of the cell to sodium and rubidium. *Plant Physiol.*, **41**, 248.

Schnepf, E. (1968). Transport by compartments. *Abhandl. Deutsch. Akad. Wissenschaften Berlin, Klasse f. Medizin*, **4**, 39–49.

Schönfelder, S. (1931). Weitere Untersuchungen über die Permeabilität von *Beggiatoa mirabilis*. Nebst kritischen Ausführungen zum Gesamtproblem der Permeabilität. *Planta (Berl.)*, **12**, 414.

Scott, G. T. (1945). The mineral composition of phosphate deficient cells of *Chlorella pyrenoidosa* during the restoration of phosphate. *J. Cell Comp. Physiol.*, **26**, 35.

Scott, G. T., and H. R. Hayward (1953a). Metabolic factors influencing the sodium and potassium distribution in *Ulva lactuca*. *J. Gen. Physiol.*, **36**, 659.

Scott, G. T., and H. R. Hayward (1953b). The influence of iodoacetate on the sodium and potassium content of *Ulva lactuca* and the prevention of its influence by light. *Science*, **117**, 719.

Scott, G. T., and H. R. Hayward (1953c). The influence of temperature and illumination on the exchange of the potassium ion in *Ulva lactuca*. *Biochim. Biophys. Acta*, **12**, 401.

Scott, G. T., and H. R. Hayward (1954). Evidence for the presence of separate mechanisms regulating potassium and sodium distribution in *Ulva lactuca*. *J. Gen. Physiol.*, **37**, 601.

Scott, R. (1954). A study on caesium accumulation by marine algae. In J. E. Johnston (Ed.), *Radioisotope Conference*, 1954. Butterworth Sci. Publ., London. pp. 373–380.

Simonis, W. (1960). Photosynthese und lichtabhängige Phosphorylierung. In W. Ruhland (Ed.), *Encyclopedia of Plant Physiology*, Vol. 5–1. Springer-Verlag, Berlin, Göttingen, Heidelberg. pp. 966–1013.

Simonis, W., and M. Mechler (1963). Action spectrum of photosynthetic phosphorylation *in vivo* by *Ankistrodesmus braunii*. *Biochem. Biophys. Res. Commun.*, **13**, 241.

Simonis, W. und W. Urbach (1963a). Über die Wirkung von Natrium Ionen auf die Phosphataufnahme und die lichtabhängige Phosphorylierung von *Ankistrodesmus braunii*. *Arch. Mikrobiol.*, **46**, 265.

Simonis, W. und W. Urbach (1963b). Untersuchungen zur lichtabhängigen Phosphorylierung bei *Ankistrodesmus braunii*. IX. Beeinflussung durch Phosphatkonzentrationen, Temperatur, Hemmstoffe, Na⁺-Ionen und Vorbelichtung. In *Microalgae and Photosynthetic Bacteria*, University of Tokyo Press, Tokyo, pp. 597–611.

Sjöstrand, F. S. (1963). A new repeat structural element of mitochondrial and certain cytoplasmic membranes. *Nature (Lond.)*, **199**, 1262.

Sjöstrand, F. S. (1967). The structure of cellular membranes. *Protoplasma*, **63**, 248.

Skierczyńska, J. (1968a). Some of the electrical characteristics of the cell membrane of *Chara australis*. *J. exp. Bot.*, **19**, 389.

Skierczyńska, J. (1968b). The electrical resistance of the tonoplast of *Chara australis*. *J. exp. Bot.*, **19**, 407.

Skou, J. C. (1957). The influence of some cations on an adenosine triphosphatase from peripheral nerves. *Biochem. Biophys. Acta*, **23**, 394.

Skou, J. C. (1960). Further investigation on a $(Mg^{++} + Na^{+})$-activated adenosine triphosphatase, possibly related to the active, linked transport of Na^+ and K^+ across the nerve membrane. *Biochim. Biophys. Acta*, **42**, 6.

Skou, J. C. (1964). Enzymatic aspects of active linked transport of Na^+ and K^+ through the cell membrane. *Progr. Biophys.*, **14**, 131.

Skou, J. C. (1965). Enzymatic basis for active transport of Na^+ and K^+ across cell membranes. *Physiol. Rev.*, **45**, 596.

Skou, J. C. (1967). Enzymatic basis for the active transport of sodium and potassium. *Protoplasma*, **63**, 303.

Skye, G. E., N. Shavit, and P. D. Boyer (1967). The catalysis by modified chloroplasts of the $P_i \rightleftharpoons ATP$, $P_i \rightleftharpoons HOH$ and $ATP \rightleftharpoons HOH$ exchange reactions in the absence of light. *Biochim. Biophys. Res. Commun.*, **28**, 724.

Smith, F. A. (1966). Active phosphate uptake by *Nitella translucens*. *Biochim. Biophys. Acta*, **126**, 94.

Smith, F. A. (1967). The control of Na uptake into *Nitella translucens*. *J. exp. Bot.*, **18**, 716.

Smith, F. A. (1968a). Rates of photosynthesis on *Characean* cells. II. Photosynthetic $^{14}CO_2$ fixation and ^{14}C-bicarbonate uptake by *Characean* cells. *J. exp. Bot.*, **19**, 207.

Smith, F. A. (1968b). Metabolic effects on ion fluxes in *Tolypella intricata*. *J. exp. Bot.*, **19**, 442.

Smith, F. A., and K. R. West (1969). The mechanism of chloride transport in giant algal cells. *Abstracts XI. Intern. Bot. Congr.*, *Seattle, Washington, USA*, 204.

Spanswick, R. M., J. Stolarek, and E. J. Williams (1967). The membrane potential of *Nitella translucens*. *J. exp. Bot.*, **18**, 1.

Spanswick, R. M., and E. J. Williams (1964). Electrical potentials and K, Na, and Cl concentrations in the cytoplasm and vacuole of *Nitella translucens*. *J. exp. Bot.*, **15**, 193.

Stein, W. D. (1967). *The Movement of Molecules across Cell Membranes*, Academic Press, London, New York.

Stoeckenius, W. (1967). Die molekulare Struktur biologischer Membranen. *Ber. Bunsenges. physik. Chemie.*, **71**, 758.

Stone, A. J. (1968). A proposed model for the Na^+ pump. *Biochim. Biophys. Acta*, **150**, 578.

Sutcliffe, J. F. (1959). Salt uptake in plants. *Biol. Rev.*, **34**, 159.

Sutcliffe, J. F. (1962). *Mineral Salts Absorption in Plants*, Pergamon Press, New York.

Taylor, A. N., and R. H. Wassermann (1967). Vitamin D_3-induced calcium-binding protein: partial purification, electrophoretic visualization, and tissue distribution. *Arch. Biochem. Biophys.*, **119**, 536.

Tillberg, J.-E., and A. Kylin (1966). Oxygen evolution and phosphorylation in *Scenedesmus* as influenced by the inhibitor-β complex from potato and by phloridzin. *Planta* (*Berl.*), **71**, 130.

Umrath, K. (1956). Elektrophysiologische Phänomene. In W. Ruhland (Ed.), *Encyclopedia of Plant Physiology*, Vol. 2. Springer-Verlag, Berlin, Göttingen, Heidelberg. pp. 747–778.

Urbach, W. und W. Simonis (1962). Wirkung von Hemmstoffen auf die lichtabhängige Phosphorylierung bei *Ankistrodesmus braunii*. In *Vorträge a.d. Gesamtgeb. d. Bot.* N.F. 1. Gustav Fischer Verlag, Stuttgart, pp. 149–156.

Urbach, W., and W. Simonis (1964). Inhibitor studies on the photophosphorylation *in vivo* by unicellular algae (*Ankistrodesmus*) with antimycin A, HOQNQ, salicylaldoxime and DCMU. *Biochem. Biophys. Res. Commun.*, **17**, 39.

Vidaver, G. A. (1966). Inhibition of parallel flux and augmentation of counterflux shown by transport models not involving a mobile carrier. *J. theor. Biol.*, **10**, 301.

Vose, J. R., and M. Spencer (1967). Energy sources for photosynthetic carbon dioxide fixation. *Biochim. Biophys. Res. Commun.*, **29**, 532.

Walker, N. A. (1955). Microelectrode experiments on *Nitella*. *Aust. J. biol. Sci.*, **8**, 476.

Walker, N. A. (1957). Ion permeability of the plasmalemma of the plant cell. *Nature (Lond.)*, **180**, 94.

Walker, N. A. (1960). The electrical resistance of the cell membranes in a *Chara* and a *Nitella* species. *Aust. J. biol. Sci.*, **13**, 468.

Warburg, O. und G. Krippahl (1958). Hill-Reaktion. *Z. Naturforschg.*, **13b**, 509.

Warburg, O. und G. Krippahl (1960). Notwendigkeit der Kohlensäure für die Chinon- und Ferricyanid-Reaktionen in grünen Grana. *Z. Naturforschg.*, **15b**, 367.

Wassermann, R. H., and A. N. Taylor (1966). Vitamin D_3-induced calcium-binding protein in chick intestinal mucosa. *Science*, **152**, 791.

Wedding, R. T., and M. K. Black (1960). Uptake and metabolism of sulfate by *Chlorella*. I. Sulfate accumulation and active sulfate. *Plant Physiol.*, **35**, 72.

Weier, T. E., H. P. Engelbrecht, A. Harrison, and E. B. Risley (1965). Subunits in the membranes of chloroplasts of *Phaseolus vulgaris*, *Pisum sativum*, and *Aspidistra* sp. *J. Ultrastructure Res.*, **13**, 92.

Weigl, J. (1967). Beweis für die Beteiligung von beweglichen Transportstrukturen (Trägern) beim Ionentransport durch pflanzliche Membranen und die Kinetik des Anionentransportes bei *Elodea* im Licht und Dunkel. *Planta (Berl.)*, **75**, 327.

West, J., and R. Hill (1967). Carbon dioxide and the reduction of indophenol and ferrycyanide by chloroplasts. *Plant Physiol.*, **42**, 819.

West, K. R., and M. G. Pitman (1967). Ionic relations and ultrastructure in *Ulva lactuca*. *Aust. J. biol. Sci.*, **20**, 901.

Wheeler, K. P., and R. Whittam (1964). Structural and enzymatic aspects of the hydrolysis of adenosine triphosphate by membranes of kidney cortex and erythrocytes. *Biochem. J.*, **93**, 349.

Whittam, R. (1962). The asymmetrical stimulation of a membrane adenosine triphosphatase in relation to active cation transport. *Biochem. J.*, **84**, 110.

Wilbrandt, W. (1967). Carrier mechanisms. *Protoplasma*, **63**, 299.

Winget, G. D., S. Izawa, and N. E. Good (1969). The inhibition of photophosphorylation by phlorizin and closely related compounds. *Biochemistry*, **8**, 2067.

Wintermanns, J. F. G. M. (1955). Polyphosphate formation in *Chlorella* in relation to photosynthesis. *Mededel. Landbouwhogesch. Wageningen*, **55**, 69.

CHAPTER 6

Light effects on carbohydrate and protein metabolism in algae

WOLFGANG KOWALLIK

Department of Botany, University of Cologne, Germany

Introduction

Influences of specific wavelengths of light on the basic composition of higher plants have been reported sporadically since 1952 (Voskresenskaja, 1952), but it was not until 1960 that similar effects were demonstrated in microorganisms. Since then a number of investigations has been devoted to that problem, but the effect is still poorly understood. This article, then, cannot be more than an attempt to summarize some of the more important data obtained thus far.

Wavelength-dependent alterations in cell composition

All developmental stages of the synchronized unicellular green alga *Chlorella pyrenoidosa* produce smaller total carbohydrate/total bound nitrogen ratios at autotrophic growth in blue light than at correspondent growth in red light, leading to equal dry mass production. In blue light that ratio varied between 1·0 and 6·0; in red light, however, it varied between 2·0 and 9·0 during the life-cycle (Pirson and Kowallik, 1960), the percentages of carbohydrate and of protein of equal dry masses from red and blue light cultures being significantly different for all stages of the cycle. After 8 hours of illumination the difference in cell composition was greatest. At that time the dry matter of blue-grown algae consisted of 50 per cent protein and of 15 per cent carbohydrate, while that of red-grown parallels contained 29 per cent protein and 38 per cent carbohydrate (Kowallik, 1962). Similar differences were present at several light intensities (Buschbohm, 1968). This demonstrates clearly that there are pronounced effects of specific spectral regions on the basic composition of *Chlorella* cells.

In the beginning of an analysis of the phenomenon, attention has been focused on the following three questions: First, do these alterations depend on specific wavelength effects in photosynthesis proper? Second, are there different effects, of blue and of red light on the final distribution of fixed

---

Figure 6.1. Spectral dependence of: (A) a change in the ratio of carbohydrate/protein in *Chlorella* in the presence of exogenous glucose, light exposure is for 16 hours, the broken line is dark control (after Kowallik, 1965); (B) is like (A), but with DCMU-poisoned photosynthesis, expressed as C/P(λ) : C/P(dark) (after Kowallik, 1966a); (C) the extra oxygen consumption of a chlorophyll-free, carotenoid-containing *Chlorella* mutant, expressed in per cent of the effect of λ 455 nm light exposure is for 20 to 30 minutes (after Kowallik, 1967); (D) a drop in apparent photosynthesis of *Chlorella*, developing within 10 to 15 minutes after the beginning of illumination, expressed as [+O_2 (15th to 20th minute)/+O_2 (0 to 5th minute)] × 100 (after Kowallik and Kowallik, 1969); (E) the extra dark oxygen uptake of *Chlorella* during 20 minutes after photosynthesis with the respective wavelength, expressed in per cent of that after λ 680 nm (after Kowallik and Kowallik, 1969); (F) the extra dark oxygen uptake of *Chlorella* after a 5 second flash of the respective wavelength (after Pickett and French, 1967); (G) RNA production in a chlorophyll-free *Chlorella* mutant, expressed in per cent of dark control, light exposure is for 24 hours (after Senger and Bishop, 1968). Each spectrum obtained at equal quantum fluxes. All figures modified.

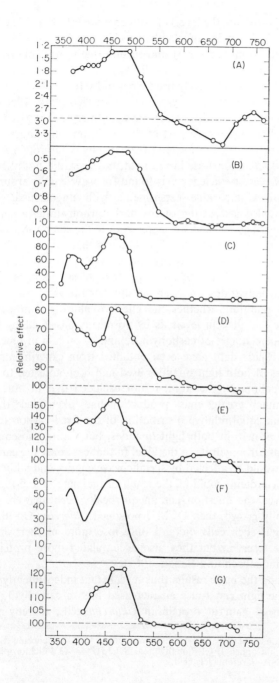

carbon? And third, are the levels of both components, that of carbohydrate as well as that of protein, affected by specific wavelengths of light?

Some clarification of the first and second questions came from experiments with glucose-fed *Chlorella* cells in which any new photosynthetic C-incorporation had selectively been blocked by 10^{-5} M DCMU†, a potent inhibitor of photosynthetic oxygen liberation (Kowallik, 1966a). If such cells were illuminated with narrow wavelength regions at equal quantum fluxes throughout the visible part of the spectrum, only those exposed to wavelengths below 580 nm lowered the carbohydrate/protein ratio below that of a dark control; those in red light, however, did not alter the dark ratio at all. Greatest efficiency was found for wavelengths around 470 nm (Figure 6.1B). Comparable experiments with unpoisoned glucose-fed cells also revealed a lowering of the dark carbohydrate/protein ratio in blue light, but they showed an increase over the dark ratio in red light (Kowallik, 1965) (Figure 6.1A). This suggests that there is only a specific effect of blue light which largely directs the above variations in cell composition independently of intact photosynthesis, perhaps by working on reserve carbohydrates which are accumulated in red light.

The third question, whether the effective blue light influences both carbohydrate and protein levels, is by no means answered by the above data on the percentages of carbohydrate and of protein in blue- and red-grown cells. These data have been obtained from experiments in which the intensities of both light qualities used had been adjusted to equal dry matter production. It, therefore, is not excluded that in blue light, for example, a much greater mass production had taken place than in red light—with an equal relative distribution of the fixed carbon into carbohydrate and protein in both light qualities, but with a subsequent much greater loss of carbohydrate in the blue. In that case only the carbohydrate metabolism would be affected by specific wavelengths, and a high recorded percentage of protein would be the result of the method and the reference used. However, the question has unequivocally been answered by using autotrophically grown, then DCMU poisoned *Chlorella*. On illumination with blue light such cells did not only lose more of their endogenous carbohydrate reserves, but they also accumulated more protein than in red light (Laudenbach and Pirson, 1969).

Summarizing the basic results thus reveals that independently of photosynthesis blue light can cause an increased loss in carbohydrate as well as an enhanced gain in protein in *Chlorella* cells, thereby producing

† Abbreviations: ADP = adenosine diphosphate, ATP = adenosine triphosphate, CMU = 3-(4-chlorophenyl)-1,1-dimethylurea, DCMU = 3-(3,4-dichlorophenyl)1,1-dimethylurea, RNA = ribonucleic acid.

pronounced differences in the basic composition of blue- or red-exposed cells.

Whether the changes in both carbohydrate and protein levels in blue light are caused by separate light reactions acting on either metabolism, or if these alterations are interdependent, cannot be deduced from the above data. A greater gain of protein might result indirectly from an enhanced loss of carbohydrate by the greater supply of energy and carbon skeletons. On the other hand, an increase in carbohydrate catabolism would be caused by the ATP consumption of a blue light-dependent protein production. In fact, it is reasonable to assume that in long-term experiments under growth conditions as described above both types of interaction are involved in establishing the final levels of carbohydrate and protein. To learn more about the initial blue light effect, we therefore have to look for more detailed results on influences of light on the consumption of carbohydrate as well as on the production of protein.

Enhanced consumption of carbohydrate in blue light

In DCMU-poisoned *Chlorella* the increased loss of cellular carbohydrate in blue light is parallelled by a strong enhancement of dark oxygen consumption and carbon dioxide release. This stimulated respiration (let us call it so because of the substrate and gas exchange, without, however, implying by that term that the reaction must be mitochondrial dark type) reaches a steady rate 10 to 15 minutes after the beginning of illumination under which it is maintained for hours. In the following darkness the enhanced rate decreases slowly to the initial dark rate. The intensity and wavelength dependences of this enhancement have first been determined with an achlorophyllous mutant of *Chlorella* (Kowallik, 1966b; Kowallik and Gaffron, 1966). Greatest efficiency was found for wavelengths around 460 nm. At this spectral region the reaction reached saturation at the very low light level of about 500 ergs/cm² second. Toward shorter wavelengths the efficiency decreased to a minimum around 400 nm followed by a second smaller maximum in near ultraviolet around 375 nm. Irradiation with yellow, red, and far red light caused no enhancement of endogenous dark respiration (Kowallik, 1967) (Figure 6.1C). A similar wavelength dependence of light-enhanced respiration has recently been noted in DCMU-poisoned green *Chlorella*, too (Kowallik, 1969a). There is also evidence that the reaction is equally present in photosynthesizing cells: Kowallik and Kowallik (1969) found a drop in apparent photosynthesis of *Chlorella* at light levels which hardly compensated for respiration. Its spectral dependence and its 10 to 15 minute induction period resemble those for light-stimulated respiration (Figure 6.1D). The conclusion at hand, that this drop in photosynthetic oxygen evolution is the result of an increased respiration during photosynthesis, is supported by the increased oxygen consumption in darkness following photosynthesis in blue light which only slowly fades away, and whose approximative action spectrum is rather similar to that of enhanced respiration in the light (Figure 6.1E). Some minor deviations are noticeable in the near ultraviolet region which, however, might be of metabolic character. Evidently this slowly ceasing blue light-stimulated oxygen consumption is a reaction independent of photosynthesis and correspondent to the, in subsequent darkness, also slowly decreasing blue light-enhanced respiration in the non-photosynthetic, achlorophyllous *Chlorella*. An after-effect of blue light on dark respiration of *Chlorella* has also been found in French's laboratory; about 1 minute after a blue light flash of 5 seconds an increase in oxygen uptake took place, which reached a maximum after 6 minutes or more and thereafter slowly faded away (Ried, 1965, 1968). The spectral response of this flash light-induced enhancement of dark oxygen uptake corresponds to those shown above (Pickett and French, 1967; Ried, personal communication) (Figure 6.1F). It appears reasonable to assume that this effect is in some way related to the steadily enhanced respiration in blue light which, however, cannot be triggered by a short flash but has to be driven by continuous irradiation.

The extended after-effect, the long induction time until respiration is fully enhanced in the light, and the observation that the acid release by *Chlorella* which occurs under conditions favouring fermentation is also enhanced by blue illumination (Kowallik, 1969b; Kowallik and Gaffron, 1967a, 1967b) indicate that the action of blue light on carbohydrate catabolism does not consist in an influence on the respiratory machinery proper, but probably in some regulation of the supply of substrate or co-factors.

Enhanced production of protein in blue light

Blue light-dependent extra protein production might result from an increased supply of reduced nitrogen, from a greater provision of amino acids, or from an enhanced formation of specific RNA species.

Blue light-dependent supply of reduced nitrogen

Although Stoy (1955, 1956) described a stimulation by blue light of the reduction of nitrate for a wheat leaf preparation, and Strotmann (1967) reported such an effect on nitrite reduction in *Chlorella*, this must be of minor or no importance on the shift in the carbohydrate/protein ratio of that alga in red or blue light, as the shift was equally present in cells grown with nitrate, with ammonium or with urea as the sole nitrogen sources (Kowallik, 1962).

Blue light-dependent production of amino acids

After 30 seconds of photosynthesis with blue light at mainly 436 nm Cayle and Emerson (1957) found that *Chlorella* cells incorporated more label from $NaH^{14}CO_3$ into glycine, serine and alanine than after photosynthesis with red light at 644 nm, given at equal quantum flux, which led to roughly the same total ^{14}C-fixation. In blue light the respective specific activities were 2·5, 1·9 and 3·0 times greater than in red light. In glycine the difference in total labelling was accompanied by a change in the distribution of ^{14}C within the molecule. In red light about 80 per cent of its total radioactivity was found in the carboxyl atom and only about 20 per cent in the α-carbon atom. In blue light, however, much more of the radioactivity was directed to the α-carbon, resulting in the distribution 60 : 40 per cent. While the former resembles the usual ^{14}C-distribution in photosynthetic phosphoglyceric acid, the latter comes closer to the uniform labelling of glycolate, which has been observed in experiments on early products of photosynthesis (Schou and coworkers, 1950).

A positive effect of blue light on the formation of serine and alanine, and in addition on aspartic and glutamic, malic and phosphoglyceric acid, has also been found by Hauschild, Nelson and Krotkov (1962a), who checked the incorporation of ^{14}C from labelled bicarbonate of *Chlorella* after 30 minutes in broad spectral bands of blue or red light. Phosphate esters and glycolic acid were the main more highly labelled compounds in red light. Although neither the specific activities nor the intramolecular ^{14}C-distribution have been determined and, in addition, the authors have mentioned the questionable reliability of the results on glycolic acid because of its volatility, the latter data shall be noted as they fit the assumption of serine production from glycolate. Such a pathway, including glycine as an intermediate, apparently exists in higher plants (Asada and Kasai, 1962; Asada and coworkers, 1965; Hess and Tolbert, 1966a, 1966b; Miflin, Marker and Whittingham, 1966; Rabson, Tolbert and Kearney, 1962; Sinha and Cossins, 1965; Tanner and Beevers, 1965; Tolbert, 1963; Tolbert and Cohan, 1953; Wang and Burris, 1963; Wang and Waygood, 1962). $NaH^{14}CO_3$-feeding experiments of comparable duration with *Scenedesmus*, another unicellular green alga, yielded more

^{14}C in glutamic and aspartic acid in blue light, while sucrose was more highly labelled in red light. In the blue-green alga *Microcystis* only glutamic acid was found to contain more ^{14}C after blue light photosynthesis, and in the photosynthetic bacterium *Chromatium* there was no difference at all in the labelling pattern of the products of photosynthesis in red or in blue light (Hauschild, Nelson and Krotkov, 1962b).

In a third group of results a specific incorporation of ^{14}C in *Chlorella*, exposed to short wavelength visible radiation, was observed in aspartic and glutamic acid. After a total of 6 minutes in light at 453 nm, including 3 minutes of preillumination without NaH^{14}CO$_3$, these compounds showed about twice as much label as after a correspondent period of photosynthesis in red light at 679 nm, adjusted to equal total ^{14}C-fixation (Ogasawara and Miyachi, 1969). In the latter more ^{14}C accumulated in sugar phosphates. Labelling of other amino acids has not separately been tested. Another important finding of these authors is that a blue-stimulated ^{14}C-incorporation in *Chlorella* was also present when photosynthesis had been suppressed by application of 5×10^{-5} M CMU, another rather selective inhibitor of photosynthetic oxygen liberation, and that in these poisoned cells additionally fixed ^{14}C also accumulated in aspartic and glutamic acid preferentially. Greatest effect was observed with light of wavelengths around 420 nm.

Blue light-dependent formation of RNA

Comparative analyses of the RNA levels at 4 hours intervals in synchronized *Chlorella* cultures in blue or in red light showed that the content of RNA was significantly higher in all developmental stages of cells grown under blue light conditions. This increased production seemed to precede protein synthesis (Kowallik, 1962). A light stimulation of the production of RNA, independent of intact photosynthesis, has been assumed by Senger and Schoser (1966) from results with DCMU-poisoned *Chlorella*; it has been recently demonstrated in an achlorophyllous mutant of that alga (Senger and Bishop, 1968). Greatest efficiency on extra RNA production was found in light around 460 nm, none at all was found in the yellow or red region of the spectrum (Figure 6.1G). This is in agreement with the wavelength dependence of the changes in the carbohydrate/protein ratios mentioned above. Also independent of photosynthesis was the light-stimulated incorporation of ^{32}P$_i$ into the RNA of a non-greening mutant of *Euglena* in which neither protochlorophyll nor chlorophyll could be detected (Zeldin and Schiff, 1968). The latter as well as Senger's RNA determinations have been performed after 24 hours of illumination.

Independence or interdependence of the blue light effects on carbohydrate and protein metabolism

Above were summarized some of the basic results on long- and short-term effects of blue light on carbohydrate and protein metabolism. We will assume a causal connexion between the rapid changes in amino acids and those in protein after extended light exposure. Let us then see if it is possible through this information to answer the question whether the initially cited changes in carbohydrate and protein content are caused by two separate actions of blue light, if they are produced by an effect of blue light on one of these main metabolic pathways only, which then in turn will influence the other secondarily, or if they depend on one basic light effect influencing both metabolic reactions independently.

Do two different blue light reactions work independently?

Comparison of the action spectra for enhanced respiration with the only one available for increased ^{14}C-incorporation into specific amino acids (Ogasawara and Miyachi, 1969) suggests the existence of two different light reactions, as the maximum activity of the former around 460 nm is well separated from that of the latter at about 420 nm. In an action spectrum for carbohydrate/protein this would lead to a levelled course around 400 nm, since at these wavelengths a relatively small loss of carbohydrate would be accompanied by a relatively great gain of protein. Such levelled spectra have been obtained with DCMU-poisoned and unpoisoned glucose-fed *Chlorella* (Kowallik, 1966a, 1965) (Figure 6.1B and 6.1A). But there are reasons to doubt the significance of this agreement: the carbohydrate/protein ratios unfortunately have been determined almost at light saturation (Kowallik, 1965), which greatly influences the relative magnitude of peaks and valleys. Further, these analyses have been performed after an illumination period of 16 hours which doubtless has resulted in a rather complex situation. Finally, Miyachi's preliminary action spectrum determination consists of measurements at four points in the blue region only, namely at 400, 420, 453, and 473 nm (Ogasawara and Miyachi, 1969). Closer spectral response analyses of the ^{14}C-incorporation into aspartic and glutamic acid might well disclose a maximum efficiency at longer wavelengths than 420 nm. As long as we do not have separate clean action spectra for the loss in carbohydrate and the gain in protein or amino acids, a conclusion on the occurrence of different blue light effects in both metabolic reactions is impossible.

*Is blue light affecting either carbohydrate breakdown or protein
biosynthesis only?*

Whether the observed alterations in carbohydrate and protein result
from an influence of blue light on either carbohydrate breakdown or
protein formation only, is equally hard to decide.

The above-mentioned assumption that a blue light-stimulated respiration might be responsible for an enhanced biosynthesis of amino acids and
of protein by the supply of more carbon skeletons and ATP is not supported
experimentally. Firstly, differences in glycine and serine content have been
found after 30 seconds exposure to blue or red light, while the earliest
indication of a blue light-stimulated respiration could not be noticed
sooner than 1 minute after a 5 second flash; complete enhancement of
respiration as well as of acid release in a nitrogen atmosphere was observed
10 to 15 minutes after the beginning of illumination. Further, a blue light-
dependent increase in nitrogenous substances has also been seen in cells
exposed to exogenous glucose (Kowallik, 1966a) in which the respiratory
rates in darkness and blue light were found to be equally high or even
slightly smaller in the light (Kowallik, unpublished).

The opposite proposal that a light-stimulated amino acid and protein
biosynthesis might activate respiration by lowering the ATP/ADP ratio
has no experimental support either, although obtained results do not exclude
this possibility. Firstly, the pronounced temporary increase in dark oxygen
uptake in *Chlorella* after a blue flash, which in some way appears related
to the steadily enhanced respiration under continuous illumination, starts
1 minute after the light exposure. Secondly, autotrophically grown cells
of *Chlorella*, carefully washed several times and resuspended in phosphate
buffer containing 10^{-5} M DCMU to prevent any new photosynthetic carbon
fixation, lost 26 per cent more of their endogenous carbohydrate reserves
during 4 hours of exposure to blue light than to red light or darkness,
while no increase in bound nitrogen could be measured. Thirdly, addition
of 10^{-3} M chloramphenicol, which might interfere with the synthesis of
new protein, did not affect the light enhancement of respiration at all
(Kowallik, unpublished). Finally, no protein production could be found
in *Chlorella* which was kept anaerobically (Nührenberg, Lesemann and
Pirson, 1968), while a blue light effect in the catabolic mechanism of that
alga could equally well be demonstrated under air or an atmosphere
of nitrogen (Kowallik, 1969b; Kowallik and Gaffron, 1967a, 1967b).

Whether an increase in protein as well as a decrease in carbohydrate
are consequences of a blue light-stimulated RNA synthesis cannot be
decided, since no data on changes in RNA are available for illumination
periods of less than 4 hours.

Is one common light reaction involved in both carbohydrate breakdown and protein biosynthesis?

Comparison of the time-course studies of increased respiration with those of extra amino acid biosynthesis in blue light allows the suggestion of one basic light reaction which influences both metabolic pathways independently of each other. As an example of such a common reaction, an effect of blue light on specific intracellular permeability barriers has been suggested in Pirson's laboratory: while the respiration of DCMU-poisoned *Chlorella* increased in blue and not in red light, some polyglucan fraction accumulated in cells exposed to the latter (Laudenbach and Pirson, 1969). Locating that fraction arbitrarily inside the chloroplast, the authors proposed an effect of short wavelength radiation on the permeability of the chloroplast envelope. The fact that until now blue light effects on carbohydrate breakdown as well as on the formation of nitrogenous material have only been found in cells containing at least rudimental plastids is in accordance with that idea.

Relevant changes in uptake affected by light as a measure for gross permeability alterations were absent in the experiments of Zeldin and Schiff (1968), as one of their chlorophyll-free *Euglena* mutants, W 3, took up amounts of radioactive phosphorus corresponding to 9×10^6 cpm in the dark and 10×10^6 cpm in the light. From these almost equal amounts, 3×10^4 cpm were found to be incorporated in the phenol-extracted RNA in darkness, but 15×10^4 cpm in the light (see page 172). There was also no evidence for a different uptake of several metabolic intermediates fed to *Chlorella* in darkness or blue light in unpublished experiments of Kowallik; and Lesemann (1966) even reports an inhibition of glucose uptake in *Chlorella* on blue irradiation.

Another possibility of such a basic action of blue light might be some regulation in glycolate metabolism, as there are indications of an involvement of that compound in an extra production of some amino acids as well as in the light enhancement of respiration. The similarities in intramolecular ^{14}C-distribution of glycine formed by *Chlorella* in blue light and of photosynthetic glycolate (Cayle and Emerson, 1957) have already been mentioned, and so has a preferred ^{14}C-incorporation into serine accompanied by a decreased labelling of glycolate (Hauschild, Nelson and Krotkov, 1962a, 1962b). The general existence of a pathway from glycolate to serine via glycine is known. Aspartic and glutamic acid have also been found preferentially labelled in blue light, and a possibility of their formation from glycolate has also been suggested (Asada and coworkers, 1965; Miflin, Marker and Whittingham, 1966). Ogasawara and Miyachi's (1969) assumption of a stimulation of phosphoenolpyruvate carboxylation by blue light as the basis for additional aspartate and glutamate ought, therefore, to be checked. All the short-term effects of blue light on amino acids observed *could* thus be related to glycolate. An involvement of that compound in light-enhanced respiration of *Chlorella* is not directly proven, but since there is much evidence that glycolate is the substrate for a 'photorespiration' in leaves of several higher plants (Zelitch, 1964) and in the highly developed green alga *Nitella* (Downton and Tregunna, 1968), its participation in blue-enhanced respiration in *Chlorella* appears possible. A comparison of conditions for and criteria of both types of light respiration may serve to balance the probability of that assumption.

Both leaves and unicellular green algae produce high amounts of glycolate in strong white light and at high oxygen partial pressure, but in contrast to leaves the algae sometimes excrete the substance in great amounts into the medium (Bassham and Kirk, 1962; Hess, Tolbert and Pike, 1967; Miller, Meyer and Tanner, 1963; Plamondon and Bassham, 1966; Prichard, Griffin and Whittingham, 1962; Tolbert, 1963; Tolbert and Zill, 1956; Warburg and Krippahl, 1960; Watt and Fogg, 1966; Whittingham, Coombs and Marker, 1967; Whittingham and Prichard, 1963; Wilson and Calvin, 1955).

An active glycolate oxidase, which is abundant in green leaves, should not allow glycolate to accumulate, and consequently the enzyme has until recently been thought to be absent from unicellular green algae (Hess and coworkers, 1965; Hess and Tolbert, 1967a; Tolbert and Hess, 1966), when its presence could be established for *Chlorella*, *Chlamydomonas* and *Ankistrodesmus*, as well as its first reaction product glyoxylate (Nelson and Tolbert, 1968; Urbach and Gimmler, 1968; Zelitch and Day, 1968). This discrepancy is explained by the finding of a strong dependence of the activity of glycolate oxidase on the developmental stage of the algae.

In leaves, glycolate accumulates in the light on application of α-hydroxy-methanesulphonates, inhibitors of glycolate oxidase (Egle and Fock, 1967; Hess and Tolbert, 1966a; Moss, 1966; Zelitch, 1958, 1959, 1965; Zelitch and Walker, 1964; and compare Tanner and Beevers, 1965). An increased excretion of glycolate was also observed with *Ankistrodesmus* exposed to 9×10^{-3} M α-hydroxy-2-pyridylmethanesulphonate (Urbach and Gimmler, 1968).

The equality of the RQ of the respiration of *Chlorella* in darkness and in the light (Kowallik, 1966b) does not necessarily exclude glycolate as a new substrate in the light, since with extracts from green leaves no complete oxidation of that compound has been found. Besides carbon dioxide formic acid also appeared as an end-product (Tolbert, Clagett and Burris, 1949; Zelitch, 1966).

High oxygen concentrations and high intensities of white light do not only increase the production of glycolate, but are also conditions that favour 'photorespiration' (Forrester, Krotkov and Nelson, 1966; Poskuta, 1968a; Tregunna, Krotkov and Nelson, 1966). Influences of different oxygen pressures have not been tested on light enhancement of respiration in *Chlorella*, whose light requirement, however, is quite low. In a chlorophyll-free mutant traces of blue radiation were already effective, and about 500 ergs/cm² second were sufficient for saturation (Kowallik, 1967). However, considering the small percentage of short wavelength radiation of most white light sources and, in addition, the shading effect of chlorophyll in green leaves this point needs closer examination.

As 'photorespiration' has not been found in non-green leaves (Hew and Krotkov, 1968) and as it disappeared when DCMU was added (Downton and Tregunna, 1968; El-Sharkawy, Loomis and Williams, 1967), it was believed to depend on running photosynthesis, a fact definitely not valid for enhanced respiration in *Chlorella* (Kowallik, 1969a; Kowallik and Kowallik, 1969). However, since CMU was found to affect stomatal opening (Zelitch, 1966), the closely related poison DCMU might act in that way and thus quite indirectly. Poskuta's (1968b) observation of a three to four times greater 'photorespiration' of spruce shoots in blue than in red light does not point to a strict connexion with photosynthesis and meets the wavelength requirement for the enhancement of respiration in *Chlorella*.

Different temperature coefficients have been found for dark respiration and for 'photorespiration' in leaves, most pronounced above 30°C (Zelitch, 1966), while an equal Q_{10} has been described for the respiration of *Chlorella* in darkness and in the light (Kowallik, 1966b). The latter, however, is only valid for short periods of time. During the first 30 minutes in each temperature the oxygen uptake in blue light was 67 per cent at 15·0°C, 59 per cent at 22·6°C and 79 per cent at 30·0°C greater than in darkness, but 68, 80 and 136 per cent respectively, when calculated for the first 90 minutes.

For the oxidation of exogenous glycolate by spinach leaf particles very low P/O ratios have been found (Zelitch and Barber, 1960). The efficiency of blue light-enhanced respiration in *Chlorella* remains to be determined.

Summing up, there are reasons to doubt that 'photorespiration' in green leaves and blue light-enhanced respiration in *Chlorella* are correspondent reactions, but considering the different organisms and methods used that possibility certainly cannot be excluded. Let us, therefore, arbitrarily consider glycolate to be a substrate and glycolate oxidase an enzyme involved in light-enhanced respiration in *Chlorella*. It is then tempting to speculate that this enzyme itself is activated by blue light, as it has been shown to be a flavoprotein (Frigerio and Harbury, 1958; Zelitch and

Ochoa, 1953) whose absorption closely resembles the action spectra for the enhancement of respiration in algae. However, illumination did not stimulate its activity in the sap of green leaves (Tolbert and Burris, 1950); that could be different *in vivo*. While a finding of Becker, Döhler and Egle (1968) of glycolate accumulation in the medium of red-exposed but none at all in that of blue-illuminated *Chlorella* might point to such an influence of light, incorporation of 30 to 36 per cent of the newly fixed carbon into glycolate in blue-adapted but almost none in red-grown *Chlamydomonas* and *Chlorella* (Hess and Tolbert, 1967b; Tolbert, 1963) speaks against it, and so does the fact that there is also an enhancement in carbohydrate breakdown under an atmosphere of nitrogen. It also should be mentioned that flavine mononucleotide-dependent glycolate oxidase has only been seen in green parts of higher plants (Kuczmak and Tolbert, 1962; Tolbert and Burris, 1950), while in achlorophyllous organs a possibly ferredoxin-dependent type has been found (Baker and Tolbert, 1967). The latter might, therefore, be expected in the yellow *Chlorella* mutant used by Kowallik (1966b); it is also assumed to be present in *Ankistrodesmus* from the action of different poisons (Urbach and Gimmler, 1968).

In case blue light leads to an increased *production* of glycolate its way of action remains obscure, the more as there is as yet no conformity on the pathway of glycolate formation. Most investigators assume that it originates from intermediates of the photosynthetic carbon cycle, preferentially from ribulosediphosphate, but also from the splitting of other sugar phosphates (Bassham, 1964; Bradbeer and Anderson, 1967; Griffith and Byerrum, 1959; Tolbert, 1963; Vandor and Tolbert, 1968; Whittingham and Prichard, 1963; Wilson and Calvin, 1955); others, however, think it to stem from a special carboxylation (Stiller, 1962; Zelitch, 1965). Such a reaction seems to occur in the photosynthetic bacterium *Rhodospirillum rubrum* (Anderson and Fuller, 1967), but from labelling kinetics is usually not accepted for leaves and green algae. If glycolate were involved in the light-enhanced respiration of the chlorophyll-free *Chlorella* mutant mentioned, its production would have to be independent of photosynthetic intermediates. That possibility seems to be established by the finding of Hess, Tolbert and Pike (1967) of a light-dependent glycolate production in *Chlorella* under an atmosphere free of carbon dioxide. These authors propose some light effect on the hydrolysis of polysaccharides.

Final remarks

On a physiological meaning of the blue light effects described nothing safe is known as yet. Only a few largely unexplained results can be mentioned, which from the wavelength and intensity requirements could some-how be related to them. So a 'healing' effect of minute amounts of blue light on fading photosynthesis of *Chlorella* in red light at 644 nm, with greatest efficiency of wavelengths around 455 nm, has been reported by Warburg and coworkers (1954a, 1954b, 1955); this has not been uncontra-dicted by others (Bassham, Shibata and Calvin, 1955; Kok, 1960). Twelve years later comparable observations were made by Terborgh (1966) with the siphonaceous green alga *Acetabularia crenulata*. In this organism oxygen production in red light decreased slowly to about 40 per cent of the initial value overnight, but was restored for 4 to 6 hours on addition of less than 10^{-8} einsteins cm^{-2} of blue radiation. The latter could be given in a single flash of down to 2·5 seconds duration. Wavelengths around 450 nm revealed greatest efficiency, those beyond 525 nm none at all. Schael and Clauss (1968) confirmed the basic result with *Acetabularia mediterranea*, and Adler (1967) with *Ankistrodesmus*. An increasing improvement of total ^{14}C-incorporation from NaH^{14}CO$_3$ of *Chlorella* in red light was found by Hauschild, Nelson and Krotkov (1964, 1965) on addition of 4 per cent blue light in the course of 90 minutes, which was preceded by the preferred labelling of aspartate, glutamate, alanine, serine and malate, mentioned earlier to occur in pure blue light. In some contrast to that, no deviation in total mass production of *Chlorella* took place during 5 days of growth in blue or red light, although the cells in the former contained about 40 per cent more protein than those in the latter (Kowallik, 1962). Schmidt-Clausen and Ziegler (1969) found light activation of NADP$^+$-dependent glyceraldehyde-3-phosphate dehydrogenase least pronounced with wave-lengths around 470 nm.

Summary

In unicellular green algae long-term exposure to blue instead of red light of equal quantum flux or productivity brings about a decrease in carbo-hydrate, an increase in protein and an enhanced production of RNA. Short-term exposure brings about a pronounced enhancement of endo-genous respiration as well as of anaerobic carbohydrate breakdown and an increased uptake of radioactive carbon from NaH^{14}CO$_3$ into some amino acids.

As far as tested, the effects are independent of intact photosynthesis. The point of attack of short wavelength visible radiation in this special response is unknown; furthermore, it is not even decided if all these effects originate from one and the same or from different basic light reac-tions. Influences on permeability barriers or on some step in protein or/

and carbohydrate metabolism have been assumed as the primary light action. Among the latter special attention has been called for an effect on glycolate, which could be a key substance in the enhanced biosynthesis of amino acids and of increased respiration as well.

Unequivocal information is also still missing on the nature of the pigment(s) involved. From the action spectra, flavins or *cis* carotenoids could be the light receptors responsible (Kowallik, 1967, 1968; Pickett and French, 1967). For both, some influence on a protein can be imagined: light-activated flavins might change the configuration of such a compound, structural or enzymic, by reducing S—S bonds (Rau, 1967), and isomerization of *cis* carotenoids might yield structural alterations.

It is beyond the scope of this article to give a complete survey of the literature on 'photorespiration' and on glycolate metabolism. As far as possible review articles have been cited, which are only extended by some more recent publications.

REFERENCES

Adler, K. (1967). Spezifische Rolle der Carotinoidabsorption bei der photosynthetischen Sauerstoffentwicklung. *Planta*, **75**, 220.

Anderson, L., and R. C. Fuller (1967). The rapid appearance of glycolate during photosynthesis in *Rhodospirillum rubrum*. *Biochim. biophys. Acta*, **131**, 198.

Asada, K., and Z. Kasai (1962). Inhibition of the photosynthetic carbon dioxide fixation of green plants by α-hydroxysulfonates, and its effects on the assimilation products. *Plant and Cell Physiol.*, **3**, 125.

Asada, K., K. Saito, S. Kitoh, and Z. Kasai (1965). Photosynthesis of glycine and serine in green plants. *Plant and Cell Physiol.*, **6**, 47.

Baker, A. L., and N. E. Tolbert (1967). Purification and some properties of an alternate form of glycolate oxidase. *Biochim. biophys. Acta*, **131**, 179.

Bassham, J. A. (1964). Kinetic studies on the photosynthetic carbon reduction cycle. *Ann. Rev. Plant Physiol.*, **15**, 101.

Bassham, J. A., and M. Kirk (1962). The effect of oxygen on the reduction of CO_2 to glycolic acid and other products during photosynthesis. *Biochem. Biophys. Res. Commun.*, **9**, 375.

Bassham, J. A., K. Shibata, and M. Calvin (1955). Quantum requirement in photosynthesis related to respiration. *Biochim. biophys. Acta*, **17**, 332.

Becker, J. D., G. Döhler und K. Egle (1968). Die Wirkung monochromatischen Lichts auf die extrazelluläre Glykolsäure-Ausscheidung bei der Photosynthese von *Chlorella*. *Z. Pflanzenphysiol.*, **58**, 212.

Bradbeer, J. W., and C. M. A. Anderson (1967). Glycollate formation in chloroplast preparations. In T. W. Goodwin (Ed.), *Biochemistry of Chloroplasts*, Vol. 2. Academic Press, London, New York. pp. 175–179.

Buschbohm, A. (1968). *Vergleichend-physiologische Untersuchungen an Algen aus Blau- und Rotlichtkulturen*, Dissertation, Göttingen.

Cayle, T., and R. Emerson (1957). Effect of wave-length on the distribution of carbon-14 in the early products of photosynthesis. *Nature*, **179**, 89.

Downton, W. J. S., and E. B. Tregunna (1968). Photorespiration and glycolate metabolism: a re-examination and correlation of some previous studies. *Plant Physiol.*, **43**, 923.

Egle, K., and H. Fock (1967). Light respiration—correlations between CO_2 fixation, O_2 pressure, and glycollate concentration. In T. W. Goodwin (Ed.), *Biochemistry of Chloroplasts*, Vol. 2. Academic Press, London, New York. pp. 79–87.

El-Sharkawy, M., A. R. Loomis, and W. A. Williams (1967). Apparent reassimilation of respiratory carbon dioxide by different plant species. *Physiol. plantarum*, **20**, 171.

Forrester, M. L., G. Krotkov, and C. D. Nelson (1966). Effect of oxygen on photosynthesis, photorespiration and respiration in detached leaves. I. Soybean. *Plant Physiol.*, **41**, 422.

Frigerio, N. A., and H. A. Harbury (1958). Preparation and some properties of crystalline glycolic acid oxidase of spinach. *J. biol. Chem.*, **231**, 135.

Griffith, T., and R. U. Byerrum (1959). Biosynthesis of glycolate and related compounds from ribose-1-^{14}C in tobacco leaves. *J. biol. Chem.*, **234**, 762.

Hauschild, A. H. W., C. D. Nelson, and G. Krotkov (1962a). The effect of light quality on the products of photosynthesis in *Chlorella vulgaris*. *Can. J. Bot.*, **40**, 179.

Hauschild, A. H. W., C. D. Nelson, and G. Krotkov (1962b). The effect of light quality on the products of photosynthesis in green and blue-green algae, and in photosynthetic bacteria. *Can. J. Bot.*, **40**, 1619.

Hauschild, A. H. W., C. D. Nelson, and G. Krotkov (1964). Concurrent changes in the products and the rate of photosynthesis in *Chlorella vulgaris* in the presence of blue light. *Naturwissenschaften*, **51**, 274.

Hauschild, A. H. W., C. D. Nelson, and G. Krotkov (1965). On the mode of action of blue light on the products of photosynthesis in *Chlorella vulgaris*. *Naturwissenschaften*, **52**, 435.

Hess, J. L., M. G. Huck, F. H. Liao, and N. E. Tolbert (1965). Glycolate metabolism in algae. *Plant Physiol.*, **40**, Suppl. XLII.

Hess, J. L., and N. E. Tolbert (1966a). Glycolate, glycine, serine, and glycerate formation during photosynthesis by tobacco leaves. *J. biol. Chem.*, **241**, 5705.

Hess, J. L., and N. E. Tolbert (1966b). Serine formation during $^{14}CO_2$-photosynthesis by algae and tobacco. *Plant Physiol.*, **41**, Suppl. XXXIX.

Hess, J. L., and N. E. Tolbert (1967a). Glycolate pathway in algae. *Plant Physiol.*, **42**, 371.

Hess, J. L., and N. E. Tolbert (1967b). Changes in chlorophyll a/b ratio and products of $^{14}CO_2$ fixation by algae grown in blue or red light. *Plant Physiol.*, **42**, 1123.

Hess, J. L., N. E. Tolbert, and L. M. Pike (1967). Glycolate biosynthesis by *Scenedesmus* and *Chlorella* in the presence or absence of $NaHCO_3$. *Planta*, **74**, 278.

Hew, C. S., and G. Krotkov (1968). Effect of oxygen on the rates of CO_2 evolution in light and in darkness by photosynthesizing and non-photosynthesizing leaves. *Plant Physiol.*, **43**, 464.

Kok, B. (1960). Efficiency of photosynthesis. In A. Pirson (Ed.), *Encyclopedia of Plant Physiology*, Vol. V/1. Springer, Berlin, Göttingen, Heidelberg. pp. 565–633.

Kowallik, U., and W. Kowallik (1969). Eine wellenlängenabhängige Atmungssteigerung während der Photosynthese von *Chlorella*. *Planta*, **84**, 141.

Kowallik, W. (1962). Uber die Wirkung des blauen und roten Spektralbereichs auf die Zusammensetzung und Zellteilung synchronisierter Chlorellen. *Planta*, **58**, 337.

Kowallik, W. (1965). Die Proteinproduktion von *Chlorella* im Licht verschiedener Wellenlängen. *Planta*, **64**, 191.

Kowallik, W. (1966a). Einfluß verschiedener Lichtwellenlängen auf die Zusammensetzung von *Chlorella* in Glucosekultur bei gehemmter Photosynthese. *Planta*, **69**, 292.

Kowallik, W. (1966b). Chlorophyll-independent photochemistry in algae. In *Energy Conversion by the Photosynthetic Apparatus. Brookhaven Symp. Biol.*, **19**, 467.

Kowallik, W. (1967). Action spectrum for an enhancement of endogenous respiration by light in *Chlorella*. *Plant Physiol.*, **42**, 672.

Kowallik, W. (1968). Uber die Wirkung von Kaliumjodid auf die lichtstimulierte endogene Atmung von Algen. *Planta*, **79**, 122.

Kowallik, W. (1969a). Der Einfluss von Licht auf die Atmung von *Chlorella* bei gehemmter Photosynthese. *Planta*, **86**, 50.

Kowallik, W. (1969b). Eine fördernde Wirkung von Blaulicht auf die Säureproduktion anaerob gehaltener Chlorellen. *Planta*, **87**, 372.

Kowallik, W., and H. Gaffron (1966). Respiration induced by blue light. *Planta*, **69**, 92.

Kowallik, W., and H. Gaffron (1967a). Enhancement of respiration and fermentation in algae by blue light. *Nature*, **215**, 1038.

Kowallik, W., and H. Gaffron (1967b). Mobilization of endogenous metabolites in algae caused specifically by blue light. *Fed. Proc.*, **26**, 861.

Kuczmak, M., and N. E. Tolbert (1962). Glycolic acid oxidase formation in greening leaves. *Plant Physiol.*, **37**, 729.

Laudenbach, B. und A. Pirson (1969). Uber den Kohlenhydratumsatz in *Chlorella* unter dem Einfluss von blauem und rotem Licht. *Arch. Mikrobiol.*, **67**, 226.

Lesemann, D. (1966). *Untersuchungen über die lichtabhängige Glukoseverwertung in Chlorella-Kulturen*, Dissertation, Göttingen.

Miflin, B. M., A. F. H. Marker, and C. P. Whittingham (1966). The metabolism of glycine and glycollate by pea leaves in relation to photosynthesis. *Biochim. biophys. Acta*, **120**, 266.

Miller, R. M., C. M. Meyer, and H. A. Tanner (1963). Glycolate excretion and uptake by *Chlorella*. *Plant Physiol.*, **38**, 184.

Moss, D. N. (1966). Glycolate as a substrate for photorespiration. *Plant Physiol.*, **41**, Suppl. XXXVIII.

Nelson, E. B., and N. E. Tolbert (1968). Glycolate excretion and metabolism in unicellular green algae. *Plant Physiol.*, **43**, Suppl. S-12.

Nührenberg, B., D. Lesemann und A. Pirson (1968). Zur Frage eines anaeroben Wachstums von einzelligen Grünalgen. *Planta*, **79**, 162.

Ogasawara, N., and S. Miyachi (1969). Effect of wavelength on ^{14}C-fixation in *Chlorella* cells. In H. Metzner (Ed.), *Progress in Photosynthesis Research*, Vol. III. Tübingen. pp. 1653–1661.

Pickett, J. M., and C. S. French (1967). The action spectrum for blue-light-stimulated oxygen uptake in *Chlorella. Proc. Natl. Acad. Sci.*, **57**, 1587.

Pirson, A. und W. Kowallik (1960). Wirkung des blauen und roten Spektralbereiches auf die Zusammensetzung von *Chlorella* bei Anzucht im Licht-Dunkel-Wechsel. *Naturwissenschaften*, **47**, 476.

Plamondon, J. E., and J. A. Bassham (1966). Glycolic acid labeling during photosynthesis with ¹⁴CO₂ and tritiated water. *Plant Physiol.*, **41**, 1272.

Poskuta, J. (1968a). Photosynthesis, photorespiration and respiration of detached spruce twigs as influenced by oxygen concentration and light intensity. *Physiol. plantarum*, **21**, 1129.

Poskuta, J. (1968b). Photosynthesis and respiration. I. Effect of light quality on the photorespiration in attached shoots of spruce. *Experientia*, **24**, 796.

Prichard, G. G., W. J. Griffin, and C. P. Whittingham (1962). The effect of CO₂ concentration, light intensity and isonicotinylhydrazide on the photosynthetic production of glycollic acid by *Chlorella. J. exptl. Bot.*, **13**, 176.

Rabson, R., N. E. Tolbert, and P. C. Kearney (1962). Formation of serine and glyceric acid by the glycolate pathway. *Arch. Biochem. Biophys.*, **98**, 154.

Rau, W. (1967). Untersuchungen über die lichtabhängige Carotinoidsynthese. II. Ersatz der Lichtinduktion durch Mercuribenzoat. *Planta*, **74**, 263.

Ried, A. (1965). Transients of oxygen exchange in *Chlorella* caused by short light exposures. *Carnegie Inst. Wash. Yearbook*, **64**, 399.

Ried, A. (1968). Interactions between photosynthesis and respiration in *Chlorella*. Types of transients of oxygen exchange after short light exposures. *Biochim. biophys. Acta*, **153**, 653.

Schael, U. und H. Clauss (1968). Die Wirkung von Rotlicht und Blaulicht auf die Photosynthese von *Acetabularia mediterranea. Planta*, **78**, 98.

Schmidt-Clausen, H. J., and I. Ziegler (1969). The influence of light quality on the activation of NADP-dependent glyceraldehyde-3-phosphate dehydrogenase. In H. Metzner (Ed.), *Progress in Photosynthesis Research*, Vol. III. Tübingen. pp. 1646–1652.

Schou, L., A. A. Benson, J. A. Bassham, and M. Calvin (1950). The path of carbon in photosynthesis. XI. The role of glycolic acid. *Physiol. plantarum*, **3**, 487.

Senger, H., and N. I. Bishop (1968). An action spectrum for nucleic acid formation in an achlorophyllous mutant of *Chlorella pyrenoidosa. Biochim. biophys. Acta*, **157**, 417.

Senger, H. und G. Schoser (1966). Die spektralabhängige Teilungsinduktion in mixotrophen Synchronkulturen von *Chlorella. Z. Pflanzenphysiol.*, **54**, 308.

Sinha, S. K., and E. A. Cossins (1965). The importance of glyoxylate in amino acid biosynthesis in plants. *Biochem. J.*, **96**, 254.

Stiller, M. (1962). The path of carbon in photosynthesis. *Ann. Rev. Plant. Physiol.*, **13**, 151.

Stoy, V. (1955). Action of different light qualities on simultaneous photosynthesis and nitrate assimilation in wheat leaves. *Physiol. plantarum*, **8**, 963.

Stoy, V. (1956). Riboflavin catalysed enzymic photoreduction of nitrate. *Biochim. biophys. Acta*, **21**, 395.

Strotmann, H. (1967). Blaulichteffekt auf die Nitritreduktion von *Chlorella. Planta*, **73**, 376.

Tanner, W. H., and H. Beevers (1965). Glycolic acid oxidase in castor bean endosperm. *Plant Physiol.*, **40**, 971.

Terborgh, J. (1966). Potentiation of photosynthetic oxygen evolution in red light by small quantities of monochromatic blue light. *Plant Physiol.*, **41**, 1401.

Tolbert, N. E. (1963). Glycolate pathway. In *Photosynthesis Mechanisms in Green Plants*. *Natl. Acad. Sci.*, *Natl. Res. Counc.*, **Publ. 1145**, pp. 648–662.

Tolbert, N. E., and R. H. Burris (1950). Light activation of the plant enzyme which oxidizes glycolic acid. *J. biol. Chem.*, **186**, 791.

Tolbert, N. E., C. O. Clagett, and R. H. Burris (1949). Products of the oxidation of glycolic acid and 1-lactic acid by enzymes from tobacco leaves. *J. biol. Chem.*, **181**, 905.

Tolbert, N. E., and M. S. Cohan (1953). Products formed from glycolic acid in plants. *J. biol. Chem.*, **204**, 649.

Tolbert, N. E., and J. L. Hess (1966). The effect of hydroxymethanesulfonates on $^{14}CO_2$ photosynthesis by algae. *J. biol. Chem.*, **241**, 5712.

Tolbert, N. E., and L. P. Zill (1956). Excretion of glycolic acid by algae during photosynthesis. *J. biol. Chem.*, **222**, 895.

Tregunna, E. B., G. Krotkov, and C. D. Nelson (1966). Effect of oxygen on the rate of photorespiration in detached tobacco leaves. *Physiol. plantarum*, **19**, 723.

Urbach, W., and H. Gimmler (1968). Stimulation of glycollate excretion of algae by disalicylidenepropanediamine and hydroxypyridinemethanesulfonate. *Z. Naturforschg.*, **23b**, 1282.

Vandor, S. L., and N. E. Tolbert (1968). Glycolate biosynthesis by isolated chloroplasts. *Plant Physiol.*, **43**, Suppl. S-12.

Voskresenskaja, N. P. (1952). Influence of the spectral distribution of light on the products of photosynthesis. (In Russian.) *Dokl. Akad. Nauk SSSR*, **86**, 427.

Wang, D., and R. H. Burris (1963). Carbon metabolism of ^{14}C-labeled amino acids in wheat leaves. II. Serine and its role in glycine metabolism. *Plant Physiol.*, **38**, 430.

Wang, D., and E. R. Waygood (1962). Carbon metabolism of ^{14}C-labeled amino acids in wheat leaves. I. A pathway of glyoxylate-serine metabolism. *Plant Physiol.*, **37**, 826.

Warburg, O. und G. Krippahl (1960). Glykolsäurebildung in *Chlorella*. *Z. Naturforschg.*, **15b**, 197.

Warburg, O., G. Krippahl, W. Schröder, W. Buchholz und E. Theel (1954a). Uber die Wirkung sehr schwachen blaugrünen Lichts auf den Quantenbedarf der Photosynthese. *Z. Naturforschg.*, **9b**, 164.

Warburg, O., G. Krippahl und W. Schröder (1954b). Katalytische Wirkung des blaugrünen Lichts auf den Energieumsatz bei der Photosynthese. *Z. Naturforschg.*, **9b**, 667.

Warburg, O., G. Krippahl und W. Schröder (1955). Wirkungsspektrum eines Photosynthese-Ferments. *Z. Naturforschg.*, **10b**, 631.

Watt, W. D., and G. E. Fogg (1966). The kinetics of extracellular glycollate production by *Chlorella pyrenoidosa*. *J. exptl. Bot.*, **17**, 117.

Whittingham, C. P., I. Coombs, and A. F. H. Marker (1967). The role of glycollate in photosynthetic carbon fixation. In T. W. Goodwin (Ed.), *Biochemistry of Chloroplasts*, Vol. 2. Academic Press, London, New York. pp. 155–173.

Whittingham, C. P., and G. G. Prichard (1963). The production of glycollate during photosynthesis in *Chlorella. Proc. Roy. Soc. B*, **157**, 366.

Wilson, A. T., and M. Calvin (1955). The photosynthetic cycle. CO_2 dependent transients. *J. Am. Chem. Soc.*, **77**, 5948.

Zeldin, M. H., and J. A. Schiff (1968). A comparison of light-dependent RNA metabolism in wild-type *Euglena* with that of mutants impaired for chloroplast development. *Planta*, **81**, 1.

Zelitch, I. (1958). The role of glycolic acid oxidase in the respiration of leaves. *J. biol. Chem.*, **233**, 1299.

Zelitch, I. (1959). The relationship of glycolic acid to respiration and photosynthesis in tobacco leaves. *J. biol. Chem.*, **234**, 3077.

Zelitch, I. (1964). Organic acids and respiration in photosynthetic tissues. *Ann. Rev. Plant Physiol.*, **15**, 121.

Zelitch, I. (1965). The relation of glycolic acid synthesis to the primary photosynthetic carboxylation reaction in leaves. *J. biol. Chem.*, **240**, 1869.

Zelitch, I. (1966). Increased rate of net photosynthetic carbon dioxide uptake caused by the inhibition of glycolate oxidase. *Plant Physiol.*, **41**, 1623.

Zelitch, I., and G. A. Barber (1960). Oxidative phosphorylation and glycolate oxidation by particles from spinach leaves. *Plant Physiol.*, **35**, 205.

Zelitch, I., and P. R. Day (1968). Glycolate oxidase activity in algae. *Plant Physiol.*, **43**, 289.

Zelitch, I., and S. Ochoa (1953). Oxidation and reduction of glycolic and glyoxylic acids in plants. I. Glycolic acid oxidase. *J. biol. Chem.*, **201**, 707.

Zelitch, I., and D. A. Walker (1964). The role of glycolic acid metabolism in opening of leaf stomata. *Plant Physiol.*, **39**, 856.

CHAPTER 7

Synchronous cultures

HARALD LORENZEN

Department of Plant Physiology, University of Göttingen, Germany

Introduction

In this chapter an attempt is made to give an up-to-date account of
ideas and experimental results of the different groups and workers engaged
in synchronous culturing of unicellular photosynthetic algae. The limited
space available for this chapter does not permit an extensive treatment of
the whole area, but the author eagerly hopes not to be blamed for omitting
major contributions in this rapidly enlarging field of research. Several
general and comprehensive reviews already written deal with important
aspects of our problem; some of them are recommended here, especially
as a source of further literature studies (Kuhl and Lorenzen, 1964; Pirson,
1962; Pirson and Lorenzen, 1966; Schmidt, 1966; Senger and Bishop,
1969; Tamiya, 1966).

For scientific purposes a general trend to synchronize the growth of microorganisms is evident. Scientists try to bring the cells of a culture into phase in such a way that they are permitted—with some restrictions— to use the whole culture to gain better information on what is happening in a 'normal' cell depending on its stage of development. It is rather difficult —perhaps impossible—to utilize isolated single cells with similar success. The synchronous cultures in a way enlarge all the events during the cell cycle to a level where successful experiments can be performed. Of course, each condition of growth in itself has great influence in modifying the physiological reactions of the cells. A question has often been raised as to whether the cells in a synchronized population are 'normal' or 'abnormal'. This, in turn, raises the question of what is a 'normal' cell. Indeed, it seems quite impossible to tell what the characteristics of a normal cell are, but the synchronization of algal cultures and maintenance of synchrony produces cells which can be called 'standardized'. There is no stress included—that means the cells are not treated with methods they cannot endure permanently—and there is a very good experimental reproducibility to compare the succeeding generations.

In this article only photosynthetic organisms are considered and, furthermore, only those methods in which the light regimen is important to induce a synchronous phasing of the cells will be discussed. In addition we have to restrict the topic mainly to methods which result in complete synchronization. That is, more than 98 per cent of the cells grow in phase during each cycle. In order to provide physiologically and morphologically well-defined cells over a long time it is important indeed that we take advantage of permanently synchronized cultures†; the cells should immediately be able to start a new cycle after they have completed one, and at any specific time during the cycles analytical or metabolical analyses should give the same data, only deviating within the experimental errors of the methods used. Thus the daughter cells ultimately produced are not different from the original ones the cycle started with.

Methods, general aspects and discussion

The most important and widely used method since 1957 to synchronize unicellular algal cultures are programmed light–dark changes combined with serial dilution (Berger, 1966; Correll and Tolbert, 1962; Döhler, 1963; Galling, 1963; Gerhardt, 1964; Guérin-Dumartrait, 1966; Kaden, 1965;

---

† Other terms used are: Dauersynchronisierung, synchrostatic technique, continuous synchronous growth, continuous synchrony, semicontinuous culture.

Knutsen, 1965; Kowallik, 1962; Lorenzen, 1957; Pirson and Lorenzen, 1958a; Ried, 1962; Ruppel, 1962; Senger, 1961; Soeder, 1964; Wanka, 1965; and others). This is true in spite of the fact that Tamiya (1966) claimed unsatisfactory results for this method. The Japanese groups of Tamiya synchronized *Chlorella* cultures some years earlier (Tamiya and coworkers, 1953) using other methods. Since 1961, Tamiya and coworkers also use a light–dark regimen together with serial dilution which they call DLD, the only real cyclic one of their different treatments tried. They employed fractionated centrifugation to concentrate cells of a certain small size. It is noteworthy to observe that after a few generations, the effect of such selection seems to be lost. Besides, this selection procedure may even have a negative effect on general metabolism, as the first cycle includes a long lag phase which means twice the cycle length of the succeeding ones. It is important to mention that using the method of serial dilution—that means reinoculation into fresh culture medium after one generation cycle to the same external starting conditions—no lag phase can be observed. The nutrient medium used by our group permits at least a ten times higher productivity than that normally observed in one synchronous growth cycle. Table 7.1 gives an impression of how far the methods used till now are away from what we would like to call an ideal synchronization.

Table 7.1. Points which should be realized under conditions of ideal synchronization

	Verified or not by the methods of Japanese (1) and German (2) authors	
	(1)	(2)
Completely synchronized	+	+
Permanently synchronized (serial dilution)	only DLD	+
Shortest possible life cycle	?	+
Stepwise nuclear divisions	−	−
Constant number of daughter cells	+	−
Sharp bursts of daughter cell release (other terms used are: sporulation, hatching, actual liberation, separation, multiple fission)	−	+
High productivity	−	+

Actually, we have to consider another important fact. If we want to gain information about the cell cycle we need to know at what part of the cycle a particular process is going on at its maximum rate, and how long it lasts.   All activities of the cell growing from a just-produced daughter cell to a ripe mother cell ready for separation are based on series of enzyme-regulated synthetic processes. When all these numerous different steps of cell activity are perfectly coordinated by optimal external conditions, the cell will grow in its shortest possible generation time, which is genetically fixed for every strain and species (for the principle of parallel pathways, see Mazia, 1961). In other words, in cells operating optimally, the time from cell division to cell division is the shortest possible one, demonstrating an inherited time factor which controls the correct temporal and spatial sequence of all events during the cell cycle (see also Rao and Engelberg, 1968). A change in the external conditions from this optimum may vary the intensity of each process to a different degree. For example, the substance production (especially that of carbohydrates) in *Chlorella pyrenoidosa* has an optimum at a temperature of 35°C (or even higher), which already blocks the cell separation (Lorenzen, 1963). At least in *Chorella pyrenoidosa* it is evident that some intervals of external conditions result in growth in the shortest possible generation time. For instance, the temperature can differ between 25° and 32°C without lengthening the cycles. The environment must not fall outside a certain critical

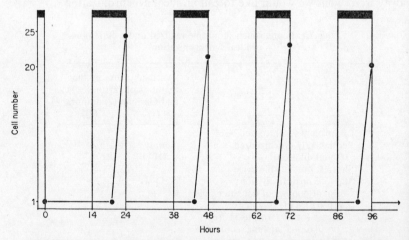

Figure 7.1. The course of cell number in a completely synchronized culture of *Chlorella pyrenoidosa*, 211–8b (Göttingen), with periodic dilution every 24 hours. It is shown that in average each cell separates in more than twenty offsprings within one cycle of 14 hours of light and 10 hours of dark. Actually, the separation is finished after 23 hours. Cell number $1 = 1 \cdot 5 \times 10^6$ cells per ml. Temperature $= 32°$C. Light intensity $= 18\,000$ lux (fluorescent tubes). Aeration with $1 \cdot 2$ per cent carbon dioxide in air. Inorganic medium, only some Fe-EDTA

range if growth and release of daughter cells are to take place in the shortest possible generation time. Generally speaking, there exists no problem to increase the length of the cycle. To measure the success of synchronizing treatments, the ratio of total cycle length to the duration of increase of cell number is often regarded as a valid parameter. One goal is that the time of real increase of cell number (the burst) should be short. That means, all cells should divide more or less simultaneously. However to record this criterion alone may be misleading and may induce great errors (Lorenzen and Schleif, 1966). To summarize, one can state the following favourable characteristics of synchronizing procedures regarding photosynthetic unicellular algae. The cells should be treated so that all are in phase (complete synchronization), all should grow in their shortest possible generation time, and all should run one cycle after the other with the same result (permanent synchronization).

Since the generation time of every cell is a genetically determined factor, it is reasonable to expect the same generation time in every cell of the clonal population in a culture. However, the external conditions also play a great role in enabling a cell to go over the cycle, and this may be the reason for the greater variability and diversity observed in fully developed cells as compared to cells at the beginning of the cycle (Komárek and Simmer, 1965). In other words, the so-called constant external conditions in synchronized cultures are actually not as constant as one likes to believe.

In Table 7.2 are given some of the methods suitable for permanent synchronization of photosynthetic microalgae; not all of them have been

Table 7.2. Possibilities to grow photosynthetic unicellular organisms in a permanently synchronized culture

Inorganic nutrient medium	Organic nutrient medium
(1) Serial dilution and light–dark change (Lorenzen, 1957)	(5) Continuous light, changes of temperature and serial dilution (Padilla and Bragg, 1968)
(2) Continuous dilution and light–dark change (Bongers, 1958)	(6) Relative short times of light only and serial dilution (Senger, 1965)
(3) Continuous light and serial dilution (this method only allows the maintainance of synchrony in already synchronized cultures for some cycles) (Baker and Schmidt, 1964)	
(4) Continuous light, changes of temperature and serial dilution (Lorenzen and Venkataraman, 1969)	

extensively attempted so far. The easiest way to start a synchronized culture is to darken a sample of a mass culture for the length of about one cell cycle. During this time all division-induced cells will divide and the onset of illumination will set the phase for the growth cycle of each cell (von Denffer, 1950; Lewin and coworkers, 1966; Pirson and Lorenzen, 1958b). Due to an intracellular endogenous time measurement, which is extinguished during such dark treatment (Pirson and Lorenzen, 1958a), even more or less grown-up cells (not division-induced!) have to start a new cycle at the beginning of illumination: they release daughter cells together with the cells starting in the stage of autospore. The number of offsprings in these cases may be higher and the size of them may be bigger.

Under certain conditions the beginning of the dark phase determines the start of cell release (Paasche, 1967; Pirson, Lorenzen and Ruppel, 1963; Soeder, Schulze and Thiele, 1967). Some attempts have been made to synchronize photosynthetic microorganisms by selecting cells of equal size. The Japanese school used to fractionate the starting population by centrifugation. Spektorov and Linkova (1962) allowed the larger cells to settle down for some hours or even a day in a non-aerated culture tube, and from the supernatant portion of the culture (but not from the surface layer!) they selected smaller cells for starting synchronous cultures. The relatively easy handling of repetitive synchronous cycles is the reason why most workers in the field do not use presynchronizing procedures (see also Fogg, 1965).

Besides the completely synchronized cultures with repetitive cycles, there are two other types which should be mentioned briefly; partially synchronized cultures (population consisting of cells with normal or double length of generation time; the cells may change from one to the other) and synchronization in groups (population consisting only of cells with double length of generation time; every burst includes one alternating group of cells). These types had already been observed in *Chlorella* by Lorenzen, 1957 (see also Pirson and Senger, 1961; Senger, 1961). There is some discussion connected with the number of daughter cells actually produced by a mother cell. The Japanese workers observed four daughter cells from more than 95 per cent of the mother cells under their experimental conditions. A regular variation from four to eight macrozoospores was reported by Göhde and Berger (1966) in *Haematococcus pluvialis*. In completely synchronized cultures where productivity is not suppressed the number of autospores formed was found to be variable. No doubt we should find conditions which allow all cells to produce exactly the same number of daughter cells in each succeeding cycle (see Table 7.1).

Continuous dilution during the light phase has been suggested to keep the supply of light and nutrients constant (Senger and Wolf, 1964). However, one must keep the fact in mind that cell size and pigment content, which change greatly during development, will influence light absorption by individual cells, even during continuous dilution. The term 'steadystate' therefore cannot be used in describing the situation of individual cells (Sorokin and Krauss, 1961). It seems to the author that due to the serial dilution after each light–dark cycle the culture is indeed reflecting the situation in the cell itself. There is strong evidence (Sorokin, 1960) that the younger cells have a higher saturation light intensity for photosynthesis than the fully developed ones. The light intensity tolerated or even needed for high productivity by daughter cells of

Table 7.3. Some species and strains already used for permanently synchronized cultures (serial dilution)

Species	Author
Anacystis nidulans (1402 + 1, Göttingen)	Lorenzen and Verkataraman (1969)
Ankistrodesmus braunii (202–70, Göttingen)	Kaden (1965)
Chlamydomonas moewusii	Bernstein (1964), Kates and Jones (1964)
Chlamydomonas reinhardii (11–32, Göttingen)	Schlösser (1966)
Chlorella ellipsoidea	Tamiya and coworkers (1961)
Chlorella pyrenoidosa (211–8b, Göttingen)	Lorenzen (1957)
Chlorella pyrenoidosa (high-temperature strain 7–11–05)	Sorokin and Myers (1957) (continuous dilution)
Chlorella pyrenoidosa (Van Niel 2.2.1)	Correll and Tolbert (1962)
Coccolithus huxleyi	Paasche (1967)
Cyclotella cryptica (Göttingen)	Werner (1966)
Ditylum brightwellii	Paasche (1968)
Euglena gracilis, var. *bacillaris*	Cook (1961a)
Euglena gracilis (strain Z)	Edmunds (1965a)
Haematococcus pluvialis	Göhde and Berger (1966)
Nitzschia turgidula	Paasche (1968)
Porphyridium sp. (637, Indiana)	Gense and coworkers (1969)
Scenedesmus sp.	Bongers (1958) (continuous dilution)
Scenedesmus obliquus (276/3a, Göttingen)	Meffert (1963)
Scenedesmus obliquus D$_3$	Senger and Bishop (1967)
Scenedesmus quadricauda (Göttingen)	Kafka (1964)

Chlorella pyrenoidosa is harmful to the grown-up cells of the same strain (Pirson and Ruppel, 1962). Schmidt and coworkers (see Schmidt, 1966) used intermittent illumination to obtain a high degree of synchrony, and then switched to continuous light which was combined further on with serial dilution. Naturally, the sharpness of synchrony was reduced as well as the productivity (Schmidt, 1966). The interval between dilutions also altered (14, 12 and 11 hours respectively), and the number of hours without increase in cell number was shortened from one cycle to the other (10, 5 and 4 hours). The advantage of this method is the continuous illumination. The optical density of the culture, of course, changed considerably. In Table 7.3 are listed the species which have so far been completely synchronized using serial dilution.

Information regarding the synchronous cell division in blue-green algae is comparatively meagre. A partial synchrony approximating Tamiya's 'light' and 'dark' cells of *Chlorella* was obtained with a blue-green alga *Chlorogloea fritschii* by Fay and Fogg (1962). Recently Lorenzen and Venkataraman (1969) reported a temperature-induced complete synchronization of cell division in *Anacystis nidulans* in continuous light (15 000 lux) by a 14-hour cycle, consisting of 8 hours at 26°C and 6 hours at 32°C, coupled with periodic dilution of the algal suspension to a constant cell number at the end of each cycle. The low temperature postponed the cell division, but did not retard the nucleic acid synthesis; the doubling occurred within 4 hours of the high temperature phase. There was a greater synthesis and accumulation of carbohydrates during the low temperature phase, while the bulk of the protein and chlorophyll was synthesized during the high temperature phase (Venkataraman and Lorenzen, 1969). Evidence for the accumulation of the photosynthetic products within the cell at 26°C has also been obtained at a submicroscopic level (Verkataraman, Amelunxen and Lorenzen, 1969).

Behaviour of cell organelles during the growth cycle and changes of synthetic capacity

During the course of every cell cycle some kind of ageing at the cellular level can be observed (Sorokin, 1964). This is due to the changes in metabolic turnover and synthetic activity. But in permanently synchronized cultures generally no ageing is observed for several weeks or even months (see, however, regarding the diatom *Ditylum brightwellii*, Eppley, Holmes and Paasche, 1967). Mostly the grown-up cells show a lower rate of photosynthesis and a reduced synthesis of essential substances, accompanied sometimes by a higher rate of respiration. This interferes with the behaviour of the cell organelles and their activities to a rather high degree. A rather weak diffuse Feulgen reaction was observed by Lorenzen (1958) and Bernstein (1964). This occurred immediately before the commencement of DNA synthesis. Tamiya and coworkers (1961), however, were unable to observe the same phenomenon with their strain, which may be due to the slower overall growth rate and also possibly to a less distinct accuracy of synchrony. Till now there is no method which can synchronize nuclear divisions of autospore-producing organisms which results in a stepwise increase of nuclear numbers (see also Table 7.1).

A few papers have been selected on the general syntheses of cell components during the cell cycle examined in *Chlorella* (Ruppel, 1962; Spektorov, Slobodskaya and Nichiporovich, 1963), *Chlamydomonas* (Bernstein, 1964) and *Euglena* (Cook, 1961a, 1961b; Edmunds, 1965a, 1965b). Information regarding cytoplasmic division, chloroplasts, pyrenoids, dictyosomes and Golgi complex of *Chlorella* at subcellular levels has been provided by Murakami, Morimura and Tahamiya (1963), Soeder (1964), Bisalputra, Ashton and Weier (1966) and Wanka (1968a). Guérin-Dumartrait (1968) examined the lamellar structures of chloroplasts during the cell development in *Chlorella pyrenoidosa*. Fischer (1969) successfully tried to describe intracellular morphometric changes in *Chlamydomonas reinhardii* synchronized according to Schlösser (1966).

Courses of photosynthesis and respiration

The mechanism of synchrony is not directly influenced by photosynthesis (Senger and Bishop, 1969). Many papers have dealt with changes in photosynthetic rates under saturation light intensity during the lifecycle. Some of the earliest articles were written by Tamiya and coworkers (1953), Sorokin (1957), Bongers (1958) and Lorenzen (1959). Generally, in organisms like *Chorella* the photosynthetic capacity of growing cells increases, reaching a maximum after a few hours, followed by a decrease in older cells even under continuous dilution. However, in *Euglena gracilis* (Cook, 1966b) the photosynthetic capacity increased linearly on a volume basis during 21 hours in a 14 : 10-cycle without dilution. A reference to a reliable index of photosynthetic activity is, however, extremely difficult. Some variations in metabolic activity, particularly those referred to cell volume, may in part be attributed to changes in water content at different stages of cell development (Schmidbauer and Ried, 1967; Yuhara and Hase, 1961). Otherwise, the larger cells have lower photosynthetic capacity than the smaller ones, as both groups were selected from the same batch culture (Sorokin, 1963). Antiparallel changes of photosynthesis and of stationary fluorescence intensity in *Chlorella* cells were described by Döhler (1963). The quantum yield of photosynthesis in synchronous cultures of *Scenedesmus* decreased, and seemed to be associated with the light reaction II (Senger and Bishop, 1967). Similar results were obtained with *Chlorella pyrenoidosa* by Govindjee, Rabinowitch and Govindjee (1968), the lowest requirement of quanta being about 10.

Respiratory changes have been examined by Sorokin and Myers (1957), Ried, Soeder and Müller (1963) and Wolf (1964). The Simonis group at Wuerzburg studied the photosynthetic phosphorylation (cyclic and non-cyclic) *in vivo* as influenced by many inhibitors, using synchronized cultures of *Ankistrodesmus braunii* and *Chlorella pyrenoidosa* (Simonis, 1964; Urbach, 1966). Gimmler and coworkers (1969) of the same group observed a maximum excretion of glycolate by *Ankistrodesmus* 5 hours after the beginning of illumination and the lowest value in the dark during the release of autospores.

Changes in nucleotides and nucleic acids, amino acids and proteins, and carbohydrates

Iwamura, Kanazawa and Kanazawa (1963), and Oh-Hama, Morimura and Tamiya (1965) gave detailed information regarding the metabolism in *Chlorella* and observed a four-fold increase of nucleotides during one cycle. Cook and Hess (1964) analysed the content of nucleotides in *Euglena* The role of DNA synthesis is often discussed as a possible trigger leading to cell division. Several experiments suggest that there is no timing of cell division by DNA formation, but the synthesis of RNA controls, via amino acids, protein and enzymes, the timing mechanism of the life cycle (Senger and Bishop, 1969). As early as 1955 Iwamura (1955) investigated the role of light on DNA metabolism in synchronized *Chlorella*. Similar investigations were continued by Ruppel (1962), Iwamura and Kuwashima (1964), Senger and Bishop (1966) and Wanka and Mulders (1967). Shen and Schmidt (1966) checked the enzymic control of nucleic acid synthesis of *Chlorella*, and more recently Schönherr (1969) gave a detailed picture of DNA-polymerase and deoxyribonuclease activities. Edmunds (1964) found the DNA synthesis restricted in *Euglena* to the last 5 hours of the light time using a 14 : 10-cycle. Kates, Chiang and Jones (1968) studied the DNA replication during synchronized vegetative growth and gametic differentiation in *Chlamydomonas*. Fetter and Altmann (1968) looked for incorporation of thymine and thymidine during DNA synthesis in *Chlorella* cells and showed that the former was incorporated ten times more rapidly than the latter. Different types of RNA were the main topics of Correll (1965a, 1965b), Galling (1965) and Enöckel (1968). Beiderbeck and Richter (1968) examined the nature of DNA-associated RNA, and the DNA content of plastids was investigated by Cook (1966a) and by Chiang and Sueoka (1967). Surprisingly high values were reported by Iwamura (1966).

Regarding the changes in amino acid content during the life-cycle of *Chlorella*, alanine, glutamic acid, glycine and leucine were found to be the most variable (Kanazawa, 1964). The amino acids were synthesized in the light phase, while protein synthesis itself was light-independent. Hare and Schmidt (1965) found a constant value of amino acid-N which was about 60 per cent of the total cellular-N throughout the synchronous growth of the high temperature strain of *Chlorella* 7–11–05. Kates and Jones (1966) found that the rate of $^{14}CO_2$-incorporation into the amino acid fraction in *Chlamydomonas* was influenced by the presence of NH_4^+ in the culture medium. Kanazawa, Yanagisawa and Tamiya (1966) isolated ethanolamine, putrescine and spermidine (aliphatic amines), which showed different behaviour during the development of *Chlorella* cells; they also reported changes in the composition patterns of basic proteins (see also Kanazawa and Kanazawa, 1968).

Since the observation of Lorenzen and Ruppel (1960) on an increased proportion of proteins in *Chlorella* during the dark phase of the cycle, many authors have confirmed this, and have explained the general activation of metabolic capacity after sufficient darkening of the cells (Lorenzen and Hesse, 1968; Soeder, Schulze and Thiele, 1966). Jones, Kates and Keller (1968) found that protein synthesis in *Chlamydomonas* occurred only in the light. The activity of *Chlorella* ribosomes in cell-free fractions was tested by Erben (1967).

Duynstee and Schmidt (1967) determined the total starch and amylose levels during two succeeding cycles in continuous light. Both substances showed a steady increase, the amylose content always being 30 per cent of the total starch level. The maximum content of oligosaccharides in *Chlorella* existed just before the release of autospores (Dedio, 1968).

According to Hülsen and Prenzel (1966), there was a better uptake of glucose into autospores as compared with mannose and fructose. Galactose, ribose and arabinose were not incorporated. Lysek and Simonis (1968) correlated the role of polyphosphates to the uptake of glucose in *Ankistrodesmus braunii* in the light and in the dark. While the growth rate of the high temperature strain of *Chlorella pyrenoidosa* with glucose in the dark was similar to that in light, the increase in cell number was only about 50 per cent of that in light (Montalvo, Carroll and Cole, 1968). Lorenzen

(1960) showed that addition of glucose (0·5 per cent) at the beginning of the light–dark cycle increased the substance production and the number of daughter cells, but the start of the release of autospores was not influenced. Problems of mixotrophic conditions were discussed extensively by Senger (1965).

Changes in enzyme activity, vitamins and pigments

Our knowledge about the changes in enzyme activity, vitamin and pigment contents during different phases of the cycle is still meagre, and it is hoped that before long a clearer understanding of these constituents will be forthcoming. Table 7.4 summarizes some available information on enzymic activities. In addition, the recent paper by Cole, Blondin and Temple (1968) establishing the occurrence of isoenzymes in *Chlorella* should also be mentioned. Morimura (1959) observed differences in the amount of vitamins in *Chlorella* during the cycle after a pretreatment of 3 days in the dark followed by centrifugation. The B_6-complex, pantothenic acid, folic acid, thiamine and riboflavin on a dry weight basis decreased during growth and increased at the final stage of ripening. P-aminobenzoic acid increased in the middle of the cycle. No significant changes were found in the case of niacin, biotin, inositol and choline. The most active stages of photosynthesis had the highest content of ascorbic acid. Under conditions of permanent synchronization, Gerhardt (1964) demonstrated a fairly good agreement between the internal content of ascorbic acid and of photosynthetic capacity, but he claimed that no direct relationship existed. While about 40 per cent of the total ascorbic acid was localized in the spinach chloroplasts, we have no information about its intracellular distribution in the algal cell. As to the pigment content, a decrease of chlorophyll concentration on a dry weight basis during the course of the life cycle of *Chlorella* was reported by Sorokin and Krauss (1962) and by Ruppel (1962). Only a relatively small amount of pigment is synthesized in the dark phase. The same was also true in the case of *Scenedesmus* (Lafeber and Steenbergen, 1967). With *Coccolithus*, Paasche (1967) observed a steady increase of chlorophyll in the light and in the dark, but with *Euglena* there is absolutely no synthesis of pigments (including carotenoids) in the dark (Cook, 1966b). In the synchronous cycle of *Mesotaenium*, about 50 per cent of the chlorophylls are synthesized in the dark (10 : 14 hours: Lorenzen, unpublished). In addition, Metzner, Rau and Senger (1965) failed to synchronize chlorophyll-less ultraviolet mutants 31 and 41 of *Chlorella*, and established partial synchrony with only the yellow mutants 10 and 11.

Enzyme	Species	Author	Maximum of activity	Remarks
Aldolase	*Chlorella pyrenoidosa* (211–8b, Göttingen)	Berger (1966)	L	Specific activities
NADP-dependent glyceraldehyde-3-phosphate-dehydrogenase			L	$(14 : \overline{10})$
Glutamic dehydrogenase			D	
Glucose-6-phosphate dehydrogenase			D	
Malate dehydrogenase			no	
Glutamic-oxaloacetic transaminase			no	
Nitrite reductase		Knutsen (1965)	L	Maximum rates of nitrite reduction $(14 : \overline{10})$
Acid and alkaline phosphatase	*Chlorella pyrenoidosa* (7–11–05)	Knutsen (1968)	L	$(14 : \overline{10})$
Aspartate transcarbamylase		Cole and Schmidt (1964)	Always increasing	Units per cell (continuous light)
Deoxythymidine monophosphate kinase		Johnson and Schmidt (1966)	Before release of daughter cells	Apparent units per cellular P (continuous light)
Deoxycytidine monophosphate deaminase		Shen and Schmidt (1966)	Increasing (stop during release)	Per cell (continuous light) $(12 : \overline{12})$
'Hatching' enzyme	*Chlamydomonas reinhardii*	Schlösser (1966)	D	?
Alanine dehydrogenase	*Chlamydomonas reinhardii* and *Chlamydomonas moewusii*	Kates and Jones (1964)	L	
Glutamate dehydrogenase			Inverse to alanine dehydrogenase	Specific activity
Alanine dehydrogenase	*Chlamydomonas reinhardii*	Kates and Jones (1967)	Increasing	Units of enzyme per volume $(12 : \overline{12})$
Glutamic dehydrogenase			D	
Phosphoenolpyruvate carboxylase			D	
Aspartate transcarbamylase			D	
Ornithine transcarbamylase			D, increase during light	

Some influences of macronutrients, micronutrients and other factors

The qualitative nature of the nitrogen source may have an important bearing on the length of the growth cycle. Due to a report of Tomova, Tinkova and Spektorov (1964) the shortest length resulted if the cells used ammonium nitrogen. On the other hand, there are strong arguments to support the assumption that synchronization by sedimentation causes a lag phase so that the shortest possible life cycle is not realized by the cells; only in this case does it seem possible that the nitrogen source influences the generation time. Meffert (1963) showed that in completely synchronized cultures of *Scenedesmus obliquus* (12 : 12 hours) the cold TCA soluble N-compounds primarily reflect the light–dark changes, but not the stages of development as was the case with the fraction soluble in hot TCA (nucleic acids) and insoluble in TCA (bulk protein). During the dark phase the nitrogen of the protein fraction increases, as had been shown by Lorenzen and Ruppel (1960) and by others later on. Schmidt (1961) obtained evidence for differences in the content of soluble nitrogen in daughter cells released in the light or in the dark. NH^+-nitrogen is preferentially absorbed compared with nitrate-nitrogen, and the cells grown with ammonia also fixed a larger quantity of phosphate. For phosphate metabolism (including polyphosphates) the reader may refer to Miyachi and Miyachi (1961), Correll and Tolbert (1962), Hase, Miyachi and Mihara (1963), Correll (1965b) and Herrmann and Schmidt (1965). Johnson and Schmidt (1963) observed a close relationship between shifts in sulphur metabolism and the induction of nuclear divisions. Hase (1962) summarized the earlier results of Japanese workers regarding the specific role of sulphur compounds in cell division with reference to light and darkness. The influences of minor elements are also important, particularly because protein synthesis is blocked by deficiency in Mg^{2+} (Galling, 1963). Mn^{2+} and Cu^{2+} were assimilated more in the ripening stage (Wey, 1964). Soeder and Thiele (1967) claimed a ten times greater need for Ca^{2+} for the liberation of daughter cells than for the process of substance production.

The release of daughter cells has been found to be delayed with increased concentration of carbon dioxide in the bubbling air during the dark phase of 16 : 8 hours of treatment (Soeder, Strotmann and Galloway, 1966; Sorokin, 1962). With nitrate as the nitrogen source, 1 per cent carbon dioxide seems to be sufficient for an optimal growth rate.

Many metabolic inhibitors were tested for their interferences with the development of the population of synchronous unicellular algae. Tamiya, Morimura and Yokota (1962) checked about fifty substances, but as the control experiment showed great variability, it is difficult to compare the different sets of experiments. A specific action of 10^{-2} M chloramphenicol inhibited the growth of the cells, although the cells divided in two small daughter cells. Guérin-Dumartrait (1966) added 3-amino-1,2,4-triazol (10^{-3} M) which inhibited the synthesis of nucleic acids, especially that of RNA, and finally blocked the cell division. This effect is not reversible by increasing the content of minor elements in the medium. No inhibition of RNA or protein syntheses was caused by 0·5 per cent colchicine (Wanka, 1965), irrespective of the stage at which it was supplied. However, the synthesis of DNA was retarded. Nevertheless, in a recent investigation the author apparently obtained polyploids of *Chlorella*, which normally is assumed to be a diploid (Wanka, 1968b).

The role of light

The specific role of light on the growth of synchronized photosynthetic unicellular algae is of particular interest. It is true that the effect of light is also dependent on the other external factors. Continuous light is to some extent harmful in *Chlorella* (Lorenzen, 1962; Pirson and Ruppel, 1962; Sorokin and Krauss, 1959), *Cyclotella* (Werner, 1966) and *Ditylum* (Paasche, 1968). The growth per hour of continuous illumination is lower per hour of light than during a light–dark change (Gorjunowa and co-workers, 1962; Pirson, Lorenzen and Ruppel, 1963; Sorokin and Krauss, 1961). Different effects of blue and red light on cell development were reported by Pirson and Kowallik (1960). In cultures with the same growth rate the actual increase in cell number is delayed by about 3 hours in blue light as compared with red light. Blue light increases the protein synthesis (Kowallik, 1962; Kowallik, Chapter 6, in this treatise), while cells grown in red light always have a higher content of total carbohydrates. The experiments of Senger and Schoser (1966) showed a maximum induction of cell division by light of 480 and 674 nm (both of 5000 erg/cm² second). Using autospores of *Chlorella pyrenoidosa* Strotmann (1967) showed a stimulated nitrite reduction in blue light which was independent of photosynthesis. Although the effects of blue light are clearly demonstrated, its precise role is not clear. Soeder, Schulze and Thiele (1966) lowered the productivity of the cells to about 50 per cent by giving light during the dark phase of only 20 to 50 lux (or less!). This low light intensity effect was independently confirmed by Lorenzen and Hesse (1968), who in addition demonstrated an endogenously ruled change in productivity of a subsequent following cycle if the synchronous autospores were left in the dark for more than the normal 10 hours. Results of the effect of ultraviolet radiation upon various physiological processes were published by Sasa (1961) for *Chlorella*, and by Cook and Hunt (1965) for *Euglena*.

Sensitive stages during the cell cycle

The qualitative and quantitative differences in the metabolic processes during the life cycle are responsible for differential sensitivity to external factors. We have some information on the effects of shocks on certain cells of a specific age, but know little about their physiological or biochemical basis. Effects of cold shocks (2 hours at 4·5°C; lower temperatures were not as harmful as 4·5° C) were reported by Lorenzen and Ruppel (1958), by Pirson, Lorenzen and Köpper (1959) and by Lorenzen (1963). The cells are most sensitive at 7 hours after dilution, and become bleached under normal conditions of growth later on. The most sensitive stage regarding heat shocks (15 minutes at 46°C) occurred as early as from 4 to 5 hours of growth (just at the beginning of effective DNA synthesis; Lorenzen, 1963). The most spectacular effect of high temperature treatment was a bleaching, but the damaging effect was reduced if the shock was given at the normal light level (15 000–20 000 lux). On the contrary, the cold shock had a pronounced consequence only when it was given at an illumination of at least 4 000 lux. Attention was also given to the response of *Chlorella* at temperatures outside its limit of normal growth (Lorenzen, 1963). Using a 16 : 8 light–dark change Ried (1962) found that the cells after 4 hours in the dark (that is about the final point of protoplast division) were killed by an increase of the total molarity of the medium from 0·016 to 0·44. But such a lethal result was not observed if the molarity of the medium was increased outside this time limit, although a delay in the development could be seen. Petroupulos (1964) inactivated the chloroplast formation in *Euglena gracilis* by ultraviolet radiation, especially at a stage just prior to cell division. A two-fold difference in the dose–response relationship of stages with low or high sensitivity resulted (see also Cook, 1966b). Davies (1965) showed different ultraviolet sensitivity of synchronized zygotes of *Chlamydomonas reinhardii*. The older ones were most sensitive. Sensitive stages effected by radiation or drugs in *Chlamydomonas* and in *Chlorella* were reported by Jacobson and Lee (1967), who concluded that the initiation of division is dependent on RNA synthesis (see also Senger and Bishop, 1966). Altmann and coworkers (1968) found *Chlorella* cells most sensitive against γ-irradiation just before onset of DNA synthesis, which was not markedly influenced in spite of blocking the cell division. The latter was blocked by some metabolic inhibitors too (Moberg, Knutsen and Goksøyr, 1968); regarding *Euglena*, see Nissen and Eldjarn (1969).

Concluding remarks

Several differences exist between the organisms producing autospores (more than two daughter cells) and those undergoing simple cell division. The burst of autospore release is rather sharp (about 10 per cent of the cell cycle), especially in *Chlamydomonas*. The duration of simple cell division in permanently synchronized cultures of *Euglena* and diatoms, for instance, is as long as about 50 per cent of the cycle length (Cook and James, 1960; Werner, 1966). A somewhat shorter time was observed by Paasche (1967) for *Coccolithus huxleyi*.

The serial dilution technique (also called Intervallverdünnung) as described by Lorenzen (1957) ensures successful and permanent synchronization; today it is also the most commonly applied method. The dilution with fresh culture medium after each cycle brings the culture back to optimum conditions of growth without including any lag phase (see also Bernstein, 1964; and compare Glooschenko and Curl, 1968).

No cell can grow without being influenced by a number of different external factors, and any artificial treatment is likely to influence the course of cellular activity during the generation cycle of an organism. However, genetic factors, such as the shortest possible generation time, are a very good absolute measure for standardized cells. The repetitive cycles with uniform cell material evidently show that no stress is included.

It is observed that the conditions of the cells in synchronous culture can be in good agreement with their natural growth rhythm (Gorjunova and coworkers, 1962; Overbeck, 1961). In addition, the view that cells in permanently synchronized cultures are living in one of the best worlds they can live in is supported by the simple handling which is applied to synchronize photosynthetic unicellular algae (Pirson and Lorenzen, 1958b). It is, however, not always necessary to use synchronous cell populations. Some experiments, especially those which do not allow the cells to grow over the full cycle, can be done satisfactorily with a starting population of standardized cells or with batch cultures. Although this chapter is too incomplete to make broad generalizations, it nevertheless serves to illuminate the importance of synchronous cultures by magnifying the activities from the cellular level.

The author is thankful to the Shiftung Volkswagenwerk for financial support and to Drs. A. Kuhl and G. S. Venkataraman for reading the manuscript. The technical assistance of Miss Gerda Wirth is gratefully acknowledged.

REFERENCES

Altmann, H., F. Fetter, E. Pfisterer, A. v. Szilvinyi und K. Kaindl (1968). Untersuchungen über die Strahlenwirkung auf die DNS-Synthese in synchronen Chlorellazellen. *Monatshefte für Chemie*, **99**, 1145.

Baker, A. L., and R. R. Schmidt (1964). Polyphosphate metabolism during nuclear division in synchronously growing *Chlorella*. *Biochim. Biophys. Acta*, **82**, 624.

Beiderbeck, R., and G. Richter (1968). On the nature of DNA-associated RNA in *Chlorella* cells. *Biochim. Biophys. Acta*, **157**, 658.

Berger, Ch. (1966). Aktivitätsänderungen einiger Enzyme synchronisierter *Chlorella*-Zellen im Licht-Dunkel-Wechsel. I. Veränderungen während des Entwicklungszyklus. *Flora, Abt. A*, **157**, 211.

Bernstein, E. (1964). Physiology of an obligate photoautotroph (*Chlamydomonas moewusii*). I. Characteristics of synchronously and randomly reproducing cells and an hypothesis to explain their population curves. *J. Protozool.*, **11**, 56.

Bisalputra, T., F. M. Ashton, and T. E. Weier (1966). Role of dictyosomes in wall formation during cell division of *Chlorella vulgaris*. *Amer. J. Bot.*, **53**, 213.

Bongers, L. H. J. (1958). Changes in photosynthetic activity during algal growth and multiplication. *Medel. Landbouwhogesch. Wageningen*, **58**, 1.

Chiang, K. S., and N. Sueoka (1967). Replication of Chloroplast DNA in *Chlamydomonas reinhardii* during vegetative cell cycle: its mode and regulation. *Proc. Natl. Acad. Sci. (Wash.)*, **57**, 1506.

Cole, F. E., J. Blondin, and L. Temple (1968). Enzyme patterns during a synchronous growth cycle of *Chlorella pyrenoidosa*. *Cell Tissue Kinet.*, **1**, 281.

Cole, F., and R. R. Schmidt (1964). Control of aspartate transcarbamylase activity during synchronous growth of *Chlorella pyrenoidosa*. *Biochim. Biophys. Acta*, **90**, 616.

Cook, J. R. (1961a). *Euglena gracilis* in synchronous division. I. Dry mass and volume characteristics. *Plant and Cell Physiol.*, **2**, 199.

Cook, J. R. (1961b). *Euglena gracilis* in synchronous division. II. Biosynthetic rates over the life cycle. *Biol. Bull.*, **121**, 277.

Cook, J. R. (1966a). The synthesis of cytoplasmatic DNA in synchronized *Euglena*. *J. Cell Biol.*, **29**, 369.

Cook, J. R. (1966b). Studies on chloroplast replication in synchronized *Euglena*. In I. L. Cameron and G. M. Padilla (Eds.), *Cell Synchrony*, Academic Press, London, New York. pp. 153–168.

Cook, J. R., and M. Hess (1964). Sulfur-containing nucleotides associated with cell division in synchronized *Euglena* cells. *Biochim. Biophys. Acta*, **80**, 148.

Cook, J. R., and W. Hunt (1965). Ultraviolet bleaching of synchronized *Euglena*. *Photochem. Photobiol.*, **4**, 877.

Cook, J. R., and T. W. James (1960). Light-induced division synchrony in *Euglena gracilis* var. *bacillaris*. *Expt. Cell Res.*, **21**, 583.

Correll, D. L. (1965a). Alkali-stable RNA fragments from *Chlorella*. *Phytochem.*, **4**, 453.

Correll, D. L. (1965b). Ribonucleic acid-polyphosphate from algae. III. Hydrolysis studies. *Plant and Cell Physiol.*, **6**, 661.

Correll, D. L., and N. E. Tolbert (1962). Ribonucleic acid-polyphosphate from algae. I. Isolation and physiology. *Plant Physiol.*, **37**, 627.

Davies, D. R. (1965). Repair mechanisms and variations in UV sensitivity within the cell cycle. *Mutation Res.*, **2**, 477.

Dedio, H. (1968). Entwicklungsabhängiger Anstau von Oligosacchariden bei *Chlorella* fusca Shihira et Krauss. *Ber. Dt. Bot. Ges.*, **81**, 359.

von Denfer, D. (1950). Die planktische Massenkultur pennater Grunddiatomeen. *Arch. Mikrobiol.*, **14**, 159.

Döhler, G. (1963). Zur Photosynthese synchron kultivierter *Chlorella pyrenoidosa*, *Planta*, **60**, 158.

Duynstee, E., and R. R. Schmidt (1967). Total starch and amylose levels during synchronous growth of *Chlorella pyrenoidosa*. *Arch. Biochem. Biophys.*, **119**, 382.

Edmunds, L. N. (1964). Replication of DNA and cell division in synchronously dividing cultures of *Euglena gracilis*. *Science*, **145**, 266.

Edmunds, L. N. (1965a). Studies on synchronously dividing cultures of *Euglena gracilis* Klebs (strain Z.). I. Attainment and characterization of rhythmic cell division. *J. Cell Comp. Physiol.*, **66**, 147.

Edmunds, L. N. (1965b). Studies on synchronously dividing cultures of *Euglena gracilis* Klebs (strain Z.). II. Patterns of biosynthesis during the cell cycle. *J. Cell Comp. Physiol.*, **66**, 149.

Enöckel, F. (1968). Änderung des tRNS- und rRNS-Gehalts von *Chlorella pyrenoidosa* während der Entwicklung. *Z. Pflanzenphysiol.*, **58**, 240.

Eppley, R. W., R. W. Holmes, and E. Paasche (1967). Periodicity in cell division and physiological behaviour of *Ditylum brightwellii*, a marine planktonic diatom during growth in light–dark cycles. *Arch. Mikrobiol.*, **56**, 305.

Erben, K. (1967). Zur quantitativen Erfassung der Proteinsynthese-Kapazität in zellfreien Fraktionen von *Chlorella*. *Z. Pflanzenphysiol.*, **57**, 329.

Fay, P., and G. E. Fogg (1962). Studies on nitrogen fixation by blue-green algae. III. Growth and nitrogen fixation in *Chlorogloea fritschii* Mitra. *Arch. Mikrobiol.*, **42**, 310.

Fetter, F., and H. Altmann (1968). Utilization of thymine and thymidine as DNA precursors in *Chlorella pyrenoidosa*. *Monatsh. Chem.*, **99**, 1076.

Fischer, B. (1969). Morphometrische Bestimmung der Zellkonstituenten von *Chlamydomonas reinhardii* Dangeard während eines Entwicklungszyklus. *Dissertation, Göttingen*.

Fogg, G. E. (1965). *Algal Cultures and Phytoplankton Ecology*, Univ. Wisconsin Press, Madison. p. 33.

Galling, G. (1963). Analyse des Magnesium-Mangels bei synchronisierten *Chlorellen*. *Arch. Mikrobiol.*, **46**, 150.

Galling, G. (1965). Uber Ribonucleinsäure in *Chlorella*. I. Die Basenzusammensetzung der RNS synchroner Chlorellen unter verschiedenen Ernährungsbedingungen. *Flora*, **155**, 596.

Gense, M.-T., E. Guérin-Dumartrait, J.-C. Leclerc, et S. Mihara (1969). Synchronisation de *Porphyridium* sp.: Evolution des quantités de pigments et de la capacite photosynthétique au cours du cycle biologique. *Phycologia*, **8**, 135.

Gerhardt, B. (1964). Untersuchungen über Beziehungen zwischen Ascorbinsäure und Photosynthese. *Planta*, **61**, 101.

Gimmler, H., W. Ullrich, J. Domanski-Kaden, and W. Urbach (1969). Excretion of glycolate during synchronous culture of *Ankistrodesmus braunii* in the presence of disalicylidenepropanediamine or hydroxypyridinemethanesulfonate. *Plant and Cell Physiol.*, **10**, 103.

Glooschenko, W. A., and H. Curl, Jr.   (1968). Obtaining synchronous cultures of algae. *Nature*, **218**, 573.

Göhde, W. und M. Berger (1966). Alternierende Vermehrungsraten in synchronisierten Kulturen von Makrozoosporen der einzelligen Grünalge *Haematococcus pluvialis*. *Naturwissenschaften*, **53**, 482.

Gorjunova, S. W., G. R. Rschnova, M. N. Ovsjannikova, W. K. Orleanskij und W. W. Kabanov (1962). Die Rolle von Synchronkulturen bei Untersuchungen der Biologie von *Chlorella* und ihre praktische Anwendung. *Mikrobiologia*, **31**, 1107.

Govindjee, R., E. Rabinowitch, and Govindjee (1968). Maximum quantum yield and action spectrum of photosynthesis and fluorescence in *Chlorella*. *Biochim. Biophys. Acta*, **162**, 539.

Guérin-Dumartrait, E. (1966). Quelques effects du 3-amino-1,2, 4-triazol sur *Chlorella pyrenoidosa* Chick. Recherches sur l'action de l'aminotriazol sur la synthèse des acides nucléiques chez des Chlorelles cultivées en cultures synchrones. *Physiol. Vég.*, **4**, 135.

Guérin-Dumartrait, E. (1968). Étude, en cryodécapage, de la morphologie des surfaces lamellaires chloroplastiques de *Chlorella pyrenoidosa*, en cultures synchrones. *Planta*, **80**, 96.

Hare, T. A., and R. R. Schmidt (1965). Nitrogen metabolism during synchronous growth of *Chlorella pyrenoidosa*. I. Protein aminoacid distribution. *J. Cell Comp. Physiol.*, **65**, 63.

Hase, E. (1962). Cell division. In R. Lewin (Ed.), *Physiology and Biochemistry of Algae*, Academic Press, London, New York. pp. 617–624.

Hase, E. S., Miyachi, and S. Mihara (1963). A preliminary note on the phosphorus compounds in chloroplasts and volutin granules isolated from *Chlorella* cells. In *Studies on Microalgae and Photosynthetic Bacteria*, Japanese Soc. of Plant Physiologists, Tokyo. pp. 619–626.

Herrmann, E. C., and R. R. Schmidt (1965). Synthesis of phosphorus-containing macromolecules during synchronous growth of *Chlorella pyrenoidosa*. *Biochim. Biophys. Acta*, **95**, 63.

Hülsen, W. und U. Prenzel (1966). Uber die Aufnahme verschiedener Zucker durch *Chlorella pyrenoidosa*. *Z. Naturforschung*, **21b**, 500.

Iwamura, T. (1955). Change of nucleic acid content in *Chlorella* cells during the course of their life-cycle. *J. of Biochem.*, **42**, 575.

Iwamura, T. (1966). Nucleic acids in chloroplasts and metabolic DNA. *Progr. Nucleic acid Res. and Molecular Biol.*, **5**, 133.

Iwamura, T., T. Kanazawa, and K. Kanazawa (1963). Nucleotide metabolism in *Chlorella*. II. Quantitative changes in component-nucleotides of the acid soluble fraction of algal cells during their life cycle. In *Studies on Microalgae and Photosynthetic Bacteria*, Japanese Soc. of Plant Physiologists, Tokyo. pp. 587–596.

Iwamura, T., and S. Kuwashima (1964). Further observations on the deoxyribonucleic acid species in *Chlorella* showing light-dependent metabolic turnover. *Biochim. Biophys. Acta*, **82**, 678.

Jacobson, B. S., and T. C. Lee (1967). Macromolecular synthesis and delayed mitotic death due to radiation and drugs in *Chlamydomonas* and *Chlorella*. *Radiation Research*, **31**, 368.

Johnson, R. A., and R. R. Schmidt (1963). Intracellular distribution of sulfur during the synchronous growth of *Chlorella pyrenoidosa*. *Biochim. Biophys. Acta*, **74**, 428.

Johnson, R. A., and R. R. Schmidt (1966). Enzymic control of nucleic acid synthesis during synchronous growth of *Chlorella pyrenoidosa*. I. Deoxythymidine monophosphate kinase. *Biochim. Biophys. Acta*, **129**, 140.

Jones, R. F., J. R. Kates, and St. J. Keller (1968). Protein turnover and macromolecular synthesis in *Chlamydomonas reinhardtii*. *Biochim. Biophys. Acta*, **157**, 589.

Kaden, J. (1965). Der Phosphatstoffwechsel synchronisierter Ankistrodesmuskulturen. *Dissertation, Würzburg*.

Kafka, R. (1964). Zur Wirkung zusätzlichen Thiamins auf Synchronkulturen einzelliger Grünalgen. *Arch. Mikrobiol.*, **49**, 1.

Kanazawa, T. (1964). Changes of amino acid composition of *Chlorella* cells during their life cycle. *Plant and Cell Physiol.*, **5**, 333.

Kanazawa, T., and K. Kanazawa (1968). Changes in composition patterns of basic proteins in *Chlorella* cells during their life cycle. *Plant and Cell Physiol.*, **9**, 701.

Kanazawa, T., T. Yanagisawa, and H. Tamiya (1966). Aliphatic amines occurring in *Chlorella* cells and changes of their contents during the life cycle of the alga. *Z. Pflanzenphysiol.*, **54**, 57.

Kates, J. R., K. S. Chiang, and R. F. Jones (1968). Studies on DNA during synchronized vegetative growth and gametic differentiation in *Chlamydomonas reinhardtii*. *Expt. Cell Res.*, **49**, 121.

Kates, J. R., and R. F. Jones (1964). Variation in alanine dehydrogenase and glutamatedehydrogenase during the synchronous development of *Chlamydomonas*. *Biochim. Biophys. Acta*, **86**, 438.

Kates, J. R., and R. F. Jones (1966). Pattern of CO_2 fixation during vegetative development and gametic differentiation in *Chlamydomonas reinhardtii*. *J. Cell Physiol.*, **67**, 101.

Kates, J. R., and R. F. Jones (1967). Periodic increases in enzyme activity in synchronized cultures of *Chlamydomonas reinhardtii*. *Biochim. Biophys. Acta*, **145**, 153.

Knutsen, G. (1965). Induction of nitrite reductase in synchronized cultures of *Chlorella pyrenoidosa*. *Biochim. Biophys. Acta*, **103**, 495.

Knutsen, G. (1968). Repressed and derepressed synthesis of phosphatases during synchronous growth of *Chlorella pyrenoidosa*. *Biochim. Biophys. Acta*, **161**, 205.

Komárek, J., and J. Simmer (1965). Synchronization of the cultures of *Scenedesmus quadricauda* (TURP.) BREB. *Biol. Plant.*, **7**, 409.

Kowallik, W. (1962). Uber die Wirkung des blauen und roten Spektralbereiches auf die Zusammensetzung und Zellteilung synchronisierter Chlorellen. *Planta*, **58**, 337.

Kuhl, A., and H. Lorenzen (1964). Handling and culturing of *Chlorella*. In D. M. Prescott (Ed.), *Methods in Cell Physiology*, Vol. 1. Academic Press, London, New York. Chap. 10, pp. 159–187.

Lafeber, A., and C. L. M. Steenbergen (1967). Simple device for obtaining synchronous cultures of algae. *Nature*, **213**, 527.

Lewin, J. C., B. E. Reimann, W. F. Busby, and B. E. Volcani (1966). Silica shell formation in synchronously dividing diatoms. In J. L. Cameron and G. M. Padilla (Eds.), _Cell Synchrony_, Academic Press, London, New York. pp. 169–188.

Lorenzen, H. (1957). Synchrone Zellteilungen von _Chlorella_ bei verschiedenen Licht-Dunkel-Wechseln. _Flora_, **144**, 473.

Lorenzen, H. (1958). Periodizität von Nuklearreaktion und Kernteilung in _Chlorella_. _Ber. Dt. Bot. Ges._, **71**, 89.

Lorenzen, H. (1959). Die photosynthetische Sauerstoffproduktion von _Chlorella_ bei langfristig intermittierender Belichtung. _Flora_, **147**, 382.

Lorenzen, H. (1960). Uber den Einfluß von Glukose auf Synchronkulturen von _Chlorella pyrenoidosa. Naturwissenschaften_, **47**, 477.

Lorenzen, H. (1962). Das Verhalten synchroner Kulturen von _Chlorella pyrenoidosa_ an der oberen Temperaturgrenze des Wachstums. _Vorträge a.d. Gesamtgebiet der Botanik, Dtsch. Bot. Ges._, **Neue Folge Nr. 1**, 231–238.

Lorenzen, H. (1963). Temperatureinflüsse auf _Chlorella pyrenoidosa_ unter besonderer Berücksichtigung der Zellentwicklung. _Flora_, **153**, 554.

Lorenzen, H. und M. Hesse (1968). Nachweis endogen bedingter Leistungsschwankungen bei _Chlorella. Z. Pflanzenphysiol._, **58**, 454.

Lorenzen, H. und H. G. Ruppel (1958). Periodische Resistenzänderung in einer synchronen Kultur von _Chlorella. Naturwissenschaften_, **45**, 553.

Lorenzen, H. und H. G. Ruppel (1960). Versuche zur Gliederung des Entwicklungsverlaufes der _Chlorella_-Zelle. _Planta_, **54**, 394.

Lorenzen, H. und J. Schleif (1966). Zur Bedeutung der kürzest möglichen Generationsdauer in Synchronkulturen von _Chlorella. Flora, Abt. A_, **156**, 673.

Lorenzen, H. und G. S. Venkataraman (1969). Synchronous cell divisions in _Anacystis nidulans_ Richter. _Arch. Mikrobiol._, **67**, 251.

Lysek, G. und W. Simonis (1968). Substrataufnahme und Phosphatstoffwechsel bei _Ankistrodesmus braunii_. I. Beteiligung der Polyphosphate an der Aufnahme von Glukose und 2-Desoxyglukose im Dunkeln und im Licht. _Planta (Berl.)_, **79**, 133.

Mazia, P. (1961). Mitosis and the physiology of cell division. In J. Brachet and A. E. Mirsky (Eds.), _The Cell_, Vol. 3. Academic Press, London, New York. pp. 77–412.

Meffert, M. E. (1963) Untersuchungen zum Stickstoff-Stoffwechsel von _Scenedesmus obliquus_ im Licht-Dunkel-Wechsel. In _Studies on Microalgae and Photosynthetic Bacteria_, Japanese Soc of Plant Physiologists, Tokyo. pp. 111–125.

Metzner, H., H. Rau und H. Senger (1965). Untersuchungen zur Synchronisierbarkeit einzelner Pigmentmangel-Mutanten von _Chlorella. Planta_, **65**, 186.

Miyachi, S., and Shizuko Miyachi (1961). Modes of formation of phosphate compounds and their turnover in _Chlorella_ cells during the process of life cycle as studied by the technique of synchronous culture. _Plant and Cell Physiol._, **2**, 415.

Moberg, S., G. Knutsen, and J. Goksøyr (1968). The 'point of no return' concept in cell division. The effects of some metabolic inhibitors on synchronized _Chlorella pyrenoidosa. Physiol. Plant._, **21**, 390.

Montalvo, F., G. Carroll, and F. E. Cole (1968). Heterotrophic metabolism in synchronous cultures of _Chlorella pyrenoidosa. Federation Proc._, **27**, 833.

Morimura, Y. (1959). Synchronous culture of *Chlorella*. II. Changes in content of various vitamins during the course of the algal life cycle. *Plant and Cell Physiol.*, **1**, 63.

Murakami, S., Y. Morimura, and A. Takamiya (1963). Electron microscopic studies along cellular life cycle of *Chlorella ellipsoidea*. In *Studies on Microalgae and Photosynthetic Bacteria*, Japanese Soc. of Plant Physiologists, Tokyo. pp. 65–83.

Nissen, P., and L. Eldjarn (1969). Differential inhibition of cell division in *Euglena gracilis* by cysteamine and cystamine. *Physiol. Plantarum*, **22**, 364.

Oh-Hama, T., Morimura, and T. Tamiya (1965). Changes in contents and oxido-reductive states of pyridine nucleotides in *Chlorella* cells during their life cycle. In *Beitr. z. Biochemie und Physiologie von Naturstoffen* (*Mothes Festschrift*), VEB G. Fischer, Jena. p. 341.

Overbeck, J. (1961). Der Tageszyclus des Phosphathaushaltes von *Scenedesmus quadricauda* (Turp.), Breb. im Freiland. *Naturwissenschaften*, **48**, 137.

Paasche, E. (1967). Marine plankton algae grown with light-dark cycles. I. *Coccolithus huxleyi*. *Physiol. Plant.*, **20**, 946.

Paasche, E. (1968). Marine plankton algae grown with light-dark cycles. II. *Ditylum brightwellii* and *Nitzschia turgidula*. *Physiol. Plant.*, **21**, 66.

Padilla, G. M., and R. J. Bragg (1968). Responses of the flagellates *Prymnesium parvum* and *Euglena gracilis* to light, dark, and synchrony-inducing temperature cycles. *J. Cell Biol.*, **39**, 101a.

Petropulos, S. F. (1964). Ultraviolet inactivation of chloroplast formation in synchronously dividing *Euglena gracilis*. *Science*, **145**, 392.

Pirson, A. (1962). Synchronanzucht von Algen im Licht-Dunkel-Wechsel. (Ein Uberblick). *Vorträge a.d. Gesamtgebeit der Botanik, Dtsch. Bot. Ges*, N.F. Nr. 1. V.S., 178–186.

Pirson, A. und W. Kowallik (1960). Wirkung des blauen und roten spektralbereiches auf die Zusammensetzung von *Chlorella* bei Anzucht im Licht-Dunkel-Wechsel. *Naturwissenschaften*, **47**, 476.

Pirson, A. und H. Lorenzen (1958a). Ein endogener Zeitfaktor bei der Teilung von *Chlorella*. *Z. Bot.*, **46**, 53.

Pirson, A. und H. Lorenzen (1958b). Photosynthetische Sauerstoffentwicklung von *Chlorella* nach Synchronisation durch Licht-Dunkel-Wechsel. *Nautrwissenschaften*, **45**, 497.

Pirson, A., and H. Lorenzen (1966). Synchronized dividing algae. *Ann. Rev. Plant Physiol.*, **17**, 439.

Pirson, A., H. Lorenzen, and A. Köpper (1959). A sensitive stage in synchronized cultures of *Chlorella*. *Plant Physiol.*, **34**, 353.

Pirson, A., H. Lorenzen und H. G. Ruppel (1963). Der Licht-Dunkel-Wechsel als synchronisierendes Prinzip. In *Studies on Microalgae and Photosynthetic Bacteria*, Japanese Soc. of Plant Physiologists, Tokyo. pp. 127–139.

Pirson, A. und H. G. Ruppel (1962). Uber die Induktion einer Teilungshemmung in synchronen Kulturen von *Chlorella*. *Arch. Mikrobiol.*, **42**, 299.

Pirson, A. und H. Senger (1961). Synchronisierungstypen bei *Chlorella* im Licht-Dunkel-Wechsel. *Naturwissenschaften*, **48**, 81.

Rao, P. N., and J. Engelberg (1968). Mitotic duration and its variability in relation to temperature in HeLa cells. *Exper. Cell Res.*, **52**, 198.

Ried, A. (1962). Entwicklungsabhängige Unterschiede in der Reaktion der Atmung von *Chlorella* auf Erhöhung der Salzkonzentration. *Ber. Dtsch. Bot. Ges.*, **74**, 431.

Ried, A., C. J. Soeder und I. Müller (1963). Uber die Atmung synchron kultivierter *Chlorella*. I. Veränderungen des respiratorischen Gaswechsels im Laufe des Entwicklungszyklus. *Arch. Mikrobiol.*, **45**, 343.

Ruppel, H. G. (1962). Untersuchungen über die Zusammensetzung von *Chlorella* bei Synchronisation im Licht-Dunkel-Wechsel. *Flora (Jena)*, **152**, 113.

Sasa, T. (1961). Effect of ultraviolet light upon various physiological activites of *Chlorella* cells at different stages in their life cycle. *Plant and Cell Physiol.*, **2**, 253.

Schlösser, U. (1966). Enzymatisch gesteuerte Freisetzung von Zoosporen bei *Chlamydomonas reinhardii* Dangeard in Synchronkultur. *Arch. Mikrobiol.*, **54**, 129.

Schmidbauer, A. und A. Ried (1967). Einfluss hypertonischer Medien auf den Stoffwechsel synchron kultivierter *Chlorella*. *Arch. Mikrobiol.*, **58**, 275.

Schmidt, R. R. (1961). Nitrogen and phosphorus metabolism during synchronous growth of *Chlorella pyrenoidosa*. *Exptl. Cell Res.*, **23**, 209.

Schmidt, R. R. (1966). Intracellular control of enzyme synthesis and activity during synchronous growth of *Chlorella*. In I. L. Cameron and G. M. Padilla (Eds.), *Cell Synchrony*, Academic Press, London, New York. pp. 189–235.

Schönherr, O. Th. (1969). Activiteit van DNA-polymerase en desoxyribonuclease in synchroon groeiende chlorella-cellen. *Diss.*, *Nijmegen*, Netherland.

Senger, H. (1961). Untersuchungen zur Synchronisierung von *Chlorella*-Kulturen. *Arch. Mikrobiol.*, **40**, 47.

Senger, H. (1965). Die teilungsinduzierende Wirkung des Lichtes in mixotrophen Synchronkulturen von *Chlorella*. *Arch. Mikrobiol.*, **51**, 307.

Senger, H., and N. I. Bishop (1966). The light-dependent formation of nucleic acids in cultures of synchronized *Chlorella*. *Plant and Cell Physiol.*, **7**, 441.

Senger, H., and N. I. Bishop (1967). Quantum yield of photosynthesis in synchronous *Scenedesmus* cultures. *Nature*, **214**, 140.

Senger, H., and N. I. Bishop (1969). Light-dependent formation of nucleic acids and its relation to the induction of synchronous cell division in *Chlorella*. In G. M. Padilla and I. L. Cameron (Eds.), *The Cell Cycle*, Academic Press, London, New York. pp. 179–202.

Senger, H. und G. Schoser (1966). Die spektralabhängige Teilungsinduktion in mixotrophen Synchronkulturen von *Chlorella*. *Z. Pflanzenphysiol.*, **54**, 308.

Senger, H. und H. J. Wolf (1964). Eine automatische Verdünnungsanlage und ihre Anwendung zur Erzielung homokontinuierlicher *Chlorella*-Kulturen. *Arch. Mikrobiol.*, **48**, 81.

Shen, S., and R. R. Schmidt (1966). Enzymic control of nucleic acid synthesis during synchronous growth of *Chlorella pyrenoidosa*. II. Deoxycytidine monophosphate deaminase. *Arch. Biochem. Biophys.*, **115**, 13.

Simonis, W. (1964). Untersuchungen zur Photosynthese-Phosphorylierung an intakten Algenzellen (*Ankistrodesmus braunii*). *Ber. Dtsch. Bot. Ges.*, **77** (5).

Soeder, C. J. (1964). Elektronenmikroskopische Untersuchungen an ungeteilten Zellen von *Chlorella* fusca Shihira et Krauss. *Arch. Mikrobiol.*, **47**, 311.

Soeder, C. J., G. Schulze, and D. Thiele (1966). A new type of 'light inhibition', occurring in synchronous cultures of *Chlorella*. *Verh. int. Ver. Limnol.*, **16**, 1595.

Soeder, C. J., G. Schulze und D. Thiele (1967). Einfluss verschiedener Kultur-bedingungen auf das Wachstum in Synchronkulturen von *Chlorella* fusca Sh. et Kr. *Arch. Hydrobiol. (Suppl.)*, **33**, 127.

Soeder, C. J., H. Strotmann, and R. A. Galloway (1966). Carbon dioxide-induced delay of cellular development in two *Chlorella* species. *J. Phycol.*, **2**, 117.

Soeder, C. J. und D. Thiele (1967). Wirkungen des Calcium-Mangels auf *Chlorella* fusca Shihira et Krauss. *Z. Pflanzenphysiol.*, **57**, 339.

Sorokin, C. (1957). Changes in photosynthetic activity in the course of cell development in *Chlorella*. *Physiol. Plant.*, **10**, 659.

Sorokin, C. (1960). Photosynthetic activity in synchronized cultures of algae and its dependence on light intensity. *Arch. Mikrobiol.*, **37**, 151.

Sorokin, C. (1962). Carbon dioxide and bicarbonate in cell division. *Arch. Mikrobiol.*, **44**, 219.

Sorokin, C. (1963). Characteristics of the process of aging in algal cells. *Science*, **140**, 385.

Sorokin, C. (1964). Aging at the cellular level. *Experientia*, **20**, 353.

Sorokin, C., and R. W. Krauss (1959). Maximum growth rates in synchronized cultures of green algae. *Science*, **129**, 1289.

Sorokin, C., and R. W. Krauss (1961). Relative efficiency of photosynthesis in the course of cell development. *Biochim. Biophys. Acta*, **48**, 314.

Sorokin, C., and R. W. Krauss (1962). Effects of temperature and illuminance on *Chlorella* growth uncoupled from cell division. *Plant Physiol.*, **37**, 37.

Sorokin, C., and J. Myers (1957). The course of respiration during the life cycle of *Chlorella* cells. *J. of Gen. Phys.*, **40**, 579.

Spektorov, K. S. und E. A. Linkova (1962). Eine neue, einfache Synchronisier-ungsmethode von *Chlorella*-Kulturen. *Doklady A.N. UDSSR*, **147**, 967.

Spektorov, K. S., G. A. Slobodskaya, and A. A. Nichiporovich (1963). Studies on photosynthesis of *Chlorella pyrenoidosa* Pringsh. 82 in a synchronous culture. In *Studies on Microalgae and Photosynthetic Bacteria*, Japanese Soc. of Plant Physiologists, Tokyo. pp. 141–149.

Strotmann, H. (1967). Blaulichteffekt auf die Nitritreduktion von *Chlorella*. *Planta (Berl.)*, **73**, 376.

Tamiya, H. (1966). Synchronous cultures of algae. *Ann. Rev. Plant Physiol.*, **17**, 1.

Tamiya, H., I. Iwamura, K. Shibata, E. Hase, and T. Nihei (1953). Correlation between photosynthesis and light-independent metabolism in the growth of *Chlorella*. *Biochim. Biophys. Acta*, **12**, 23.

Tamiya, H., Y. Morimura, and M. Yokota (1962). Effects of various anti-metabolites upon the life cycle of *Chlorella*. *Arch. Mikrobiol.*, **42**, 4.

Tamiya, H., Y. Morimura, M. Yokota, and R. Kunieda (1961). Mode of nuclear division in synchronous cultures of Chlorella: comparison of various methods of synchronization. *Plant and Cell Physiol.*, **2**, 383.

Tomova, N. E., Linkova, and K. Spektorov (1964). The effect of different nitrogen sources on the growth and development of a synchronous culture of *Chlorella pyrenoidosa* Pringsheim 82. *C.r. de l'Acad. bulgare des Sciences*, **17**, 757.

Urbach, W. (1966). Die Wirkung von Sauerstoff auf den Elektronen-Transport der Photosynthese. *Ber. Dtsch. Bot. Ges.*, **79**, 107.

Venkataraman, G. S., F. Amelunxen and H. Lorenzen (1969). Note to the fine structure of *Anacystis nidulans* during its synchronous growth. *Arch. Mikrobiol.*, **69**, 370.

Venkataraman, G. S., and H. Lorenzen (1969). Biochemical studies on *Anacystis nidulans* during its synchronous growth. *Arch. Mikrobiol.* (In the press).

Wanka, F. (1965). The use of colchicine in investigation of the life cycle of *Chlorella. Arch. Microbiol.*, **52**, 305.

Wanka, F. (1968a). Ultrastructural changes during normal and colchicine-inhibited cell division of *Chlorella. Protoplasma*, **66**, 105.

Wanka, F. (1968b). Durch Colchicinbehandlung erzeugte polyploide *Chlorella*-Stämme. *Ber. Dtsch. Bot. Ges.*, **81**, 38.

Wanka, F., and P. F. M. Mulders (1967). The effect of light on DNA synthesis and related processes in synchronous cultures of *Chlorella. Arch. Mikrobiol.*, **58**, 257.

Werner, D. (1966). Die Kieselsäure im Stoffwechsel von *Cyclotella cryptica*, Reimann, Lewin und Guillard. *Arch. Mikrobiol.*, **55**, 278.

Wey, M. (1964). Determination of contents of manganese and copper during the life cycle of *Chlorella* cells by neutron activation analysis. *Hua Hsueh*, **4**, 165.

Wolf, G. (1964). Die Atmung von *Chlorella* im Verlauf der Zellentwicklung. *Dissertation, Göttingen*.

Yuhara, T., and E. Hase (1961). A brief note on the absorption of water by *Chlorella* cells during the process of their life cycle. *J. Gen. Appl. Microbiol.*, **7**, 70.

CHAPTER 8

Photomotion of microorganisms and its interactions with photosynthesis

WILHELM NULTSCH

Department of Botany, University of Marburg, Germany

Introduction

Movement of photosynthetic microorganisms may be influenced by light in several different ways, either resulting in changes of the speed of movement, called photokinesis, or leading to distinct patterns in spatial arrangements and distribution of the organisms, called phototaxis. Phototactic reactions can be further divided into two types, namely topic and phobic. In phototopotaxis the orientation of the movement is related to the direction of the incident light beam, whereas photophobotactic responses are caused by sudden changes in light intensity or exposure to a steep light gradient. All these reactions, summarized in the term 'photomotion' by Wolken and Shin (1958), can be positive or negative. Consequently, we have to distinguish the following reaction types:

(1) positive photokinesis, i.e. an acceleration of movement by light or, in organisms which are immotile in the dark, even an excitation of the movement by light;

(2) negative photokinesis, i.e. a decrease in the speed of movement by light, eventually resulting in immotility;

(3) positive phototopotaxis, i.e. a movement towards the light source;

(4) negative phototopotaxis, i.e. a movement away from the light source;

(5) positive photophobotaxis, i.e. a reversal of the direction of movement caused by a sudden decrease in light intensity or exposure to a steep light gradient; it results in an accumulation of the organisms in a bright light field projected on a surrounding field of less light intensity ('light trap'); and

(6) negative photophobotaxis, caused by a sudden increase in light intensity. As a result, the brighter light field becomes empty.

Phototaxes and photokinesis in bacteria and algae have been reviewed by Bendix (1960), Clayton (1959, 1964), Halldal (1962, 1964) and Haupt (1959, 1965, 1966).

Photokinesis

Photokinetic phenomena were described for the first time by Strasburger (1878). He observed that green swarmspores of *Ulothrix zonata*, *Botrydium granulatum* and *Haematococcus lacustris*, as well as colourless swarmers of the phycomycete *Chytridium vorax*, soon become immotile in light (*Lichtstarre*), whereas in dark the motility is maintained for a period of from several days to a fortnight. This behaviour is an example of negative photokinesis, according to the fore-mentioned terminology. Positive photokinesis has been observed by Engelmann (1882, 1883, 1888) in some species of purple bacteria. These organisms become immotile in the dark the sooner the less oxygen is present (*Dunkelstarre*). After a short irradiation (some seconds to a few minutes) they become motile again, and the speed of movement increases with increasing light intensity. After a new darkening, movement is maintained for a shorter or longer period of time, and then the bacteria become immotile again. Engelmann (1882), who first used the term 'photokinesis', concluded that 'a pool of any substance may be produced by irradiation which is necessary for movement and is consumed in the dark'. Under continuous illumination, especially at high intensities, the purple bacteria also become immotile (negative photokinesis, *Lichtstarre*).

More detailed investigations on photokinesis have been carried out by Bolte (1920). Contrary to Strasburger, she failed to demonstrate any effect of light on motility of colourless microorganisms such as *Euglena*, *Polytoma* and *Chilomonas curvata*. *Euglena proxima*, a green organism, also displayed photokinetic indifference. In most of the green flagellates, however, light was active in the sense of positive photokinesis, and only a few species of the genera *Phacus*, *Lepocinclis* and *Chlamydomonas* became immotile in light through the loss of their flagella. Bolte found that in ten of the species investigated the motility depended on the presence of carbon dioxide. She therefore suggested correlations between positive photokinesis and photosynthesis.

Further observations and investigations on photokinesis revealed contradictory results. While Mainx (1929) in *Volvox*, *Eudorina*, *Gonium* and other species, and Haxo and Clendenning (1953) in gametes of *Ulva* failed to observe any kinetic effect of light, Holmes (1903), Oltmanns (1917) and Mast (1926) found an increase of velocity with increasing light intensity in *Volvox*. After maximum speed had been reached, a further increase of light intensity had no more photokinetic effect, not even a negative one.

More extensive investigations have been carried out by Luntz (1931a, 1931b, 1932). In some organisms which were immotile in the dark (*Volvox*, *Eudorina* and *Chlamydomonas*), he evaluated the 'threshold of motility' (*Bewegungsschwelle*), i.e. the lowest light intensity needed to cause movement. He found that the presentation time (t) depended on light intensity (I). Thus, at lower intensities the photokinetic effect is a function of light quantity ($I \times t$), i.e. the reciprocity law is valid. Under continuous illumination, however, the photokinetic effect seems to depend only on light intensity (I).

Photokinesis of blue-green algae has been investigated by Nultsch (1962a, 1962c) in several species of the genus *Phormidium*. In these organisms, which are also motile in the dark, the speed of movement increases with increasing light intensities, reaching a maximum at about 2000 lux ('photokinetic optimum'). By a further increase of light intensity the speed of movement is decreased, and beyond 30 000 lux the photokinetic effect becomes negative, i.e. the speed of the movement in light is slower than in the dark. The zero threshold of photokinesis is estimated by Nultsch at 0·01–0·02 lux. Positive photokinesis in diatoms has been observed by Nultsch (1956), and a kinetic effect of light on the movement of the desmid *Micrasterias denticulata* has been pointed out by Neuscheler (1967b). Photokinesis in the purple bacterium *Rhodospirillum rubrum* has been recently reinvestigated by Throm (1968).

Action spectra

The most frequently used method to evaluate the photokinetic effect is to measure the speed of movement at different wavelengths and light intensities. Usually the speed of single organisms is measured microscopically,

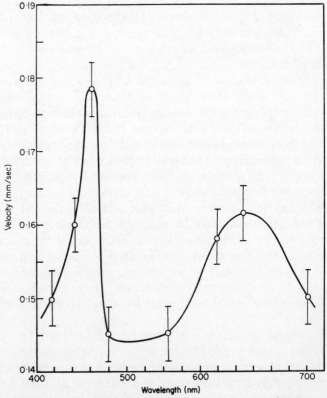

Figure 8.1. Action spectrum of photokinesis in *Euglena* (Wolken and Shin, 1958)

but in slowly moving organisms, as in blue-green algae, it is more convenient to measure the speed of a population spreading over an agar plate
(Nultsch, 1962a). A null method has been used by Luntz (1931a, 1931b,
1932), who evaluated the threshold values of motility at different wavelengths and intensities in some organisms which are immotile in the dark
(see above). Finally, an indirect method has been devised by Throm (1968).
In photophobotactic experiments he measured the increase of optical
density in a 'light trap' caused by photophobotactic accumulation of

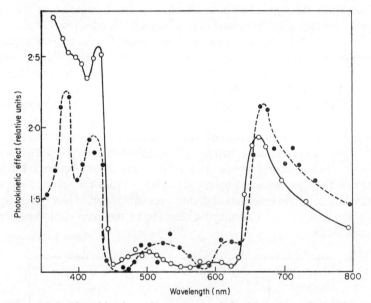

Figure 8.2. Photokinetic action spectra of the blue-green algae *Phormidium uncinatum* (- ● - - ● -) and *Phormidium* sp. (–○–○–) (Nultsch,
1962c)

positive-reacting organisms (see Figure 8.6, page 226), yielding a continuous
record of the time course of the response. The slope of its linear part depends
on the speed of the organisms collecting in the 'light trap', and can be used
as an index of photokinesis.

Luntz (1931a) presented rough action spectra of positive photokinesis in
Eudorina, *Pandorina* and *Chlamydomonas*, each of which showed a maximum at 492 nm and were consistent with the phototactic action spectrum
of the same organism. The action spectrum of positive photokinesis in
Euglena, as measured by Wolken and Shin (1958), has one sharp maximum

8

in the blue region at 460 nm and another lower and broader one in the red region of the spectrum between 600 and 650 nm (Figure 8.1).

Photokinetic action spectra of blue-green algae have been measured by Nultsch (1962a, 1962c). He found that both the action spectra of the phycoerythrin containing *Phormidium uncinatum* and the blue-green form *Phormidium* sp., which only contains phycocyanin, are very similar in the visible region, showing maxima between 430 and 440 nm as well as between 670 and 680 nm (Figure 8.2). These maxima coincide with the absorption maxima of chlorophyll *a* within the living cell, while the biliproteins are not as active as the light absorption would suggest. Irradiation of the main absorption region of the carotenoids is only slightly effective, if at all. From these results Nultsch concluded that chlorophyll *a* is the main photoreceptor of photokinesis in the visible region. As the ultraviolet region of the spectrum is photokinetically effective, too, this effectiveness must be due to another photoreceptor of unknown chemical structure.

In diatoms only some incidental observations have been reported by Nultsch (1956). Recently, however, an action spectrum of photokinesis in *Nitzschia* sp. has been measured (Nultsch, unpublished) which also displays two regions of maximum activity, the first in the red region between 660 and 700 nm, and a second smaller one in the blue region between 420 and 440 nm. The light absorbed by carotenoids has little or no effect. Thus, the photokinetic action spectrum of diatoms resembles that of blue-green algae. In both these groups chlorophyll *a* seems to be the main photoreceptor of photokinesis.

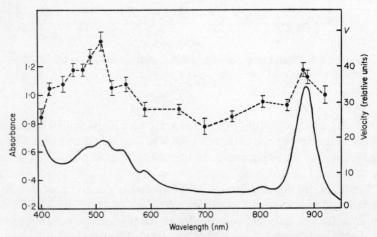

Figure 8.3. Photokinetic action spectrum (- ● -- ● -) and *in vivo* absorption spectrum (——) of *Rhodospirillum rubrum* (Throm, 1968)

Recently, Throm (1968) has measured the action spectrum of photokinesis in the purple bacterium *Rhodospirillum rubrum* (Figure 8.3). As the *Rhodospirilla*, under the chosen conditions, were also motile in the dark, the photokinetic effect consists only in an increase of the speed. The action spectrum resembles both the absorption spectrum of the cells and the photosynthetic action spectrum (Clayton, 1953a; Duysens, 1952; Manten, 1948a, 1948b; Thomas, 1950), but the relative height of the two peaks at shorter and longer wavelengths is inverse in the two spectra. The peak at 510 nm and the shoulder at 550 nm points out a stronger participation of spirilloxanthin in photokinesis than in photosynthesis. As action spectra of photophobotaxis in *Rhodospirillum rubrum* also differ in the region of carotenoid absorption (Clayton, 1959), a reexamination of the action spectra is desirable in order to obtain more information of the role of carotenoids in these photic reactions.

Effect of inhibitors

The photokinetic action of light absorbed by photosynthetic pigments has led to the conclusion that photokinesis may be linked with photosynthesis. However, as in blue-green algae the effectiveness of far red light, compared with that of shorter wavelengths, is far out of proportion to its effectiveness in photosynthesis (Duysens, 1952), photokinesis seems to be coupled mainly with the photosynthetic photosystem I (see Halldal, Chapter 2, of this treatise), which is sufficient to drive photosynthetic phosphorylation (Kok and Hoch, 1961). This idea gave rise to a series of investigations on the effect of photosynthetic inhibitors (Nultsch, 1965; Nultsch and Jeeji-Bai, 1966), uncouplers (Nultsch, 1966a, 1966b, 1967) and redox systems (Nultsch, 1968a) on photokinesis in blue-green algae.

Photokinesis is markedly impaired by DCMU (3-(3,4-dichlorophenyl)-1,1-dimethylurea), a well-known inhibitor of the non-cyclic electron transport and coupled phosphorylation. Less than 50 per cent of the DCMU-resistant movement in light is due to oxidative phosphorylation, as shown by experiments carried out in an argon atmosphere (Nultsch and Jeeji-Bai, 1966). As anaerobic phosphorylation does not play an important role in the energy supply of movement, the remaining oxygen-independent and DCMU-resistant part of movement is obviously due to cyclic phosphorylation. Indeed, it is sensitive to uncouplers which more or less specifically inhibit cyclic photophosphorylation as antimycin *a* (Nultsch and Jeeji-Bai, 1966; Tagawa, Tsujimoto and Arnon, 1963) and desaspidin (Arnon, Tsujimoto and McSwain, 1965; Baltscheffsky and de Kiewiet, 1964; Gromet-Elhanan and Arnon, 1965; Gromet-Elhanan and Avron, 1966;

Table 8.1. Effect of desaspidin (2×10^{-5} M) and DCMU (5×10^{-5}) on photokinesis in white light and at 714 nm (6000 erg/cm² second) under argon

	Photokinetic effect (%)	
Treatment	White light	714 nm
Control	100	100
Desaspidin	82	56
DCMU	52	76

Nultsch, 1968b; Tsujimoto, McSwain and Arnon, 1966). Experiments with far red light, which only excites photosystem I, confirm these conclusions. As shown in Table 8.1, under an argon atmosphere the photokinetic effect of far red light is more strongly inhibited by desaspidin than in white light, which excites both photosystems I and II. These findings are consistent with the results of investigations on the effect of antimycin *a*, desaspidin and DCMU upon ³²P incorporation in intact cells of the green alga *Ankistrodesmus braunii*, reported by Urbach and Simonis (1964, 1967).

Investigating the effect of redox systems on the motility of blue-green algae, Nultsch (1968a) has found that every substance which traps electrons from the electron transport chain of photosynthesis inhibits the coupled photosynthetic phosphorylation and, hence, photokinesis. The redox systems with the most negative E_0' values (viologens) are most effective in inhibiting photokinesis. A second smaller inhibition maximum lies between $+ 0.06$ and $+ 0.08$ V (see Figure 8.10, page 234). Obviously, redox systems which are able to trap electrons from the cyclic electron transport are most effective in inhibition of photokinesis.

Photokinesis of *Rhodospirillum rubrum* is also sensitive to uncouplers and some inhibitors of photosynthetic electron transport, with the exception of DCMU (Throm, 1968). Obviously, even in purple bacteria, photokinesis is coupled with photophosphorylation.

Effect of ATP

Based on these results, Nultsch (1967, 1968, 1969) has concluded that the acceleration of movement by light is the result of an additional ATP supply from cyclic and non-cyclic photosynthetic phosphorylation, whereas movement in the dark is maintained by ATP produced in oxidative phosphorylation and, to a very small extent, by anaerobic phosphorylation in glycolysis. However, because the ATP produced by non-cyclic phosphorylation is consumed to a large extent by carbon dioxide fixation, under normal conditions, i.e. in air, cyclic photophosphorylation should be the main energy source of photokinesis. In order to confirm this hypothesis, the effect of ATP on movement has been studied repeatedly.

Attempts to increase the speed of the movement by externally added ATP in blue-green algae and diatoms have not been successful (Nultsch, unpublished). Probably ATP cannot penetrate the cytoplasmic membrane. In *Rhodospirillum rubrum* Clayton (1958) also failed to demonstrate any effect of ATP on movement in short-time experiments. Recently, Throm (1968) has repeated these experiments. In accordance with Clayton's results he could not detect any effect of ATP during the first few minutes of the experiment, but with increasing time the movement in the dark was markedly accelerated by 10^{-3} M ATP to about the same speed as in the light (200 lux). This suggests that externally supplied ATP acts in the ATPase system of flagella after an adaptation time. Moreover, in these samples no further increase of the speed by irradiation could be observed, because the apparatus of movement was energetically saturated. Thus, light and ATP are equivalent in their ability to accelerate the movement. Nultsch and Throm (1968) have interpreted these results as indicating that the photokinetic effect of light is due to an additional energy supply from photophosphorylation. Obviously, the substance predicted by Engelmann (see page 215) is ATP.

Summarizing the results of all investigations on photokinesis in purple bacteria, blue-green algae and diatoms it seems to be admissible to accept the suggestion that photokinesis is a function of photosynthetic phosphorylation as a general concept. The action spectrum of photokinesis in *Euglena* does not permit a clear interpretation because of its unsatisfactory resolution. Therefore, more detailed investigations on photokinesis in flagellates are necessary.

Photophobotaxis

Photophobotactic responses are caused by temporal changes in light intensity (I), dI/dt, and do not show any relation to the direction of the incident light beam. Phobotactic responses elicited by an abrupt decrease of light intensity are called positive, while responses caused by an increase of intensity are called negative. A temporal light stimulus is of course perceived by an individual when it passes the boundary between two fields of different light intensities, e.g. when it quickly moves from light to dark and vice versa. In this case the spatial gradient of light intensity is transformed into a temporal one. Consequently, positively reacting organisms gather in a light spot projected onto a preparation which is exposed to lower light intensities or is kept in the dark (Engelmann's 'light trap'), because they are prevented from leaving it but not from entering it. Negatively reacting organisms leave the light field but are not able to enter it. Thus the light zone becomes empty.

At first sight there seem to be some similarities between photophobotaxis and photokinesis: both reactions depend on the intensity of undirected light and both reactions are coupled with a change in the speed of movement. However, photokinesis is a long-time effect of light, and the change of speed does not depend on a distinct ratio dI/dt. Moreover, the direction of movement is not changed during alteration of light intensity, provided that no phobotactic reaction is elicited simultaneously. On the other hand, a photophobotactic reaction is caused by a short-time change in light intensity which is strongly dependent on a distinct ratio dI/dt, and the individuals show only a transient change of speed, usually a stop, followed by a resumption of movement in a new, mostly opposite, direction.

Modes of photophobotactic reactions

The mode of a photophobotactic reaction, i.e. the behaviour of an individual after photic stimulation, depends on the morphology of the organism on the one hand and the mechanism of movement on the other. In many cases, as in some purple bacteria (*Rhodospirillum, Thiospirillum*), blue-green algae and diatoms, the phobotactic response is simply a stop of movement, normally followed by a reversal of swimming or gliding direction.

In the unipolarly flagellated purple bacterium *Thiospirillum*, which, under optimal conditions, can swim in either direction parallel to its long axis, the reversal ensues very quickly as a result of an abrupt reversal of beating direction of the flagella, accompanied by a 'snapping over' (like an umbrella) of the flagella. As has been shown by Buder (1915), this 'snapping over' is an active response, which also ensues in individuals kept by small particles in the culture medium. Sometimes, under less optimal conditions, the movement in one direction is preferred. In this case, in response to the stimulus, *Thiospirillum* behaves more like *Chromatium* (see below), i.e. one light stimulus is followed by two reversals, resulting in a movement in the preferred direction whereby the orientation of the long axis of the cell may be changed a little.

In the bipolarly flagellated purple bacterium *Rhodospirillum rubrum*, which does not show any polarity in the direction of movement, the flagella at both ends of the cell normally 'snap over' simultaneously (Figure 8.4) in

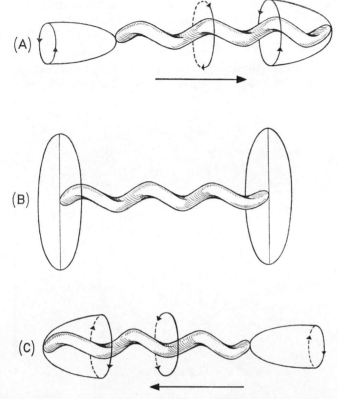

Figure 8.4. Photophobotactic reaction of bipolarly flagellated spirilla, illustrating the 'snapping over'. (Changed after Metzner, 1920)

response to the stimulus (Metzner, 1920). However, a very strong stimulus can induce a series of repeated reversals of up to ten and more.

In diatoms and blue-green algae the gliding movement is much slower than the swimming movement of purple bacteria, and normally a slow-down precedes the stop (Drews, 1957, 1959; Nultsch, 1956, 1962b). The 'rest period' lasts one or a few seconds, and then the cells or filaments resume gliding in the opposite direction. Sometimes in the range of the threshold value the phototactic reaction can be overpowered. In this case the phobotactic response is only indicated by a transient slow-down, or by a complete stoppage of the filaments, but the movement is resumed after the 'rest period' in the same direction as before (Harder, 1920).

A more complicated type of photophobotactic reaction is realized in the unipolarly flagellated purple bacterium *Chromatium*, which shows a strong polarity in movement. Usually it swims with the flagellum at the rear pole (forward direction). In response to a light stimulus, it performs a typical shock reaction (the *Schreckbewegung* of Engelmann, 1882), i.e. it goes back ten- to twenty-fold of its cell-length with an increased speed, as if it recoils from an invisible wall. This reversal is only the result of a change in the rotating direction of the flagellum (Figure 8.5), like a reversal of the turning

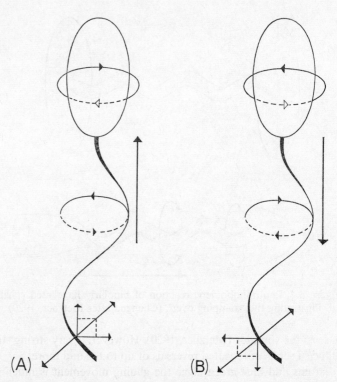

Figure 8.5. Mechanism of movement in *Chromatium*. (A) Normal forward movement. (B) Reversal during photophobotactic response. (Changed after Buder, 1915)

direction of a propeller without 'snapping over'. Then the *Chromatium* pauses for a second or fractions of a second; this is followed by a resumption of forward movement, the direction of which may be more or less divergent from the former one due to changes in the orientation of the long axis of the cell during the reversal or, by Brownian movement, during the 'rest period' (Buder, 1915). In other cases, however, the individuals stagger or turn at the same place after the reversal (Buder, 1919), whereby of course the orientation of the cell axes is changed.

In *Euglena*, a green flagellate with a flagellum at the anterior part of the swimming cell, the photophobotactic response results in a sharp turn of the cell, due to an abrupt sideward beating of the flagellum to the ventral side, i.e. the side opposite to the stigma. Consequently, the cell deviates towards the dorsal side. Thus, the direction of deviation depends on the morphology of the cell and is independent of the direction of the incident light beam. The direction of the resumed movement depends of course on the strength of the turning impulse, which for its part quantitatively depends on the strength of stimulus, i.e. the change of light intensity. Similar responses have been observed in dinoflagellates by Metzner (1929). This type of reaction is the so-called 'shock reaction' or 'motor reflex' (Jennings, 1904). No differences have been observed in the behaviour of cells reacting either positively or negatively.

Methods of investigation

The simplest and most frequently used method to investigate photophobotactic reactions is the direct observation of single organisms. For this purpose pulses of increased or decreased light are applied to the cells, and their individual tactic responses are observed microscopically. A quantitative evaluation is possible in such experiments if the numbers of responding and non-responding individuals are counted. A dose–response curve may be obtained in this way by plotting the percentage of reacting organisms against ΔI.

Another more convenient way is the 'light-trap' method of Engelmann (1882). In this method a light field is projected by a slit onto a preparation of microorganisms on an agar layer, a microscopic slide or in a cuvette. As mentioned above, positively reacting organisms gather in the light trap. The number of organisms collected in the light field depends on the photophobotactic sensitivity, and can be evaluated by measuring the changes in optical density. The same method is practicable to evaluate the discrimination threshold, using two light fields of different intensities, and also to measure action spectra, using monochromatic light. A one-light-beam method has been used by Nultsch (1962b) and Throm (1968), and a double-beam method by Lindes, Diehn and Tollin (1965) and Diehn (1969) in the so-called 'phototaxigraph'. In slowly moving organisms, as in blue-green algae and diatoms, the accumulations are stable for some minutes and can therefore be measured photometrically at the end of the experiment (Nultsch, 1962b), while in quickly moving organisms, as in flagellates and purple bacteria, the optical density must be measured during the experiment (Diehn, 1969; Throm, 1968). If the increase of the optical density is recorded continuously, a time course of the phobotactic response is obtained (Figure 8.6). Sooner or later the time course reaches the saturation level. This is constant for a longer period of time, because the number of organisms which enter the 'light trap' equals the number which leave it. The relative height of the saturation level depends on the photophobotactic sensitivity and the effectiveness of the irradiation respectively, while the slope of the linear part of the curve depends on the speed of movement (see Figure 8.6). Finally, a null method can be used by establishing a border between two fields of different light intensities or wavelengths (Manten, 1948a, 1948b; Schlegel, 1956; Schrammeck, 1934).

However, some error and misinterpretation is possible using the 'light-trap' method (Clayton, 1957, 1964). On the one hand, positive or negative phototaxis can be simulated by photokinetic effects. For instance, if micro-organisms have lost their motility in the dark, the light in the projected field

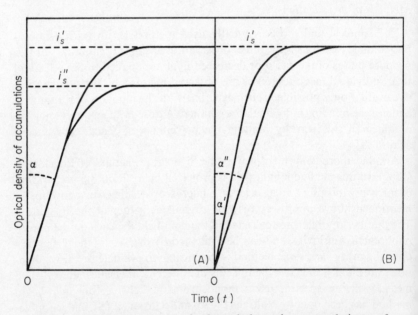

Figure 8.6. Time courses of photophobotactic accumulations of *Rhodospirillum rubrum.* (A) Two curves with different saturation levels (i'_s, i''_s), indicating differences in photophobotactic activity. (B) Two curves with the same saturation level (i'_s), but with a different slope in the linear part of the time course (α', α''), indicating different speeds of movement (Throm, 1968)

can arouse motility again (positive photokinesis). As a result, some of the organisms will leave the light field. This vacation of the light spot, which is solely due to photokinesis, develops a pattern which suggests negative photophobotaxis. Conversely, accumulation of organisms in the light field can be due to a negative photokinesis, because the individuals become immotile in bright light. Finally, the organisms accumulate in the light trap the faster the more quickly they swim. Therefore, only the saturation level, and not the slope of the recorded time course, can be used as a measure of the photophobotactic response, as has been shown by Throm (1968).

On the other hand, photophobotaxis can be superimposed by photo-topotaxis (see below), as organisms in the vicinity of the light field may be enticed topotactically by laterally scattered light. A phobotactic accumulation or vacation can also be the result of positive or negative chemotaxis. Therefore, the most reliable way is to observe microscopically the behaviour of individuals which are subjected to temporal light stimuli.

Influence of light intensity

Most of the photophobotactically reacting microorganisms investigated so far respond positively in dim light and negatively in bright light. The intensity ranges of positive and negative reactions are usually separated by a smaller or broader range of intensities, in which the sense of reaction is inverted. At this 'inversion intensity' the organisms do not respond photophobotactically at all. It has also been called the 'indifference zone' (Buder, 1917), 'neutral point' (Luntz, 1932) or 'change point' (Mainx, 1929) of phototaxis.

The prerequisite of any phobotactic reaction is that the stimulus exceeds the threshold value, i.e. the least perceptible change in light intensity from I_0 to $I_0 \pm \Delta I$. The ratio $\Delta I / I_0$ is called the discrimination threshold. If $-\Delta I$ equals I_0 or, in other words, if the organism comes from light to dark, the least intensity causing an accumulation of organisms in the light trap is called the zero threshold. In addition, the stimulus, i.e. the decrease (increase) of light intensity in positive (negative) phobotaxis, must last a minimum time, called the presentation time. Both presentation time and threshold values are functions of phototactic sensitivity. Therefore, they depend of course on the species of organism on the one hand and on environmental factors on the other.

In purple bacteria the discrimination threshold, $\Delta I / I_0$, is 0·03 to 0·05 and is constant over a wide range of intensities, as 10^5-fold in *Rhodospirillum rubrum* (Clayton, 1953a; Schrammeck, 1934; but 100-fold in Schlegel, 1956), 100-fold in *Chromatium vinosum* and *Chromatium okeni* and ten-fold in *Thiospirillum jenense* (Schlegel, 1956). Thus, in the forementioned ranges of light intensities the Weber–Fechner law is valid in purple bacteria, provided that the intensity is changed very quickly. The zero threshold, evaluated by Schlegel (1956), is about 0·01 lux in *Chromatium vinosum* and *Chromatium okeni*, and 0·02–0·03 lux in *Thiospirillum jenense*. The presentation time is very short (about a second), but depends of course on the strength of the stimulus. Vacations of the light field due to negative phobotaxis have been observed by Schlegel (1956) in *Chromatium vinosum* at 36 500 lux, in *Chromatium okeni* at intensities > 15 000 lux and in *Thiospirillum jenense* at intensities > 1200 lux. In *Rhodospirillum rubrum* Schlegel failed to observe negative phobotaxis up to 36 500 lux, and the vacations of the light spot, reported by Clayton (1957), may be due to photokinesis (Clayton, 1959, 1964).

In the blue-green alga *Phormidium uncinatum* the zero threshold has been found at about 0·1 lux by Drews (1959) and Nultsch (1962b), while in other *Phormidium* species the threshold values were 1 and 20 lux (Nultsch, 1962c) The density in the light field increases with increasing light intensity due to the accumulation of organisms. The maximum density, i.e. the 'optimum' of photophobotaxis in the sense of Mainx and Wolf (1940), is reached at 5000 lux in *Phormidium uncinatum* (Nultsch, 1962b) and at 10 000 lux in *Phormidium* sp. (Nultsch, 1962c). With a further increase of intensity the density of accumulations decreases.

Negative photophobotaxis has been reported by Schmid (1923) in *Oscillatoria jenensis*, which seems to respond only negatively. On the other hand, Harder (1920), Drews (1959) and Nultsch (1962b, 1962c) have never observed negative reactions in species of the genera *Phormidium*, *Oscillatoria* and *Nostoc*, even when they used very high light intensities (50 000 lux, Nultsch, 1962b). Vacations of the light fields in the presence of some redox systems (methylene blue, thionin and others), reported by Nultsch (1968a), may be of a photokinetic nature. *Anabaena variabilis*, investigated by Drews, did not show any phobotactic response at all. The discrimination threshold, most recently evaluated by Nultsch and Haeder (unpublished), is about 0·05, i.e. of the same order of magnitude as in purple bacteria.

Detailed investigations on presentation time have been carried out by Harder (1920). The time of darkening which induces a phobotactic response is strongly dependent on the duration and intensity of the foregoing illumination. Provided that the product of intensity and time of the foregoing illumination is constant, the presentation time of darkening is the longer the higher the foregoing light intensity.

In diatoms, zero threshold intensities have been measured by Heidingsfeld (1943) and Nultsch (1956). While the threshold value in *Navicula radiosa* is about 50 lux (after Heidingsfeld, cited by Nultsch), Nultsch has found values of 3–5 lux in five species of the genera *Navicula*, *Nitzschia* and *Amphora*. As in blue-green algae, the density of accumulations in the light spot was the stronger the higher the light intensity. The presentation time has not yet been investigated in diatoms. Vacations of the light field observed by Heidingsfeld in white light and by Nultsch in red light are probably due to photokinesis, but more detailed investigations of these problems are necessary. Only the negative reactions of *Nitzschia stagnorum* at intensities > 2500 lux seem to be true phototactic responses.

Most of the flagellates investigated so far react positively in dim light and negatively in bright light, as reported by numerous authors (see the review of Haupt, 1959). In some species, as in *Chlamydomonas* (Oltmanns, 1917), negative reactions have never been observed. This may be due to the fact that the intensities used in older investigations were too low. Zero and discrimination thresholds and inversion intensities either have not been measured exactly or are not reliable with regard to the optical equipment and the experimental conditions used. Furthermore, there are considerable difficulties in interpreting many results of former investigations, because the investigators did not separate phobotaxis and topotaxis methodically (see Clayton, 1964), or the results were distorted by such factors as photokinesis and chemotaxis (see above). Finally, threshold values and inversion intensity depend on internal and external factors, as will be shown later in this chapter (page 244).

Action spectra

Since the classical work of Engelmann (1883), Molisch (1907) and Buder (1919) it is known that radiation in purple bacteria is photophobotactically active according to its absorption by the photosynthetic pigments (bacterio-chlorophyll and carotenoids). Later, it was found by several authors (Clayton, 1953a; Duysens, 1952; Manten, 1948a, 1948b; Thomas, 1950) that the action spectra of positive photophobotaxis (Figure 8.7) and photosynthesis in *Chromatium* and *Rhodospirillum* are essentially identical, apart from some inconsistencies in the carotenoid absorption region, which may be due to changes in the carotenoid content of purple bacteria with the increasing age of cultures (Van Niel, Goodwin and Sissins, 1956). Based on these results, the forementioned authors have concluded that the photophobotactic responses in purple bacteria are caused by sudden changes in the rate of photosynthesis. Action spectra of negative photophobotaxis have not yet been measured.

In blue-green algae the first investigations with coloured light were carried

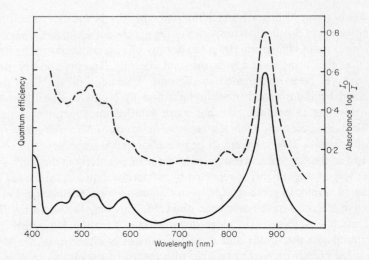

Figure 8.7. Photophobotactic action spectrum (——) and *in vivo* absorption spectrum (– – –) of *Rhodospirillum rubrum*. (After Clayton, 1953a, from Haupt, 1966)

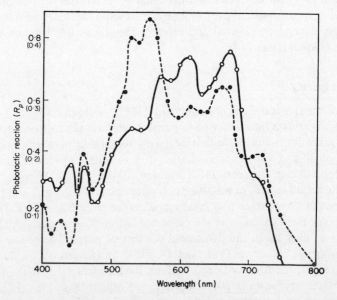

Figure 8.8. Photophobotactic action spectra of the blue-green algae *Phormidium uncinatum* (- ● - - ● -) and *Phormidium* sp. (–○–○–). (Nultsch, 1962c)

out by Dangeard (1910, 1911). He found red and far red light to be photo-phobotactically more active than blue light. These observations have essentially been confirmed by Drews (1959) who used cut-off and inter-ference filters. Action spectra have been measured by Nultsch (1962b, 1962c) in some species of the genus *Phormidium* containing abundant amounts of either phycoerythrin or phycocyanin (Figure 8.8). As the main maxima of photophobotactic activity coincide with the absorption maxima of *C*-phycoerythrin and *C*-phycocyanin respectively, the function of bili-proteins as photoreceptors in photophobotaxis seems to be proved. More-over, in all species investigated a second maximum exists at about 680 nm, which corresponds to the red absorption maximum of chlorophyll *a in vivo*, while the phobotactic activity of blue light is far out of proportion to its absorption by chlorophyll *a*. Altogether, the action spectra of photophobo-taxis resemble those of photosynthesis (Duysens, 1952; Haxo and Norris, 1953). In *Phormidium uncinatum* the action spectrum of photophobotaxis shows a striking similarity with that of $^{14}CO_2$-incorporation (Nultsch and Richter, 1963). In this connexion the activity of far red light between 700 and 750 nm deserves to be mentioned (Figure 8.9).

Action spectra of diatoms have not yet been measured. Heidingsfeld

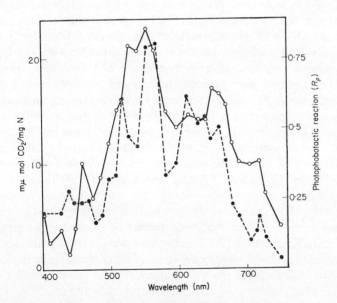

Figure 8.9. Action spectra of photophobotaxis (–○–○–) and $^{14}CO_2$ incorporation (- ● -- ● -) in *Phormidium uncinatum* (Nultsch and Richter, 1963)

(1943) observed photophobotactic responses in blue and red light. Using interference filters, Nultsch (1956) found the whole spectrum between 398 and 705 nm to be more or less phobotactically active, whereas light of wavelengths > 739 nm was inactive. In red light both accumulations and vacations with sharp borders on the one hand and diffuse aggregations in the light field on the other have been observed. We are not able to give a clear interpretation of these confusing results. At least in some cases the vacations seem to be of a photokinetic nature (see above), and the diffuse accumulations could be the result of positive chemotaxis towards oxygen produced by photosynthesis (aerotaxis). Thus, in diatoms correlations between photosynthesis and photophobotaxis are not proved, but are not out of the question.

Although in flagellates topotaxis and phobotaxis have not been clearly distinguished and methodically separated by the former investigators, as mentioned above, it seems to be proved from the action spectra that chlorophyll is not the photoreceptor because of the inefficiency of red light. Engelmann (1882) has observed accumulation maxima of *Euglena* in the range between 470 and 490 nm of a microspectrum. Mast (1917) and Oltmanns (1917) have also found maximum activity at these wavelengths, in positive and in negative reactions as well, with an additional subsidiary maximum at 405 nm (Oltmanns, 1917). The spectral phototactic sensitivity of *Phacus* and *Trachelomonas* is reported to be the same (Mast, 1917). From the experimental conditions one has to conclude that Mast has investigated the topotactic response, whereas Engelmann and Oltmanns, using the 'light-trap' method, have probably investigated photophobotaxis. However, since the light laterally scattered by the organisms under certain circumstances can act as a secondary light source (see above), a superposition of the phobotactic response by topotaxis cannot be excluded.

This possible source of error, of course, must always be considered in interpretation of phototactic action spectra. In *Euglena*, Bünning and Schneiderhöhn (1956) and Bünning and Gössel (1959) found different action spectra of negative photophobotaxis which depended on the pre-illumination of the organisms, whereas Wolken and Shin (1958) found only quantitative differences in the action spectra of *Euglena* cells adapted to light and dark respectively. As Bünning has never observed positive phobotactic reactions in his *Euglena* strain, he could not compare the action spectra of positive and negative photophobotaxis. Unlike Bünning and his co-workers, who found only light of short wavelengths (up to about 540 nm) active, Wolken and Shin (1958) have reported phototactic activity of red light. However, from the experimental conditions it seems possible that the red-light effect is due to photokinesis. Recently, action spectra of positive and negative photophobotaxis in *Euglena* have been measured by Diehn (1969), using the phototaxigraph (see above). The action spectrum of positive phobotaxis shows two distinct peaks at 375 and 480 nm, while that of negative phobotaxis exhibits major sharp peaks at 450 and 480 nm and a minor one at 412 nm. A major broad band of negative phototactic activity appears below 400 nm when the plane of polarized light is parallel to the long axis of the organisms. If it is perpendicular to the long axis, no negative phototaxis occurs below 425 nm. Different action spectra in polarized and non-polarized light have also been measured by Wolken and Shin (1958).

Although all findings concerning the effect of coloured light on photo-phobotaxis in flagellates, with the exception of the observations of Wolken and Shin (1958), agree in the one point that only blue and blue-green light and, as far as has been investigated, near ultraviolet are active, we do not know the chemical nature of the photoreceptor. The reason for this is the striking inconsistencies in the position of the peaks in the action spectra published. Carotenoids and flavoproteins are in discussion, but mainly with respect to phototopotaxis (see below).

In colourless forms of *Euglena* with and without a stigma, Gössel (1957) found the maximum activity at 410 nm. Provided that these action spectra represent the absorption spectrum of the photoreceptor, neither carotenoids nor flavoproteins come into the question. On the other hand, based on these results, a distortion of phototactic action spectra by photosynthesis, presumed by Halldal (1958), can certainly be excluded. Finally, from Diehn's (1969) experiments we must conclude that positive and negative photo-phobotactic reactions are mediated by different photoreceptors. Thus, most of the reported experiments have necessarily to be repeated under more exact experimental conditions in order to obtain reliable information of the chemical nature of the photoreceptors in photophobotaxis of flagellates.

Correlations between photophobotaxis and photosynthesis

The striking similarity between the action spectra of photosynthesis and of photophobotaxis in purple bacteria led Manten (1948a, 1948b) to the conclusion that photophobotactic reactions are caused by sudden changes in the rate of photosynthesis. Thomas and Nijenhuis (1950) found that the saturation levels of photosynthesis and of photophobotaxis are reached at the same light intensity. Clayton's (1953b) experiments gave evidence that, even if photosynthesis is light-saturated, a sudden change in light intensity can cause a transient change in the steady state of photosynthesis and, hence, initiate a phototactic response. As a general hypothesis, Links (1955) has suggested that the phobotactic response is caused by a sudden decrease

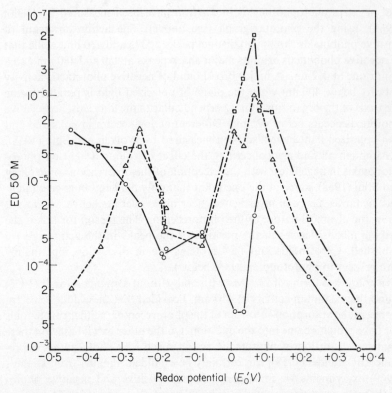

Figure 8.10. Effect of redox systems on photokinesis (–○–○–), dark movement (- △- - △-) and photophobotaxis (- □ - ·· - □ -) in *Phormidium uncinatum* (Nultsch, 1968a)

in the energy supply, i.e. in the ATP supply from photophosphorylation in the case of phototaxis. The investigations on photophobotactic action spectra in blue-green algae (Nultsch, 1962b, 1962c; Nultsch and Richter, 1963) support Manten's idea of correlations between photophobotaxis and photosynthesis. However, because of the fact that the action spectrum of photokinesis resembles that of the first light reaction whereas the action spectrum of photophobotaxis is more like that of the second one in biliprotein-containing organisms, Nultsch concluded that both photokinesis and photophobotaxis are coupled with photosynthesis, but in different ways and at different points. After he had shown that photokinesis is linked to photophosphorylation (see above), he rejected the hypothesis of Links (1955), because it is unlikely that two reactions which are both coupled with the same photoprocess have different action spectra.

Therefore, Nultsch and Jeeji-Bai (1966) investigated the effect of photo-synthetic inhibitors on the photophobotactic response. As they found that photophobotaxis is more sensitive to inhibitors of the non-cyclic electron transport than to those of the cyclic one, they concluded that photophobotaxis is coupled with the non-cyclic electron transport. Because of its sensitivity to DCMU, which is low compared with other inhibitors (Nultsch, 1965), and because of the strong activity of the light absorbed by biliproteins, they presumed the coupling point to be located in the region of the accessory pigments—but before the DCMU block (Duysens, 1963, 1964). In further experiments Nultsch (1966a, 1966b, 1967) has studied the effect of un-couplers on photophobotaxis. Most of the uncouplers investigated, with the exception of dinitrophenol and atebrin, are less effective on photophobotaxis than on photokinesis. However, the interpretation of these results meets with serious difficulties because of the strong inhibitory effect of un-couplers on the dark movement. Of course, no photophobotactic responses can occur when the movement in the dark is completely inhibited.

Recently, further attempts have been made by Nultsch (1966a, 1968a) to locate the position of the linkage between photophobotaxis and photo-synthesis by measuring the inhibitory effect of numerous redox systems on photophobotactic responses of *Phormidium uncinatum*. The redox potentials (E_0') of these substances covered the range between -0.44 and $+0.36$ V. As a measure of the inhibitory effect the ED 50, i.e. the concentration in M that inhibited photophobotactic reactions to an extent of 50 per cent of the control, has been used. As shown in Figure 8.10, redox systems with E_0' values between ±0 and $+0.1$ V most strongly inhibit the phobotactic responses. The maximum inhibition is shown by thionin ($E_0' + 0.064$ V). A second smaller maximum lies below -0.25 V. In this range methyl-viologene ($E_0' - 0.44$ V) is the most effective substance. Based on these results Nultsch suggested that photophobotaxis is coupled with the electron flow along the electron transport chain of photosynthesis, and that the linkage is not restricted to the second light reaction, although the regions of electron transfer from the two light reactions to the next electron acceptors are most sensitive to externally added redox systems and the maximum sensitivity has been found at E_0' values of the first electron acceptors of the second light reaction. Therefore, the action spectrum resembles that of oxygen evolution and carbon dioxide fixation, but is quite different from the action spectrum of photokinesis, which mainly represents that of cyclic photophosphorylation.

This conception is supported by recent studies of Throm (1968) and Nultsch and Throm (1968) on the effect of inhibitors on photophobotaxis in *Rhodospirillum rubrum*. They have shown that uncouplers which inhibit photokinesis do not influence photophobotaxis, whereas inhibitors of the photosynthetic electron flow, with the exception of DCMU, inhibit both photophobotaxis and photokinesis. Moreover, concentrations of ATP which considerably increase the speed of the movement have no significant effect on the photophobotactic response.

These results are convincing arguments against the hypothesis of Links (1955), and favour the conception that photophobotaxis, at least in purple bacteria and blue-green algae, is linked to the photosynthetic electron transport, whereas photokinesis is coupled with photophosphorylation. Accordingly, one has to conceive that the photophobotactic responses are induced by transient changes in the steady state of the photosynthetic electron flow caused by sudden changes in light intensity and, hence, in the redox state of the cell. Since correlations between photosynthesis, ion uptake (MacRobbie, 1965) and bioelectric potentials (Schilde, 1966) are well documented by investigations in several algae, one could imagine that changes in the steady state of the electron transport are transformed into bioelectric potential changes, possibly action potentials, which are quickly conducted by the cell membrane. In this way, the excitation could be transmitted to the locomotor apparatus, causing a reversal of the beating direction of the flagella (Clayton, 1959). Such correlations between action potentials and flagellar activity have been observed by Hisida (1959) in *Noctiluca*.

Although correlations between photophobotaxis and photosynthesis in blue-green algae and purple bacteria seem to be well established, we are far from an understanding of the mechanism of photophobotactic responses in flagellates. Of course, the photoreceptor pigment which absorbs violet and blue light could also induce bioelectric potentials, but no observation exists to favour such a hypothesis.

Phototopotaxis

The term phototopotaxis designates a phototactic reaction type in which the response is related to the direction of the incident light beam. In other words, phototopotaxis results in a movement either towards the light source (positive) or away from it (negative). This does not necessarily mean that the phototopotactic orientation is the result of a steering act, or a succession of steering acts. On the contrary, there are rather different types of topotactic reactions developed in microorganisms which depend on morphology of the organisms as well as on the mechanisms of the movement.

Modes of phototopotactic reactions

A widely spread (probably the phylogenetic original) type of photo-topotactic orientation is a result of trial and error motion. This type is realized in blue-green algae of the family Oscillatoriaceae (Drews, 1957, 1959), in diatoms (Nultsch, 1956) and probably in the purple bacterium *Rhodospirillum rubrum*, which, according to the recent investigations of Throm (1968), can also react phototopotactically. Under diffuse light the diatoms and the filaments of the Oscillatoriaceae normally show an alternating forward and backward movement, like a shunting-engine, without preferring any direction. As the cells and filaments do not exhibit any morphological polarity, we cannot use the terms 'forward' and 'backward' in the strict sense of the words. The reversals of the movement ensue autonomously, even if their frequency seems to be influenced by external factors. The orientation of the movement in the individuals of a population is, of course, random. After switching on the unilateral actinic light no steering reaction, i.e. no active change in the direction of the movement, occurs. However, as has been shown by Nultsch (1956) and Drews (1959), the rhythm of autonomous reversals is considerably changed. Consequently, in positively reacting individuals which are in a more or less parallel position to the light beam, the ways towards the light source are prolonged, while the ways away from it are shortened (Figure 8.11), and vice versa in the

Light

Figure 8.11. Scheme of positive phototopotactic movement in diatoms of the *Navicula* type and blue-green algae of the *Phormidium* type. The arrow indicates the light direction. (Changed after Nultsch, 1956, and Drews, 1959)

negative reaction. The frequency of reversals is decreased in this way. In organisms which move about perpendicularly to the light beam, the unilateral light has no topotactic effect at all. The effect of unilateral light on rhythm and rate of reversals is the stronger the more acute the angle between the long axis of the organism and the direction of the light beam. As the direction of movement in blue-green algae and diatoms is never absolutely straight, all individuals of a population come, sooner or later, for a shorter or longer period of time, into a more or less 'parallel' position and approach the light source by degrees in positive reactions or withdraw from it in negative reactions. As a result, the organisms aggregate at either the front or back side of the culture vessel.

Blue-green algae of the family Nostocaceae, e.g. some species of the genus *Anabaena* and *Cylindrospermum*, display phototopotactic reactions *sensu strictu*, i.e. an active orientation towards the light source by means of a steering mechanism (Drews, 1959). Contrary to the Oscillatoriaceae, the filaments of *Anabaena* do not rotate around their long axis. Under unilateral illumination most of the filaments form a U, the bent part of it being orientated towards the light source and the symmetrical axis of it being parallel to the light beam. Both halves of the U-shaped filament glide with the same speed. When the light direction is changed the bent 'tip' begins to turn until the U is again directed to the light source. Straight filaments, perpendicularly orientated to the light direction, begin to bend after the onset of illumination. Obviously, this bending is due to a different behaviour of the shadowed and the illuminated sides of the filaments, comparable to phototropic responses.

Euglena and other flagellates also exhibit true phototopotactic responses so far as they react on the direction of the incident light and not on a spatial gradient of light intensity. This has been shown by Buder (1917) in experiments with parallel and convergent light beams. But contrary to *Anabaena*, the *Euglena* cells do not steadily curve into the direction of the unilaterally incident light beam after the onset of illumination. Instead, the steering act consists of several steps, each of which brings the cell into a position in which the angle between the long axis and the direction of the rays is more acute than before. Every small change in the direction of movement is caused by a slight sideward beating of the flagellum, like a photophobotactic reaction on a small stimulus. This happens just at the moment when the stigma of the rotating cell which is located at the dorsal side (see above) passes between the flagellar basis and the light source; this shadows the paraflagellar body which is considered to be the photoreceptor (Vávra, 1956). The result is a slight deviation to the dorsal side, i.e. to the light source, in the case of positive reaction. Thus, in *Euglena* the phototopotactic reactions seem to be composed of numerous small photophobotactic reactions induced by periodical shadowing of the photoreceptor. These take place until the flagellated end of the cell points to the light source. In this position, from morphological reasons, the stigma cannot shadow the paraflagellar body.

However, it has been emphasized by Halldal (1958) that numerous phototactic flagellates exist which handle the steering act perfectly well, although they lack a stigma and, consequently, cannot react by means of a periodic shading mechanism. Therefore, *Euglena* must be regarded as a unique type which represents an exception rather than a model of phototopotactic orientation in flagellates.

In desmids, such as *Closterium* and *Micrasterias*, which move slowly over solid substrates by excretion of mucilage, phototopotactic reactions have been observed by Stahl (1880), Klebs (1885) and Aderhold (1888). Recently, the phototactic behaviour of *Micrasterias* has been reinvestigated by Bendix (1960) and Neuscheler (1967a, 1967b). The *Micrasterias* cells, which produce mucilage only at their rear poles, exhibit an oscillating movement, resulting in meander traces of mucilage. From time to time the cells overturn due to a change of polarity, interrupting the traces of mucilage. Under unilateral illumination the oscillations decrease and the traces are straighter and are directed towards the light source. The cells orientate their long axes parallel to the incident light beam, in the case of a positive reaction, and perpendicular to it, in the case of phototactic indifference.

Methods of investigation

In older investigations phototopotactic reactions have been estimated only qualitatively. The investigators illuminated the vessels containing the test organisms unilaterally and observed whether individuals accumulated at the 'positive' or 'negative' side of the vessel. Semiquantitative experiments in *Euglena* have been carried out by Bünning and Schneiderhöhn (1956) and Gössel (1957), who evaluated the threshold intensities of phototaxis at different wavelengths and thus measured action spectra.

Methods for quantitative evaluation of phototopotactic responses of flagellates have been devised by Halldal (1958) and Feinleib and Curry (1967). A detailed description of their instruments and methodical details is given by Hand and Davenport (see Chapter 9 of this treatise).

A monitoring system similar to that described by Feinleib and Curry (1967) has been used by Throm (1968) to measure the phototopotactic response of the purple bacterium *Rhodospirillum rubrum*. Unlike Feinleib and Curry, however, he mounted the cuvette equipped with two photocells at both its ends in an angle of 45° to the light beam; in this way it serves simultaneously as both a measuring and actinic beam. A further difference is that at the beginning of the experiment Throm let the cell suspension flow continuously through the cuvette. If the suspension is homogeneous the recorded baseline is nearly straight, because the bacteria are mechanically prevented by the stream of the medium from aggregating at one side of the cuvette or the other. Immediately after the flow of the suspension has been stopped, the organisms begin to move to one side of the cuvette, producing a difference in output between the two photocells. The resulting time course of the response recorded by this system is nearly linear for several minutes. Its slope depends on the quality and quantity of the stimulus.

Diatoms and blue-green algae of the family Oscillatoriaceae, transferred onto a slide or an agar plate, normally spread circularly or elliptically from the inoculation spot under diffuse light. Under unilateral illumination this spreading area is shifted to the light source in the case of positive phototopotaxis. The shifting is the larger the stronger the phototactic effect of the incident light (Drews, 1959; Nultsch, 1956). The extent of shifting, corrected to eliminate any distortion by photokinesis, has been used as an index of the phototopotactic response by Nultsch (1961). To measure phototopotactic responses and action spectra in the desmid *Micrasterias*, Neuscheler (1967a, 1967b) has devised a double-beam method in which the actinic light beam enters the cuvette rectangularly to the reference beam of constant intensity and wavelength. As a measure of phototopotactic orientation he used tg α, being the angle between the direction of movement and the direction of the reference beam. If the two light beams have the same wavelength the law of resultants is valid, i.e. the orientation depends on the sum of the physical vectors. However, this is not true for two beams of different colours, e.g. blue versus red. In this case around $\alpha = 45°$, if the blue light intensity is doubled, tg α is increased by the factor 1·15–1·35.

Intensity dependence of phototopotactic reactions

Phototopotactically reacting flagellates investigated so far respond positively in dim light and negatively in bright light (Bancroft, 1913; Buder, 1917; Halldal, 1958; Mainx, 1929). As in photophobotaxis, the intensity ranges of positive and negative reactions are separated by a smaller or broader indifference zone (Buder, 1917), in which the sense of response is inverted and the cells display a random orientation in movement. The same seems to be true in blue-green algae (Nultsch, 1961, 1962c), in which inversion intensities between 1000 and 10 000 lux have been found. In diatoms, negative reactions have been investigated by Heidingsfeld (1943) in *Navicula radiosa*, while Nultsch (1956) observed only an indifferent behaviour in *Nitzschia stagnorum* at higher intensities (> 2500 lux). Recently, Throm (1968) has shown that purple bacteria (*Rhodospirillum rubrum*), which are reported not to have any phototopotaxis at all (Buder, 1919), exhibit clear, though slight, topotactic responses under unilateral illumination. They react positively and negatively as well; the inversion intensity is about 18 000 lux. Studying the phototopotactic reaction of desmids, Stahl (1880) observed that at low intensities the *Closterium* cells orientated their long axes parallel to the light beam and then moved slowly towards the light source. At inversion intensities, the cells stopped and orientated their long axes perpendicularly to the incident light beam. At high intensities the cells displayed negative reactions. These observations have been confirmed by several authors (Aderhold, 1888; Pringsheim, 1912; Kol, 1927) and, recently, by Neuscheler (1967a, 1967b) in *Micrasterias denticulata*.

Correlations between threshold values and inversion intensities of photo-topotaxis in flagellates have been investigated by Mainx (1929) and Mainx and Wolf (1940). They found that factors which increase the inversion intensity decrease the threshold values of positive phototopotaxis. In diatoms zero threshold intensities of phototopotaxis have been measured by Heidingsfeld (1943) and Nultsch (1956). According to Heidingsfeld the threshold value in *Navicula radiosa* lies between 10 and 30 lux, while Nultsch found threshold values at 1–3 lux in some species of the genera *Navicula*, *Amphora* and *Nitzschia*. Several blue-green algae of the genera *Phormidium* and *Oscillatoria* investigated by Drews (1959) and Nultsch (1962c) displayed threshold values of between 1 and 30 lux. Nultsch (1962c), who also measured the optima of phototopotactic reactions, found them at 50–200 lux in several *Phormidium* species. The zero threshold value of the desmid *Micrasteria* is reported to lie at very low light intensities, between 10^{-4} and 10^{-5} lux, while the inversion intensity is about 4000 lux (Neu-scheler, 1967b).

It will be shown later (page 244) that threshold values and inversion intensities of phototopotaxis, just as in photophobotaxis, depend on internal and external factors.

Action spectra

Action spectra are of the greatest importance in answering the question as to whether or not positive and negative reactions on the one hand and phototopotaxis and photophobotaxis on the other are mediated by the same photoreceptors and reaction mechanisms. In purple bacteria this question is undecided because the phototopotactic action spectrum of *Rhodospirillum rubrum* is not yet known. The same is the case with diatoms. Although Nultsch (1956) has shown that in phototopotaxis, unlike in photophobotaxis, only light of wavelengths shorter than 550 nm is active, neither topotactic nor phobotactic action spectra have been measured so far in these organisms.

In blue-green algae the first investigations on the phototopotactic effect of coloured light were carried out by Pieper (1915). He found that filaments of the genus *Oscillatoria* reacted positively at wavelengths longer than 575 nm and negatively at wavelengths shorter than 500 nm. Between 500 and 575 nm positive responses occurred in dim light and negative ones in bright light. Drews (1959), using cut-off and interference filters, observed positive reactions in the range between 400 and 610 nm in *Phormidium uncinatum* and between 400 and 720 nm in *Oscillatoria mougeotii*. In *Anabaena variabilis* and *Cylindrospermum* positive reactions occurred in orange-red light, whereas violet and blue light were ineffective. Action spectra have been measured by Nultsch (1961, 1962c) in *Phormidium autumnale* and *Phormidium uncinatum* (Figure 8.12). The action spectra of these species, the

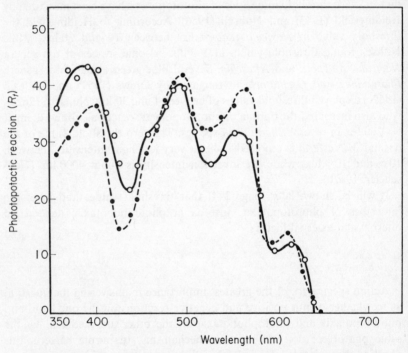

Figure 8.12. Phototopotactic action spectra of the blue-green algae *Phormidium autumnale* (–O–O–) and *Phormidium uncinatum* (- ● - - ● -) (Nultsch, 1961, 1962c)

pigment content of which is very similar, are essentially the same, showing maxima in the absorption region of carotenoids (490–500 nm) and phycoerythrin (560–570 nm). A small subsidiary maximum at 615 nm may be due to phycocyanin, only small amounts of which are present in both species. Therefore, Nultsch has concluded that both carotenoids and biliproteins are the photoreceptors of phototopotaxis in blue-green algae. Since radiation between 350 and 410 nm is only slightly absorbed by these pigments, the phototopotactic effect of the violet and near ultraviolet must be due to another photoreceptor of unknown chemical structure. Chlorophyll *a* can be excluded because of the inefficiency of red light > 650 nm and the minimum in the action spectrum between 430 and 440 nm (see Figure 8.12).

Action spectra of phototopotaxis in the desmid *Micrasterias* have been measured by Bendix (1960) and Neuscheler (1967b), both showing maximum activity in blue and red light. However, Bendix has not considered the possible distortion of the action spectrum by photokinesis. Therefore, we are not able to decide whether the action spectrum represents phototopotaxis or photokinesis, or both. Neuscheler has clearly separated phototopotaxis and photokinesis methodically. Thus, the action spectrum measured by him really represents phototopotaxis. As it demonstrates the greatest effectiveness at short wavelengths between 370 and 470 nm and in the red between 650 and 666 nm, chlorophyll may be an effective light absorber for phototopotaxis in this organism. However, the broad range of acticity in the blue and near ultraviolet requires that at least one more pigment is involved in the light absorption at shorter wavelengths.

Although the discussion of action spectra in flagellates meets the difficulties mentioned above, because of methodical uncertainties on the one hand and some inconsistencies in the results reported by several authors on the other, the action spectra of phototopotaxis have one feature in common: only light of shorter wavelengths in the visible region is active. Most of the flagellates investigated so far show a main maximum of phototactic sensitivity between 475 and 495 nm and a secondary one between 425 and 435 nm (Halldal, 1958). Only the dinoflagellate *Prorocentrum micans* exhibits maximum sensitivity around 570 nm. Action spectra of positive and negative topotaxis measured in the same organism and under the same or at least comparable conditions are essentially identical (Halldal, 1958, 1961). Investigating the effect of ultraviolet on phototaxis of the Volvocale *Platymonas subcordiformis*, Halldal (1961) has found reaction maxima around 220, 275 and 335 nm.

Conclusions

Although numerous papers dealing with phototopotaxis of unicellular algae have been published, little is known about the mechanisms of phototopotactic orientation and the photoreceptors involved. In *Euglena* it has been suggested that the phototopotactic orientation is the result of numerous successive phototactic reactions which are caused by periodical shadowing of the paraflagellar body by the eye spot. In this case, the action spectrum of phototopotaxis should be composed of the absorption spectra of the photoreceptor and the shadowing pigments as well. However, as mentioned above (page 238), the behaviour of *Euglena* under unilateral illumination must be regarded as a unique type of phototopotactic orientation, which cannot be generalized.

In the case of a positive reaction in other flagellates, the cells orientate their long axes so that the photoreceptor at the flagellar basis is illuminated directly. Whenever the photoreceptor is shaded by any part of the cell, the direction of movement is adjusted (Halldal, 1958). Because of its spectral sensitivity at shorter wavelengths of the visible spectrum, the photoreceptor is suggested to be a carotenoprotein (Halldal, 1961, 1963), although flavoproteins are not out of the question.

In the case of a trial-and-error motion of blue-green algae and diatoms we have no idea of how the effect of light on rhythm and the rate of reversals may come about, although it seems to be proved that in blue-green algae of the genus *Phormidium* carotenoids and biliproteins function as photoreceptors. Nothing is yet known about the mechanism of phototopotactic orientation in Nostocaceae and desmids.

Influence of internal and environmental factors on phototaxis

Since Strasburger's classical work (1878) it is known that phototactic sensitivity depends on the momentary disposition of the organisms, the so-called phototaktische Stimmung. This concerns topotaxis and phobotaxis as well. The disposition is influenced by internal and external factors, and a change in the phototactic disposition (Stimmungsänderung) results in a change of inversion intensities and threshold values as well.

Internal factors

One of the most important internal factors which determine the phototactic disposition is the age of the individuals. Mainx (1929) and Luntz (1931a) have reported that in *Volvox* the inversion intensity of phototopotaxis is the lower the older the colonies. Moreover, the females display lower inversion intensities than the vegetative colonies (Oltmanns, 1892). In several green algae, as in *Monostroma, Hydrodictyon, Acetabularia* and *Ulva* (Hämmerling, 1944; Haxo and Clendenning, 1953; Mainx, 1929; Strasburger, 1878), the gametes react positively, whereas the planozygotes show negative topotactic reactions shortly after copulation. Furthermore, the phototactic reactivity can also depend on the age of cultures and individuals respectively, as has been found by Rothert (1901) and Mast (1926).

Another internal factor which can influence the phototactic behaviour is the pigment content of the organisms. This seems to be evident, as the pigments are either the photoreceptors or shadow the photoreceptor region (see above).

In *Euglena* the phototactic reactivity shows circadian fluctuations which are controlled by the 'physiological clock' (Bruce and Pittendrigh, 1956; Diehn and Tollin, 1966; Pohl, 1948). In addition, *Euglena* also displays a diurnal rhythm in motility (Brinkmann, 1966), which in some cases may superimpose the endodiurnal rhythms of phototactic sensitivity. Endo-diurnal changes in the speed of movement in the light and in the dark, as well as in phototopotaxis, have also been observed in *Micrasterias denticulata* by Neuscheler (1967b).

Environmental factors

External factors which influence the phototactic disposition are light, temperature, carbon dioxide concentration, concentration and the ratio of certain ions and other chemical factors.

Although the effect of light has been investigated most frequently its mode of action is far from being clear because of contradictory results reported by numerous authors. While Strasburger (1878) and Oltmanns (1892, 1917) have observed that the 'phototactic disposition' is increased by a longer preillumination and vice versa, Jacobsen (1910), Mast (1918), Mainx (1929) and Mainx and Wolf (1940) found the contrary to be true. Moreover, while in Mast's (1919) experiments only light of shorter wave-lengths (violet, blue and green) was active, Terry (1906) observed a strong effect of red light on the galvanotactic disposition, which usually changes in the same sense as the phototactic one (Mast, 1927). Brucker (1954) has concluded from his investigations on the phototactic behaviour of *Lepo-cinclis texta* that the effect of light is an indirect one and may be due to the change in the carbon dioxide concentration of the medium caused by photosynthesis. Unlike Brucker, Halldal (1960) did not find that the photo-tactic disposition depended on the carbon dioxide tension of the medium. A more direct effect of light by photosynthesis has been presumed by Nultsch (1962a), who found that in the blue-green alga *Phormidium uncinatum* the phototactic sensitivity is decreased by preillumination in bright light.

The most important chemical factor that influences phototactic sensi-tivity and disposition is the concentration of certain ions in the medium. As has been shown by Halldal (1957, 1958, 1959, 1960), the phototactic behaviour of *Platymonas subcordiformis* is determined by a distinct $Mg^{2+}/$

Ca^{2+} ratio. If it is greater than six, the *Platymonas* cells react positively; if it is less than six, they respond negatively; if it equals six, the movement displays random orientation. Moreover, the phototactic behaviour also depends on the Ca^{2+}/K^+ ratio, while in media containing Mg^{2+} and K^+ ions the cells react positively at any Mg^{2+}/K^+ ratio. Based on these findings Halldal has concluded that light influences the phototactic disposition indirectly by changing the intracellular ratio of forementioned cations. Adjusting the cells close to the 'inversion intensity' by using a distinct ion ratio, Halldal has found that in *Platymonas* the − to + transformation is brought about by violet, blue and green light (400–540 nm), as well as by red light (660–685 nm), while in the + to − transformation light of wavelengths between 555 and 660 nm is active. These action spectra for phototactic disposition, which must not be confused with action spectra of phototaxis, point out a possible participation of chlorophyll in light absorption and, hence, correlations with photosynthetic processes (Clayton, 1964).

The effects of other chemical agents as inhibitors, uncouplers and redox systems have already been discussed in the foregoing chapters.

Little is known about the effect of temperature. Mast (1918, 1919) and Jacobsen (1910) reported that the inversion intensity is increased by increasing temperature in some volvocales and vice versa. On the other hand, Mainx (1929) has found in *Gonium*, *Eudorina* and *Synura* that the inversion intensity is decreased by an increase and a decrease of temperature as well. Therefore, more detailed investigations are necessary to clarify the effect of environmental factors on the phototactic sensitivity and disposition.

REFERENCES

Aderhold, R. (1888). Beitrag zur Kenntnis richtender Kräfte bei der Bewegung niederer Organismen. *Jena. Z. Naturw.*, **22** (N.F. **15**), 310.

Arnon, D. J., H. Y. Tsujimoto, and B. O. McSwain (1965). Photosynthetic phosphorylation and electron transport. *Nature*, **207**, 1367.

Baltscheffsky, H., and D. Y. de Kiewiet (1964). Existence and localization of two phosphorylation sites in photophosphorylation of isolated spinach chloroplasts. *Acta chem. scand.*, **18**, 2406.

Bancroft, F. W. (1913). Heliotropism, differential sensibility and galvanotropism in *Euglena*. *J. exp. Zool.*, **15**, 383.

Bendix, S. (1960). Pigments in Phototaxis. In M. B. Allen (Ed.), *Comparative Biochemistry of Photoreactive Systems*, Academic Press, New York and London. pp. 107–127.

Bolte, E. (1920). Uber die Wirkung von Licht und Kohlensäure auf die Beweglichkeit grüner und farbloser Schwärmer. *Jb. wiss. Bot.*, **59**, 287.

Brinkman, K. (1966). Temperatureinflüsse auf die circadiane Rhythmik von *Euglena gracilis* bei Mixotrophie und Autotrophie. *Planta*, **70**, 344.

Bruce, V. G., and C. S. Pittendrigh (1956). Temperature independence in a unicellular 'clock'. *Proc. natn Acad. Sci.*, **42**, 676.

Brucker, W. (1954). Beiträge zur Kenntnis der Phototaxis grüner Schwärmzellen. *Arch. Protistenk.*, **99**, 294.

Buder, J. (1915). Zur Kenntnis des *Thiospirillum jenense* und seiner Reaktion auf Lichtreize. *Jb. wiss. Bot.*, **56**, 529.

Buder, J. (1917). Zur Kenntnis phototaktischer Richtungsbewegungen. *Jb. wiss. Bot.*, **58**, 105.

Buder, J. (1919). Zur Biologie des Bakteriopurpurins und der Purpurbakterien. *Jb. wiss. Bot.*, **58**, 525.

Bünning, E. und I. Gössel (1959). Ergänzende Versuche über die phototaktischen Aktionsspektren von *Euglena*. *Arch. Mikrobiol.*, **32**, 319.

Bünning, E. und G. Schneiderhöhn (1956). Uber das Aktionspektrum der phototaktischen Reaktionen von *Euglena*. *Arch. Microbiol.*, **24**, 80.

Clayton, R. K. (1953a). Studies in the phototaxis of *Rhodospirillum rubrum*. I. Action spectrum, growth in green light, and Weber Law adherence. *Arch. Mikrobiol.*, **19**, 107.

Clayton, R. K. (1953b). Studies in the phototaxis of *Rhodospirillum rubrum*. II. The relation between phototaxis and photosynthesis. *Arch. Mikrobiol.*, **19**, 125.

Clayton, R. K. (1957). Patterns of accumulation resulting from taxes and changes in motility of microorganisms. *Arch. Mikrobiol.*, **27**, 311.

Clayton, R. K. (1958). On the interplay of environmental factors affecting taxis and motility in *Rhodospirillum rubrum*. *Arch. Mikrobiol.*, **29**, 189.

Clayton, R. K. (1959). Phototaxis of purple bacteria. In W. Ruhland (Ed.), *Handbuch der Pflanzenphysiologie*, Vol. 17/1. Springer, Berlin, Göttingen, Heidelberg. pp. 371–387.

Clayton, R. K. (1964). Phototaxis in microorganisms. In A. C. Giese (Ed.), *Photophysiology*, Vol. II. Academic Press, New York, London. pp. 51–77.

Dangeard, P. A. (1910). Phototactisme, assimilation, phénoméne de coissance. *Bull. Soc. bot.*, *France*, **57**, 315.

Dangeard, P. A. (1911). Sur les conditions de l'assimilation chlorophylienne chez les Cyanophycee. *C.R.A. Sc.*, *Paris*, 152.

Diehn, B. (1969). Action spectra of the phototactic responses in *Euglena*. *Biochim. biophys. Acta*. (In the press.)

Diehn, B., and G. Tollin (1966). Phototaxis in *Euglena*. II. Physical factors determining the rate of phototactic response. *Photochem. Photobiol.*, **5**, 523.

Drews, G. (1957). Die phototaktischen Reaktionen einiger Cyanophyceen. *Ber. dt. bot. Ges.*, **70**, 259.

Drews, G. (1959). Beiträge zur Kenntnis der phototaktischen Reaktionen der Cyanophyceen. *Arch. Protistenk.*, **104**, 27.

Duysens, L. N. M. (1952). Transfer of excitation energy in photosynthesis. *Diss.*, *Utrecht*.

Duysens, L. N. M. (1963). Photosynthesis mechanisms in green plants. *Nat. Acad. Sci.-Nat. Research Council*, **Publ. 1145.**

Duysens, L. N. M. (1964). Photosynthesis. *Progr. in Biophys.*, **14**, 1.

Engelmann, Th. W. (1882). Uber Licht- und Farbenperception niederster Organismen. *Pflügers Archl ges. Physiol.*, **29**, 387.

Engelmann, Th. W. (1883). Bacterium photometricum. *Pflügers Arch. ges. Physiol.*, **30**, 95.

Engelmann, Th. W. (1888). Die Purpurbakterien und ihre Beziehungen zum Licht. *Bot. Ztg.*, **46**, 661.

Feinleib, M. E. H., and G. M. Curry (1967). Methods for measuring phototaxis of cell populations and individual cells. *Physiologia Pl.*, **20**, 1083.

Gössel, I. (1957). Über das Aktionsspektrum der Phototaxis chlorophyllfreier Euglenen und über die Absorption des Augenflecks. *Arch. Mikrobiol.*, **27**, 288.

Gromet-Elhanan, Z., and D. I. Arnon (1965). Effect of desaspidin on photosynthetic phosphorylation. *Pl. Physiol.*, **40**, 1060.

Gromet-Elhanan, Z., and M. Avron (1966). Desaspidin: non-specific uncoupler of photophosphorylation. *Pl. Physiol.*, **41**, 1231.

Halldal, P. (1957). Importance of calcium and magnesium ions in phototaxis of motile green algae. *Nature*, **179**, 215.

Halldal, P. (1958). Action spectra of phototaxis and related problems in Volvocales, *Ulva*-Gametes and Dinophyceae. *Physiologia Pl.*, **11**, 118.

Halldal, P. (1959). Factors affecting light response in phototactic algae. *Physiologia Pl.*, **12**, 742.

Halldal, P. (1960). Action spectra of induced phototactic response changes in *Platymonas*. *Physiologia Pl.*, **13**, 726.

Halldal, P. (1961). Ultraviolet action spectra of positive and negative phototaxis in *Platymonas subcordiformis*. *Physiologia Pl.*, **14**, 133.

Halldal, P. (1962). Taxes. In R. A. Lewin (Ed.), *Physiology and Biochemistry of Algae*, Academic Press, New York, London. pp. 583–593.

Halldal, P. (1963). Zur Frage des photoreceptors bei der Topophototaxis der Flagellaten. *Ber. dr. bot. Ges.*, **76**, 323.

Halldal, P. (1964). Phototaxis in Protozoa. In S. H. Hunter (Ed.), *Biochem. and Physiol. of Protozoa*, Vol. 3. Academic Press, New York, London. pp. 277–297.

Hämmerling, J. (1944). Zur Lebensweise, Fortpflanzung und Entwicklung verschiedener Dasycladaceen. *Arch. Protistenk.*, **97**, 7.

Harder, R. (1920). Uber die Reaktionen freibeweglicher Organismen auf plötzliche Änderungen der Lichtintensität. *Z. Bot.*, **12**, 353.

Haupt, W. (1959). Die Phototaxis der Algen. In W. Ruhland (Ed.), *Handbuch der Pflanzenphysiologie*, Vol. 17/1. Springer, Heidelberg. pp. 318–370.

Haupt, W. (1965). Perception of environmental stimuli orienting growth and movement in lower plants. *A. Rev. Pl. Physiol.*, **16**, 267.

Haupt, W. (1966). Phototaxis in plants. *Int. Rev. of Cytol.*, **19**, 267.

Haxo, F. T., and K. A. Clendenning (1953). Photosynthesis and phototaxis in *Ulva lactuca* gametes. *Biol. Bull.*, **105**, 103.

Haxo, F. T., and P. S. Norris (1953). Photosynthetic activity of phycobilins in some red and blue-green algae. *Biol. Bull.*, **105**, 374.

Heidingsfeld, J. (1943). Phototaktische Untersuchungen an *Navicula radiosa*. *Diss.*, *Breslau*.

Hisida, M. (1959). Membrane resting and action potentials from a protozoan *Noctiluca scintillans*. *Misaki Marine Biol. Stat.*, *Kanagawaken*.

Holmes, S. J. (1903). Phototaxis in *Volvox*. *Biol. Bull.*, **4**, 319.

Jacobsen, H. C. (1910). Kulturversuche mit einigen niederen Volvocaceen. *Z. Bot.*, **2**, 145.

Jennings, H. S. (1904). *Contributions of the Study of the Behaviour of Lower Organisms*, Carnegie Inst., Publ. XVI.

Klebs, G. (1885). Uber die Bewegung und Schleimbildung der Desmidiaceen. *Biol. Zbl.*, **5**, 353.

Kok, B., and G. Hoch (1961). Spectral changes in photosynthesis. In W. D. McElroy and B. Glass (Eds.), *Light and Life*, Johns Hopkins Press, Baltimore, Maryland. pp. 397–416.

Kol, E. (1927). Uber die Bewegung durch Schleimbildung einiger Desmidiaceen aus der Hohen Tátra. *Folia cryptog.*, **1**, 435.

Lindes, D. A., B. Diehn, and G. Tollin (1965). Phototaxigraph: recording instrument for determination of the rate of response of phototactic microorganisms to light of controlled intensity and wavelength. *Rev. scient. Instrum.*, **36**, 1121.

Links, J. (1955). I. A hypothesis for the mechanism of (phobo-) chemotaxis. II. The carotenoids, steroids, and fatty acids of *Polytoma uvella*. *Diss.*, *Leiden*.

Luntz, A. (1931a). Untersuchungen über die Phototaxis. I. Die absoluten Schwellenwerte und die relative Wirksamkeit von Spektralfarben bei grünen und farblosen Einzelligen. *Z. vergl. Physiol.*, **14**, 68.

Luntz, A. (1931b). II. Lichtintensität und Schwimmgeschwindigkeit bei *Eudorina elegans*. *Z. vergl. Physiol.*, **15**, 652.

Luntz, A. (1932). III. Die Umkehr der Reaktionsrichtung bei starken Lichtintensitäten und ihre Bedeutung für eine allgemeine Theorie der photischen Reizwirkung. *Z. vergl. Physiol.*, **16**, 204.

MacRobbie, E. A. C. (1965). The nature of the coupling between light energy and active ion transport in *Nitella translucens*. *Biochim. biophys. Acta*, **94**, 64.

Mainx, F. (1929). Untersuchungen über den Einfluss von Aussenfaktoren auf die phototaktische Stimmung. *Arch. Protistenk.*, **68**, 105.

Mainx, F., and H. Wolf (1940). Reaktionsintensität und Stimmungsänderung in ihrer Bedeutung für eine Theorie der Phototaxis. *Arch. Protistenk.*, **93**, 105.

Manten, E. (1948a). Phototaxis, phototropism and photosynthesis in purple bacteria and blue-green algae. *Diss.*, *Utrecht*.

Manten, A. (1948b). Phototaxis in the purple bacterium *Rhodospirillum rubrum*, and the relation between phototaxis and photosynthesis. *Antonie van Leeuwenhoek*, **14**, 65.

Mast, S. O. (1917). The relation between spectral colour and stimulation in the lower organisms. *J. exp. Zool.*, **22**, 472.

Mast, S. O. (1918). Effects of chemicals on reversion in orientation to light in the colonial form, *Spondylomorum quaternarium*. *J. exp. Zool.*, **26**, 503.

Mast, S. O. (1919). Reversion in the sense of orientation to light in the colonial forms, *Volvox globator* and *Pandorina morum*. *J. exp. Zool.*, **27**, 367.

Mast, S. O. (1926). Reactions to light in *Volvox*, with special reference to the process of orientation. *Z. vergl. Physiol.*, **4**, 637.

Mast, S. O. (1927). Response to electricity in *Volvox* and the nature of galvanic stimulation. *Z. vergl. Physiol.*, **5**, 739.

Metzner, P. (1920). Die Bewegung und Reizbeantwortung der bipolar begeisselten Spirillen. *Jb. wiss. Bot.*, **59**, 325.

Metzner, P. (1929). Bewegungstudien an Peridineen. *Z. Bot.*, **22**, 225.

Molisch, H. (1907). *Die Purpurbakterien nach neuen Untersuchungen*, Gustav Fischer, Jena.

Neuscheler, W. (1967a). Bewegung und Orientierung bei *Micrasterias denticulata* Bréb. im Licht. I. Zur Bewegungs- und Orientierungsweise. *Z. Pflanzenphysiol.*, **57**, 46.

9

Neuscheler, W. (1967b). Bewegung und Orientierung bei *Micrasterias denticulata* Bréb. im Licht. II. Photokinesis und Phototaxis. *Z. Pflanzenphysiol.*, **57**, 151.

Nultsch, W. (1956). Studien über die Phototaxis der Diatomeen. *Arch. Protistenk.*, **101**, 1.

Nultsch, W. (1961). Der Einfluss des Lichtes auf die Bewegung der Cyanophyceen. I. Photo-topotaxis von *Phormidium autumnale*. *Planta*, **56**, 632.

Nultsch, W. (1962a). Der Einfluss des Lichtes auf die Bewegung des Cyanophyceen. II. Photokinesis bei *Phormidium autumnale*. *Planta*, **57**, 613.

Nultsch, W. (1962b). Der Einfluss des Lichtes auf die Bewegung der Cyanophyceen. III. Photophobotaxis von *Phormidium uncinatum*. *Planta*, **58**, 647.

Nultsch, W. (1962c). Phototaktische Aktionsspektren von Cyanophyceen. *Ber. dt. bot. Ges.*, **75**, 443.

Nultsch, W. (1965). Light reaction systems in Cyanophyceae. *Photochem. Photobiol.*, **4**, 613.

Nultsch, W. (1966a). Correlations between photosynthesis, phototaxis and photokinesis in blue-green algae. In J. B. Thomas and J. C. Goedheer (Eds.), *Currents in Photosynthesis*, A. D. Donker, Rotterdam, The Netherlands. pp. 421–429.

Nultsch, W. (1966b). Über den Antagonismus von Atebrin und Flavinnucleotiden im Bewegungs- und Lichtreaktionsverhalten von *Phormidium uncinatum*. *Arch. Mikrobiol.*, **55**, 187.

Nultsch, W. (1967). Untersuchungen über den Einfluss von Entkopplern auf die Bewegungsaktivität und das phototaktische Reaktionsverhalten blaugrüner Algen. *Z. Pflanzenphysiol.*, **56**, 1.

Nultsch, W. (1968). Einfluss von Redox-Systemen auf die Bewegungsaktivität und das phototaktische Reaktionsverhalten von *Phormidium uncinatum*. *Arch. Mikrobiol.*, **63**, 295.

Nultsch, W. (1969). Effect of desaspidin and DCMU on photokinesis of blue-green algae, *Photochem. Photobiol.*, **10**, 119.

Nultsch, W. und Jeeji-Bai (1966). Untersuchungen über den Einfluss von Photosynthese-Hemmstoffen auf das phototaktische und photokinetische Reaktionsverhalten blaugrüner Algen. *Z. Pflanzenphysiol.*, **54**, 84.

Nultsch, W. und G. Richter (1963). Aktionsspektren des photosynthetischen $^{14}CO_2$-Einbaus von *Phormidium uncinatum*. *Arch. Mikrobiol.*, **47**, 207.

Nultsch, W., and G. Throm (1968). Equivalence of light and ATP in photokinesis of *Rhodospirillum rubrum*. *Nature*, **218**, 697.

Oltmanns, F. (1892). Über die photometrischen Bewegungen der Pflanzen. *Flora (Jena)*, **75**, 183.

Oltmanns, F. (1917). Über Phototaxis. *Z. Bot.*, **9**, 257.

Pieper, A. (1915). Die Phototaxis der Oscillarien. *Diss., Berlin.*

Pohl, R. (1948). Tagesrhythmus im phototaktischen Verhalten der *Euglena gracilis*. *Z. Naturf.*, **3b**, 367.

Pringsheim, E. G. (1912). Kulturversuche mit chlorophyllführenden Mikroorganismen. I. Die Kultur von Algen in Agar. *Beitr. Biol. Pfl.*, **11**, 305.

Rothert, W. (1901). Beobachtungen und Betrachtungen über taktische Reizerscheinungen. *Flora (Jena)*, **88**, 371.

Schilde, C. (1966). Zur Wirkung des Lichts auf das Ruhepotential der grünen Pflanzenzelle. *Planta*, **71**, 184.

Schlegel, H. G. (1956). Vergleichende Untersuchungen über die Lichtempfindlichkeit einiger Purpurbakterien. *Arch. Protistenk.*, **101**, 69.

Schmid, G. (1923). Das Reizverhalten, die Kontraktilität und das osmotische Verhalten der *Oscillatoria jenensis*. *Jb. wiss. Bot.*, **62**, 328.

Schrammeck, J. (1934). Untersuchungen über die Phototaxis der Purpurbakterien. *Beitr. Biol. Pfl.*, **22**, 315.

Stahl, E. (1880). Uber den Einfluss von Richtung und Stärke der Beleuchtung auf einige Bewegungserscheinungen im Pflanzenreich. *Bot. Ztg.*, **38**, 297.

Strasburger, E. (1878). *Wirkung des Lichtes und der Wärme auf Schwärmsporen*, Jena.

Tagawa, K., H. Y. Tsujimoto, and D. J. Arnon (1963). Role of chloroplast ferredoxin in the energy conversion process of photosynthesis. *Proc. natn. Acad. Sci.*, **49**, 567.

Terry, O. P. (1906). Galvanotropism of *Volvox*. *Am. J. Physiol.*, **15**, 235.

Thomas, J. B. (1950). On the role of the carotenoids in photosynthesis in *Rhodospirillum rubrum*. *Biochim. Biophys. Acta*, **5**, 186.

Thomas, J. B., and L. E. Nijenhuis (1950). On the relation between phototaxis and photosynthesis in *Rhodospirillum rubrum*. *Biochim. Biophys. Acta*, **6**, 317.

Throm, G. (1968). Untersuchungen zum Reaktionsmechanismus von Phobotaxis und Kinesis an *Rhodospirillum rubrum*. *Arch. Protistenk.*, **110**, 313.

Tsujimoto, H. Y., B. O. McSwain, and D. J. Arnon (1966). Differential effects of desaspidin on photosynthetic phosphorylation. *Pl. Physiol.*, **41**, 1376.

Urbach, W., and W. Simonis (1964). Inhibitor studies on the photophosphorylation *in vivo* by unicellular algae (*Ankistrodesmus*) with antimycin A, HOQNO, salicylaldoxime and DCMU. *Biochem. biophys. Res. Commun.*, **17**, 39.

Urbach, W., and W. Simonis (1967). Further evidence for the existence of cyclic and non-cyclic photophosphorylation *in vivo* by means of desaspidin and DCMU. *Z. Naturf.*, **22b**, 537.

Van Niel, C. B., T. W. Goodwin, and M. E. Sissins (1956). The nature of the changes in carotenoid synthesis in *Rhodospirillum rubrum* during growth. *Biochem. J.*, **63**, 408.

Vávra, J. (1956). Ist der Photoreceptor eine unabhängige Organelle der Eugleniden? *Arch. Mikrobiol.*, **25**, 223.

Wolken, J. J., and E. Shin (1958). Photomotion in *Euglena gracilis*. I. Photokinesis. II. Phototaxis. *J. Protozool.*, **5**, 39.

CHAPTER 9

The experimental analysis of phototaxis and photokinesis in flagellates

WILLIAM G. HAND
Department of Biology, Occidental College, Los Angeles

DEMOREST DAVENPORT
Department of Biological Sciences, University of California, Santa Barbara

---

'A spot of dull stagnation, without light or power of movement ...'
—*Tennyson*

Introduction

In this chapter it is not our intention to present an inclusive review of advances in the study of photoorientation and photomigration in unicellular organisms. Such information may be found in the comprehensive reviews of Bendix (1960), Clayton (1964), Halldal (1964), Haupt (1966) and Jahn and Bovee (1966). Here we will emphasize the experimental analysis of two types of behaviour, phototaxis and photokinesis, and the methods used in their investigation. There are few physiological and behavioural experiments which contribute to our understanding of the internal mechanisms governing these two types of behaviour. We would hope to point out a number of potential problems which may be amenable to investigation, the solution of which should enable us ultimately to support (or disprove) with hard experimental data much currently accepted theory about this internal mediation, a great part of which rests upon

253

circumstantial evidence. It is our belief that even in organisms of as small a size as some nannoplankters, the current explosion of knowledge and techniques in electrophysiology, immuno- and pigment chemistry, electron microscopy, membrane phenomena, etc., presages a much deeper understanding of the physiological basis for overt behaviour in motile microorganisms.

At the outset we must define as precisely as possible the phenomena with which we are concerned. We are quite aware of the fact that even the briefest perusal of the extensive literature dealing with the overt responses of motile microorganisms to light will indicate immediately that no two schools of investigators can agree on the definition of terms. However, students of photoorientation and photomigration in Metazoa have come to accept some rather precise distinctions over the years. Fraenkel and Gunn's (1940) publication brought about the virtual abandonment by zoologists of the use of the word 'taxis' for almost any sort of relatively simple behaviour which results in an organism being displaced toward or away from a stimulus scource. Therefore, in accordance with Gunn, Kennedy and Pielou (1937), Kennedy (1945) and Fraenkel and Gunn (1961) we shall adhere to the following definition: *taxes* are directed reactions with a single source of stimulation, in which the long axis of the body is oriented in line with the source and locomotion toward or away from it. Classically, as this definition implies, such reactions are thought to result from the unequal stimulation of symmetrically arranged sensors on a bilaterally symmetrical organism, allowing simultaneous *comparison* of intensities of stimulation on the two sides (see Fraenkel and Gunn, 1961, page 76). However, at least theoretically, such directed reactions may be effected via a single receptor which can make successive comparisons of stimulus intensities in time. Obviously that part of the definition demanding that the long axis of the organism be oriented in line with the source is frequently inapplicable for protista: such forms as *Amoeba* are not bilaterally symmetrical. But the critical point here is that taxes are *directed* responses demanding that comparison be made, either between a source and its background, between two sources of differing intensity, or between successive intensities of a single source in time. In spite of the fact that there is no experimental evidence as yet that comparison either in space or time can be accomplished within a single cell, we have no choice but to assume that it can. There is a wealth of experimental evidence establishing the effectiveness of evaluative behaviour of this sort in bringing about orientation and aggregation in Metazoa. We may therefore feel free to use the terms *positive phototaxis* and *negative phototaxis* in protistan behaviour, but will restrict their use to behaviour which would appear from circumstantial evidence to depend upon comparisons made in space or time.

Kineses are 'variations in generalized, undirected, random, locomotory activity due to variations in intensity of stimulation. Such variations can be of two kinds, namely changes in linear velocity' (orthokineses) . . . 'and changes in rate of change of direction or angular velocity' (klinokineses)—(Gunn, Kennedy and Pielou, 1937). It is important to note that these responses bear no relation to the *direction* of stimulation (even though they may result in differential distribution in a gradient). Hence, to avoid the directional connotations of the adjectives 'positive' and 'negative' (Kennedy, 1945) kineses in which activity is directly proportional to intensity of stimulation are known as *direct* kineses, while those in which activity is inversely proportional to stimulus intensity are known as *inverse* kineses. For example, if an organism increases its speed of swimming with a rise in general light intensity, it shows a direct orthokinetic response to light. If an organism decreases its rate of randomly directed turning ('goes straighter') under these conditions, it shows an inverse klinokinetic response to light. There is adequate experimental evidence (for example, Gunn, 1937) to show that differential distribution in a gradient may result from a simple orthokinesis. Pill bugs go fast in dry air and slow in moist air. Through the resulting inactivity they are 'trapped' in very moist regions. Mathematical analysis based on experiments where directional cues are absent has shown that this behavioural phenomenon is entirely adequate to explain differential distribution in a gradient, without the necessity of attributing to the organism an ability to sense the *direction* of the gradient.

The existence of klinokinesis as a distinct behavioural phenomenon has long been established by experimentation (Davenport, Camougis and Hickok, 1960; Ullyott, 1936; Wigglesworth, 1941), and a high order of specificity in some chemokineses (Davenport, Camougis and Hickok, 1960; Dethier, 1957) suggests that this behaviour may serve an adaptive function in bringing about aggregation at some point in a gradient. However, there is as yet no satisfactory *experimental* evidence which is supported by mathematical analysis to show that a klinokinetic response, with or without sensory adaptation, is effective in bringing about displacement up or down a gradient, the experiments of Ullyott (1936) and the theoretical treatment of Rohlf and Davenport (1969) notwithstanding. This is because of the difficulty, if not the impossibility, of designing experimental protocol in which directional cues are completely eliminated.

We believe that the establishment of these distinctions by Fraenkel, Gunn and their colleagues did much to clarify the thinking of workers concerned with the control of behaviour in the lower Metazoa, and we urge that they be adopted by students of protistan behaviour as well. As late as 1964 Clayton said: 'Phototaxis is a motor response elicited by light; that is, by a temporal change in light intensity or a non-uniform field of illumination.' In 1965 Lindes, Diehn and Tollin said; 'Phototaxis is defined as the movement of organisms in response to a light stimulus.' Both of these definitions include at least four types of demonstrably different behavioural responses. Haupt (1966) says: 'Phototaxis is the orientation of free-moving organisms to light'; the use of the words '*orientation to*' restricts the concept of taxis, as it should be, to movements dependent upon the reception of directional cues. But in discussing his concept of the phenomenon of 'phobophototaxis', which is, to say the least, an uncomfortable mouthful,† he says the phenomenon is due to 'changes of light intensity in time and independent of the direction of light'. We would urge that 'taxes' be terminologically restricted to behaviour dependent upon directional cues and the word 'kineses' be used for behaviour which is not so dependent.We wish to stress that kineses to light can only be demonstrated by experimental results in which responses bear no relation to the direction of a gradient or source; if they bear such a relation a tactic component may be present. We feel that the good editor of this volume falls into a kind of 'semantic light trap' when he uses the word 'phobophototaxis' for a gathering of cells in a gradient according to light intensity (Halldal, 1964). He is correct that this is a reaction distinct from his 'topophototaxis' (which is, by our definition, simply phototaxis), but the gathering may result from two different phenomena: an ability to sense and use the direction that clearly exists in the gradient and the purely mathematical phenomenon of the accumulation of cells at a point where they move slower and/or make randomly directed turns more frequently. It is quite true that there is always a directional cue in a gradient, but whether the cells *use* the cue or not is another matter, as Fraenkel and Gunn (1940) demonstrated with the pill bugs. This is the reason why we believe that efforts should be made in every case where differential displacement of organisms under stimulation is being considered to separate *experimentally* behaviour which is dependent upon the receipt of directional cues from behaviour which is not so dependent. 'Shock reactions' (Halldal, 1964; Haupt, 1966) and the 'phobotactic' response (turning) of *Euglena* as described by Oltmanns (1917) and Bünning and Tazawa (1957) we must define as kineses, since they have been experimentally demonstrated to be

---

† Haupt (1966): 'All influences of light on motion, viz. topophototaxis (phototopotaxis), phobophototaxis (photophobotaxis) and photokinesis have been called 'photomotion' by Wolken and Shin (1958) . . . παῦσαι!

responses dependent upon a change in intensity without a directional cue.

In any case, we feel that the tendency to proliferate terminology should be reversed, that cumbersome word-combinations should be abandoned and, above all, that defining terms should be restricted to phenomena for which there is solid experimental evidence which ultimately must be based on observation in detail of the movements of the individual cell, as well as on an understanding of the intracellular mechanisms controlling the behavioural phenomena. In our own work, for example, *Gyrodinium* exhibits a sudden complete cessation of movement when radiant light intensity is sharply increased. This phenomenon, an akinesis, may be physiologically similar to the 'shock reaction' discussed by Halldal and Haupt, and is perhaps thus akin to the 'phobophototaxis' of the purple bacteria (for example, see Buder, 1915; Clayton, 1964; Engelmann, 1882). We have not created a new 'kinesis', but have merely called this a 'stop response' because it is the simplest way to describe what happens. It will be time enough to assign terms for these phenomena in accordance with the above definitions when internal mechanisms are thoroughly understood.

However, as we shall see below, it becomes imperative to determine the extent to which taxes and kineses each *play a part in effecting* the aggregation which may be observed in any species. Perhaps the best demonstration of the manner in which aggregation can be brought about by a mechanism not demanding a directional cue is presented by the above-cited investigations of photokinesis in purple bacteria. In these organisms a drop in radiant intensity causes a reversal of their direction of movement. As the superb films made by Norbert Pfennig at Göttingen show, this phenomenon is entirely adequate to explain the aggregation of these bacteria in a light spot, without the necessity of assigning to them an ability to sense the directional cue which they must receive when they encounter the steep gradient of the light–dark interface. Yet we feel that it is going to be quite difficult, in those organisms which do respond in an oriented way to a directional light cue, to determine the relative effectiveness of taxes or kineses in bringing about aggregation in a zone of different intensity such as a spot. This will be the case, at least until such a time as one can, by monitoring the movements of individual cells, compare mathematically linear velocities and rates (and manner) of change of direction *inside* the spot, *just inside* the light–dark interface, *at* the light–dark interface, *just outside* the interface and *at some distance* from the spot. The behaviour of a cell which exhibits both a taxis and a kinesis should differ in each one of these regions. We cannot understand the relative role of these behaviours until what the cell does in each region has been quantitatively analysed.

Finally, as is well known, when we investigate the behaviour of Metazoa and in particular that unlearned behaviour serving to bring the organism into an optimal environment for its activities (predator to prey, parasite to host, male to female, etc.), it is always necessary to assemble tables of data composed of numerical evaluations of individual responses from which statistical analyses can be made. It is not necessary to point out that to date, because of technical difficulties involved in monitoring the behaviour of the individual motile microorganism, very few data of this sort are available. Careful analysis of the movement of the individual cell in gradients or interfaces has not been made in such a way as to enable us to say what happens to velocity or rate (and manner) of turning under such circumstances. This is the only way in which we can reach an understanding of the manner in which changes in the external environment effect behaviour or the manner in which reception or transfer of such information within the cell is effected. There is every indication that such studies will soon be possible, even with the smallest microorganisms.

Methods

In the discussion of methods we will frequently quote precisely the description of apparatus used by certain investigators in their own words.

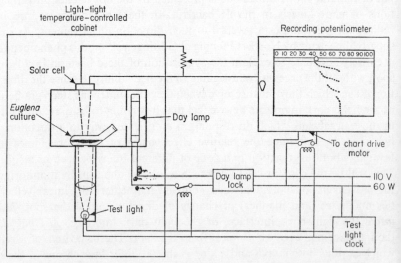

Figure 9.1. Diagram of an apparatus used to test and record accumulation responses of cells to light. (From Bruce and Pittendrigh, 1956)

We believe that this will help the reader to determine the characteristics, advantages and limitations of specific instruments in their systems.

In general, methods used to investigate the behaviour of flagellates fall into two categories: the 'mass movement' or 'optical density' method and the 'individual cell' method, in which conventional optics either alone or with electronic elements are used.

'Mass movement' method

All such systems utilize the principle of the measurement of optical density of cell suspensions. The most successful early use of this system is seen in the work of Pohl (1948). Later Bruce and Pittendrigh (1956) utilized this to determine the presence of a circadian rhythm in the photic response of *Euglena*. The system is described by Bruce and Pittendrigh (see Figure 9.1):

> 'Approximately 10 ml of culture are contained in a 50-mm-diameter Carrel flask. A microscope lamp directly below the culture throws a vertical beam of light about 7 mm in diameter (the test light) on to a Silicon Solar Cell. An adjustable fraction of the voltage output of the Solar Cell is recorded without amplification on a Brown Strip Chart Electronik Potentiometer.
>
> 'A so-called "day lamp" consisting of a 4-watt fluorescent lamp is located laterally to the culture and provides illumination for a day–night cycle. In operation the test lamp has generally been turned on for 30 minutes every 2 hours, and the Chart Recorder only operated when the test lamp was on; thus the test records generally consist of half-hour "responses" at 2-hour intervals. The day lamp is always off when the test lamp is on. The culture vessel is mounted in a black lucite chamber, and the whole assembly is kept in a light-tight, temperature-controlled cabinet.'

A more recent innovation of this method is presented by Lindes, Diehn and Tollin (1965). These authors establish the following design criteria:

(1) that the apparatus should provide permanent records of measurements of phototaxis over extended periods of time under controlled conditions,
(2) that intensity and spectral quality of the stimulating light should be controllable, and
(3) that simultaneously with stimulus, exposure of the organisms to randomly scattered background light of controlled intensity and wavelength should be possible.

Again a photoelectric method is used in which changes in optical density are recorded, but in this case the organisms accumulate in a beam of actinic light passing at right angles through the sample. Figure 9.2 shows

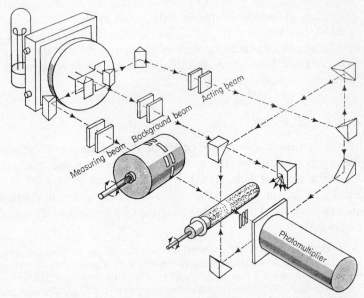

Figure 9.2. Diagram of the 'phototaxigraph'. (From Lindes, Diehn and Tollin, 1965)

the optical configuration of the 'phototaxigraph'. Note that three beams of light are utilized: a measuring (monitoring) beam, a background beam and a actinic beam. The actinic beam is the stimulating beam. It is split so that opposing beams of equal intensity illuminate a specific and limited zone in the cuvette containing the sample. It is important to recognize that the arrangement of opposing beams of equal intensity is a wise one if one wishes to study the accumulation of cells in regions of higher light intensity, for these provide a zone of light more intense than the scattered illumination in the surrounding medium but with minimized intensity gradients and directional effects. An effort appears to have been made here to create a 'light trap' in which the machinery of what has been called 'phobotaxis' can be studied. The background beam to be seen in the figure provides the desired scattered light for general background illumination of the sample. The measuring (monitoring) beam ($\sim$800 nm) elicits no overt response from *Euglena* (although it could conceivably have physiological effects). The beam is caused by the rotating chopper to pass through the sample in three discrete entities: a central measuring beam passes through that zone in the sample illuminated by the actinic beam (but at right angles to it), and two reference beams on either side of the axis of the measuring beam pass through zones illuminated only by the scattered light of the background beam. These serve to generate the 'blank' signal. The authors say:

> 'Measuring and reference beams alternately impinge upon the cathode of a photomultiplier tube. After electronic processing of the signals, an output is obtained and recorded which is a measure of the difference in the numbers of organisms inside and outside the stimulated region of the suspension as a function of time.
>
> 'The cell (cuvette) is rotated slowly (10 rev/min) to avoid settling of the organisms and to minimize attachment of the organisms to the sample cell walls. A stainless steel thermistor temperature probe is immersed into the suspension through the hollow shaft of the rotating motor. The temperature of the suspension is controlled thermostatically to within $\pm 0.5°c$ and can be varied from $0°c$ to $50°c$. Micro-organism suspensions of the same initial optical density (O.D.$_{800}$ = 0.5) are used in all runs.'

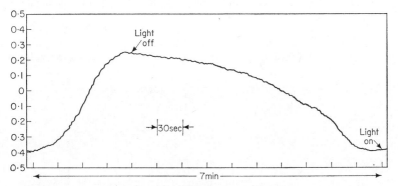

Figure 9.3. A typical 'phototaxigram'. (From Lindes, Diehn and Tollin, 1965)

The recorder trace of a typical phototaxis experiment apparatus may be seen in Figure 9.3. The figure shows that after the initial stimulus from the actinic light there is a 'lag period' followed by a steeply ascending curve, the slope of which the authors consider a measure of 'the intrinsic rate of phototaxis'. The curve reaches a peak of maximum amplitude which 'is a measure of the population density of organisms in the stimulated zone under near steady-state conditions, i.e. when almost as many organisms are leaving the stimulated zone as are entering it.' A second lag period occurs between removal of the stimulus and cessation of phototactic activity, and the descending slope of the curve (after 'light off') is a measure of the 'intrinsic motility of the organisms, independent of the phototactic effect'. The baseline indicates uniform distribution in the cuvette.

With this apparatus, in which all light sources allow control of wavelength and intensity, where similarly controllable background light is continuously provided and where *T* is controlled, it is obvious that important studies on mass displacement under light stimulation can be carried out with many types of organisms. Almost complete automation makes possible the continuing collection of data on the relation to such displacement of temperature, stimulus intensity and wavelength, culturing regimes, (photoperiod), etc.

Feinleib and Curry (1967) have described a somewhat similar but in many respects simpler system (Figure 9.4):

'The swimming chamber is a Coleman cylindrical spectrophotometer cuvette, 19 mm in diameter, and cut to a length of 80 mm. It has been filled to one fifth of its depth with black paraffin, shaved down to a flat surface. At the start of an experiment, the cuvette is filled to the top with 15 ml of algal suspension, and is covered with a No. 1 cover glass fastened with silicone grease.

'The cuvette is placed in a horizontal position in front of a pair of selenium barrier-layer photovoltaic cells (G. E. 8PVIAAF). These are connected in a comparison circuit of the type recommended by General Electric for use

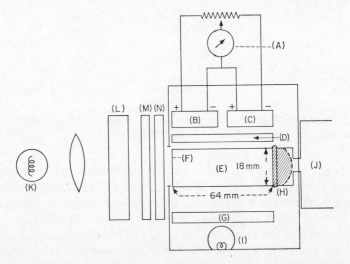

Figure 9.4. Schematic diagram of apparatus used in the population method, viewed from above. (A) Amplifier–galvanometer. (B) Photocell 1. (C) Photocell 2. (D) Red filter. (E) Algal suspension. (F) No. 1 round covership. (G) Red filter. (H) Black paraffin. (I) Tungsten filament lamp. (J) 60 rpm motor. (K) 300 W tungsten filament lamp. (L) Water filter. (M) Neutral density filter. (N) Blue-green filter. (From Feinleib and Curry, 1967)

where linear response is desired (G.E. booklet C-690). The voltage difference between the two photocells (mV) is measured with a Kin Tel 204A electronic galvanometer-amplifier, and is plotted continuously against time on a Leeds and Northrup Speedomax type G recorder (Model S60 000 series). An R-C filter with a time constant of 0·1 second has been incorporated to eliminate short term fluctuations which appear as noise on the recording.

'A monitoring beam passes through the side walls of the cuvette, traverses the suspension and impinges on the photocells. Dim red light is used for this purpose because it has been found to cause practically no phototactic movement. The light source is a 6 V lamp (G.E. W 63 bulb), run off a voltage regulator, and it is used in combination with a Corning II.R. 2-60 red cut-off filter, which absorbs visible radiation below ca. 620 nm. A second filter of the same type has been placed directly over the photocell faces, so that they are shielded from any stray light of shorter wavelengths. Each photocell has an 18 mm² area exposed to the light coming through the suspension. These two areas are separated by a distance of 28 mm. As shown, the photocell face extends up to the end wall of the cuvette at front and back.'

If we compare the Lindes, Diehn and Tollin system with that of Feinleib and Curry, we see that the former measures the difference between absorption at a point in the stimulus field and that at a point outside the field, while the latter continually measures the difference in absorption between two points in the field, one point closer to the light source and the other further away. The essential difference between the two systems is that that of Lindes, Diehn and Tollin measures movements in and out of a stimulus beam (i.e. movements which are perpendicular to it), while that of Feinleib and Curry monitors orientational swimming toward or away from the stimulus source (i.e. movements which are parallel with the stimulus beam). It should be pointed out that in both experimental situations directional cues are present. Cell aggregations interpreted as tactic in nature may possibly be effected by both taxes and kineses.

There remains the elegant modification of the technique of Manten (1948) made by the editor of this volume in his work on action spectra of 'phototaxis' (Halldal, 1958). This method consists in the illumination of a vessel from one side by a reference beam uniform in intensity and from the

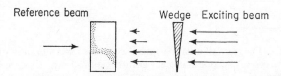

Figure 9.5. The distribution of photoresponsive cells in a vessel illuminated from one side by a reference light of uniform intensity, and from the other by an intensity graded light. This accumulation may be traced or photographed through a microscope. (From Halldal, 1958)

other side by an exciting beam having an intensity gradient (Figure 9.5).
In the words of the author:

'When the two lights are properly chosen in intensity and wavelength,
the organisms will, at high intensity of the exciting beam, swim toward it
and will collect in a certain region at the wall nearest the exciting light
source. At lower intensities of the exciting beam, they will be more attracted
by the reference beam and will collect at the wall nearest this light source.
In between, there will be a certain region where the organisms will swim
by random motion. When the intensity of the exciting beam and its gradient
are known, $W\lambda$ can be computed from the position of the organisms showing
random motion.' (This is defined as the phototactic effect of light of wavelength λ_a with respect to the reference light, from which the smallest incre-
ment of light necessary for 'topotaxis' may be precisely determined.)

'Another method of measuring the action spectrum of phototaxis in a
single exposure is to replace the monochromatic exciting beam with a projected spectrum having an intensity gradient at a right-angle to the wavelength scale (Figures 9.6 and 9.7).

'Under these light conditions the organisms will collect on the side of the
vessel illuminated by the spectrum wherever this light is more effective
than the uniform reference beam. If the spectrum is adjusted to have equal
numbers of quanta along any horizontal line, and the intensity gradient at
a right angle to the wavelength scale is linear from full intensity to zero, the
regions of random motion will form a curve representing the action spectrum
of topo-phototaxis for the organism.

'The organisms which have collected at the wall illuminated by the spectrum are made visible by shutting off both the reference beam and the

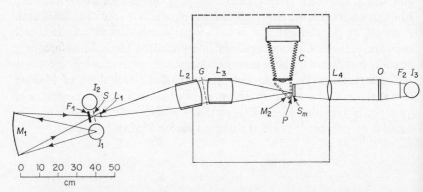

Figure 9.6. Diagram of an apparatus used in the determination of action
spectra from a cell population. Legend is as follows: C: camera; F_1:
heat filter; F_2: colour filter; G: grating; I_1: 'white' light source; I_2:
mercury lamp; I_3: 'white' reference light; L_1: lens; L_2: lens; L_3: lens;
L_4: lens; M_1: spherical mirror; M_2: removable mirror; O: opal glass;
P: glass plate containing mercury lines of 405, 436, 546 and 578 nm;
S: slit; S_m: sample. (From Halldal, 1958)

PLATE III

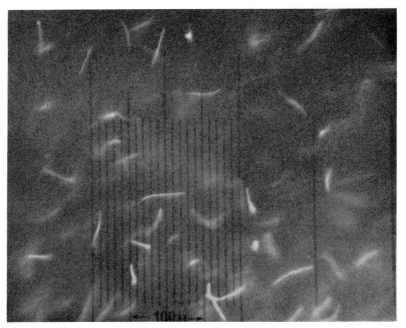

Figure 9.9. One-fifth second dark-field exposure showing swimming tracks of *Chlamydomonas* cells in dim red light. (From Feinleib and Curry, 1967)

PLATE IV

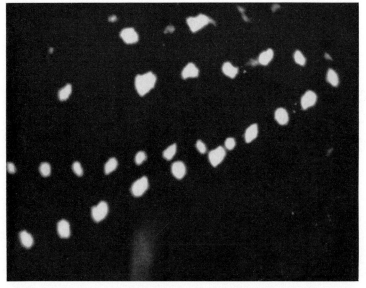

Figure 9.12. Typical 'tracks' of the dinoflagellate *Gyrodinium* as photographed through the flying spot scanning device

projected spectrum and then illuminating these organisms by intense white light from above. This was done by sending a narrow beam of light as indicated by the arrows and the dotted lines in Figure 9.7. Due to the Tyndall effect, the organisms show up as light spots against a dark background. This pattern can either be traced or photographed.'

The 'mass movement' method is well suited to the measurement of the responses of *populations* of cells to a given set of environmental conditions (as, for example, laboratory studies of vertical or horizontal migrations in gradients). However, in the majority of experiments in which this method has been used the measurement of cell accumulation has been made without resolving the manner in which this accumulation was effected. We submit that the method generally gives no better information on the role of taxes and kineses in mass movements than the observation of mass movement of moths to a light source can give evidence of the physiological and behavioural events triggered in the moth by the light. As yet, no worker has

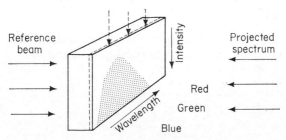

Figure 9.7. A pictorial representation of the action spectrum measurement produced by the apparatus described in Figure 9.6. (After Halldal, 1958)

succeeded in combining the conventional mass movement technique with one which simultaneously allows him, in a single experiment, to observe changes in linear velocity, rate of change of direction, or simple directed responses. To the behavioural physiologist, a response is the property of an individual, not a population.

A further weakness of the mass movement method should be stressed. The measurement of such movements *takes time*; some methods employ a paper chart recorder that yields data on accumulation over time. Responses are not only dependent upon the nature but also the *duration* of the stimulus. For example, as stimulus duration increases, adaptation of the receptor or the receptor–effector chain may occur. The altering of these units in the cell will change the response of the cell with respect to the stimulus (Hand, Forward and Davenport, 1967; Mayer, 1968). Hence monitoring an accumulation of cells over time means that one may be monitoring a *change* in physiological state. This makes conclusions concerning internal mechanisms difficult. Such mechanisms should be investigated as nearly as possible to the time when they can be said to be in a 'steady' or 'baseline' state.

For these reasons the authors find it difficult to support some of the conclusions drawn by investigators using the mass movement method—in particular those concerning internal mechanisms which determine the nature of responses. As an example, as Clayton (1964) has already so clearly pointed out, it would appear to be entirely possible that the derived action spectrum for aggregation in particular species may be made up of the combined spectra for several behavioural components effecting aggregation. However, there is no doubt that the method has served and will continue to serve as a valuable primary technique in the determination of factors that effect mass displacement.

'*Individual cell*' method

The oldest system, of course, involves the optical observation of individual cell movement with the light microscope. Events are then recorded

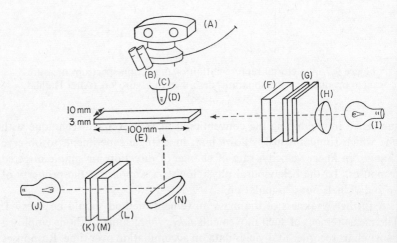

Figure 9.8. Schematic diagram of apparatus used in photomicrographic system. (A) 35 mm camera. (B) Ocular 10×. (C) AO Spencer trinocular microstar microscope. (D) Objective 20×. (E) Lucite chamber with the black patch. (F) Water filter. (G) Neutral density filter. (H) Blue-green filter. (I) 7·5 V tungsten filament lamp stimulus light. (J) 7·5 V tungsten filament lamp. (K) Red filter. (L) Water filter. (M) Monitoring light. (N) Mirror. (From Feinleib and Curry, 1967)

by standard photomicrographic techniques. Modifications of this technique have been used with valuable results by many workers to study the behaviour of flagellates, motile gametes, etc. (for example, Brokaw, 1958; Gebauer, 1930; Mast, 1911; Rothschild and Swann, 1949). A more recent one is that employed by Feinleib and Curry (1967) to study the behaviour of *Chlamydomonas* (Figure 9.8). In their own words:

'The algal suspension is placed in a lucite chamber, 100 mm × 10 mm × 3 mm (in depth), and is covered with a strip made from a No. 1c over glass, care being taken to exclude air bubbles. Red light for observation and photography is directed at the chamber from below, while a blue-green stimulus light is focused to cover the end wall. The stimulus beam is adjusted so that it diverges only slightly. The light source for each of these is a 7·5 V tungsten-filament microscope illuminator. The monitoring light is filtered through water and through a Corning H.R. 2-60 red cut-off filter; the stimulus beam is passed through a water filter and through a Corning 4-96 blue-green filter.

'A chamber several mm in depth was judged to be the most suitable for these measurements for the following reason. Since the stimulus light comes from the side, a slide and cover slip arrangement would make the light path very difficult to determine; in a deep chamber, on the other hand, the cells would be difficult to see and to photograph. Even with the 3 mm chamber, it is necessary to use a modified version of dark-field illumination in order to obtain sufficient contrast for photography. By means of a microscope mirror, the red light is directed obliquely at a 3 mm² black patch pasted on the bottom of the chamber. The edge of the patch has been placed at the centre of the microscope field and its position has been adjusted so that the cells are brilliantly illuminated. The cells are simultaneously observed and photographed using an AO Spencer 35 mm photomicrographic camera (Model 635) in combination with an AO Spencer Microstar trinocular microscope and Kodak Plus-X film.'

With this system one can obtain very accurate information about a cell which directly pertains to phototaxis and photokinesis as defined above. For instance, one can measure the velocity, or rate of change of direction of the cell in question, as well as determine its orientation, with all this information coming from a single photograph (Figure 9.9, Plate III). In our estimation this method and procedure are relatively simple, efficient and well suited to the task of obtaining information about the movement of large numbers of individual cells in a short time. However, the system is not designed to give detailed information concerning the sequence of behavioural events that occurs when an individual cell responds to a particular stimulus. Only a system that permits the monitoring of individual cell movements at high magnification can accomplish this.

An interesting modification of the 'individual cell' method which is applicable to the study of taxes and kineses is the electronic technique used by Rikmenspiel and Van Herpen (1957) in their investigations of sperm behaviour. Through the use of dark-field illumination and a microscope objective, an image of a sperm cell is produced. A diaphragm with an aperature of approximately 100 μ is placed in the image plane. A photomultiplier (EMI 5371), positioned so as to face the diaphragm, receives a flash of light whenever a sperm cell passes the aperture, this flash of light being produced by the sperm cell head. The resulting photomultiplier response is then recorded with an oscilloscope or paper chart recorder. The basic design of the system is seen in Figure 9.10.

This system is very good if one is content with the determination of response capability as a function of the number of cells migrating toward or away from a light stimulus or of linear velocity under light stimulation.

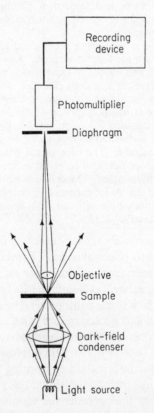

Figure 9.10. Diagram of the apparatus used by Rikmenspiel and colleagues to determine the motility of bull spermatozoa. (From Rikmenspiel and Van Herpen, 1957)

It does not, of course, permit the discernment of the sequence of behavioural events which constitute the response of an individual cell.

Photomicrographic techniques present certain problems in the protocol of experiments, particularly those designed for the elucidation of behaviour and its physiological control. First of all there is the problem of field illumination, of particular importance in the investigation of phytoflagellates. This must be of a wavelength and intensity which does not alter cell behaviour with respect to a controlled stimulus beam. If field illumination is known to affect cells, this factor must be accounted for in the experimental design. As an example, we question the use of red illumination by Feinleib and Curry. Although in the case of *Chlamydomonas* red illumination does not produce an overt behavioural response, it may have a profound effect on the cell's responses to other wavelengths (Forward and Davenport, 1968; Hand, Forward and Davenport, 1967). Intensity of illumination should be kept as low as possible, since secondary physiological effects at a given wavelength at high intensity may be absent at lower ones. In examining the criteria for field illumination, we discover what may be the major weakness in the photomicrographic system. The quality of the field illumination is dependent upon the spectral sensitivity of the monitoring sensor, whether it be human or an instrument, as well as upon the opacity of the objects being examined at the particular wavelength chosen. Unfortunately, most photomicrographic techniques require rather high light levels for good resolution of an object like an algal cell.

A further factor in photomicrography is the limitation of field size which accompanies higher levels of magnification in the conventional compound microscope. This demands restriction of the movements of the experimental subject within limited areas and with cover slips, etc. Ideally one would wish to be able to monitor the movements of very small microorganisms in a relatively large unrestricted field, in much the same way as a limnologist might wish to monitor from an altitude of a thousand feet the movements of a large number of trout swimming above the illuminated sandy bottom of a lake. The latter should be entirely possible, for with a scanning laser beam one can now resolve a matchhead on the ground from an airplane flying at Mach 2 at 500 feet. Similar techniques which allow the monitoring of the movements of numbers of cells, while at the same time giving high enough resolution so that the chain of events in the response of the individual can be recorded, should ultimately be available to the investigator of the microworld. Of course, data obtained by such techniques can readily lend themselves to computerization.

In the light of the above criteria we may examine two systems currently being used by the authors, which resolve some, but by no means all, of the problems.

The electronic system of Hand, Collard and Davenport (1965) was expressly designed to monitor the movements of large numbers of motile microorganisms at varying levels of magnification. Resolution was frankly sacrificed to facilitate manipulation of preparations and the recording of the position of organisms. The field illuminating system (scanner) permits

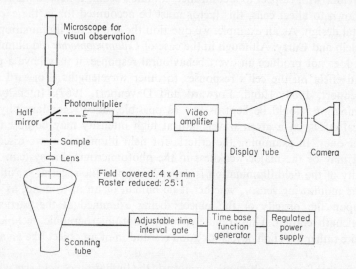

Figure 9.11. Diagram of the 'flying spot' scanning device. (From Hand, Collard and Davenport, 1965)

great reduction in intensity; the investigator does not view the cells directly but secondarily, through an electronically amplified primary monitoring system. The instrument (Figure 9.11) is best described by the authors as follows:

'A gated square sweep (scan) of uniform low intensity is generated on a Dumont Type 304 H oscilloscope mounted vertically. The gated sweep is accomplished by a generating circuit which is capable of producing several sweep frequencies. The scan frequency is further controllable by emphasizing a certain section of the generated sweep with accurate regularity. This is accomplished by the incorporation of a blanking circuit which can also be adjusted. By adjusting the interplay between these two controls, as well as the oscilloscope intensity, it is possible to obtain gated scans with intervals of one scan per ten minutes to fifty scans per second.

'The gated scan is transmitted optically to a microscope stage above the oscilloscope via a 1·0-mm camera lens mounted below the stage. The position of this lens can be adjusted by a fine-adjustment so that the scan may be focused on a prepared slide of the organism to be observed. The area of the test preparation scanned is a 4-mm square, while the depth of field is less than 0·5-mm. The magnification of the subject may be changed by manipulating the horizontal and vertical sweep controls of the vertically mounted oscilloscope. The field scanned remains the same.

'The scan is transferred by a mirror to a type 1 P 21 photomultiplier tube and associated preamplifier unit. Changes in light intensity within the scan, caused by objects (such as moving organisms) between the scan and the photomultiplier tube, are registered by that tube as changes in electrical potential. The potential changes are amplified and transmitted to a second Dumont 304 H oscillscope mounted horizontally. This scope in turn displays a "negative" image of the intensity differences in the generated scan, i.e. the discrete areas of reduced illumination in the original scan are finally displayed as light "blips" on a dark field, each blip recording the organism's position at the moment it is scanned. Resolution of the system is fine enough for it to perceive relatively opaque organisms larger than 15–20 μ and to distinguish the external morphological features of many forms in the 50–300 μ range. By coupling a Dumont type 450 A oscilloscope camera fitted with a Robot 'Recorder 24' 35-mm camera-back to the display scope, track images of moving organisms may be photographically recorded as they occur; a time exposure on Kodak Tri X Pan film records the positional images or blips for a chosen number of scans.

'With scan rate as well as power of magnification known, information necessary for linear velocity determination is available. We determine field magnification by photographing, at the beginning of each test series, a calibrated guide of known dimension; for this purpose a lucite slide inscribed with parallel lines 1 mm apart was used.

'The organisms being scanned were stimulated by light from a Bausch and Lomb No. 33-86-02 Grating Monochromator. This instrument was adjusted to give a 20 nm band-pass between 400 and 700 nm. Intensity was controlled with an Optical Coating Laboratory continuously graded neutral density filter. Stimulus duration and interval were controlled by a simple revolving blanking wheel in which a notch of the required size was cut. The wheel was driven by a Bodine type NSH 128 motor. Wheel velocity was controlled by a rheostat. A stimulus marker was operated in conjunction with the revolving disc.'

A typical 'track' recorded in such an apparatus is shown in Figure 9.12, Plate IV; as can be seen it is similar to those obtained by the photomicrographic technique of Feinleib and Curry (Figure 9.9). There is one major difference, however. Because the scan may be 'strobed', information concerning linear velocity is not dependent on camera shutter speed. Instead, it is calculated by measuring the distance between several 'blips', the number of blips used being determined by the scan rate.

There are some major shortcomings in this system. First, the scanning beam is of a fixed wavelength band, this being determined by the emission characteristics of the scanning tube. It was purely fortuitous that in the particular case of the cells which the authors chose to investigate (Hand, Collard and Davenport, 1965) the monitoring scanner appeared to have no effect on the cells either directly or indirectly. It is important to note that in a system of this sort the scanning source consists of a rapidly moving spot of light which actually illuminates the cell many times per second rather than bathing the cell in continuous light. This reduces the overall time that experimental organisms are illuminated, and coupled with the low intensity, further contributes to the neutrality of the monitoring system. A second and perhaps the greatest shortcoming of the system is its poor resolving capability, the highest level of resolution being not better than 30–40 μ. This is due to the nature of the scanner, the resolution capability of which is determined by the number of scan lines. Some effort has been made by the authors to investigate ways in which spot size, and hence distance between scan lines, could be decreased, so that in effect very small organisms would be less likely to 'fall in the cracks' between the scan lines. To date this effort has not been successful. One could decrease the distance between the scan lines optically. However, in so doing the size of the field which could be monitored would also be reduced, with obvious results in so far as the observation of large numbers of cells or a single large organism is concerned. To resolve this difficulty one would need a tube with a spot size smaller than that of standard tubes commercially available at this time. It has been proposed that one replace the cathode ray scanning tube with a 'cold' mechanically scanning microbeam laser. But here one would face the great difficulty of designing a reliable high-speed mechanical scan.

The above-described electronic scanning technique is, then, well suited to record the position of fairly large microorganisms and to monitor their movements, but, like the method of Feinleib and Curry, does not allow simultaneous observation and monitoring of such organisms at high levels of magnification or resolution.

Largely as a result of our inability to improve the above system in such a way as to make it possible to study organisms of a 2–10 μ size range, we have recently designed and assembled a system which combines the advantages of the best photomicrographic and electronic systems, while at the same time providing optimally efficient elements for rapid data collection.

We employ the basic elements of photomicrographic technique along with closed-circuit television in a somewhat more sophisticated modification of the instrumentation described by Davenport, Wright and Causley (1962). The entire apparatus is composed of standard commercially available optical and electronic units. The basic optical instrument is a

Nikon Model M Inverted microscope with all magnifications from 20×
to oil, with phase-contrast and dark field. In this instrument the illuminator
is some 8 inches from the microscope stage, which practically eliminates
the effects of heat on a preparation. The large unencumbered open stage
permits free manipulation of preparations, perhaps the greatest advantage
of the instrument. One can, for example, observe a field at low magnification as much as several millimeters on a side, while the movements of
organisms barely visible in the microscope can be monitored by the video
camera. The positions of the condenser, well above a preparation, and the
nose-piece beneath the stage allow the greatest freedom in the manipulation
of preparations. With the use of the high-dry long-focus lens one may bring
a preparation into focus at a magnification of 400× without the necessity
of employing an especially thin microslide or restricting the movements of
organisms by cover-slips, etc. In this stage of the instrumentation one has,
therefore, all the advantages of a superb and versatile optical system, particularly suited to the manipulation of free and unencumbered experimental
subjects.

The electronic elements of this system consists of a COHU Electronics
Model 3000 television camera, with camera control (Model 3900) and
monitors. The camera 'views' the optical image through the monocular
ciné attachment of the microscope. It and the television monitor are
especially designed for the high resolution necessary for closed-circuit
instructional television.

There are many advantages in this system. Because the television camera
is sensitive to a broad spectrum of wavelengths, the experimentor may
select for field illumination any wavelength which he knows will not
affect his experimental subjects by merely placing a filter in the microscope
illuminator. Stimulus beams may be introduced to an experimental
preparation, as in the previously described flying-spot system, parallel to
the plane of the microscope stage and at right angles to the field illuminator
beam, without in any way affecting the television image of the organism.
Brilliance and contrast of the image on the monitor screen may be modified
in a number of ways in accordance with particular ends; one may do this
with the optical diaphragms, condensers and field illuminator of the
microscope, the sensitivity control of the video camera or the light–dark
contrast of the monitor. Phase or dark-field can be used to attain particular ends. With the former, internal cellular structures may give one cues
as to direction of movement, in such a way that changes in direction can
be precisely monitored. With the latter, which gives image reversal on the
television screen (light tracks against a black background), one can collect
data with ease by exposing film for a known interval to the brilliant tracks

moving on the television screen. This film can be processed and simple measurements made to give data on linear velocity, rate of change of direction or temporal change in numbers of cells present. By the use of a video tape-recorder (Ampex Model 7500), the investigator can record and store responses of cells and replay these at his convenience. This not only allows him to select a particular part of the record for data analysis but also to observe a particular response repetitively if he wishes. In addition, he can slow down the tape in such a way that the elements of a behavioural act become more obvious to him. In practice an 8-inch monitor is used for data retrieval, while a 25-inch monitor in the same circuit permits the subjective observation of cells at great magnification. There is no doubt whatever that the increase in size of the cells as they appear on the 25-inch screen enhances the investigator's perception of the events of a response. The image of the cell on such a screen, particularly when phase-contrast is used, becomes of paramount importance when one wishes to analyse the position of subcellular units (flagella, etc.) with respect to orientational movement, direction of stimuli, etc.

Finally, it is now possible to facilitate data collection from such a system by the use of a digital computer. A television tape presents in one small area a mass of information that is almost beyond reckoning, with pictures coming off at a rate of thirty times per second. Clearly, if we fed so much information at this rate into a computer, we would soon produce a state of information overload. The answer is to employ a video preprocessor which selectively strips off that portion of the picture in which we are interested, and thus reduces the inflow of data to the computer. John O. B. Greaves of the Department of Electrical Engineering at Santa Barbara has designed and constructed such a device for us, which we have christened the 'bugwatcher'. This unit essentially allows us to send, at variable rates depending upon the speed of movement of the organisms, two kinds of information to the computer from the tape: first, the position of organisms and, second, their outline in the form of x–y coordinates of points on the margin. To programme studies on linear velocity, net displacement, rate of change of direction, etc., clearly only the former is necessary, in which case organisms are taped at low optical magnifications. For studies of change in orientation of the individual cell, however, the latter is most useful, in which case experiments are taped at higher levels of magnification. We can, then, store in the mathematical memory of a digital computer a record of movement stored on video tape, whether it be of numbers of individuals in a suspension or of one individual or its parts. We can ask the computer to return these movements for our visual review on a Tektronics 611 storage scope.

For a more detailed and technical description of the above apparatus, see Davenport and coworkers, 1970.

Discussion

At the beginning of this chapter we defined the types of responses of flagellates which may occur as a result of directional and non-directional stimulation. We may now discuss certain findings in terms of these definitions, and pose resulting questions which are open to investigation.

We may first ask whether orientation to a directional light stimulus (phototaxis) occurs in like fashion in all unicellular organisms exhibiting a light response. To date the majority of efforts have been concentrated on *Euglena*, with occasional studies of other forms. In *Euglena* the mechanism of orientation is still a matter of conjecture, but the most widely accepted view is still that proposed by Mast (1911), with some modifications. This hypothesis assumes the existence of a single photoreceptor which determines the direction of a light stimulus by measuring the intensity of that stimulus at two points in time through the effects of a second organelle, the stigma or 'eye-spot', which periodically shades the photoreceptor as the *Euglena* swims. The period of this intermittant shading is governed by the rotation rate of the *Euglena*. The theory asserts that the alternating light intensity effects a response in the receptor and that with each alternation there is a correction in course until the receptor either receives continuous illumination or is completely shaded by the stigma at all times. Halldal (1958) believes that in forms which reverse the sign of their response, corrections in course are made which keep the sensor illuminated when cells are responding positively, and shaded when they are responding negatively.

One would expect an orientation mechanism of this type to be slow, with the time from reception of the stimulus until final orientation dependent upon several factors:

(1) The 'state' of the receptor, i.e. its threshold at the time of stimulation. This may vary, for as demonstrated by Bruce and Pittendrigh (1956) and other workers the response level varies with the time of day, age of culture, etc.

(2) The rate of rotation of the cell which may or may not be a function of swimming velocity. The rotation rate will determine the rate of shading.

(3) The position of the cell relative to stimulus direction at the time of reception of the stimulus. An interesting question arises here. If we assume that the *direction* of rotation of a cell remains constant,

will the direction of orientation to the stimulus always be constant with respect to the rotation direction, or will it vary depending upon the position of the cell with reference to stimulus direction?

(4) The intensity of the stimulus, which should in turn affect the intensity of the resulting 'shock reaction'. This, of course, assumes that the amount of turn made at each 'shock' is dependent upon stimulus intensity. Quite possibly this response is independent of intensity. If this is the case, then threshold should be determinable.

(5) Any latencies (energetic or otherwise) involved in the transmission chain between the receptor and the 'motorium' effecting the response.

Plausible as the above theory of orientation is, we are still by no means certain that a flagellate photoreceptive organelle cannot monitor direction *per se*. In this connexion Halldal (1959) has contributed some important information. In an ingenious experiment with *Platymonas* and other flagellates he supported the observations of Buder (1919) that photomigration is independent of intensity. Figure 9.13 shows the experi-

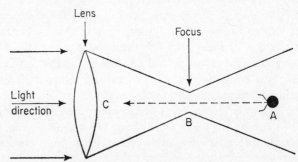

Figure 9.13. The behaviour of a positively phototactic algal cell in a focused beam of light. (After Halldal, 1959)

mental protocol whereby this was demonstrated. It can be seen that under these conditions a moving flagellate responding positively to unidirectional stimulation moves from point A, where there is relatively low radiant intensity, through the focal point B, where there is relatively high radiant intensity, and then on toward the source into a region C of relatively low intensity again. It is clear that in this course the organism moves directly toward the stimulus source regardless of the fact that in so doing it moves from a point where there are relatively few quanta per unit area of its surface, through a region where there are relatively many and on into a region where there are relatively few again. Under these conditions there would, incidentally, appear to be no effective 'kinesis component' in the aggregation which occurs near the source. If there were such a component there would surely be some differential aggregation at the 'focal point' or the two 'lower intensity' regions. This is an extraordinarily interesting phenomenon. The elucidation of the mechanism at the subcellular level would in our opinion constitute as important an advance as could be made in our understanding of the control of photomigration in the Protista. We still do not know whether, under the conditions of Halldal's experiment, the cells are moving after orientation along a preprogrammed pathway or are continually making midcourse correction. Halldal (unpublished) believes the latter.

What about other flagellates? Does their response suggest that orientation is controlled by temporal changes in intensity at the receptor? Recent studies in our laboratories suggest that there may be a different mechanism in some forms. In the dinoflagellate *Gyrodinium* the initial response to a directional light stimulus is quite different from that of *Euglena*. Upon stimulation of a sufficient intensity, the cells cease moving very soon after the onset of the stimulus (200–500 milliseconds, dependent upon intensity). They remain in a stationary position for several seconds, again depending upon stimulus intensity. The first movement of the cells following this 'stop' response is a rapid shift of position in the cell's axis, so that the anterior end (in this case the anterior is opposite the longitudinal flagellum) faces directly toward the stimulus origin. Only after this manoeuvre has been completed does the cell resume its normal forward travel. On the basis of this observation, i.e. that the cell makes one complete course correction without interruptions, it would appear that the organism may be capable of monitoring the direction and/or the intensity of the stimulus at only one point in time, but from two points in space. Classically, this type of response has been attributed to organisms that have two or more receptor organs which feed information into some sort of decision-making system. This becomes an inviting possibility when one examines the gross anatomy of biflagellate forms such as dinoflagellates. The two flagella of a form such as *Gyrodinium* are located at different points and operate in different planes. In *Euglena* with its single flagellum the photoreceptor is thought to be located on the flagellum or near it. Perhaps a dinoflagellate has two photoreceptors with which spatial comparison can be made, while in *Euglena* comparison of intensity in time must be made by the receptor, effecting orientation in such a way that the cell simply turns so many degrees each time the organelle is 'shocked' or stimulated.

Let us now briefly consider what is known about the photoreceptor and associated structures. In *Euglena* the paraflagellar swelling is the accepted photoreceptor. This small organelle (200–500 Å) is located at the base of the flagellum and is in the logical place to be shaded by the stigma, which lies to the side of the flagellum. Unpublished work by Diehn, Tollin and their coworkers suggests that the photopigment responsible for phototaxis and photokinesis is probably a flavin, possibly riboflavin. Their findings are supported by action spectra analysis of the behaviour of *Euglena* in ultraviolet light. Recent work (Diehn, 1969) using polarized light suggests that two perpendicularly oriented and physically separated pigment systems may exist in the shading structure (stigma). One system screens the light between 400 and 550 nm and probably has its molecules arranged with their long axes parallel to the long axis of the organism, while a short wavelength system, with its molecules arranged at right angles to the long axis of the cell, screens the shorter wavelengths (from 400 nm down). Could not such a geometric arrangement make possible orientation based on the analysis of stimulus direction rather than intensity? The majority of the pigments making up the stigma have been identified as lutein (51 per cent) and cryptoxanthin (32 per cent), with a trace of β carotene and some unknown fraction (Batra and Tollin, 1964).

Halldal (1958) has made the most extensive survey of other forms with respect to action spectra. Most of the forms investigated (dinoflagellates, *Platymonas* and *Ulva* gametes) exhibit similar action spectra in the blue region of the spectrum between 450 and 490 nm, with one exception (the dinoflagellate *Prorocentrum*), which has its peak in the yellow-orange (570 nm). No attempt was made to determine an action spectrum in the ultraviolet except in *Platymonas*. Halldal (1961) ruled out the possibility of riboflavin as the photoreceptive pigment in *Platymonas* on the basis of the ultraviolet action spectra, and concluded that a carotenoprotein was the photoreceptive pigment complex involved in photic orientation. He also predicted the involvement of a carotenoid in the phototactic process in other forms.

As a result of recent work by Forward and Davenport (1968) we must now consider still another class of pigments as being involved in photoresponses; these are the phytochromes. In the dinoflagellate *Gyrodinium* a mediation of photoreceptor sensitivity is observed by a phytochrome-type pigment system. It is not known at this time whether the pigment system is active at the photoreceptor level or whether it is secondarily active somewhere in the receptor–effector chain.

Finally, we must direct our attention to the nature of information transfer and propagation in the receptor–effector complex. Intracellular information transfer may take the form of an electrical potential generated by some photochemical reaction at the receptor site. Electrical potentials have been measured in dinoflagellate cells (Chang, 1960; Eckert, 1965) and have been used to mediate cellular responses (Eckert, 1965). It is possible that such electrical events are utilized by the cell to cause deviations in flagellar activity resulting in orientation. As Jahn (1963) hypothesizes, the initial response of the photopigment to light may result in a chain reaction of conduction of electrons across the receptor through a poised redox state. The potential could then cause a bending of the flagellum via the flagellum's piezoelectric properties (Jahn and Bovee, 1968). Such piezoelectric states appear to be common in biological fibres (Shamos and Lavine, 1967) and may also be present in flagella. The overall mechanism for orientation using this model is presented by Jahn and Bovee (1968). They state that because the structure of the flagellum is 'quasi-crystalline' it may be piezoelectric. If so, the voltage change generated by the change in the photoreceptor pigment state may cause the flagellum to bend. The straightening of the flagellum then results from its inherent elasticity. In straightening, the sign of the charge is reversed. The bend, then, could be proportional to the charge, which may be proportional to the stimulus. If this is the case, then it can be examined experimentally by standard dose–response methods or by more novel means, such as the analysis of polar particles in the medium which align themselves along the flagellum on the basis of that organelle's charge. One may well ask whether this proposed model can explain all observed behaviour, as, for example, our 'stop' response in *Gyrodinium*. Could this be the result of an excessive electrical potential generated by a totally activated receptor?

The role of kineses in flagellate behaviour has received little attention. Feinleib and Curry (1967) have shown that the swimming rate in *Chlamydomonas* does not change with light intensity, while Wolken and Shin (1958) have shown that in *Euglena* the swimming rate increases with intensity. We know of no studies with flagellates in which the relation of rate of change of direction to intensity has been investigated. Detailed studies are obviously needed.

In concluding this chapter we may express our belief that there will turn out to be as great a diversity in the physiological mechanisms controlling behaviour in the 'acellular' protists as there is among Metazoa; indeed there may even be greater diversities for once information-carrying systems were evolved in the Metazoa, they appeared to follow a somewhat common evolutionary pathway. This may not be true in the Protista, but we do not have enough information to reach a conclusion as yet. We must ultimately be able to make a comparison of the modes of action of the fast-acting stimulus–response chains triggered by light and other stimuli in bacteria, plant and animal flagellates, ciliates, amoeboid cells and motile gametes. We believe that the most rapid advances in our understanding

of the chains involved will be made by following the approach recommended
by Clayton (1964) who studied *Rhodospirillum* 'from the point of view of
the nerve physiologist' and showed that 'the "phototactic" response shows
nearly all the "traditional" features of nerve excitation: refractoriness and
recovery, rhythmicity . . . accommodation . . . and the Gildemeister effect'.
Further, he says that by definition 'phototaxis' demands 'a photoreceptor
and an effector that alter the nature of . . . movement. A mechanism that
bridges the photoreceptive act and the resulting motor response is also
implicated. A full understanding of phototaxis will therefore require that
the operations of the receptor, the mediating system and the effector be
understood'. Clayton clearly appreciated the necessity of breaking down
the chain of physiological events which occur from the reception of the
stimulus to completed orientation and photomigration into its component
links. We believe that a number of discrete and probably physiologically
differing links will be demonstrated to make up the whole chain. Clearly,
a legion of problems remain to be clarified.

REFERENCES

Batra, P. P., and G. Tollin (1964). Phototaxis in *Euglena* I. Isolation of the eye-
    spot granules and identification of the eye-spot pigments. *Biochim. et Biophys.
    Acta.*, **79**, 371.
Bendix, S. W. (1960). Phototaxis. *Botan. Rev.*, **26**, 145.
Brokaw, C. J. (1958). Chemotaxis of bracken spermatozoids—implications of
    electrochemical orientation. *J. Exp. Biol.*, **35**, 197.
Bruce, V. G., and C. S. Pittendrigh (1956). Temperature independence in a
    unicellular 'clock'. *Proc. Nat. Acad. Sci.*, **42**, 676.
Buder, J. (1915). Zur Kennetnis des *Thiospirillum jenense* und seiner Reaktionen
    auf Lichtreize. *Jahrb. Wiss. Botan.*, **56**, 529.
Buder, J. (1919). Zur kenntnis der phototaktischen Richtungsbewegungen.
    Jahrb. Wiss. Bot., **58**, 105.
Bünning, E. und M. Tazawa (1957). Uber die negativ-phototaktische Reaktion
    von *Euglena. Arch. Mikrobiol.*, **27**, 306.
Chang, J. J. (1960). Electrophysiological studies of a non-luminescent form of
    the dinoflagellate *Noctiluca miliaris. J. Cell. Comp. Physiol.*, **56**, 33.
Clayton, R. K. (1964). Phototaxis in microorganisms. In A. C. Giese (Ed.),
    Photophysiology, Vol. II. Academic Press, New York, London. p. 51.
Davenport, D., G. Camougis, and J. F. Hickok (1960). Analyses of the behavior
    of commensals in host-factor. I. A hesionid polychaete and a pinnotherid crab.
    An. Beh., **8**, 209.
Davenport, D., G. J. Culler, J. O. B. Greaves, R. Forward, and W. G. Hand
    (1970). The investigation of the behavior of microorganisms by computerized
    television. *I.E.E.E. Trans. Biomed. Eng.*, July (In the press).
Davenport, D., C. A. Wright, and D. Causley (1962). Technique for the study of
    the behavior of motile microorganisms. *Science*, **135**, 1059.

Dethier, V. G. (1957). Communication by insects: the physiology of dancing. *Science*, **125**, 331.

Diehn, B. (1969). Two perpendicularly oriented pigment systems involved in phototaxis of *Euglena*. *Nature*, **221**, 366.

Eckert, R. (1965). Bioelectric control of bioluminescence in the dinoflagellate *Noctiluca*. *Science*, **147**, 1140.

Engelmann, T. W. (1882). Uber Licht-und Farbenperseption niederster Organismen. *Pflugers Arch. Ges. Physiol.*, **29**, 387.

Feinleib, M. E. H., and G. M. Curry (1967). Methods for measuring phototaxis of cell populations and individual cells. *Physiol. Plant.*, **20**, 1083.

Forward, R., and D. Davenport (1968). Red and far-red light effects on a short-term behavioral response of a dinoflagellate. *Science*, **161**, 1028.

Fraenkel, G. S., and D. L. Gunn (1940). *The Orientation of Animals; Kineses, Taxes and Compass Reactions*, Clarendon Press, Oxford.

Fraenkel, G. S., and D. L. Gunn (1961). *The Orientation of Animals; Kineses, Taxes and Compass Reactions*, Dover, New York.

Gebauer, H. (1930). Zur Kennetnis der Galvanotaxis von *Polytoma uvella* und einigen anderen Volvocineen. *Beitr. Biol. Pflanz.*, **18**, 463.

Gunn, D. L. (1937). The humidity reactions of the wood-louse, *Porcellio scaber* Latrielle. *J. Exper. Biol.*, **14**, 178.

Gunn, D. L., J. S. Kennedy, and D. P. Pielou (1937). Classification of taxes and kineses. *Nature*, **140**, 1064.

Halldal, P. (1958). Action spectra of phototaxis and related problems in Volvocales, *Ulva*-gametes and Dinophyceae. *Physiol. Plant.*, **11**, 118.

Halldal, P. (1959). Factors affecting light response in phototactic algae. *Physiol. Plant.*, **12**, 742.

Halldall, P. (1961). Ultraviolet action spectra of positive and negative phototaxis in *Platymonas subcordipronis*. *Physiol. Plant.*, **14**, 133.

Halldal, P. (1964). Phototaxis in Protozoa. *Biochem. and Physiol. of Protozoa*, Vol. 3. Academic Press, New York, London. pp. 277–296.

Hand, W. G., P. A. Collard, and D. Davenport (1965). The effects of temperature and salinity change in swimming rate in the dinoflagellates, *Gonyaulax* and *Gyrodinium*. *Biol. Bull.*, **128**, 90.

Hand, W. G., R. Forward, and D. Davenport (1967). Short-term photic regulation of a receptor mechanism in a dinoflagellate. *Biol. Bull.*, **133**, 150.

Haupt, W. (1966). Phototaxis in plants. *Int. Rev. Cytology*, **19**, 267.

Jahn, T. L. (1963). A possible mechanism for the amplifier effect in the retina. *Vis. Res.*, **3**, 25.

Jahn, T. L., and E. C. Bovee (1966). Motile behaviour of protozoa. In T. T. Chen (Ed.), *Research in Protozoology*, Vol. I. Pergamon Press, Oxford. pp. 39–198.

Jahn, T. L., and E. C. Bovee (1968). Locomotive and motile response in *Euglena*. In D. Buetow (Ed.), *The Biology of Euglena*, Vol. I. Academic Press, New York, London. pp. 45–108.

Kennedy, J. S. (1945). Classification and nomenclature of animal behaviour. *Nature*, **156**, 754.

Lindes, D. A., B. Diehn, and G. Tollin (1965). Phototaxigraph: recording instrument for determination of the rate of response of phototactic microorganisms to light of controlled intensity and wavelength. *Rev. Scient. Instrum.*, **36**, 1721.

10

Manten, A. (1948). Unpublished Ph.D. Thesis, University of Utrecht, Netherlands.

Mast, S. O. (1911). *Light and Behavior of Organisms*, Wiley, New York. 410 pp.

Mayer, A. M. (1968). *Chlamydomonas*: adaptation phenomena in phototaxis.
Nature, 217, 875.

Oltmanns, F. (1917). Uber phototaxis. *Z. Bot.*, 9, 257.

Pohl, R. (1948). Tagesrhythmus im phototaktischen Verhalten der *Euglena
gracilis*. *Z. Naturforsch.*, 3b, 367.

Rikmenspiel, R., and G. Van Herpen (1957). Photoelectric and cinematographic measurements of the motility of bull sperm cells. *Phys. in Med. and
Biol.*, 21, 54.

Rohlf, F. J., and D. Davenport (1969). Simulation of simple models of animal
behavior with a digital computer. *J. Theoret. Biol.*, 23, 400.

Rothschild, Lord, and M. M. Swann (1949). The fertilization reaction in the
sea-urchin egg. *J. Exp. Biol.*, 26, 164.

Shamos, M. H., and L. S. Lavine (1967). Piezoelectricity as a fundamental
property of biological tissues. *Nature*, 213, 267.

Ullyott, P. (1936). The behavior of *Dendrocoelum lactum*. I—Responses at light-
and-dark boundaries. II—Responses in non-directional gradients. *J. Exp.
Biol.*, 13, 253.

Wigglesworth, V. B. (1941). The sensory physiology of the human louse *Pediculus
humanus corporis* De Geer (Anoplura). *Parasit.*, 33, 67.

Wolken, J. J., and E. Shin (1958). Photomotion in *Euglena gracilis*. I. Photokinesis. II. Phototaxis. *J. Protozool.*, 5, 39.

CHAPTER 10

Light-oriented chloroplast movements

WOLFGANG HAUPT

Department of Biology, University of Erlangen-Nürnberg, Germany

EKKEHARD SCHÖNBOHM

Department of Biology, University of Marburg, Germany

Introduction

Since the discovery of chloroplast orientation movements in light more than 100 years ago, it has generally been accepted that in light of low and medium intensity chloroplasts move in such a way as to expose as large an area to light as possible (positive response), which results in conditions most favourable for photosynthesis; in contrast, in high intensity light they move in such a way as to expose as little of their area as possible (negative response), which results in protection against too much light absorption (Stahl, 1880). In most cases it can be observed, therefore, that chloroplasts move to the most or least illuminated places in the cell

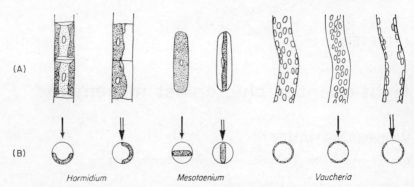

Figure 10.1. Chloroplast arrangements in some algae, showing surface views (A) and cross-sections (B). Low intensity and high intensity light are shown by single and double arrows respectively; in addition, a specific dark arrangement is shown in *Vaucheria*. (After Haupt, 1963; Senn, 1908)

respectively. These places are differently located in relation to the light direction (for example, see Figure 10.1), depending upon the cell morphology and other conditions; accordingly, several 'types' of orientation have been established by Senn (1908), and the chloroplast arrangements of these types have been given special names, e.g. *epistrophe*, *diastrophe*, *peristrophe*, *parastrophe*, etc. The use of these names, however, seems to be more of historical interest; for the sake of simplicity, we shall restrict ourselves in this review to the easily understandable terms 'dark arrangement', 'low intensity arrangement' and 'high intensity arrangement', and, accordingly, the respective movements, viz. 'dark movement', 'low intensity movement' and 'high intensity movement' (Haupt, 1959a).

Light intensities which are perceived by different plants as 'low' or as 'high' intensities differ from species to species. In addition, this perception depends on environmental and internal factors. But as an order of magnitude, we can consider daylight without direct sun as 'low intensity light' and, consequently, direct sunlight as 'high intensity light'. However, the distinction 'low' and 'high' intensity *sensu strictu* is correct only as long as we are dealing with white light. In coloured light there are cases where both the light quality and its intensity will determine whether high or low intensity movement will occur. It might then be more correct to use the terms 'positive' and 'negative' response. However, this is a formalism of little importance.

As far as we know, several main steps of light-oriented chloroplast movements are similar or even identical in higher and lower plants, and investigations on algae have contributed much to our knowledge on this topic. Some of these algae are microorganisms *sensu strictu*, e.g. the unicellular *Mesotaenium* or the small filamentous *Hormidium*. On the other hand, the large filamentous alga *Mougeotia* shares so many details and specialities with *Mesotaenium* that it is quite impossible to deal with *Mesotaenium* without also regarding *Mougeotia*. This may justify the unusual widening of the term 'microorganism' in this chapter. Occasionally, the filamentous alga *Vaucheria* has also to be included in our review. Because of the general uniformity in the responses, reference will be made to comparable facts in higher plants.

If movement is induced and oriented by light, we can divide the reaction chain into at least three steps: light perception, physical or chemical processes following light perception, and movement of the chloroplast. These three steps will be treated in the following sections. In the first step, we are interested in the nature of the photoreceptor pigment, its localization and, eventually, association with cell structures, and, furthermore, the mechanism of perception of light direction. In the second step we hope to obtain information on metabolic pathways involved in the response. The third step deals mainly with the mechanics of the movement, which, however, has hardly been understood until now.

Besides the orientation movements there are also, in some diatoms (e.g. *Striatella*, *Biddulphia*), light-induced movements without any relationship to the light direction. Here also we may distinguish between low intensity and high intensity arrangements with the above-mentioned biological meaning. However, the arrangements are determined only by the cell morphology; in high intensity light all chloroplasts are clustered around the nucleus in the centre of the cell, whereas in low intensity light they are distributed in the peripheral portion of the cell (Höfler, 1963; Senn, 1919).

Perception of light

The nature of the photoreceptor

One of the fundamental goals in all analyses on photic responses is the identification of the photoreceptor. Such analyses normally involve action spectra determinations. For chloroplast motion, action spectra have been measured in *Vaucheria*, *Mesotaenium* and *Mougeotia* (Figure 10.2). In *Vaucheria*, as in some higher plants (*Funaria*, *Selaginella* and *Lemna*), the action spectra of both response types are identical (Fischer-Arnold, 1963; Haupt and Schönfeld, 1962; Mayer, 1964; Zurzycki, 1962b, 1967a). Moreover, there is a strong similarity between the action spectra of *Vaucheria* and of higher plants (Haupt, 1963). This suggests uniformity of the primary processes of chloroplast orientation in nearly all plants (for

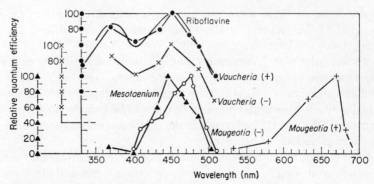

Figure 10.2. Action spectra of chloroplast orientation movements. These spectra demonstrate high intensity movement for *Mesotaenium*, whereas for *Mougeotia* and *Vaucheria* low (+) and high (−) intensity movements are indicated. The absorption spectrum of riboflavin is given for comparison. Ordinates for *Vaucheria* and riboflavin are shifted. (After Haupt, 1959b, 1963; Haupt and Gärtner, 1966; Schönbohm, 1963)

exceptions, see below). The overall shape of these action spectra closely resembles that of the absorption spectrum of flavins, from which it may be concluded that flavins act as photoreceptive pigments. This is supported in higher plants by experiments with iodide ions which reversibly inhibit the response and are supposed to quench the energy transfer from excited flavins to secondary processes (Mayer, 1966; Seitz, 1967b). However, this is not as conclusive as it seems to be, because in *Mougeotia* iodide has been shown to inhibit dark processes of the reaction chain rather than the photoprocess (Schönbohm, 1967b, 1969a). Nevertheless, even if we do not have conclusive evidence that a flavin is involved in the photoprocess, no experimental data favour any other yellow pigment.

In contrast to *Vaucheria* and higher plants, the action spectra of both response types in *Mesotaenium* and *Mougeotia* are different from each other: high intensity movement is induced by blue light, but low intensity movement mainly by red light. As can be seen from Figure 10.2, the high intensity spectra roughly resemble the spectra of the other plants, except for a shift of the peak in *Mougeotia* and a lack of a pronounced near ultra-violet peak in *Mesotaenium* (Haupt and Gärtner, 1966; Schönbohm, 1963). The low intensity response, however, is induced by phytochrome, as can be supposed from the action spectra (Dorscheid, 1966; Haupt, 1959b) and, especially, as can be demonstrated by the far-red reversibility in both algae (Haupt, 1958; Haupt and Thiele, 1961).

It should be added that the phytochrome of *Mesotaenium* is slightly different from that of higher plants, judging from a shift of the absorption peaks to shorter wavelengths measured *in vitro* (Taylor and Bonner, 1967); in *Mougeotia* the same conclusion could be derived from calculations based upon physiological results (Haupt, 1970).

Thus we come to a rather unified conclusion: with the exception of the more complicated systems in *Mougeotia* and *Mesotaenium* (see also page 292), both types of response generally seem to be mediated by a yellow pigment.

Localization of the photoreceptor in the cell

It is essential to know whether the photoreceptor system of chloroplast orientation is localized to the chloroplast itself (this will then by definition be phototaxis), or if it is confined to the cytoplasm. It can be demonstrated in two different ways that the cytoplasm normally is the site of the photoreceptor.

(1) Irradiations with microbeams of small areas of the cell result in chloroplast orientation whether or not the chloroplasts are hit. This has been found by Fischer-Arnold (1963) for a blue light response in *Vaucheria* (and equally well in the higher plant *Selaginella* by Mayer, 1964) and is also true for a phytochrome response in *Mougeotia* (Bock and Haupt, 1961).

(2) Polarized light has a very specific effect, as the response depends upon the orientation of the electric vector relative to cell morphology (Figure 10.3). This may be explained by assuming a dichroic arrangement of photoreceptors. Since this well-defined dichroism is found irrespective of the momentary chloroplast arrangement, the photoreceptors cannot be localized at the chloroplasts. Furthermore, the existence of dichroism would not correspond with the photoreceptor's localization in the overall cytoplasm which undergoes streaming.

From these observations and conclusions, the photoreceptor is assumed to be localized in the cortical cytoplasm, perhaps associated with the plasmalemma. This conclusion has been arrived at for blue light responses—*Hormidium* (Scholz, 1970), the high intensity response of *Mesotaenium* (Haupt and Gärtner, 1966), and higher plants (Haupt and Weisenseel, 1967; Mayer, 1964; Zurzycki, 1967b)—as well as for phytochrome responses—*Mesotaenium* (Haupt and Thiele, 1961) and *Mougeotia* (Haupt, Köhler and Müller, 1960). For exceptions, see below.

In addition to its bearing on the localization, the dichroic orientation

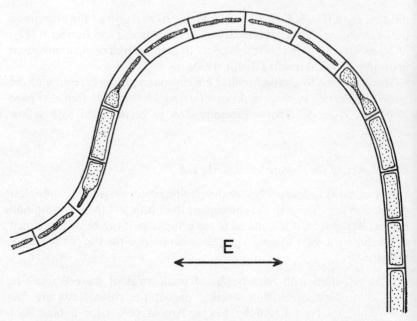

Figure 10.3. Low intensity response in *Mougeotia* after irradiation with polarized red light, the electrical vector of which is indicated by the double arrow. (After Haupt, Köhler and Müller, 1960)

of the photoreceptors is by itself a very interesting problem. As far as generalizations are possible, it has been stated in all cases that the dichroic orientation is parallel to the cell surface. This is true for both the blue and the red light responses. This could also be specified in *Mougeotia* by demonstrating that the phytochrome molecules are oriented along a screw pattern in the cell periphery (Haupt and Bock, 1962). However, some data and calculations strongly favour the assumption that this peripheral parallel orientation of phytochrome molecules in *Mougeotia* is valid only for the red absorbing form (P_{660}), whereas the far-red absorbing form (P_{730}) might have its dichroic orientation normal to the surface (Haupt, 1970). If this is the case, photobiological studies on *Mougeotia* might be an excellent approach to the analyses of the phytochrome system in general.

Perception of light direction

In order to induce light-oriented responses, not only must light perception take place, but the perception of light direction must be an additional

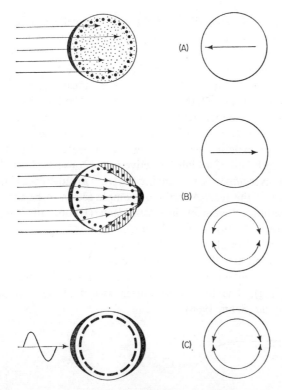

Figure 10.4. Perception of light direction (arrows at left) by a cylindrical cell. Photoreceptor molecules are indicated by dots or dashes near the surface, absorption by black thickening at the periphery. The arrows at right point from stronger to lower absorption, thus showing the absorption gradient. (A) illustrates the shadow principle, (B) the lens and by-passing principles and (C) the dichroic principle (see also Figure 10.6). (After Haupt, 1965)

requirement. Since chloroplasts orientate themselves in relation to light absorption in the cytoplasm, we have to postulate an absorption gradient in the cytoplasm, resulting from unidirectional irradiation. The question, therefore, is to find the transformation of light direction into spatial differences of absorption (absorption gradient).

Theoretically, four different possibilities exist for this transformation (Figure 10.4; see also Haupt, 1965), which all seem to have some bearing on the responses in question.

(1) Light which passes through a cell is attenuated by absorption. In this way the rear of the cell is exposed to less light than the front.

This kind of perception of light direction will be referred to in the following as the 'shadowing principle'; an example will be given later.

(2) An inverse gradient to that effected by the 'shadowing principle' results if, by differences of refractive indexes of the surrounding medium and the cell, the latter acts as a collecting lens, thus focusing the light to the rear (referred to as the 'lens principle' in the following). The 'lens principle' or 'lens effect' can be demonstrated by the inversion experiment: replacing air or water as the surrounding medium by oil of a high refractive index converts the collecting lens into a diverging lens, thus giving the front an advantage in absorption instead of the rear. A striking example seems to be the chloroplast orientation movement in *Hormidium* (Haupt and Scholz, 1966) (Figure 10.5). For complications, see below.

(3) The lens effect also results in the light by-passing some parts of the cell behind the flanks, thus establishing roughly a front-and-rear to flank gradient (referred to as the 'by-passing principle' in the following). The light direction is measured by this principle in *Vaucheria* (Senn, 1908).

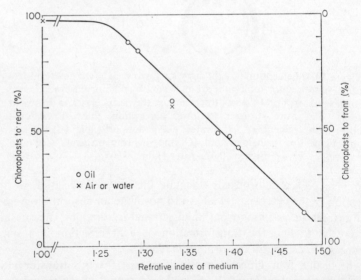

Figure 10.5. Light-induced chloroplast movement of *Hormidium* immersed in oils of different refractive indexes; controls irradiated in air or water respectively. Percentage of cells in which the chloroplast moves to the rear or to the front is shown in the ordinates (after Haupt and Scholz, 1966)

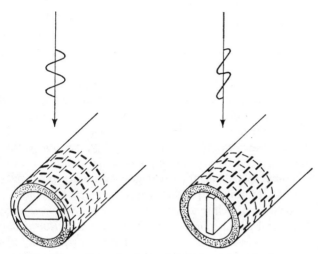

Figure 10.6. Absorption of polarized light by oriented photoreceptors in *Mougeotia*. Heavy and thin dashes show photoreceptors which can or cannot absorb the polarized light due to their orientation. The resulting absorption is indicated by dots near the periphery; the absorption gradient is shown by arrows. The electrical vector of the light is indicated by the waves. (After Haupt, 1960)

(4) Apart from attenuation and light refraction, an absorption gradient must also arise if dichroic photoreceptors are oriented parallel to the cell surface (Figure 10.6). At the front and rear, all photoreceptors are in an orientation to absorb unpolarized light, whereas at the flanks, only part of the photoreceptors are in the right orientation; this results in less absorption at the flanks, i.e. in establishing a front-and-rear to flank gradient (referred to as the 'dichroic principle' in the following). The 'dichroic principle' can be demonstrated by differential action of polarized light, as is well known in *Mesotaenium* (Haupt and Thiele, 1961) and *Mougeotia* (Haupt, 1960; see also Figure 10.3).

Some complications in perception of light or light direction

The four principles used to measure the light direction can easily be distinguished as long as only one principle is followed by one cell. In some cases, however, two or more of these principles are used in combination. A further complication may arise if different types of photoreceptive pigments are simultaneously involved in chloroplast orientation. Examples for both possibilities will be given below.

In the low light intensity response of *Hormidium* the chloroplast moves, with a high degree of probability, to the brightest region of the cell, i.e. to the rear, by means of the lens principle. This is true in air, water and oils of low refractive indexes. In oils of high refractive indexes, however, the chloroplast usually moves to the area facing the light, which area now is brightest illuminated (see the inversion experiment above). In oils of a well-defined medium refractive index, no advantage of absorption should be expected in any region of the cell. Nevertheless, the chloroplast goes to either the front or the rear with equal probability, but not to the flanks. This must be explained by a front-and-rear to flank gradient in addition to the (now compensated) front to rear gradient. Since polarized light is effective only if vibrating perpendicularly to the cell axis rather than if parallel, the dichroic principle seems to be followed by the cell in addition to the lens principle. Still more complications arise in *Hormidium* if the light intensity is increased, but is still working as low intensity light. The chloroplast, then, always goes to the front, independent of the refractive index of the surrounding medium. This is consistent only with the shadowing principle. Thus, in one cell, either the lens principle is working without interference with the shadowing effect, or the shadowing principle is operating without interference with the lens effect, depending on light intensity. One way to explain such responses is to assume two different sites of photoreceptors in the cell. One group should then be arranged in the cortical cytoplasm (the lens and the dichroic principles), and the other should be closely associated with the chloroplast (shadowing principle). Experimental data further suggest that these two groups of differently localized photoreceptors also represent different photoreceptive pigments; they may be separated by the difference in their responses to coloured and polarized light, to metabolic inhibitors and to pretreatments (Scholz, 1970).

In the low intensity orientation in *Mesotaenium*, chlorophyll is suggested to function as a photoreceptor in addition to phytochrome (Dorscheid and Wartenberg, 1966). It is, however, difficult to understand how absorption in the flat chloroplast can result in a front-and-rear to flank gradient necessary to turn the chloroplast from the profile to the plane position. But if chlorophyll were indeed able to act as a photoreceptor in chloroplast orientation, its dichroic orientation in the chloroplast could possibly be judged from experiments in *Mesotaenium* (Dorscheid and Wartenberg, 1966), as well as in *Hormidium* (Scholz, 1970).

In *Mougeotia* some data from analyses on the low intensity orientation suggest two different photoreceptor systems which do not seem to have different sites. In addition to the red light effect mediated by phytochrome, blue light is also effective. Under certain conditions, the photoreceptive system excited by blue light seems to operate independently of the phytochrome system. This can especially be shown if irradiation takes place in a low temperature. At 2°C no response occurs in red light, even if given continuously, while continuous blue light induces movement with ease (Weisenseel, 1968b). However, though this might best be interpreted by assuming two photoreceptor systems, another interpretation will be presented below (page 297).

Another complication has been found in the high intensity movement in *Mougeotia*. From action spectra analyses it had originally been concluded that a yellow pigment is involved, while phytochrome is the light absorber for the low intensity response only. However, if simultaneously irradiated with red and blue light from different directions (Figure 10.7), the chloroplast is oriented by red rather than by blue light, and this red light effect is far-red reversible and hence a phytochrome effect. The only effect of blue light, in this case, seems to be to change the chloroplast arrangement to a high intensity response oriented by phytochrome; to operate in this way,

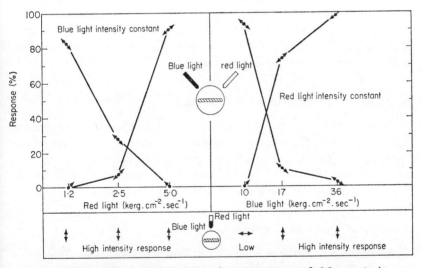

Figure 10.7. High intensity chloroplast movement of *Mougeotia* in double beam irradiation with blue and red light. The upper part of the figure shows the orientation of the profile to either blue or red light, depending on the intensities; the ordinate shows the percentage of cells which respond according to the double arrows (chloroplast orientation). The lower part shows the orientation to both light beams irradiating the plane; low or high intensity response is indicated only qualitatively by the double arrows (see above). (After Schönbohm, 1967a)

blue light must be irradiated with sufficient intensity, but the direction of the blue light beam has no bearing on the orientation (Schönbohm, 1967a). The action spectrum of this 'tonic' effect resembles closely the original action spectrum of the blue light mediated response (see Figure 10.2); the effect is restricted to the blue and near ultraviolet region of the spectrum and therefore reflects light absorption in a yellow pigment (Schönbohm, 1970). Strong blue light alone, then, results in high intensity movement because it is absorbed simultaneously by two pigments:

(1) the phytochrome which establishes the gradient, and
(2) the unknown yellow pigment which determines the type of response.

There is good evidence to indicate that this interaction of phytochrome and a yellow pigment is also working in the high intensity movement of *Mesotaenium* (Gärtner, 1970). It is not yet clear in which way the 'tonic' action of the yellow pigment has to be understood; it might have some direct interaction with the phytochrome and/or it might contribute to the energy supply for the movement (Schönbohm, 1969a). However, this cannot be the key effect for triggering the high intensity response, as will be seen later (page 307). In these latter cases we were no longer dealing with the true primary processes of our responses; this question, therefore, will be referred to in more detail on page 299.

Secondary processes following photoperception

Though our knowledge of the processes which follow the perception of light is still more fragmentary than our knowledge of the perception of light, there are two approaches to a preliminary understanding of such processes: the influence of temperature on the light requirement for the response and the role of the metabolism in the response.

Influence of temperature on the light requirement

Between light intensities which act as 'low intensities' and those which act as 'high intensities', we can find an intermediate intensity which forces the chloroplasts with equal probability to either of both response types. This 'indifference intensity' is neither absolutely fixed in each species nor in each specimen, but depends remarkably on the temperature (Senn, 1908); it is lower in low temperatures and higher in high ones. This can be demonstrated not only by plotting intensity–response curves at different temperatures (as has been done mainly in higher plants by Weisenseel,

1968a) but also convincingly, by keeping an intermediate light intensity constant and varying the temperature. If, then, the appropriate light intensity is chosen, this variation of temperature results in chloroplast movement, i.e. low intensity movement with increasing temperature and high intensity movement with decreasing temperature. The result of this experiment, originally performed in higher plants, can be seen in Figure 10.8 for *Mougeotia* (Weisenseel, 1968a; Weisenseel and Haupt, 1967);

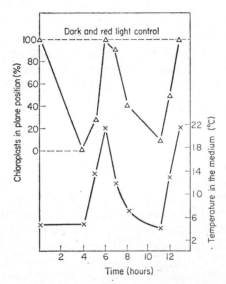

Figure 10.8. Chloroplast movement in *Mougeotia* in constant blue light intensity, caused by changing the temperature. The experimentally changed temperature in the medium is shown on the lower right ordinate, while the resulting chloroplast position is shown on the upper left ordinate. Increasing and decreasing percentages of plane position indicate low intensity and high intensity movements respectively. (After Weisenseel and Haupt, 1967)

the same result has also been obtained for *Mesotaenium* (Gärtner, 1970). Incidentally, it should be mentioned that only blue or white light is suited for this kind of experiment, because red light alone can only result in low intensity movement in *Mesotaenium* and *Mougeotia* and, therefore, no indifference intensity can exist.

To explain this temperature effect on the indifference intensity, it must be kept in mind that the primary photochemical processes are expected to be temperature independent, which in favourable systems has indeed been demonstrated (Mugele, 1962). The effect in question must be concerned, therefore, with secondary processes. The following hypothesis seems to be reasonable (Weisenseel, 1968a). As an effect of the absorption gradient, a biochemical gradient might be established which results in the orientation response. By thermodynamic reasons this biochemical gradient is not stable, but is trying to level out. This levelling out should be a temperature-dependent process. In continuous light, therefore, this biochemical gradient is a steady state between two antagonistic processes: its temperature-independent establishment by the absorption gradient and its temperature-dependent decay. From this it is evident that at lower temperatures lower light intensities are needed to keep a certain steady state.

This hypothesis is supported by experiments with intermittent irradiations. In the high intensity response of *Mesotaenium* and *Mougeotia*, continuous blue light can be replaced effectively by short irradiations with dark periods of some seconds to about 1 minute in between. The longer the dark periods, the less the response. However, to obtain the same response, the dark periods can be prolonged if the temperature is lowered (Gärtner, 1970; Mugele, 1962). Here, the temperature-dependent decay of each light effect can be demonstrated clearly.

In contrast to this need for a steadily renewed gradient in blue light mediated responses, the phytochrome induced response of *Mougeotia* needs only one short induction. However, here also something is lost after a short time, as can be seen from the red far-red antagonism: if an induction by red is followed by far-red not more than a few minutes later, no response takes place (Haupt, 1959b); this demonstrates that here also secondary processes require the steady presence of a gradient. The main difference seems to be that, in a blue light photoreceptor, absorption results in an excited state which lasts for only fractions of a second, and which, therefore, has to be continued or repeated more quickly than the decay of the physiological gradient; phytochrome is converted, however, by a single red absorption to a rather stable form (P_{730}) which maintains the physiological gradient, at least during a great part of the response in *Mesotaenium* (Haupt and Thiele, 1961) or even during the whole response in *Mougeotia* (Haupt, 1959b). However, this happens in *Mougeotia* only if the response is not retarded very strongly by a low temperature, which would cause the active form (P_{730}) to become lost before it could trigger the response (Mugele, 1962).

The role of the metabolism

Besides the effect of temperature on the light requirement, there is also a marked effect on the speed of movement. Agreeing with the results in higher plants, the movements of both response types of *Mougeotia* are temperature dependent, the Q_{10} being in the range of 2 to 4 below 20°c and approaching unity in higher temperatures (Mugele, 1962; Weisenseel, 1968b). This suggests that, at least in a certain range of temperature, biochemical processes are rate limiting. Nothing can be said, however, from those very unspecific treatments, about the nature of the processes in question. Before trying to solve this problem with more specific means, we have to add a few complications concerning temperature effects.

(1) Lowering the temperature below 5°c in *Mougeotia* retards the low intensity movement as expected if irradiation is achieved with continuous blue light; if red light is used, however, no movement is possible at all, as stated above. This points to either a temperature effect on viscosity which is blue light reversible or to different primary and secondary processes after blue versus red irradiation with different temperature requirements (Weisenseel, 1968b).

(2) Viscosity effects of temperature may also be involved in the contraction of the chloroplast of *Mougeotia* which occurs in a combination of low temperature and high intensity blue light (Weisenseel, 1968b). Because of the drastic change of cytology under these conditions, the retardation of movement of such contracted chloroplasts by cold must not be explained only by influences on metabolism and viscosity.

(3) The low intensity movement of *Mesotaenium* eventually shows Q_{10} values below unity (Dorscheid, 1966). This points to a complicated interaction of two or more temperature-dependent processes, acting antagonistically against each other. More information is needed in this case.

To go into more detail of the role of metabolism, inhibitors have to be used which more or less specifically block well-defined steps of the metabolism. In this way it is possible to relate different types of movement to different steps of metabolism and, especially, to get information about the energy source for the movement. This has been done, as well as in some of the higher plants (viz. *Lemna* by Zurzycki, 1965; *Vallisneria* by Seitz, 1967b), mainly in *Mougeotia* (Fetzer, 1963; Schönbohm, 1967b, 1969a, 1969b). The following information is therefore nearly restricted to this alga.

In principle, three main sources of energy are available in a photosynthetic cell: glycolysis, respiration and photosynthesis, or, more exactly, substrate-level phosphorylation, oxydative phosphorylation and photosynthetic phosphorylation. Of course, all plants which have a typical dark arrangement of chloroplasts are restricted to the dissimilatory energy sources when performing dark movement. However, *Mougeotia* also has to be restricted to only these sources if low intensity movement is induced by a short irradiation and is continued in darkness (in this case, the energy of the inducing light can be neglected with regard to photosynthesis). In contrast, all three energy sources should be available in this low intensity movement if it is caused by continuous light rather than by a short induction. The same is also valid, of course, for the high intensity movement which always needs longer lasting irradiations. Thus, we have to ask which one of the theoretically possible energy sources contribute in fact to the three types of movement.

(a) *Low intensity movement in the dark after a short induction* ('*induction type of low intensity movement*')

Inhibitors of glycolysis are without effect upon the movement as long as they are not applied in toxic concentrations. Inhibitors of respiration, however, are very effective, and even an inhibition of the respiration of

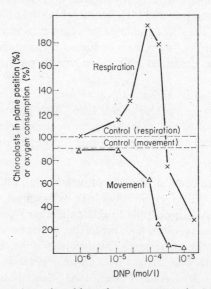

Figure 10.9. Low intensity chloroplast movement in *Mougeotia*, continuing in the dark after a short induction, as influenced by 2,4-dinitrophenol (DNP); oxygen consumption is given for comparison. Controls without inhibitors are shown by the dashed lines. The ordinate shows the percentage of chloroplasts reaching the low intensity position and the oxygen consumption measured as a percentage of the control. (After Fetzer, 1963)

only 50 per cent by cyanide or azide abolishes the movement completely (Fetzer, 1963). Thus, in contrast to the shuttle streaming of slime moulds (Kamiya, 1959), this chloroplast movement utilizes the energy from the oxydative rather than from the anaerobic part of the dissimilation.

More information was obtained from the finding that uncouplers of the oxydative phosphorylation also very effectively inhibit the movement, showing, at the same time, the increase of oxygen consumption which is typical for this uncoupling (Figure 10.9). Thus, ATP from the oxydative phosphorylation seems to be the energy source for the induction type of low intensity movement (Fetzer, 1963).

(*b*) *Low intensity movement in continuous light* ('*continuous light type*')

In the continuous light type the speed of movement is much more rapid than in the induction type. This suggests that additional energy is provided by photosynthesis. Indeed, application of inhibitors of photosynthesis reduces the speed to that of the induction type; the percentage of response which finally will be reached is not influenced. Inhibition of photosynthesis and respiration at the same time abolishes the movement completely, which, however, can be restored by application of ATP in addition to the inhibitors (Figure 10.10). Thus, the continuous light type makes use of photophosphorylation in addition to the oxydative phosphorylation, with the result of a more rapid movement. At the same time this shows that the energy delivering system is rate limiting for the speed of movement (Schönbohm, 1969a).

These results are supported by similar findings for higher plants (Zurzycki, 1965), as well as by some evidence that light absorption in chlorophyll is important for some low intensity responses in *Mesotaenium* (Dorscheid and Wartenberg, 1966), *Hormidium* (Scholz and Haupt, 1968) and, perhaps, *Vaucheria* (Fischer-Arnold, 1963).

(c) *High intensity movement*

Neither inhibitors of glycolysis nor of respiration have a marked effect on the high intensity movement. Instead, this response can easily be abolished by inhibitors of photosynthesis (Fetzer, 1963). This is true for inhibitors like 3-(3,4-dichlorophenyl) 1-,1-dimethylurea (DCMU) or Salicylaldoxim which block the electron transport chain and the photophosphorylation, and it is also true for those inhibitors which block the photophosphorylation without inhibiting the electron transport (e.g. atebrine). Prolonged irradiation time does not reverse this inhibition, but the addition of ATP can partly restore the response (Schönbohm, 1969a). Thus, mainly ATP from photophosphorylation seems to contribute to the

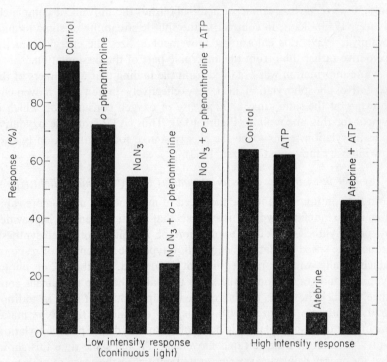

Figure 10.10. Chloroplast movement in *Mougeotia* under the influence of metabolic inhibitors. *o*-phenanthroline: 5×10^{-5} mol/l; NaN$_3$: 5×10^{-4} mol/l; atebrine: 10^{-4} mol/l; ATP: 10^{-3} mol/l (left) and 5×10^{-4} mol/l (right). (After Schönbohm, 1969a)

energy supply for the high intensity movement; whether cyclic or non-cyclic phosphorylation is more important cannot be stated with certainty.

One experiment by Fetzer (1963) seems to contradict the assumption that photosynthetic phosphorylation is the main energy source for the high intensity movement. In *Mougeotia*, this movement is strongly inhibited under anaerobic conditions. However, this does not necessarily have to be caused by reduced respiration, as recovery of response in air takes an unexpectedly long time, indicating cell injury. If this injury is of a general nature, chloroplast movements would also be involved.

As can be seen from the continuous light type of low intensity movement and from the high intensity movement, light may have a dual function: it may trigger the energy in the correct way and supply the energy by photosynthesis. In the high intensity movement, however, there seems to be a third function of light, at least in *Mougeotia*. This becomes obvious from the finding that orientation to red light absorbed in the phytochrome is possible provided there is additional irradiation with blue light, the direction of which has no bearing on the direction of response (see page 294). This blue light cannot act primarily as a source of photosynthetic energy, because this effect should be obtained equally well by red light or experimentally added ATP. Instead, there seems to be a step in the reaction chain which depends on blue light. This directs the chloroplast orientation to a low or a high intensity response type.

The mechanics of light-oriented chloroplast movement

With regard to the mechanics of chloroplast movement, one interesting problem is the question of whether the movement is 'active' or 'passive' (Senn, 1908). This means either a movement of chloroplasts relative to the whole cytoplasm, which may then be regarded as being motionless, or a movement of part of the cytoplasm, which takes with it the included organelles, among which are the chloroplasts.

Several hypotheses have been suggested (for example, Zurzycki, 1962a) which will not be referred to here, although some of the observations on lower plants contribute to this problem. The most detailed investigation has been undertaken by Fischer-Arnold (1963) in *Vaucheria* (Figure 10.11).

If chloroplast movement in this alga in low or high light intensity is carefully followed, it can easily be seen that not only chloroplasts are rearranged by light, but also other cytoplasmic organelles and inclusions (for example, oil drops; Fischer-Arnold, 1963) as well. Furthermore, if only a small area of the filament is irradiated by a spot of light, a strong accumulation of cytoplasm plus chloroplasts plus inclusions can be observed at this place, which eventually leads to occupation of the whole space of the vacuole.

This strongly points to the chloroplasts being taken passively with the moving cytoplasm. This is supported by another observation: light does not induce the movement of the chloroplasts, but only modifies this movement which goes on autonomously in a longitudinal direction in darkness or in diffuse light. This movement is not restricted to the chloroplasts, but again comprises masses of cytoplasm with other inclusions. Even though not all of the chloroplasts move in the same direction, at least groups of them always move together and change speed and direction together, thus showing the underlying general cytoplasmic movement. However, chloroplasts sometimes change their relationship from one to another moving group, thus showing the steadily changing structure of the cytoplasm.

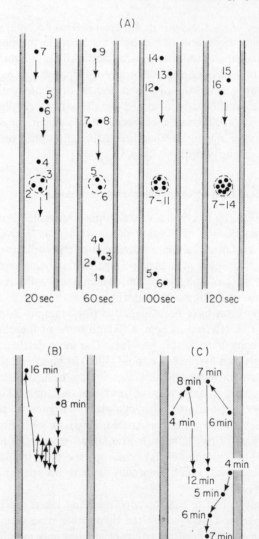

Figure 10.11. Movement of some single chloroplasts in *Vaucheria*. (A) Low intensity movement, irradiation with light spot (dashed); four consecutive positions after beginning of irradiation. Individual chloroplasts are numbered. Arrows show direction of overall streaming of cytoplasm. (B) High intensity movement, total irradiation and movement of one single chloroplast at different times after beginning of irradiation. (C) Low intensity movement starting from high intensity arrangement; movement of three individual chloroplasts as in (B), but after turning off high intensity light. (After Fischer-Arnold, 1963)

Next, the problem arises of how the accumulation of the cytoplasm with chloroplasts in the light gradient is brought about in *Vaucheria*. Several cases have to be distinguished: low intensity movement as a result of partial or total irradiation, or high intensity movement.

(1) Low intensity movement as a result of partial irradiation. If only a small area of the 'cell' is irradiated with a light spot, the autonomous cytoplasmic streaming in the neighbourhood is quite uninfluenced and goes on in the longitudinal direction. No 'attraction' takes place in the irradiated area. But whenever, after a lag phase of some seconds to a few minutes, a chloroplast enters this area, it stops moving; this is indicative of stopping the movement of the whole cytoplasm at this place. Thus, only randomly entering chloroplasts are prevented from escaping the 'light trap'. However, this has nothing to do with classical 'light trap' experiments in photophobotactic organisms, because it is not the change of light intensity in time which acts as the stimulus, but the spatial longitudinal difference in light intensity, the speed of cytoplasmic streaming being otherwise independent of the overall irradiation.

(2) Low intensity movement as a result of unidirectional total irradiation. If, by unidirectional irradiation, a front-and-rear to flank gradient of light absorption is established, the longitudinally streaming cytoplasm is gradually diverted in order to accumulate in the stronger absorbing front and rear regions. The overall pattern of streaming, however, is unchanged.

(3) High intensity movement as a result of unidirectional total irradiation. The strong light arrangement, i.e. accumulation at the less irradiated flanks, is reached in a more complicated way than the low intensity arrangements. During a lag phase of about 10 minutes, no response can be observed at all. In a second phase, after that, the streaming system is disturbed fundamentally, and the chloroplasts change with high frequency speed and direction of movement, without covering a net distance in the longitudinal direction. When, by these small movements, the chloroplasts have approached the flanks, a more uniform streaming is resumed which finally, in a somewhat oblique direction, brings the chloroplasts to the flanks where normal streaming is continued, although with reduced speed. Here also it can be seen that the front and rear areas are emptied not only by the chloroplasts, but also by the bulk of cytoplasm.

It should be noted that by this arrangement the distribution of cytoplasm is strongly unbalanced and tries to level out as soon as the absorption gradient of high intensity light is lost. Therefore, after the light has been turned off, chloroplasts and cytoplasm immediately move back to the front and rear areas in direct oblique ways, resulting, after a few minutes, in the random distribution of cell contents which are typical for darkness.

Very similar results concerning the patterns of movement in low and high intensity light and in the dark have been obtained in higher plants (Zurzycka and Zurzycki, 1957). It has been concluded that these different patterns demonstrate different biophysical states of the cytoplasm, e.g. viscosity. This latter assumption agrees very well with the measurements taken in higher plants, viz. the increase and decrease in viscosity by low and high intensity light respectively, in *Lemna* (Zurzycki, 1960), *Funaria* (Voerkel, 1934) and *Vallisneria* (Seitz, 1967a). In the latter plant the identity of the photoreceptors for viscosity change, chloroplast rearrangement and, in addition, light-enhanced overall protoplasmic streaming (photodinesis) has also been shown (Seitz, 1967a).

In spite of this interesting information, the mechanics of movement on a molecular base remain an open question. Very little is contributed on chloroplast-containing microorganisms; however, to complete this picture, the reader will shortly be referred to investigations on higher algae, viz. Characeae, where the bulk of protoplasm ('endoplasm') is continuously moving along a fixed outer cytoplasmic layer ('ectoplasm'). Here, the motive force has to be localized at the interphase between both layers, and is supposed to be generated by contractile protein fibrils (Kamiya, 1959; Zurzycki, 1962a). Though such fibrils have indeed been found in Characeae (Jarosch, 1958), no reference can be given to similar observations in plants and cells showing light-oriented chloroplast movements. However, the SH-blocking substance *p*-chloromercuribenzoate (PCMB), which inhibits the protoplasmic streaming in higher plants (Abe, 1964), has also been used with good success to inhibit the chloroplast movement in *Mougeotia*, whether it occurs in continuous light or in darkness after a short induction. Thus, this effect of PCMB cannot be explained only by its inhibiting function on photosynthesis. In the same way as streaming in higher plants, the chloroplast movement response of *Mougeotia* can be restored by applying cystein in addition to PCMB (Schönbohm, 1969b).

Thus, indirect evidence is given to contractile protein fibrils underlying the motive force in *Mougeotia*. Considering the fact that the edges of the single chloroplast extend to the ectoplasm, one would assume that the fibrils are localized at the interphase between the chloroplast and the ectoplasm. However, as a complication, another kind of movement has to be considered in *Mougeotia* besides that of the well-known turning around the longitudinal axis: the chloroplast can also move from its central position to one side of the cell (Senn, 1908). This movement which occurs in some extreme conditions (Haupt, 1970; Wirth, 1970) has not yet been thoroughly investigated.

REFERENCES

Abe, S. (1964). The effect of *p*-chloromercuribenzoate on rotational protoplasmatic streaming in plant cells. *Protoplasma*, **58**, 483.

Bock, G. und W. Haupt (1961). Die Chloroplastendrehung bei *Mougeotia*. III. Die Frage der Lokalisierung des Hellrot-Dunkelrot-Pigmentsystems in der Zelle. *Planta*, **57**, 518.

Dorscheid, T. (1966). Untersuchungen über die Schwachlichtbewegung des *Mesotaenium*-Chloroplasten. *Dissertation der Math. Nat. Fakultät der Univ. des Saarlandes*.

Dorscheid, T. und A. Wartenberg (1966). Chlorophyll als Photoreceptor bei der Schwachlichtbewegung des *Mesotaenium*-Chloroplasten. *Planta*, **70**, 187.

Fetzer, J. (1963). Über die Beteiligung energieliefernder Stoffwechselprozesse an den lichtinduzierten Chloroplastenbewegungen von *Mougeotia*. *Z. Bot.*, **51**, 468.

Fischer-Arnold, G. (1963). Untersuchungen über die Chloroplastenbewegung bei *Vaucheria sessilis*. *Protoplasma*, **56**, 495.

Gärtner, R. (1970). Die Bewegung des *Mesotaenium*-Chloroplasten im Starklichtbereich. *Z. Pflanzenphysiol.*, **62** (In the press).

Haupt, W. (1958). Hellrot-Dunkelrot-Antagonismus bei der Auslösung der Chloroplastenbewegung. *Naturwiss.*, **45**, 273.

Haupt, W. (1959a). Chloroplastenbewegung. In W. Ruhland (Ed.), *Encyclopedia of Plant Physiology*, Vol. 17-1. Springer Verlag, Berlin. pp. 278–317.

Haupt, W. (1959b). Die Chloroplastendrehung bei *Mougeotia*. I. Über den quantitativen und qualitativen Lichtbedarf der Schwachlichtbewegung. *Planta*, **53**, 484.

Haupt, W. (1960). Die Chloroplastendrehung bei *Mougeotia*. II. Die Induktion der Schwachlichtbewegung durch linear polarisiertes Licht. *Planta*, **55**, 465.

Haupt, W. (1963). Photoreceptorprobleme der Chloroplastenbewegung. *Ber. Dtsch. Bot. Ges.*, **76**, 313.

Haupt, W. (1965). Perception of environmental stimuli orienting growth and movement in lower plants. *Ann. Rev. Plant Physiol.*, **16**, 267.

Haupt, W. (1970). Über den Dichroismus von Phytochrom 660 und Phytochrom 730 bei *Mougeotia*. *Z. Pflanzenphysiol.*, **62** (In the press).

Haupt, W. und G. Bock (1962). Die Chloroplastenbewegung bei *Mougeotia*. IV. Die Orientierung der Phytochrom-Moleküle im Cytoplasma. *Planta*, **59**, 38.

Haupt, W. und R. Gärtner (1966). Die Chloroplastenorientierung von *Mesotaenium* in starkem Licht. *Naturwiss.*, **53**, 411.

Haupt, W., G. Köhler und D. Müller (1960). Chloroplastenbewegung in polarisiertem Licht. *Naturwiss.*, **47**, 113.

Haupt, W. und A. Scholz (1966). Nachweis des Linseneffektes bei der Chloroplastenorientierung von *Hormidium flaccidum*. *Naturwiss.*, **53**, 388.

Haupt, W. und I. Schönfeld (1962). Über das Wirkungsspektrum der 'negativen Phototaxis' der *Vaucheria*-Chloroplasten. *Ber. Dtsch. Bot. Ges.*, **75**, 14.

Haupt, W. und R. Thiele (1961). Chloroplastenbewegung bei *Mesotaenium*. *Planta*, **56**, 388.

Haupt, W. und M. Weisenseel (1967). Chloroplastenbewegung bei *Lemna trisulca* in polarisiertem Licht. *Naturwiss.*, **54**, 48.

Höfler, K. (1963). Zellstudien an *Biddulphia titiana* Grunow. *Protoplasma*, **56**, 1.

Jarosch, R. (1958). Protoplasmafibrillen der Characeen. *Protoplasma*, **50**, 93.

Kamiya, N. (1959). Protoplasmic streaming. In *Protoplasmatologia*, VIII, 3a. Springer Verlag, Wien.

Mayer, F. (1964). Lichtorientierte Chloroplasten-Verlagerung bei *Selaginella martensii*. *Z. Bot.*, **52**, 346.

Mayer, F. (1966). Lichtinduzierte Chloroplasten-Verlagerung bei *Selaginella martensii*. Untersuchungen zur Identifizierung des Photoreceptors durch Anwendung von Quencher-Substanzen. *Z. Pflanzenphysiol.*, **55**, 65.

Mugele, F. (1962). Der Einfluss der Temperatur auf die lichtinduzierte Chloroplastenbewegung. *Z. Bot.*, **50**, 368.

Scholz, A. (1970). Chloroplastenbewegung bei *Hormidium flaccidum*. *Dissertation der Nat. Fakultät der Universität Erlangen-Nürnberg*.

Scholz, A. und W. Haupt (1968). Lichtorientierte Chloroplastenbewegung bei der Grünalge *Hormidium flaccidum*. *Naturwiss.*, **55**, 186.

Schönbohm, E. (1963). Untersuchungen über die Starklichtbewegung des *Mougeotia*-Chloroplasten. *Z. Bot.*, **51**, 233.

Schönbohm, E. (1967a). Die Bedeutung des Gradienten von Phytochrom 730 und die tonische Wirkung von Blaulicht bei der negativen Phototaxis des *Mougeotia*-Chloroplasten. *Z. Pflanzenphysiol.*, **56**, 282.

Schönbohm, E. (1967b). Die Hemmung der positiven und negativen Phototaxis des *Mougeotia*-Chloroplasten durch Jodid-Ionen. *Z. Pflanzenphysiol.*, **56**, 366.

Schönbohm, E. (1969a). Untersuchungen über den Einfluss von Photosynthesehemmstoffen und Halogeniden auf die Starklicht- und Schwachlichtbewegung des Chloroplasten von *Mougeotia* spec. *Z. Pflanzenphysiol.* **60**, 255.

Schönbohm, E. (1969b). Die Hemmung der lichtinduzierten Bewegung des *Mougeotia*-Chloroplasten durch *p*-chlormercuribenzoat (Versuche zur Mechanik der Chloroplastenbewegung). *Z. Pflanzenphysiol.*, **61**, 250.

Schönbohm, E. (1970). In preparation.

Seitz, K. (1967a). Wirkungsspektren für die Starklichtbewegung der Chloroplasten, die Photodinese und die lichtabhängige Viskositätsänderung bei *Vallisneria spiralis* ssp. *torta*. *Z. Pflanzenphysiol.*, **56**, 246.

Seitz, K. (1967b). Eine Analyse der für die lichtabhängigen Bewegungen der Chloroplasten verantwortlichen Photoreceptorsysteme bei *Vallisneria spiralis* ssp. *torta*. *Z. Pflanzenphysiol.*, **57**, 96.

Senn, G. (1908). *Die Gestalts- und Lageveränderungen der Pflanzen-Chromatophoren*, Wilhelm Engelmann, Leipzig.

Senn, G. (1919). Weitere Untersuchungen über Gestalts- und Lageveränderungen der Chromatophoren. *Z. Bot.*, **11**, 81.

Stahl, E. (1880). Über den Einfluss von Richtung und Stärke der Beleuchtung auf einige Bewegungserschelnungen im Pflanzenreich. *Bot. Ztg.*, **38**, 257.

Taylor, A. O., and B. A. Bonner (1967). Isolation of phytochrome from the alga *Mesotaenium* and liverwort *Sphaerocarpos*. *Plant Physiol.*, **42**, 762.

Voerkel, S. H. (1934). Untersuchungen über die Phototaxis der Chloroplasten. *Planta*, **21**, 156.

Weisenseel, M. (1968a). Vergleichende Untersuchungen zum Einfluss der Temperatur auf lichtinduzierte Chloroplastenverlagerungen. 1. Die Wirkung verschiedener Lichtintensitäten auf die Chloroplastenanordnung und ihre Abhängigkeit von der Temperatur. *Z. Pflanzenphysiol.*, **59**, 56.

Weisenseel, M. (1968b). Vergleichende Untersuchungen zum Einfluss der Temperatur auf lichtinduzierte Chloroplastenverlagerungen. II. Die statistische Bewegungsgeschwindigkeit der Chloroplasten und ihre Abhängigkeit von der Temperatur. *Z. Pflanzenphysiol.*, **59**, 153.

Weisenseel, M. und W. Haupt (1967). Die Temperaturabhängigkeit der Chloroplastenbewegung bei *Mougeotia* sp. *Naturwiss.*, **54**, 145.

Wirth, H. (1970). Wachstum, Entwicklung und Bewegung von *Mougeotia* unter dem Einfluss des Lichtes. *Dissertation der Nat. Fakultät der Universität Erlangen-Nürnberg.*

Zurzycka, A., and J. Zurzycki (1957). Cinematographic studies on phototactic movements of chloroplasts. *Acta Soc. Bot. Polon.*, **26**, 177.

Zurzycki, J. (1960). Studies on the centrifugation of chloroplasts in *Lemna trisulca*. *Acta Soc. Bot. Polon.*, **29**, 385.

Zurzycki, J. (1962a). The mechanisms of the movements of plastids. In W. Ruhland (Ed.), *Encyclopedia of Plant Physiology*, Vol. 17-2. Springer Verlag, Berlin. pp. 940–978.

Zurzycki, J. (1962b). The action spectrum for the light dependent movements of the chloroplasts in *Lemna trisulca* L. *Acta Soc. Bot. Polon.*, **31**, 489.

Zurzycki, J. (1965). The energy of chloroplast movements in *Lemna trisulca* L. *Acta Soc. Bot. Polon.*, **34**, 637.

Zurzycki, J. (1967a). Properties and localization of the photoreceptor active in displacements of chloroplasts in *Funaria hygrometrica*. I. Action spectrum. *Acta Soc. Bot. Polon.*, **36**, 133.

Zurzycki, J. (1967b). Properties and localization of the photoreceptor active in displacements of Chloroplasts in *Funaria hygrometrica*. II. Studies with polarized light. *Acta Soc. Bot. Polon.*, **36**, 143.

CHAPTER 11

The photoresponses of fungi

M. J. CARLILE

Department of Biochemistry, Imperial College of Science and Technology, London, S.W.7

This chapter is dedicated to the memory of Robert M. Page (1919–1968), distinguished for his elegant experiments on the fungus Pilobolus, especially on its photoresponses.

Introduction

Some years ago the present author published a review (Carlile, 1965) in which an attempt was made to mention and assess all major work on fungus photobiology. In the present chapter, examples of the various photoresponses will be considered, but some arguments stated in full in the earlier publication will not be repeated, and fewer species will be mentioned. Instead, some important work carried out since the earlier review will be discussed, as will photoresponses in yeasts, acrasiales (cellular slime moulds) and myxomycetes (non-cellular slime moulds), groups not considered in the earlier publication. The present chapter and earlier review are thus complimentary, differing in presentation and to some extent in content.

Numerous references to research papers on the photoresponses of fungi are cited in the author's earlier review (Carlile, 1965), the textbook by Cochrane (1958) and the literature guide compiled by Marsh, Taylor and Bassler (1959).

Light and vegetative growth

There are numerous reports of light influencing vegetative growth of fungi, with both liquid and agar media, but few detailed investigations. Duration of lag phase, growth rate and total yield may all be influenced, and whether in a particular species light is found to stimulate or inhibit growth often depends upon the medium employed (Carlile, 1965; Cochrane, 1958; Page, 1965). Occasionally, the reported inhibition or stimulation may have been an indirect consequence of illumination, as a result of increased temperature or of photochemical effects on the medium. Weinhold and Hendrix (1963), Holloman (1966) and Barnett (1968) found, with some media and species, that illumination of media *prior* to inoculation could bring about subsequent growth inhibition. The studies of Weinhold and Hendrix and of Holloman both indicated that the formation of peroxides was involved, but in Holloman's experiments it is clear that the peroxides did not act directly on the fungus, but caused chemical changes in the medium. Barnett attributed the adverse effects of light to the destruction of pyridoxine. The stimulation or depression of growth of fungi reported as a consequence of illumination is often small, so it is important that the above-mentioned indirect effects are considered in the design of experiments.

The most detailed recent work on the depression of fungal growth by visible light has been carried out with the yeast *Saccharomyces cerevisiae*. Ehrenberg (1966a) found some effect with blue light (400–500 nm) at

$400\mu W/cm^2$ but not with longer wavelengths. Most of her work was carried out at rather higher intensities. At 44 000 $\mu W/cm^2$ marked inhibition of oxygen uptake was observed, but carbon dioxide output actually increased after a short period. It was shown (Ehrenberg, 1966b) that this stimulation of fermentation was an indirect consequence of the inhibition of aerobic respiration, through the operation of the Pasteur effect. At still higher light intensities (95 000 $\mu W/cm^2$) inhibition of fermentation occurred. The magnitude of these effects depends markedly on the phase of growth of the culture (Ehrenberg, 1968). The above results are in substantial agreement with those of an earlier study by Matile and Frey-Wyssling (1962) who found that light at 13 000 $\mu W/cm^2$ severely inhibited oxygen uptake but not carbon dioxide output, and that some growth inhibition occurred at intensities as low as 500 $\mu W/cm^2$. Yeast grown under anaerobic conditions and then transferred to air is particularly susceptible to the effect of light until adaptation has occurred, intensities of 1000 to 1500 $\mu W/cm^2$ severely inhibiting oxygen uptake, growth and ultimately carbon dioxide production (Guerin and Sulkowski, 1966; Sulkowski and coworkers, 1964). This work, however, and that of Elkind and Sutton (1957a, 1957b), who found that a mutant strain of *S. cerevisiae* which was deficient in respiratory enzymes was rapidly killed under aerobic conditions by visible light (46 000 $\mu W/cm^2$), may be more relevant to the study of protective mechanisms against photoinactivation than to the photoresponses discussed in this chapter.

Light has also been shown to depress growth and oxygen uptake in plasmodia of the myxomycete *Physarum polycephalum* (Daniel, 1966). The response is rapid and reversible, changes in the rate of oxygen uptake being detectable within 15 seconds of changed illumination. The depression in oxygen uptake has been demonstrated with isolated mitochondria as well as with intact plasmodia.

The most detailed study of growth stimulation in fungi by light remains that of Cantino and his coworkers on the watermould *Blastocladiella emersonii*. Cantino and Horenstein (1956) found that light stimulated both growth and carbon dioxide fixation, and suggested that the effect was produced by the stimulation of an 'SKI' (succinate-α-ketoglutarate isocitrate) cycle. It was subsequently shown (Cantino and Horenstein, 1959) that growth stimulation, which did not occur in the absence of carbon dioxide or bicarbonate, could be duplicated by an equimolar mixture of succinate and glyoxylate (believed to be products of the SKI cycle) and that the effective wavelengths were in the range 400 to 500 nm. Further studies (Cantino, 1959; Cantino and Goldstein, 1967; Cantino and Horenstein, 1957; Cantino and Turian, 1961; Goldstein and Cantino,

1962; Turian and Cantino, 1959) supported the scheme proposed, and revealed other metabolic consequences of illumination. Cantino (1965) showed that light influenced the intracellular location of a haemoprotein, thought to be a form of cytochrome *b*, and provided a useful list of metabolic, developmental and growth parameters influenced by light. The study remains unique in the extent to which the metabolic consequences of illumination have been explored, but the initial locus, or perhaps loci, of light action remains undiscovered.

Stimulation of growth by the longer ultraviolet wavelengths has also been reported. Leach (1961) found that shaken liquid cultures of *Helminthosporium oryzae* (Fungi Imperfecti) exposed to ultraviolet radiation (310 to 400 nm; 232 μW/cm^2) gave higher yields than control cultures in darkness. Brandt (1967) found that whereas the growth rate of colonies of *Verticillium albo-atrum* (Fungi Imperfecti) on agar media in darkness gradually declined (i.e. showed 'staling'), colonies exposed to ultraviolet radiation (320 to 400 nm) maintained their original growth rate. This he attributed to a suppression of the synthesis of a staling agent (DMF, diffusible morphogenetic factor) he had earlier studied.

Phototropism

Phototropism (light-oriented growth) is widespread in the fungi (Banbury, 1959; Carlile, 1965; Page, 1968; Thimann, 1967). Positive phototropism (growth towards light) is common in developing reproductive structures, and has a role in getting spores past obstacles and into the open to facilitate their dispersal (Buller, 1934; Ingold, 1953). Negative phototropism (growth away from light) has been observed in the germination of the spores of some plant pathogens, and may assist their penetration of host tissues. Various phototropic systems will now be considered.

The sporangiophore of PHYCOMYCES

The photoresponses of the sporangiophore of the zygomycete *Phycomyces blakesleeanus* have been intensively studied by biophysicists interested in the analysis of stimulus–response systems. The topic has recently been reviewed briefly by Castle (1966a) and at length by Bergman and coworkers (1969) in a review containing much hitherto unpublished work. The present account will therefore be brief and will cite only a few references. Some discussion is, however, essential in view of the relevance of these studies to our understanding of photoresponses in other fungi.

The sporangiophore is at first merely a stout erect hypha with a tapering tip, but later carries at its apex a spherical package of spores, the sporangium. Almost all study of phototropism has been confined to the final phase of growth, when the sporangiophore elongates at about 3 mm per hour, the growing zone extending for about 3 mm immediately below the sporangium. The sporangiophore responds to illumination from one side by bending towards that side at a rate of about 5 degrees per minute, after a latent period of about 6 minutes. This positive phototropic response takes place within the growing zone; perception of the phototropic stimulus is also confined to the growing zone. Phototropic curvature will continue as long as the growing zone is subjected to asymmetrical illumination.

The intervening steps between unilateral illumination and phototropic curvature are known in outline, but are not understood in detail. A horizontal beam of light entering the sporangiophore from one side is focused on the far side, the sporangiophore acting as a cylindrical converging lens. The consequence of this 'lens effect' is that unilateral illumination acts on the *far* side of the sporangiophore rather than on the near side, and brings about an increased growth rate on the far side. The growth rate on the near side of the sporangiophore is diminished—an indirect consequence of growth promotion on the far side, perhaps through competition for a limited supply of an essential metabolite. The increased growth rate on the far side relative to the near side results in the apex of the sporangiophore being swung towards the light. A clear demonstration that curvature away from the region of greatest light action occurs was provided by Banbury (1952), who illuminated one side of the growing zone of a sporangiophore with a grazing beam of light, and found that curvature towards the side in darkness took place.

The reality of the lens effect has been confirmed in a variety of ways. It is not understood exactly how the lens effect brings about a greater effect on the far side than on the near side, but it is clear that for the lens effect to operate the sporangiophore must act as a converging lens and that attentuation of light during its passage across the sporangiophore should not be so large as to counteract the effect of focusing. If the sporangiophore can be made to act as a diverging lens, or if its absorption can be increased, then the lens effect should not operate, unilateral light should promote growth on the near side instead of on the far side, and the sporangiophore should bend away from the light. Negative phototropism has in fact been achieved by both these approaches. Immersion in a liquid of high refractive index (e.g. liquid paraffin) causes the sporangiophore to act as a diverging lens, and unilateral illumination under these conditions results in negative phototropism. Unilateral ultraviolet radiation of less than 300 nm will also bring about negative phototropism, as these

11

wavelengths are so strongly absorbed by the sporangiophore that attenuation wholly counteracts the effects of focusing.

Phototropism is a response to asymmetrical illumination. Sporangiophores illuminated *symmetrically* will respond to an increase in light intensity (whether a step-up in intensity or a flash of light) by a *transient* increase in the growth rate, termed the light-growth response. A decrease in light intensity evokes a transient decrease in the growth rate, distinguished by some workers as the dark-growth response. An important feature of the light-growth response (including in this term the dark-growth response) is adaptation—within about 15 minutes the growth rate returns to a value that is independent of light intensity. The light-growth response would appear in itself to have no significance in the life of the organism, but to be an expression of the mechanisms which underlie phototropism, responding to the stimulus of a temporal instead of a spatial change in illumination. Models proposed to account for phototropism should therefore also account for the details of the light-growth response.

Recently, such a model has been devised by Castle (1966b). (An alternative earlier model developed by Delbrück and Reichardt, 1956, is mentioned later, see page 320.) It is suggested that a material M is supplied to the growing zone at a constant rate and is then converted to a product P, the concentration of which determines the growth rate. The conversion of M into P can be either by a light-dependent path or a light-independent path. Hence an increased light intensity would lead initially to an increased supply of P through increased activity of the light-dependent path, and hence increased growth. Subsequently, however, depletion of M would lead to a reduced rate of formation of P, and hence a reduced growth rate. Thus an increase in growth rate followed by adaptation, as in the light-growth reaction, would be observed. To account for phototropic curvature it is necessary to assume that the near and far sides of the sporangiophore are competing for the same supply of M. If this is so, then the side having the more active light-dependent path will form the greater amount of P and will show faster growth. Castle's paper should be consulted for the detailed formulation of the model and demonstration of its agreement with the consequences of various illumination programmes.

Major gaps in our knowledge of the photoresponses of the *Phycomyces* sporangiophore are that the nature and location of the photoreceptor molecules are uncertain and that almost nothing is known of the biochemical steps brought about by light action; attempts to study the latter topic have encountered great difficulties and have proved rather unrewarding (Shropshire and Bergman, 1968). The very similar action spectra

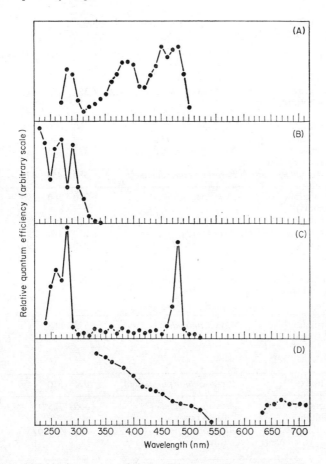

Figure 11.1. Action spectra. (A) Growth response of sporangiophores of *Phycomyces blakesleeanus* (Delbrück and Shropshire, 1960). (B) Induction of sporulation (conidia) of *Pleospora herbarum* (Leach and Trione, 1966). (C) Inhibition of terminal phase of sporulation (conidia) of *Pleospora herbarum* (Leach, 1968). (D) Induction of sporulation of *Physarum nudum* (Rakoczy, 1965). Scales for relative quantum efficiencies are arbitrary, but indication of effective intensities is given in Table 11.3

obtained for phototropism by Curry and Gruen (1959) and for phototropism and the light-growth response (Figure 11.1A and Table 11.1 A, B) by Delbrück and Shropshire (1960) have been interpreted by some as favouring β-carotene and by others a flavoprotein as the photoreceptor. The recent *in vivo* absorption measurements by Zankel, Burke and Delbrück (1967) indicate that the bulk of the β-carotene conspicuously present in

Table 11.1. Detailed action spectra

Reaction	Organism	Maxima (nm)[a]				Upper limit (nm)
(A) Light-growth response and phototropism[b]	*Phycomyces blakesleeanus*	280	385	*455*	485	—
(B) Phototropism[c]	*Phycomyces blakesleeanus*	280	370	*445*	470	525
(C) Phototropism[d]	*Pilobolus kleinii*		360	*450*		520
(D) Sporulation (conidia)[e]	*Trichoderma viride*		*380*	440		525
(E) Suppression of circadian rhythm[f]	*Neurospora crassa*		375	*465*	485	515
(F) Carotenoid synthesis[g]	*Fusarium aquaeductuum*		380	*450*	475	520
(G) Sporulation (pycnidia)[h]	*Ascochyta pisi*	230	260	*290*		360
(H) Sporulation (conidia)[i]	*Alternaria dauci*	*230*	260	280		360
(I) Sporulation (conidia)[i]	*Pleospora herbarum*	230	*270*	290		360
(J) Sporulation (perithecia)[i]	*Pleospora herbarum*	230		*290*		380
(K) Inhibition of terminal phase of sporulation[j]	*Pleospora herbarum*	*280*	480			weak beyond 510
(L) Sporulation[k]	*Physarum nudum*	*333*	668			713 (580–622 also inactive)
(M) Phototaxis[l]	*Dictyostelium purpureum*	*425*	550			600
(N) Reversal 450 nm inhibition of sporulation[m]	*Alternaria solani* (wild type)	*600*	725?			—
	Alternaria solani (mutant)	575	*650–675*			—

a The most effective wavelength is italicized; some shoulders mentioned by authors are omitted; horizontal lines divide reactions into groups with similar action spectra
b Delbrück and Shropshire (1960)
c Curry and Gruen (1959)
d Page and Curry (1966)
e Gressel and Hartmann (1968)
f Sargent and Briggs (1967)
g Rau (1967a)
h Leach and Trione (1965)
i Leach and Trione (1966)
j Leach (1968)
k Rakoczy (1965)
l Francis (1964)
m Lukens (1965)

the growing zone cannot be the photoreceptor, a conclusion supported by the study of mutants having abnormal pigmentation (Meissner and Delbrück, 1968). Consideration of the effectiveness of polarized light in promoting photoresponses (Haupt and Buchwald, 1967; Jaffe, 1960) suggests that the photoreceptor molecules are dichroic and have a peripheral location adjacent to the hyphal wall and a periclinal orientation (i.e. axis of maximal absorption parallel to the cell surface), but the location as well as the nature of the photoreceptor remains uncertain. However, these problems are now being attacked by a wide range of methods, including the production and study of mutants with abnormal pigmentation or photoresponses (Bergman and coworkers, 1969), so progress may be anticipated.

The sporangiophore of PILOBOLUS

As with *Phycomyces*, the sporangiophore of the zygomycete *Pilobolus* is capable of positive phototropism both before and after sporangium formation. An important recent study of the phototropism of young sporangiophores (i.e. before sporangium formation) of *Pilobolus kleinii* is that of Page and Curry (1966). It was found that a beam of light grazing one side of the photosensitive zone caused curvature of the sporangiophore towards the side in darkness, and that unilateral illumination with wavelengths of less than 300 nm caused negative instead of positive phototropism. Hence, as in *Phycomyces*, positive phototropism can result from growth stimulation on the far side of the sporangiophore through the operation of a lens effect. An apparent anomaly is that unilateral illumination of sporangiophores immersed in liquid paraffin did not lead to reversal of phototropism. It was found, however, that the refractive index of the sporangiophore in the photosensitive zone (the terminal 50 μ) of the sporangiophore was so high that reversal would not be expected. This indicates that while phototropic reversal in liquid paraffin can be regarded as evidence for the operation of a lens effect, non-reversal does not necessarily establish its absence. This invalidates Paul's conclusion (cited by Banbury, 1959) that the lens effect is absent in young sporangiophores of *Pilobolus kleinii* and *P. spherosporus*, although it is present in *P. crystallinus*; the absence of phototropic reversal in liquid paraffin in the former two species is presumably merely a consequence of higher refractive indices in the photosensitive zone. Page and Curry (1966) obtained an action spectrum (Table 11.1 C) for phototropism in *P. kleinii* which was substantially similar to that of *Phycomyces*. The action spectrum and *in vivo* absorption measurements indicate that the carotenoids conspicuously present in the sporangiophore are not the photoreceptor pigment; the most highly pigmented region, moreover, lies below the photosensitive zone.

Although phototropism by bending is well established for *Pilobolus kleinii,* young sporangiophores of this species can under some conditions respond by a different mechanism (Page, 1962; Page and Curry, 1966). Sporangiophores transferred from darkness to unilateral light have been observed to cease elongation and subsequently to resume growth from a new growing zone located on the side of the sporangiophore nearest the light source and a little below the former apex (Page, 1962). Moreover, exposure to equal phototropic stimuli from two sides sometimes leads to branching, each branch growing towards one light source (Page and Curry, 1966). These observations suggest relocation of the growing zone and 'phototropism by bulging' (see page 319) instead of 'phototropism by bending' through differential growth. Further study of the circumstances leading to the changed behaviour are clearly needed.

The phototropism of *P. kleinii* after sporangium formation and the development of the subsporangial swelling characteristic of the genus was studied in detail by Buller (1934) and more recently by Page (1962). A special feature of the phototropism of this stage is that if the sporangiophore is exposed to two light sources separated by more than 10 degrees, orientation is towards one of the light sources and not to an intermediate position; this is a consequence of the optics of the subsporangial swelling (Buller, 1934).

The sporangiophore of THAMNIDIUM

The positive phototropic response of the sporangiophore of the zygomycete *Thamnidium elegans* has been investigated by Lythgoe (1961). Bending towards the light would seem to be due to stimulation of growth on the far side of the sporangiophore, as sporangiophores immersed in liquid paraffin showed negative phototropism. Transfer from darkness to symmetrical illumination apparently produces a transient fall in growth rate. A more detailed study of the photoresponses of this interesting organism is clearly required.

The conidiophore of ASPERGILLUS GIGANTEUS

The phototropic response of the large conidiophores of *Aspergillus giganteus* (Fungi Imperfecti) has recently been studied by Trinci and Banbury (1968). During curvature towards the light, increased growth rate on the far side of the conidiophore and reduced growth rate on the near side occurs. Since negative phototropism of conidiophores immersed in liquid paraffin or exposed to unilateral ultraviolet radiation (280 nm) was observed, curvature appears to be due to stimulation of growth on the far side of the conidiophore through the operation of the lens effect. Conidiophore growth in this species is dependent on light, the growth rate falling gradually towards zero in darkness (over the course of about one day), but recovering to its former value fairly rapidly (a few hours) if returned to light.

Germinating spores of rusts and BOTRYTIS CINEREA

Negative phototropism is shown by the germinating spores of some plant pathogens. In *Botrytis cinerea* (Fungi Imperfecti) germtube emergence takes place on the far side of spores exposed to unilateral light, and from the illuminated region of partially illuminated spores (Jaffe and Etzold, 1962). Hence unilateral light stimulates growth (in the form of germtube emergence) on the far side of the spore. Experiments with polarized light established that the photoreceptor molecules had an anticlinal orientation (i.e. axes of maximal absorption perpendicular to the cell surface), and were located at or near the inner surface of the spore wall. Greater light absorption by photoreceptors at the far side of the spore is a consequence of the focusing of light entering the spore and the anticlinal orientation of the photoreceptor molecules. In Jaffe and Etzold's study germination took place in an aqueous medium; *B. cinerea* spores and perhaps those of rusts (Cochrane, 1960) can, however, germinate in the absence of liquid water and may well do so in the humid layer of air adjacent to the leaf surface. Under these circumstances refraction of incident light will be greater than that computed by Jaffe and Etzold and germtube emergence will probably occur still more markedly towards the far pole of the spore.

Negative phototropism of the emergent germtube takes place in *B. cinerea* and in some, but not all, rust species (Gettkandt, 1954). In the rust *Puccinia triticina*, which was studied in detail, immersion in liquid paraffin caused phototropic reversal. Gettkandt concluded correctly that this implied that the normal negative phototropism was the result of light acting on the far side of the germtube, but incorrectly (in the opinion of the present author) that light was acting by retarding growth. In germtubes —and in fact in almost all hyphae except sporangiophores and a few conidiophores—growth is confined to the hemispherical apex of the hypha. Under these circumstances tropism results from bulging of the apex and not from bending consequent upon differential subapical growth (Carlile, 1966); negative tropism hence results from growth stimulation of the far side of the apex. Indeed, germtube phototropism in *P. triticina* may well be controlled in a way similar to that envisaged for germtube emergence in *B. cinerea* by Jaffe and Etzold.

Other phototropic systems

A variety of other phototropic responses have been observed in the fungi. A few which would seem to merit further investigation are mentioned here.

Positive phototropism of hyphae of *Penicillium expansum* (Fungi Imperfecti) was observed by Carlile, Dickens and Schipper (1962). Etzold (1961) reported positive phototropism of hyphae of '*Penicillium glaucum*'— a species not currently recognized by authorities on the genus (Raper and Thom, 1949), but which may well have been *P. expansum*—and also observed the unusual response of growth orientation perpendicular to the vibration plane of polarized light. Etzold also found the stolons of the zygomycete *Rhizopus nigricans* to be positively phototropic. Positive phototropism of germtubes and hyphae has been found in the zygomycete *Empusa* (Sawyer, 1929). Negative phototropism of vegetative hyphae has been found in *Aspergillus restrictus* (Welty and Christensen, 1965).

The phototropism of multihyphal systems has received little study. One of the more extensive studies concerns the coremia of several species of *Penicillium* (Carlile and coworkers, 1961, 1962; Carlile, Dickens and Schipper, 1962); the positive phototropism of the coremia of *Penicillium isariaeforme* was found to result from the positively phototropic response of the constituent hyphae. The phototropism of the fruit bodies of basidiomycetes probably has a similar basis (Carlile, 1965).

General comments on phototropism

Phototropic responses may be brought about either by differential subapical growth and subsequent bending (well known in *Phycomyces* and higher plants) or by bulging of an apical growing zone. This latter process is clearly important in the fungi. Whether the response is positive or negative depends on the optics of the growing zone, maximal light action on the far side of the organ due to a 'lens effect' being widespread. The ways in which positive and negative phototropism can be brought about by light stimulating growth are indicated in Table 11.2. Theoretically it should be possible for phototropism to be brought about by light depressing growth; whether this in fact occurs is unknown.

The model for phototropism in *Phycomyces* developed by Castle (1966b) may be applicable to other fungi; on the other hand, the phototropic behaviour of young sporangiophores of *Pilobolus kleinii* (Page, 1962; Page and Curry, 1966) and of *Thamnidium elegans* (Lythgoe, 1961) are difficult to reconcile with the concept of a single reaction promoted by light. An earlier model (Delbrück and Reichardt, 1956) envisaged a fast

Table 11.2. The relationship between light stimulation and growth and phototropic response

Sign of phototropic response	Position of growing zone	
	Intercalary	Apical
Positive	Maximal growth stimulus on the far side through operation of the lens effect	Maximal growth stimulus on the near side
	Examples: sporangiophores of *Phycomyces*[a] and *Pilobolus kleinii*,[b] conidiophores of *Aspergillus giganteus*[c]	Examples: sporangiophores of *Pilobolus kleinii*[a] under some circumstances, probably hyphae of some *Penicillium* spp.[e],[f]
Negative	Maximal growth stimulus on the near side	Maximal growth stimulus on the far side through operation of the lens effect
	Examples: artificially produced by elimination of the lens effect in sporangiophores of *Phycomyces*[a] and *Pilobolus kleinii*,[b] conidiophores of *Aspergillus giganteus*[c]	Examples: Germ tubes of *Botrytis cinerea*,[g] probably *Puccinia triticina* also[h]

a Shropshire (1962)
b Page and Curry (1966)
c Trinci and Banbury (1969)
d Page (1962)
e Etzold (1961)
f Carlile, Dickens and Schipper (1962)
g Jaffe and Etzold (1962)
h Gettkandt (1954)

light reaction stimulating growth and a slower adapting reaction reducing the growth rate; such a model may be applicable to these other systems (Carlile, 1965).

Most phototropic responses in the fungi are found on adequate investigation to be promoted only by wavelengths of less than 530 nm. A notable exception which deserves further investigation is the positive phototropism of conidiophores of the zygomycete *Entomophthora coronata* (or *Conidiobolus villosus*), in which sensitivity extends to about 650 nm (Page, 1965, 1968; Page and Brungard, 1961).

Phototaxis

Positive phototaxis occurs in the aquatic phycomycetes, having been reported for zoospores of *Phytophthora cambivora* (Petri, 1925) and for a few chytrids (Kusano, 1930; Strasburger, 1878; Wager, 1913). Electron microscope studies have revealed structures reminiscent of algal photoreceptor organelles in *Nowakowskiella profusa* (Chambers, Marcus and

Willoughby, 1967) and *Monoblepharella* sp. (Fuller and Reichle, 1968); these workers did not, however, demonstrate any phototactic responses in these organisms. Negative phototaxis of myxomycete plasmodia was long ago reported (Baranetzki, 1876). Further studies of the phototaxis of both myxomycetes (Rakoczy, 1963) and aquatic phycomycetes are clearly needed.

The positive phototaxis of the pseudoplasmodium of some members of the acrasiales has received more attention, the most detailed studies being those of Francis (1964) and Bonner and Whitfield (1965). The latter workers observed phototaxis in species having large pseudoplasmodia (*Dictyostelium purpureum*, *Polysphondylium pallidum* and *Polysphondylium violaceum*), but not in those with small pseudoplasmodia (*Acytostelium leptosomum* and *Protostelium mycophaga*). Within the species *Dictyostelium purpureum* it was found that large pseudoplasmodia were more sensitive to phototactic stimuli than small ones, and that the smallest pseudoplasmodia studied did not respond. Samuel (1961) had earlier failed to find phototaxis in solitary amoebae of *Dictyostelium mucoroides*, the large pseudoplasmodia of which are phototactic. Francis (1964) found by means of illumination with a grazing beam of light that phototactic sensitivity in *Dictyostelium discoideum* was limited to the apex of the pseudoplasmodium, and that illumination of one side of the apex led to curvature to the side in darkness. Subsequently, Bonner and Whitfield (1965) showed phototactic reversal on immersion in liquid paraffin with *D. purpureum*, so it is clear that a 'lens effect' brings about maximum action on the far side of the pseudoplasmodium. Francis (1964) suggested that light brought about curvature by inducing a more rapid hardening of the slime sheath on the far side of the apex. Rough action spectra (Table 11.1 M) for the phototactic response in *D. purpureum* and *D. discoideum* were obtained by Francis, the most effective wavelengths being at the blue end of the visible spectrum.

Morphogenetic effects of light

Ultraviolet induction of sporulation and related effects

It has been known since the work of Stevens (1928) that it is the ultraviolet component of daylight that induces sporulation in some fungi. For many years this phenomenon attracted little attention, although the initiation of sporulation by the longer ultraviolet wavelengths was confirmed by a number of workers, including Ramsey and Bailey (1930), McCallan and Chan (1944), Charlton (1953) and Carlile (1956). Recently, however, more detailed studies have been carried out by Leach (for a review of the topic by this worker, see Leach, 1967b) who has studied the effect of temperature, visible light and ultraviolet radiation on sporulation in several species; these factors had earlier been considered separately, but their relationships to each other had not been thoroughly investigated. Leach's investigations have led to a major clarification of the topic and will now be considered in more detail.

Reproduction is initiated in many fungi by exposure to ultraviolet radiation. Leach and Trione (1965, 1966) have obtained action spectra for both perithecium production (a sexual process) and conidium initiation (Figure 11.1 B) in the ascomycete *Pleospora herbarum* (in the absence of sexual reproduction, often referred to as *Stemphylium botryosum*), for conidium initiation in *Alternaria dauci* (Fungi Imperfecti) and for pycnidium formation in *Ascochyta pisi* (Fungi Imperfecti). All these action spectra are essentially similar (Table 11.1 G to J). The longest effective wavelengths are in the region 360 to 380 nm, and a peak in effectiveness occurs at about 290 nm in *Pleospora herbarum*, for example, irradiation at 290 nm is about 5000 times as effective as at 360 nm. All shorter wavelengths down to the shortest tested (230 nm) are effective, but there were differences in relative effectiveness between species. Many substances absorb at these wavelengths, and details of the action spectra may here reflect secondary effects rather than the absorption spectrum of the photoreceptor. This is likely, as it has been shown for *Ascochyta pisi* (Leach, 1962), *Pleospora herbarum* (Leach, 1963) and *Alternaria chrysanthemi* (Leach, 1964) that whereas with the longer ultraviolet wavelengths (313, 334 and 366 nm) high dosages may be used to stimulate sporulation, at shorter wavelengths (280 nm or less) only relatively low dosages can be employed without causing severe inhibition. For this reason, it is best to employ lamps emitting only the longer ultraviolet wavelengths for routine sporulation induction (Leach, 1967b). Leach (1967a) showed that some Fungi Imperfecti that need ultraviolet wavelengths for sporulation (e.g. *Helminthosporium catenarium*, *Fusarium nivale* and *Cercosporella herpotrichoides*) will sporulate satisfactorily under continuous ultraviolet irradiation. In other species (e.g. *Alternaria dauci*, *Alternaria tomato* and *Pleospora herbarum*) ultraviolet radiation will bring about the inductive phase of sporulation (conidiophore development), but the terminal phase (conidium production) requires a period of darkness. In *Alternaria dauci* and *Alternaria tomato* the best results were obtained when the twelve hours of exposure to ultraviolet radiation was at a high temperature (e.g. 30 to 35°c) and the subsequent dark period was at a low temperature (18 or 20°c). Leach classified the two groups of fungi as 'constant temperature sporulators' and 'diurnal sporulators' respectively. However, even some of the 'constant temperature sporulators' seem to sporulate most effectively when exposed to an alternation of a 'warm day' and a 'cool night'.

Recently, Leach (1968) has obtained an action spectrum (Figure 11.1 C) for the inhibition by light of the terminal phase of asexual sporulation in *Pleospora herbarum*. All wavelengths tested (240 to 650 nm) were active, but the longest (520 to 650 nm) were only weakly so. Peaks of effectiveness occurred at 280 and 480 nm. Earlier workers had obtained less detailed action spectra for the inhibition of the terminal phase of sporulation in *Alternaria tomato* (Aragaki, 1962) and *Alternaria solani* (Lukens, 1963). For *Alt. solani* it was found that photoinhibition of the terminal phase of sporulation could be prevented by treatment with flavin mononucleotide prior to irradiation (Lukens, 1963) or reversed by subsequent exposure to red light (Lukens, 1965).

It has been proposed (Leach, 1965; Trione, Leach and Mutch, 1966) that one step in sporulation in many fungi is the production of a sporogenic substance termed, from the absorption maximum of crude preparations, P 310. In some fungi that will sporulate in darkness, such as one strain of *Ascochyta pisi* on a special medium, the substance has been detected in dark-grown cultures. In fungi requiring ultraviolet irradiation for sporulation (such as *Asc. pisi* under most circumstances) applied P310 can induce sporulation in dark-grown cultures. The substance has been obtained, for example, from irradiated cultures of *Asc. pisi* and when applied to dark-grown cultures of *Asc. pisi* and *Pleospora herbarum* has induced pycnidia in the former and both conidia and perithecia in the latter. In *Stemphylium solani* which normally requires ultraviolet radiation for sporulation, it has been found (Sproston and Setlow, 1968) that sporulation can be induced by treatment with ergosterol. Lesser, but still considerable, sporulation was obtained by using the carrier solutions, such as 5 per cent dimethyl-sulphoxide or 2 per cent ethanol, without added ergosterol. It was suggested that free ergosterol was required for sporulation and that the efficacy of ultraviolet irradiation and the partial efficacy of solvents alone was due to the release of bound ergosterol. The approaches of both groups of workers seem promising, and perhaps further studies will lead to the integration of their conclusions.

Brandt and Reese (1964) showed that ageing cultures of *Verticillium albo-atrum* produced a diffusible morphogenetic factor (DMF) which promoted the production of microsclerotia and melanin and inhibited sporulation and hyphal growth. Irradiation with the longer ultraviolet wavelengths, however, inhibited microsclerotia and melanin formation (Brandt, 1964), and promoted sporulation (Roth and Brandt, 1964) and the maintenance of high hyphal growth rates (Brandt, 1967). The

suppression of DMF formation by irradiation was demonstrated by Brandt
and Reese (1964), providing a possible explanation for at least some of the
effects of ultraviolet radiation. Treatment of cultures with catechol (a
precursor of the non-nitrogenous melanins) permitted microsclerotium
and melanin formation to occur even under continuous ultraviolet irradiation (Brandt, 1965).

Effect of visible light on sporulation

In many fungi, as described in the previous section, only ultraviolet
radiation is capable of inducing sporulation, visible radiation being ineffective. In the present section, well-substantiated examples of the promotion of sporulation by visible light will be discussed.

In many strains of *Trichoderma viride* (Fungi Imperfecti) conidia are not
produced in darkness. Sporulation may, however, be brought about by
brief exposure to light. Gressel and Hartmann (1968) determined the
action spectrum (Table 11.1 D) in the range 350 to 1100 nm by means of
30-second exposures to a range of light intensities. Peaks of effectiveness
were demonstrated at approximately 380 and 440 nm and no response was
shown at wavelengths longer than about 525 nm. Galun and Gressel
(1966) showed that photoinduced sporulation in *T. viride* could be prevented by 5 hours of exposure to 5-fluorouracil at concentrations having
little effect on growth. Gressel and Galun (1967) extended this study.
They found that conidiophore production began about 3 hours after
irradiation, and extensive branching of conidiophores occurred at 6 to
8 hours. Branching was complete at 12 hours, at which time the first
conidia were observed. Sporulation was completed in 24 hours. Exposure
to high concentrations of fluorouracil for 1 hour at the time of irradiation
or for 7 to 8 hours later, prevented conidium production, but exposure
when conidiophore branching was complete (11 to 12 or 16 to 17 hours
after irradiation) did not do so. The incorporation of fluorouracil into
various categories of RNA was demonstrated, as well as prevention of the
effects of fluorouracil by simultaneous application of uracil. It was concluded that the photoinduction of sporulation was RNA mediated.

The photoinduction of sporulation in *T. viride* is an example of a light
requirement for the initiation of sporulation. Alternatively, light may be
needed not for the initiation of the process but for its completion, or there
may be light requirements for several phases in the process, including
perhaps both the earliest and latest stages. Page (1956) showed that light
was required to induce trophocyst formation, the first step in sporulation
in the zygomycete *Pilobolus kleinii*. He found that exposures as short as

0·1 seconds gave some and 100 seconds gave copious trophocyst product-
tion, but that only wavelengths of less than 510 nm were effective. For the
next phase of development (sporangiophore emergence and elongation) to
occur, a second period of illumination preceeded by a period of darkness
was needed. Light was also required for inducing the final phase of
sporulation, sporangium development. In a comparative study of the
light requirements of different members of the genus, Jacob (1954) showed
that *Pilobolus gracilis* resembled *P. kleinii* in its requirements, *P. crystallinus*
and *P. umbonatus* needed light only for initiating the final phase of the
process (sporangium development), and that other species, such as
P. sphaerosporus had no light requirement for sporulation.

The basidiomycete *Sphaerobolus stellatus* requires light (Alasoadura,
1963) for both the initiation and maturation of the sporophore, a process
that takes in all from 2 to 3 weeks at 20°c and terminates in the discharge
of the spore mass after a final 24 hour period in which light is not required.
During approximately the first 10 days of sporophore development, only
wavelengths of less than 500 nm are effective, and fairly high intensities
(1000 lux) are needed. At about this stage the red end of the spectrum,
hitherto ineffective, becomes active and the intensities required diminish
greatly, about 1 lux being adequate. Ingold and Nawaz (1967) showed that
during this phase red light is more effective than blue and that more rapid
maturation of sporophores occurs at 640 or 720 nm than at 400 or 460 nm.
If cultures in which sporophore discharge is occurring are transferred to
darkness, discharge ceases after 24 hours; this is due to the development of
sporophores still in their final light-sensitive phase being arrested. Ingold
(1968) showed that exposure of such cultures to yellow light (585 nm;
100 lux)for 30 minutes led to discharge 24 hours later, but that a comparable
exposure to blue light (about 450 nm) had no stimulatory effect. He also
found that the stimulatory effect of 30 minutes of exposure to yellow light
was annulled if followed by 30 minutes of exposure to blue light. Thus, as
the development of the sporophore of *Sphaerobolus* proceeds, it would
seem that a second photoreceptor sensitive to the longer visible wavelengths appears, and that effects mediated by this system can be antagonized
by shorter wavelengths.

As discussed above, light is essential in many species for the induction or
completion of sporulation. In such species, and others in which light is
not essential for sporulation, the duration and intensity of illumination
may be an important factor in determining the form of the reproductive
structures. For example, the coremia of *Penicillium claviforme* can be
produced in darkness, but their number, distribution, form and size are
influenced by illumination (Carlile and coworkers, 1961; Piskorz, 1967b).

In this species, and in *Penicillium isariaeforme* in which appropriate conditions of illumination can produce coremia of a variety of bizarre forms (Carlile and coworkers, 1962; Piskorz, 1967a), the morphogenetic effects of light would seem to be the consequence of the interaction of inductive effects with phototropic responses of individual hyphae. Transfer of the coremia of *P. isariaeforme* (Carlile and coworkers, 1962) or the giant conidiophores of *Aspergillus giganteus* (Trinci and Banbury, 1967) to continuous darkness, terminates their elongation after about 24 hours. In contrast, transfer of the zygomycetes *Pilobolus crystallinus* (Page, 1962) and *Phycomyces blakesleeanus* (Banbury, 1959) to darkness, results in sporangiophores many times their usual length. In *Thamnidium elegans* the effects of light on sporangiophore development are complex (Lythgoe, 1961) and require further investigation, but in general it would seem that light promotes sporangium formation, whereas the production of numerous small sporangioles on lateral branches can proceed in darkness. In *Choanephora cucurbitarum* sporangium formation is not influenced by light, condium formation requires a period of light followed by a period of darkness (Barnett and Lilly, 1950) and zygospore production is inhibited at low temperatures by continuous illumination (Barnett and Lilly, 1956). In *Botrytis cinerea* (Paul, 1929) and other *Botrytis* spp. light promotes conidium formation and inhibits sclerotium development. Whether these are interrelated or distinct effects is not clear, but the matter could well be resolved by more detailed study, perhaps utilizing predominantly conidial and predominantly sclerotial strains.

Literature on the effect of light on yeast sporulation includes reports of stimulation, indifference and inhibition, and is often contradictory and difficult to evaluate. The situation has, however, been clarified by the demonstration (Kelley and Gay, 1969) that different wavelengths may have opposite effects, and that strains of a single species may respond differently. Enhancement of sporulation by red light (approximately $1000 \mu W/cm^2$) was found in one strain of *Saccharomyces carlsbergensis*, and depression of sporulation by blue light (intensities from 100 to 14 000 $\mu W/cm^2$) was found in both *S. carlsbergensis* and *S. cerevisiae*. As aerobic respiration is essential for yeast sporulation, the latter effect may well be mediated through the blue light depression of aerobic respiration discussed on page 310; the red effect is an additional instance in fungi of red sensitivity and of opposing effects of long and short wavelengths.

Light is essential for sporulation in many myxomycetes, as was demonstrated by Gray (1938), who found a light requirement in many, but not all, of the species he investigated. The most detailed study of the effects of different wavelengths is that of Rakoczy (1965) who obtained an action spectrum (Figure 11.1 D) for *Physarum nudum* using exposure periods of 12 hours. At 333 nm, the shortest wavelength tested, $4 \mu W/cm^2$ caused some sporulation, and intensities of above $80 \mu W/cm^2$ inhibited sporulation. Similar effects were observed with longer ultraviolet wavelengths. At 417 nm at least $200 \mu W/cm^2$ was necessary to induce sporulation, but no inhibitory effects were found at higher intensities. Similar but weaker stimulatory effects were found at wavelengths up to 541 nm. No stimulatory effects were observed at 560, 580, 602 and 622 nm, and in a later study Rakoczy (1967) showed some of these wavelengths to be inhibitory, partially or completely annulling the stimulatory effects of simultaneous irradiation with blue or red light. Between 632 and 713 nm stimulatory effects were observed at intensities above $240 \mu W/cm^2$, but inhibitory effects were observed at high intensities. No stimulation was observed in the range 724–823 nm. In practice the blue end of the visible spectrum is most effective in promoting sporulation, as the inhibitory effects observed with ultraviolet wavelengths are absent.

Other less detailed studies are in substantial agreement with the above study. Daniel and Rusch (1961) found that the blue end of the visible spectrum was effective in promoting the sporulation of *Physarum polycephalum*. Leith (1956) in a study of sporulation in *Didymium nigripes* found near ultraviolet and blue light to be effective at low intensities, red light at high intensities and green light without effect or inhibitory. Nair and Zabka (1965) reached essentially similar conclusions in a study on *Didymium iridis*; in addition, they concluded that the stimulatory effects of 'near red' (650 to 700 nm) could be annulled by subsequent exposure to 'far red' (690 to 750 nm).

Effect of light on spore discharge

Sporulation terminates in spore liberation, which in many fungi takes the form of active discharge. In some species discharge takes place predominantly during the daytime, in others at night (Ingold, 1965). The alternation of light and darkness seems to be the main factor in determining the time of discharge, as is indicated by a study (Walkey and Harvey, 1966) on twenty-four pyrenomycetes—members of the ascomycetes producing perithecia. When subjected under otherwise constant conditions to alternate periods of 12 hours of light and 12 hours of darkness, half of the species discharged their spores during the light periods and half wholly or predominantly discharged during darkness.

Light may bring about periodic spore liberation in a variety of ways. In *Pilobolus kleinii* (Jacob, 1961) and *Sphaerobolus* (Alasoadura, 1963), the timing of discharge is not directly determined by light, but indirectly through the light requirements of an earlier step in sporulation. In three species studied by Ingold and Marshall (1963), light influenced the timing of spore discharge more directly. In *Sordaria fimicola* they found that transfer from darkness to light led to a massive increase in discharge in a few hours, accounting for a daytime maximum. In *Sordaria verruculosa* a much longer lag between illumination and discharge leads to nocturnal discharge. In *Hypoxylon fuscum* illumination was found to rapidly depress spore discharge, leading also to maximum discharge in darkness. Light may also act, as in *Daldinia concentrica* (Ingold and Cox, 1955), by bringing an endogenous 24-hour (i.e. circadian) rhythm of discharge into phase with night and day.

The interaction of light with rhythmic processes will be discussed further in the next section.

Effect of light on spore germination

There are many reports of light either stimulating or inhibiting spore germination, but few detailed studies. Sussman (1965) lists many reports of light-stimulated spore germination; more detailed investigation of these might well prove rewarding.

Perhaps the most detailed study is that of Hebert and Kelman (1958) on the germination of the resting sporangia of the chytrid *Physoderma maydis*. They found light to be essential, and only the blue end of the spectrum to be effective. Recently, there have been reports of light stimulating the germination of oospores of members of the economically important genus *Phytophthora* (peronosporales), responsible for many serious plant diseases. Leal and Gomez-Miranda (1965) found that in *Phytophthora heveae*, light following a period of darkness was required for germination of oospores; Berg and Gallegly (1966) found that blue light (400–480 nm) and to a lesser extent other wavelengths, particularly far red (700–1000 nm), stimulated germination in *P. heveae* and several other species.

Inhibition of uredospore germination in rusts by light has often been reported. Givan and Bromfield (1964) studied retardation of spore germination by light in *Puccinia recondita*, and discussed possible explanations for the conflicting results obtained by earlier workers.

Relationships between light and rhythmic processes

Rhythmic processes are widespread in the fungi, and are commonly influenced by light. The most extensively studied rhythms are the periodicities in spore discharge (Ingold, 1965) and the production of concentric zones of differing form in colonies on agar media (Jerebzoff, 1965). Such zonation may result from alternate zones of sparse and dense mycelium, or of sterile and fertile hyphae.

The determination of zonation in a fungal colony may be wholly exogenous. Thus, if light is essential for sporulation, and newly formed hyphae are capable of responding for a few hours only, then diurnal alternation of light and darkness will lead each day to the production of a zone of spores and a zone of sterile mycelium. Periodicity in spore discharge may also be exogenously determined in a similar way (see page 328). Exogenously determined rhythms are a direct consequence of the periodic fluctuation of one or more environmental factors, such as light, and cease in a constant environment.

Rhythmic spore discharge or zonation in colony form can also be endogenously determined, and is able to persist for long periods under constant environmental conditions. Such rhythms may have a natural period closely approaching 24 hours under a wide variety of conditions, including high and low temperatures, and are then known as circadian rhythms, regarded as the overt expression of a 'physiological clock' (Bünning, 1964; Cumming and Wagner, 1968). Such rhythms are widespread in plants, animals and microorganisms including fungi. Endogenous rhythms with a period that may differ markedly from 24 hours also occur in the fungi, and are more susceptible to influence by temperature (Cumming and Wagner, 1968; Jerebzoff, 1965). Light may have relatively little effect on such rhythms, as with that shown by the 'clock' mutant of the ascomycete *Neurospora crassa* (Berliner and Neurath, 1965), or may lead to their suppression and replacement by a light-controlled exogenous rhythm as found in *Alternaria tenuis* (Jerebzoff, 1965).

Circadian rhythms may well be widespread in the fungi, but as yet only a few have been studied in detail. The *Neurospora crassa* mutant 'timex', when grown under constant conditions in continuous darkness, exhibits a rhythm of alternate zones of dense and sparse sporulation with a periodicity of 22·7 hours at 25°c (Sargent, Briggs and Woodward, 1966); the length of the period varies little over a wide range of temperatures. In an appropriate light–dark regime (9 hours light; 15 hours darkness), entrainment of the rhythm to 24 hours occurs, a single brief exposure to light can shift the phase of the rhythm, and the rhythm is suppressed by continuous light. The action spectrum (Table 11.1 E) for this suppression was determined by Sargent and Briggs (1967); only the blue end of the spectrum was effective. A circadian rhythm of sporulation by another *N. crassa* mutant 21863, *prolineless* or 'patch' studied earlier by Brandt (1953), Stadler (1959) and Pittendrigh and coworkers (1959), was similarly affected by light.

Circadian rhythms of spore discharge have been studied in *Pilobolus sphaerosporus* (Bruce, Weight and Pittendrigh, 1960; Schmidle, 1951; Uebelmesser, 1954) and in *Daldinia concentrica* (Ingold and Cox, 1955). These too, as seems characteristic of circadian rhythms, persisted in continuous darkness but not continuous light, and, within limits, could be entrained to a variety of light regimes. In *Pilobolus sphaerosporus* a single high intensity light flash of 0·0005 seconds was shown to bring about a shift in phase.

It appears that light is the environmental variable that has the greatest effect on circadian rhythms, and that the nature of these effects is similar in a wide variety of organisms.

Light and pigmentation

Light affects fungal pigmentation in a variety of ways. Melanin synthesis, for example, is stimulated by light in the basidiomycete *Aureobasidium pullulans* (Lingappa, Sussman and Bernstein, 1963) but inhibited in a black mutant of *Neurospora crassa* (Schaeffer, 1953). A topic of particular interest is carotenoid pigmentation in relation to light.

Light-induced carotenoid synthesis

Light has been shown to be essential for carotenoid synthesis in many fungi and to enhance it in others (Carlile, 1965). The most detailed studies have been carried out on *Neurospora crassa* and *Fusarium aquaeductuum*.

The mycelium, but not the conidia, of *Neurospora crassa* require light for the production of carotenoids; only colourless polyenes and trace amounts of coloured carotenoids are formed in darkness (Zalokar, 1954). An action spectrum for the response was obtained by Zalokar (1955), who found that only wavelengths of less than 520 nm were effective. One minute of exposure to light is sufficient to produce near maximal pigmentation. The addition of actidione, an inhibitor of protein synthesis, will prevent pigmentation if it is added up to 10 minutes after the onset of illumination, but not later (Rau, Lindemann and Rau-Hund, 1968). Carotenoid production begins 40 minutes after the start of illumination and is complete in about 6 hours.

Similar enhancement of carotenoid production by light was found in *Fusarium aquaeductuum* (Eberhard, Rau and Zehender, 1961). In this organism also, a detailed action spectrum (Table 11.1 F) established that only wavelengths of less than 520 nm are effective (Rau, 1967a). Carotenoid production could be induced not only by light but also by means of a 2-

hour exposure to mercuribenzoate (either *p*-chloro or *p*-hydroxy), although other mercuric compounds or SH-group poisons were not effective (Rau, 1967b). Both light-induced and mercuribenzoate-induced carotenoid synthesis could be prevented by means of actidione if applied soon after induction.

Rilling (1962, 1964) studied light-induced carotenoid synthesis in *Mycobacterium* sp. He concluded that the process consisted of three phases—photoinduction, the synthesis of one or more carotenogenic enzymes and carotenoid synthesis itself. The studies of Zalokar and Rau and coworkers suggest that light-induced carotenoid biosynthesis in fungi consists of a similar sequence of phases. Recently Rau (1969) has established that photoinduction can occur under anaerobic conditions in both *Neurospora crassa* and *Fusarium aquaeductuum*, although the amount of pigment subsequently synthesized is less than when induction takes place under aerobic conditions. His results suggest that 'oxygen does not directly participate in the primary photochemical event, but functions as an electron acceptor to keep the photoreceptor in the proper oxidation state'.

The role of carotenoids in relation to light

Carotenoids are widespread in the fungi. They are particularly common in photosensitive organs and in spores (which commonly being dispersed by air currents are liable to prolonged exposure to light), and often, as indicated above, they appear as a consequence of illumination.

Prolonged exposure to light can bring about the death of cells through oxidations mediated by light-absorbing sensitizers, either exogenously applied dyes or endogenous pigments, such as chlorophyll, porphyrins and flavins. In bacteria it has been established that carotenoids can provide protection against this effect, although the way in which the pigment acts remains unsolved (Krinsky, 1968). The circumstances and location of carotenoid pigmentation in fungi suggest a similar role in these organisms also, and recent experiments support this possibility. Goldstrohm and Lilly (1965) have shown a close correlation between carotenoid content of cells of the basidiomycete *Dacryopinax spathularia*, varied by means of differing periods of growth in light and their survival when exposed to sunlight or to intense artificial light. Macmillan, Maxwell and Chichester (1966) find that various circumstances leading to low carotenoid content in the yeast *Rhodotorula glutinis* result in decreased survival of cells sensitized by toluidine blue and exposed to laser illumination (30 000 μW; 633 nm). The treatments employed in the experiments on *Dacryopinax* and on *Rhodotorula* aimed at influencing carotenoid content may well have affected many aspects of cellular physiology, but the correlation between carotenoid content and sensitivity to harmful illumination remains suggestive of a protective role for these pigments.

Action spectra and photoreceptors

Detailed action spectra have been obtained for a variety of responses in a range of fungi (see Figure 11.1 and Table 11.1), and less detailed studies on the effectiveness of different wavelengths have been reported for many other species.

Many studies (see Figure 11.1 A and Table 11.1 A to F) have shown sensitivity limited to wavelengths of less than about 520 nm, with blue light (450 nm) particularly effective, and a further peak in the near ultraviolet region (about 370 nm). Subsidiary peaks or 'shoulders' on either side of the major peak are apparent at about 430 and 480 nm in some studies (Curry and Gruen, 1959; Delbrück and Shropshire, 1960; Rau, 1967a; Sargent and Briggs, 1967). Action spectra of this sort are widespread, occurring not only in fungi but also in bacteria (Batra and Rilling, 1964), algae (Kowallik, 1967; Pickett and French, 1967) and higher plants (Shropshire and Withrow, 1958; Thimann and Curry, 1960).

Since the absorption spectra of riboflavin and β-carotene are very similar, such action spectra can be interpreted as indicative of either a carotenoid or a flavoprotein photoreceptor. The action spectrum peak at 370 nm, an apparent embarrassment to supporters of the carotenoid hypothesis, can be explained as being due to *cis* carotenoids (Thimann and Curry, 1960). The 'shoulders' at 430 and 480 nm, apparent in some action spectra, are absent in the absorption spectra of free flavins; they are, however, present in many flavoproteins (Massey and Ganther, 1965), and hence do not constitute evidence against the flavin hypothesis. As an action spectrum may for a variety of reasons (Delbrück and Shropshire, 1960) differ slightly from the absorption spectrum of the photoreceptor, it seems unlikely that the controversy will be resolved by action spectrum studies, valuable though these are.

Formerly, the presence of carotenoids in many photosensitive organs was cited as being indicative of a carotenoid photoreceptor, and the continued photosensitivity of albino mutants, or cultures with carotenoid levels reduced in other ways, as evidence against the hypothesis (Carlile, 1965). Recently, however, it has been demonstrated that the photoreceptor in the very sensitive (Table 11.3) *Phycomyces* sporangiophore must constitute only a minute fraction of the pigment present, with an *in vivo* absorption undetectable by present methods (Meissner and Delbrück, 1968); therefore both classes of argument have lost much of their force. It seems likely that the controversy will only be resolved when the immediate biochemical consequences of illumination have been elucidated. Meanwhile, in view of the intimate involvement of flavoproteins in electron transfer paths, and their *in vitro* photosensitivity, the present author favours the flavoprotein hypothesis.

Table 11.3. Effective light intensities.

Reaction	Organism	Intensity $(\mu W/cm^2)^a$	Comment
Inhibition of growth and respiration[b]	*Saccharomyces cerevisiae*	400–95 000	Range employed
Growth stimulation[c]	*Helminthosporium oryzae*	230	—
Phototropism[d]	*Phycomyces blakesleeanus*	3×10^{-7}	Threshold
		0·04–20	Normal range
Phototaxis[e]	*Dictyostelium purpureum*	0·1–0·3	Threshold
Sporulation (conidium initiation)[f]	*Pleospora herbarum*	50	Standard stimulus at 290 nm
Sporulation (inhibition of terminal phase)[g]	*Pleospora herbarum*	100–170	Standard stimulus at 280 nm
Sporulation (conidium initiation)[h]	*Trichoderma viride*	6	Half-maximum response at 447 nm
Sporulation (induction)[i]	*Physarum nudum*	4	Threshold at 333 nm
Sporulation (inhibition)[i]	*Physarum nudum*	200	Threshold at 417 nm
		80	Threshold at 333 nm
Suppression of circadian rhythm[j]	*Neurospora crassa*	$1·8 \times 10^{-2}$	Threshold at 465 nm
Carotenoid biosynthesis[k]	*Fusarium aquaeductuum*	55–115	Employed in action spectrum determinations

a Calculated, in some instances, from data expressed differently
b Ehrenberg (1966a)
c Leach (1961)
d Bergman and coworkers (1969)
e Francis (1964)
f Leach and Trione (1966)
g Leach (1968)
h Gressel and Hartmann (1968)
i Rakoczy (1965)
j Sargent and Briggs (1967)
k Rau (1967a)

An action spectrum (see Figure 11.1 B and Table 11.1 G to J) indicative of a different photoreceptor is that for the induction of sporulation in many ascomycetes and Fungi Imperfecti (Leach and Trione, 1965, 1966). Only ultraviolet radiation is effective in bringing about this response, as wavelengths longer than 330 nm are effective only in high dosages and those longer than 360 to 380 nm are inactive. There is a peak of effectiveness at about 290 nm. Shorter wavelengths are also effective, but here the action spectrum is unlikely to represent closely the absorption spectrum of the photoreceptor.

The widespread occurrence in fungi of the 'blue' and 'ultraviolet' photoresponses discussed above has been firmly established for some years, whereas at the time of the author's previous review (Carlile, 1965) there was little convincing evidence for the occurrence of 'red' photoreceptors in fungi. Recently, however, it has become clear that some fungi do respond to wavelengths longer than 520 nm. These longer wavelengths in some instances have the same effect as blue light. Page and Brungard (1961) showed that phototropic sensitivity in *Entomophthora coronata* (or *Conidiobolus villosus*) extended to 650 nm, although the organism is about twenty-five times as sensitive at 405 nm as at 640 nm (Page, 1968). Leach (1968) found that wavelengths as long as 650 nm inhibited the terminal phase of sporulation in *Pleospora herbarum*, although they were far less effective than shorter wavelengths. Rakoczy (1965) found stimulation of sporulation in the myxomycete *Physarum nudum* by light with wavelengths 632 to 713 nm, as well as by wavelengths of less than 541 nm. The longer wavelengths may on the other hand act in opposition to the effects of shorter wavelengths. Lukens (1965) found that the inhibitory effect of blue light (450 nm) on the terminal phase of sporulation in *Alternaria solani* could be reversed by longer wavelengths, peaks of activity being observed at 600 nm in the wild type and at 575 and 650 to 675 nm in a mutant. Ingold (1968) reported that the effect of yellow light (585 nm) in promoting sporulation in *Sphaerobolus* was reversible by blue light (450 nm), and Ingold and Oso (1969) reported that the immediate spore discharge in *Ascobolus* caused by transfer from darkness to blue light (e.g. 440 nm) can be presented by simultaneous exposure to yellow light (580 nm). In some myxomycetes it has been found that the stimulating effects of blue and near red light (650–700 nm) in promoting sporulation can be reversed by far red (690–750 nm) light (Nair and Zabka, 1965) or by green light (Rakoczy, 1967). Further studies, including the determination of action spectra, on the effects of the red end of the visible spectrum are clearly needed.

The mode of action of light

There is little detailed information as to how light brings about fungal photoresponses; there are, however, a few general comments that can be made.

It is likely that the various observed responses fall roughly into two categories, which for convenience can be termed 'direct' and 'indirect' responses. 'Direct' responses are those in which an effect can be observed in seconds or in minutes, and which are essentially quantitative in nature. Examples are changes in respiration rate, in direction of movement (phototaxis) or in the spatial distribution of growth (phototropism). Changes in enzyme activity and metabolic rates will be involved, but not (as, for example, in morphogenesis) the synthesis of enzymes previously absent. Such responses should hence be uninfluenced by inhibitors of RNA or protein synthesis. Some preliminary results of Trinci and Border (personal communication) are consistent with this suggestion; they find that the threshold time for phototropism of the *Phycomyces* sporangiophore is unaffected by concentrations of uracil analogues that inhibit sporangium formation.

'Indirect' responses are those which are not observed for some hours, and which clearly involve novel biosynthesis entailing the production of enzymes not hitherto present. Examples are photoinduced sporulation and carotenoid biosynthesis. In *Trichoderma viride* (see page 325) light-induced sporulation takes about 24 hours to complete, can be inhibited by fluorouracil and appears to be RNA mediated (Gressel and Galun, 1967). Light-induced carotenoid synthesis (see page 331) also takes several hours to complete, involves enzyme biosynthesis and can be prevented by actidione inhibition of protein synthesis (Rau, Lindemann and Rau-Hund, 1968). Such indirect responses are of great interest, but only the earlier phases prior to the production of novel enzymes are strictly speaking 'photobiological'.

The early phases of both direct and indirect responses have received little study. Respiratory enzymes commonly have light-absorbing prosthetic groups (flavins and porphyrins) and the present author (Carlile 1962, 1965) has tentatively suggested that an early step in a fungal photoresponse may be increased or diminished activity in one or more electron-transfer paths. As mentioned earlier (page 311), Daniel (1966) has shown a very rapid (within 15 seconds) and reversible depression of oxygen uptake by plasmodia and isolated mitochondria of the myxomycete *Physarum polycephalum*. Fairly rapid (10 to 15 minutes) light-stimulated oxygen uptake, with an action spectrum suggestive of a flavoprotein photoreceptor, has been demonstrated in the alga *Chlorella* (Kowallik, 1967; Pickett and French, 1967). Similar studies on the respiration of light-sensitive fungi could prove to be rewarding.

REFERENCES

Alasoadura, S. O. (1963). Fruiting in *Sphaerobolus* with special reference to light. *Ann. Bot.*, **27**, 123.

Aragaki, M. (1962). Quality of radiation inhibitory to sporulation of *Alternaria tomato*. *Phytopathology*, **52**, 1227.

Banbury, G. H. (1952). Physiological studies in the Mucorales. Part I. The phototropism of sporangiophores of *Phycomyces blakesleeanus*. *J. exp. Bot.*, **3**, 77.

Banbury, G. H. (1959). Phototropism of lower plants. In W. Ruhland (Ed.), *Encyclopedia of Plant Physiology*, Vol. 17-1. Springer, Berlin. pp. 530–578.

Baranetzki, J. (1876). Influence de la lumière sur les plasmodia des Myxomycetes. *Mém. Soc. natn. Sci. nat. math.*, *Cherbourg*, **19**, 321.

Barnett, H. L. (1968). The effects of light, pyridoxine and biotin on the development of the mycoparasite *Gonatobotryum fuscum*. *Mycologia*, **60**, 244.

Barnett, H. L., and V. G. Lilly (1950). Influence of nutritional and environmental factors upon asexual reproduction of *Choanephora cucurbitarum* in culture. *Phytopathology*, **40**, 80.

Barnett, H. L., and V. G. Lilly (1956). Factors affecting the production of zygospores by *Choanephora cucurbitarum*. *Mycologia*, **48**, 617.

Batra, P. P., and H. C. Rilling (1964). On the mechanism of photoinduced carotenoid synthesis: aspects of the photoinductive reaction. *Archs Biochem. Biophys.*, **107**, 485.

Berg, L. A., and M. E. Gallegly (1966). The effect of light on oospore germination in species of *Phytophthora*. *Phytopathology*, **56**, 583.

Bergman, K., P. V. Burke, E. Cerdá-Olmedo, C. N. David, M. Delbrück, K. W. Foster, E. W. Goodell, M. Heisenberg, G. Meissner, M. Zalokar, D. S. Dennison, and W. Shropshire (1969). Phycomyces. *Bact. Rev.*, **33**, 99.

Berliner, M. D., and P. W. Neurath (1965). The band forming rhythm of *Neurospora* mutants. *J. cell. comp. Physiol.*, **65**, 183.

Bonner, J. T., and F. E. Whitfield (1965). The relation of sorocarp size to phototaxis in the cellular slime mold, *Dictyostelium purpureum*. *Biol. Bull.*, **128**, 51.

Brandt, W. H. (1953). Zonation in a prolineless strain of *Neurospora*. *Mycologia*, **45**, 194.

Brandt, W. H. (1964). Morphogenesis in *Verticillium*: effects of light and ultraviolet radiation on micro-sclerotia and melanin. *Can. J. Bot.*, **42**, 1017.

Brandt, W. H. (1965). Morphogenesis in *Verticillium*: reversal of the near-UV effect by catechol. *BioScience*, **15**, 669.

Brandt, W. H. (1967). Influence of near-ultraviolet light on hyphal elongation in *Verticillium*. *Mycologia*, **59**, 736.

Brandt, W. H., and J. E. Reese (1964). Morphogenesis in *Verticillium*: a self-produced, diffusible morphogenetic factor. *Am. J. Bot.*, **51**, 922.

Bruce, V. G., F. Weight, and C. S. Pittendrigh (1960). Resetting the sporulation rhythm in *Pilobolus* with short light flashes of high intensity. *Science*, **131**, 728.

Buller, A. H. R. (1934). *Researches on Fungi*, Vol. 6. Longmans, Green and Co., London.

Bünning, E. (1964). *The Physiological Clock*, Academic Press, London, New York.

Cantino, E. C. (1959). A role for glycine in light-stimulated development and phosphorus metabolism in the mold *Blastocladiella emersonii. Devl. Biol.*, 1, 396.

Cantino, E. C. (1965). Intracellular distribution of ^{14}C during sporogenesis in *Blastocladiella emersonii.* Effect of light on hemoprotein. *Arch. Mikrobiol.*, 51, 42.

Cantino, E. C., and A. Goldstein (1967). Citrate-induced citrate production and light-induced growth of *Blastocladiella emersonii. J. gen. Microbiol.*, 46, 347.

Cantino, E. C., and E. A. Horenstein (1956). The stimulatory effect of light upon growth and CO_2 fixation in *Blastocladiella* I. The SKI cycle. *Mycologia*, 48, 777.

Cantino, E. C., and E. A. Horenstein (1957). The stimulatory effect of light upon growth and CO_2 fixation in *Blastocladiella.* II. Mechanism at an organismal level of integration. *Mycologia*, 49, 892.

Cantino, E. C., and E. A. Horenstein (1959). The stimulatory effect of light upon growth and carbon dioxide fixation in *Blastocladiella.* III. Further studies, *in vivo* and *in vitro. Physiologia Pl.*, 12, 251.

Cantino, E. C., and G. Turian (1961). A role for glycine in light-stimulated nucleic acid synthesis by *Blastocladiella emersonii. Arch. Mikrobiol.*, 38, 272.

Carlile, M. J. (1956). A study of the factors influencing non-genetic variation of a strain of *Fusarium oxysporum. J. gen. Microbiol.*, 14, 643.

Carlile, M. J. (1962). Evidence for a flavoprotein photoreceptor in *Phycomyces. J. gen. Microbiol.*, 28, 161.

Carlile, M. J. (1965). Photobiology of fungi. *A. Rev. Pl. Physiol.*, 16, 175.

Carlile, M. J. (1966). The orientation of zoospores and germ-tubes. In M. F. Madelin (Ed.), *The Fungus Spore*, Butterworths, London. pp. 175–187.

Carlile, M. J., J. S. W. Dickens, E. M. Mordue, and M. A. A. Schipper (1962). The development of coremia. II. *Penicillium isariaeforme. Trans. Br. mycol. Soc.*, 45, 457.

Carlile, M. J., J. S. W. Dickens, and M. A. A. Schipper (1962). The development of coremia. III. *Penicillium clavigerum* with observations on *P. expansum* and *P. italicum. Trans. Br. mycol. Soc.*, 45, 462.

Carlile, M. J., B. G. Lewis, E. M. Mordue, and J. Northover (1961). The development of coremia. I. *Penicillium claviforme. Trans. Br. mycol. Soc.*, 44, 129.

Castle, E. S. (1966a). Light responses of *Phycomyces. Science*, 154, 1416.

Castle, E. S. (1966b). A kinetic model for adaptation and the light responses of *Phycomyces. J. gen. Physiol.*, 49, 925.

Chambers, T. C., K. Marcus, and L. G. Willoughby (1967). The fine structure of the mature zoosporangium of *Nowakowskiella profusa. J. gen. Microbiol.*, 46, 135.

Charlton, K. M. (1953). The sporulation of *Alternaria solani* in culture. *Trans. Br. mycol. Soc.*, 36, 349.

Cochrane, V. W. (1958). *Physiology of fungi*, Wiley, New York.

Cochrane, V. W. (1960). Spore germination. In J. G. Horsfall and A. E. Dimond (Eds.), *Plant Pathology: An Advanced Treatise*, Vol. 2, Academic Press, London New York. pp. 167–202.

Cumming, B. G., and E. Wagner (1968). Rhythmic processes in plants. *A. Rev. Pl. Physiol.*, 19, 381.

Curry, G. M., and H. E. Gruen (1959). Action spectra for the positive and negative phototropism of *Phycomyces* sporangiophores. *Proc. natn Acad. Sci. U.S.A.*, 45, 797.

Daniel, J. W. (1966). Light-induced synchronous sporulation of a myxomycete—the relation of initial metabolic changes to the establishment of a new cell state. In I. L. Cameron and A. M. Padilla (Eds.), *Cell Synchromy*, Academic Press, London, New York. pp. 117–152.

Daniel, J. W., and H. P. Rusch (1961). Method for inducing sporulation of pure cultures of the myxomycete *Physarum polycephalum*. *J. Bact.*, **83**, 234.

Delbrück, M., and W. Reichardt (1956). System analysis for the light growth reactions of *Phycomyces*. In D. Rudnick (Ed.), *Cellular Mechanisms in Differentiation and Growth*, Princeton University Press, New Jersey. pp. 3–44.

Delbrück, M., and W. Shropshire (1960). Action and transmission spectra of *Phycomyces*. *Pl. Physiol.*, **35**, 194.

Eberhard, D., W. Rau, and C. Zehender (1961). Über den Einfluss des Lichtes auf die Carotinoidbildung von *Fusarium aquaeductuum*. *Planta*, **56**, 302.

Ehrenberg, M. (1966a). Wirkungen sichtbaren Lichtes auf *Saccharomyces cerevisiae*. I. Einfluss verschiedener Faktoren auf die Höhe des Lichteffektes auf Wachstum und Stoffwechsel. *Arch. Mikrobiol.*, **54**, 358.

Ehrenberg, M. (1966b). Wirkungen sichtbaren Lichtes auf *Saccharomyces cerevisiae*. II. Die Gärung unter dem Einfluss von Pasteur-Effekt und Licht-Effekt. *Arch. Mikrobiol.*, **55**, 26.

Ehrenberg, M. (1968). Die Höhe des Lichteffektes auf Wachstum und Stoffwechsel von *Saccharomyces cerevisiae* in Abhängigkeit vom Phasenstatus der Vorkulturzellen. *Arch. Mikrobiol.*, **61**, 20.

Elkind, M. M., and H. Sutton (1957a). Lethal effect of visible light on a mutant strain of haploid yeast. I. General dependencies. *Archs Biochem. Biophys.*, **72**, 84.

Elkind, M. M., and H. Sutton (1957b). Lethal effect of visible light on a mutant strain of haploid yeast. II. Absorption and action spectra. *Archs Biochem. Biophys.*, **72**, 96.

Etzold, H. (1961). Die Wirkungen des linear polarisierten Lichtes auf Pilze und ihre beziehungen zu den tropistischen Wirkungen des einseitigen Lichtes. *Expl. Cell Res.*, **25**, 229.

Francis, D. W. (1964). Some studies on phototaxis of *Dictyostelium*. *J. cell. comp. Physiol.*, **64**, 131.

Fuller, M. S., and R. Reichle (1968). The fine structure of *Monoblepharella* sp. zoospores. *Can. J. Bot.*, **46**, 279.

Galun, E., and J. Gressel (1966). Morphogenesis in *Trichoderma*: suppression of photoinduction by 5-fluorouracil. *Science*, **151**, 696.

Gettkandt, G. (1954). Zur Kenntnis des Phototropismus der Keimmyzelien einiger parasitischer Pilze. *Wiss. Z. Martin-Luther-Univ. Halle-Wittenb.*, **3**, 691.

Givan, C. V., and K. R. Bromfield (1964). Light inhibition of uredospore germination in *Puccinia recondita*. *Phytopathology*, **54**, 116.

Goldstein, A., and E. C. Cantino (1962). Light-stimulated polysaccharide and protein synthesis by synchronized, single generations of *Blastocladiella emersonii*. *J. gen. Microbiol.*, **28**, 689.

Goldstrohm, D. D., and V. G. Lilly (1965). The effect of light on the survival of pigmented and nonpigmented cells of *Dacryopinax spathularia*. *Mycologia*, **57**, 612.

Gray, W. D. (1938). The effect of light on the fruiting of myxomycetes. *Am. J. Bot.*, **25**, 511.

Gressel, J., and E. Galun (1967). Morphogenesis in *Trichoderma*: photoinduction and RNA. *Devl. Biol.*, **15**, 575.

Gressel, J. B., and K. M. Hartmann (1968). Morphogenesis in *Trichoderma*: action spectrum of photoinduced sporulation. *Planta*, **79**, 271.

Guerin, B., et E. Sulkowski (1966). Photoinhibition de l'adaptation respiratoire chez *Saccharomyces cerevisiae*. 1. Variations de la sensibilite a l'inhibition. *Biochim. biophys. Acta*, **129**, 193.

Haupt, W., und M. Buchwald (1967). Die Orientierung der Photorezeptor—Moleküle im Sporangienträger von *Phycomyces*. *Z. Pflphysiol.*, **56**, 20.

Hebert, T. T., and A. Kelman (1958). Factors influencing the germination of resting sporangia of *Physoderma maydis*. *Phytopathology*, **48**, 102.

Holloman, D. W. (1966). Reducing compounds and the growth of *Phytophthora infestans*. *J. gen. Microbiol.*, **45**, 315.

Ingold, C. T. (1953). *Dispersal in Fungi*, Clarendon Press, Oxford.

Ingold, C. T. (1965). *Spore Liberation*, Clarendon Press, Oxford.

Ingold, C. T. (1968). Fruiting in *Sphaerobolus*: an effect of yellow light reversed by blue. *Nature*, **219**, 1265.

Ingold, C. T., and V. J. Cox (1955). Periodicity of spore discharge in *Daldinia*. *Ann. Bot.*, **19**, 201.

Ingold, C. T., and B. Marshall (1963). Further observations on light and spore discharge in certain pyrenomycetes. *Ann. Bot.*, **27**, 481.

Ingold, C. T., and M. Nawaz (1967). Sporophore development in *Sphaerobolus*: effect of blue and red light. *Ann. Bot.*, **31**, 469.

Ingold, C. T., and B. A. Oso (1969). Light and spore discharge in *Ascobolus*. *Ann. Bot.*, **33**, 463.

Jacob, F. (1954). Die Rolle des Lichtes im Entwicklungsgang der Sporangienträger von *Pilobolus*—Arten. *Wiss. Z. Martin-Luther-Univ. Halle-Wittenb.*, **4**, 125.

Jacob, F. (1961). Die Steuerung der tagesperiodischen Sporangienträgerbildung bei *Pilobolus kleinii* und *umbonatus*. *Flora*, **154**, 329.

Jaffe, L. F. (1960). The effect of polarized light on the growth of a transparent cell: a theoretical analysis. *J. gen. Physiol.*, **43**, 897.

Jaffe, L. F., and H. Etzold (1962). Orientation and locus of tropic photoreceptor molecules in spores of *Botrytis* and *Osmunda*. *J. Cell Biol.*, **13**, 13.

Jerebzoff, S. (1965). Growth rhythms. In G. C. Ainsworth and A. S. Sussman (Eds.), *The Fungi*, Vol. 1. Academic Press, London, New York., pp. 625–645.

Kelly, M. S., and J. L. Gay (1969). The action of visible radiation on the formation and properties of *Saccharomyces* ascospores. *Arch. Mikrobiol.*, **66**, 259.

Kowallik, W. (1967). Action spectrum for an enhancement of endogenous respiration by light in *Chlorella*. *Pl. Physiol.*, **42**, 672.

Krinsky, N. I. (1968). The protective function of carotenoid pigments. In A. C. Giese (Ed.), *Photophysiology*, Vol. 3, Academic Press, London, New York, pp. 123–195.

Kusano, S. (1930). The life history of *Synchytrium fulgens* Schroet, with special reference to its sexuality. *Jap. J. Bot.*, **5**, 35.

Leach, C. M. (1961). The sporulation of *Helminthosporium oryzae* as affected by exposure to near ultraviolet radiation and dark periods. *Can. J. Bot.*, **39**, 705.

Leach, C. M. (1962). The quantitative and qualitative relationship of ultraviolet and visible radiation to the induction of reproduction in *Ascochyta pisi*. *Can. J. Bot.*, **40**, 1577.

Leach, C. M. (1963). The qualitative and quantitative relationship of mono-chromatic radiation to sexual and asexual reproduction of *Pleospora herbarum*. *Mycologia*, **55**, 151.

Leach, C. M. (1964). The relationship of visible and ultraviolet light to sporulation of *Alternaria chrysanthemi*. *Trans. Br. mycol. Soc.*, **47**, 153.

Leach, C. M. (1965). Ultraviolet-absorbing substances associated with light-induced sporulation in fungi. *Can. J. Bot.*, **43**, 185.

Leach, C. M. (1967a). Interaction of near-ultraviolet light and temperature on sporulation of the fungi *Alternaria*, *Cercosporella*, *Fusarium*, *Helmintho-sporium* and *Stemphylium*. *Can. J. Bot.*, **45**, 1999.

Leach, C. M. (1967b). The light factor in the detection and identification of seed-borne fungi. *Proc. int. Seed Test Ass.*, **32**, 565.

Leach, C. M. (1968). An action spectrum for light inhibition of the 'terminal phase' of photosporogenesis in the fungus *Stemphylium botryosum*. *Mycologia*, **60**, 532.

Leach, C. M., and E. J. Trione (1965). An action spectrum for light induced sporulation in the fungus *Ascochyta pisi*. *Pl. Physiol.*, **40**, 808.

Leach, C. M., and E. J. Trione (1966). Action spectra for light-induced sporula-tion of the fungi *Pleospora herbarum* and *Alternaria dauci*. *Photochem. Photo-biol.*, **5**, 621.

Leal, J. A., and B. Gomez-Miranda (1965). The effect of light and darkness on the germination of the spores of certain species of *Phytophthora* on some synthetic media. *Trans. Br. mycol. Soc.*, **48**, 491.

Leith, H. (1956). Die Wirkung des Grünlichtes auf die Fruchtkörperbildung bei *Didymium eunigripes*. *Arch. Mikrobiol.*, **24**, 91.

Lingappa, Y., A. S. Sussman, and I. A. Bernstein (1963). Effect of light and media upon growth and melanin formation in *Aureobasidium pullulans* (De By.) Arn. (=*Pullularia pullulans*). *Mycopath. Mycol. appl.*, **20**, 109.

Lukens, R. J. (1963). Photoinhibition of sporulation in *Alternaria solani*. *Am. J. Bot.*, **50**, 720.

Lukens, R. J. (1965). Reversal by red light of blue light inhibition of sporulation in *Alternaria solani*. *Phytopathology*, **55**, 1032.

Lythgoe, J. N. (1961). Effect of light and temperature on growth and develop-ment in *Thamnidium elegans* Link. *Trans. Br. mycol. Soc.*, **44**, 199.

McCallan, S. E. A., and S. Y. Chan (1944). Inducing sporulation of *Alternaria solani* in pure culture. *Contr. Boyce Thomson Inst. Pl. Res.*, **13**, 323.

Macmillan, J. D., W. A. Maxwell, and C. O. Chichester (1966). Lethal photo-sensitization of microorganisms with light from a continuous-wave gas laser. *Photochem. Photobiol.*, **5**, 555.

Marsh, B., E. E. Taylor, and L. M. Bassler (1959). A guide to the literature on certain effects of light on fungi: reproduction, morphology, pigmentation, and phototropic phenomena. *Pl. Dis. Reptr. No.* 261.

Massey, V., and H. Ganther (1965). On the interpretation of the absorption spectra of flavoproteins with special reference to D-amino acid oxidase. *Biochemistry*, **4**, 1161.

Matile, P., und A. Frey-Wyssling (1962). Atmung und Wachstum von Hefe im Licht. *Planta*, **58**, 154.

Meissner, G., and M. Delbrück (1968). Carotenes and retinal in *Phycomyces* mutants. *Pl. Physiol.*, **43**, 1279.

Nair, P., and G. G. Zabka (1965). Light quality and sporulation in myxomycete with special reference to a red, far-red reversible reaction. *Mycopath. Mycol. appl.*, **26**, 123.

Page, R. M. (1956). Studies on the development of asexual reproductive structures in *Pilobolus. Mycologia*, **48**, 206.

Page, R. M. (1962). Light and the asexual reproduction of *Pilobolus. Science*, **138**, 1238.

Page, R. M. (1965). The physical environment for fungal growth. 3. Light. In G. C. Ainsworth and A. S. Sussman (Eds.), *The Fungi*, Vol. 1. Academic Press, London, New York., pp. 559–574.

Page, R. M. (1968). Phototropism in fungi. In A. C. Giese (Ed.), *Photophysiology*, Vol. 3. Academic Press, London, New York., pp. 65–90.

Page, R. M., and J. Brungard (1961). Phototropism in *Conidiobolus*, some preliminary observations. *Science*, **134**, 733.

Page, R. M., and G. M. Curry (1966). Studies on the phototropism of young sporangiophores of *Pilobolus kleinii. Photochem. Photobiol.*, **5**, 31.

Paul, W. R. C. (1929). A comparative morphological and physiological study of a number of strains of *Botrytis cinerea* Pers. with special reference to their virulence. *Trans. Br. mycol. Soc.*, **14**, 118.

Petri, L. (1925). Osservazioni biologiche sulla '*Blepharospora cambivora*'. Reprinted from *Rep. Inst. Sup. Agrario e Forestale*, Ser. 2a, i. *Rev. appl. Mycol.*, **10**, 122.

Pickett, J. M., and C. S. French (1967). The action spectrum for blue-light-stimulated oxygen uptake in *Chlorella. Proc. natn Acad. Sci.*, **57**, 1587.

Piskorz, B. (1967a). Investigations on the formation of coremia. 1. Action of light on the formation of coremia in *Penicillium isariaeforme. Acta. Soc. Bot. Pol.*, **36**, 123.

Piskorz, B. (1967b). Investigations on the action of light on the growth and development of *Penicillium claviforme* Bainer. *Acta. Soc. Bot. Pol.*, **36**, 677.

Pittendrigh, C. S., V. G. Bruce, N. S. Rosensweig, and M. L. Rubin (1959). A biological clock in *Neurospora. Nature* **184**, 169.

Rakoczy, L. (1963). *Acta Soc. Bot. Pol.*, **32**, 393.

Rakoczy, L. (1965). Action spectrum in sporulation of the slime-mold *Physarum nudum* Macbr. *Acta Soc. Bot. Pol.*, **34**, 97.

Rakoczy, L. (1967). Antagonistic action of light in sporulation of the myxomycete *Physarum nudum. Acta Soc. Bot. Pol.*, **36**, 153.

Ramsey, C. M., and A. A. Bailey (1930). Effects of ultraviolet radiation upon sporulation in *Macrosporium* and *Fusarium. Bot. Gaz.*, **89**, 113.

Raper, K. B., and G. Thom (1949). *A Manual of the Penicillia*, Williams and Wilkins, Baltimore.

Rau, W. (1967a). Untersuchungen über die lichtabhängige Carotinoidsynthese. I. Das Wirkungsspektrum von *Fusarium aquaeductuum. Planta*, **72**, 14.

Rau, W. (1967b). Untersuchungen über die lichtabhängige Carotinoidsynthese. II. Ersatz der Lichtinduktion durch Mercuribenzoat. *Planta*, **74**, 263.

Rau, W. (1969). Untersuchungen über die lichtabhängige Carotinoidsynthese. IV. Die Rolle des Sauerstoffs bei der Lichtinduktion. *Planta*, **84**, 30.

Rau, W., I. Lindemann, und A. Rau-Hund (1968). Untersuchungen über die lichtabhängige Carotinoidsynthese. III. Die Farbstoffbildung von *Neurospora crassa* in Submerskultur. *Planta*, **80**, 309.

Rilling, H. C. (1962). Photoinduction of carotenoid synthesis of a *Mycobacterium* sp. *Biochim. biophys. Acta*, **60**, 548.

Rilling, H. C. (1964). On the mechanism of photoinduction of carotenoid synthesis. *Biochim. biophys. Acta*, **79**, 464.

Roth, J. N., and W. H. Brandt (1964). Influence of some environmental factors on hereditary variation in monospore cultures of *Verticillium albo-atrum*. *Phytopathology*, **54**, 1454.

Samuel, E. D. (1961). Orientation and rate of locomotion of individual amoebas in the life cycle of the cellular slime mold *Dictyostelium mucoroides*. *Devl. Biol.*, **3**, 317.

Sargent, M. L., and W. R. Briggs (1967). The effects of light on a circadian rhythm of conidiation in *Neurospora*. *Pl. Physiol.*, **42**, 1504.

Sargent, M. L., W. R. Briggs, and D. O. Woodward (1966). Circadian nature of a rhythm expressed by an invertaseless strain of *Neurospora crassa*. *Pl. Physiol.*, **41**, 1343.

Sawyer, W. H. (1929). Observations on some entomogenous members of the Entomophthoraceae in artificial culture. *Am. J. Bot.*, **16**, 87.

Schaeffer, P. (1953). A black mutant of *Neurospora crassa*. *Archs. Biochem. Biophys.*, **47**, 359.

Schmidle, A. (1951). Die Tagesperiodizitat der asexuellen Reproduction von *Pilobolus sphaerosporus*. *Arch. Mikrobiol.*, **16**, 80.

Shropshire, W. (1962). The lens effect and phototropism of *Phycomyces*. *J. gen. Physiol.*, **45**, 949.

Shropshire, W., and K. Bergman (1968). Light induced concentration changes of ATP from *Phycomyces* sporangiophores: a re-examination. *Pl. Physiol.*, **43**, 1317.

Shropshire, W., and R. B. Withrow (1958). Action spectrum of phototropic tip-curvature of *Avena*. *Pl. Physiol.*, **33**, 360.

Sproston, T., and R. B. Setlow (1968). Ergosterol and substitutes for the ultraviolet radiation requirement for conidia formation in *Stemphylium solani*. *Mycologia*, **60**, 104.

Stadler, D. R. (1959). Genetic control of a cyclic growth pattern in *Neurospora*. *Nature*, **184**, 170.

Stevens, F. L. (1928). Effects of ultraviolet radiation on various fungi. *Bot. Gaz.*, **86**, 210.

Strasburger, E. (1878). Wirkung des Lichts und der Wärme auf Schwärmsporen. *Jena. Z. Naturw.*, **12**, 551.

Sulkowski, E., B. Guerin, J. Defaye, and P. P. Slonimski (1964). Inhibition of protein synthesis in yeast by low intensities of visible light. *Nature*, **202**, 36.

Sussman, A. S. (1965). Physiology of dormancy and germination in the propagules of cryptogamic plants. In A. Lang (Ed.), *Encyclopedia of Plant Physiology*, Vol. 15. Springer, Berlin. pp. 933–1025.

Thimann, K. V. (1967). Phototropism. In M. Florkin and E. H. Stoltz (Eds.), *Comprehensive Biochemistry*, Vol. 27. Elsevier, Amsterdam. pp. 1–29.

Thimann, K. V., and G. M. Curry (1960). Phototropism and phototaxis. In M. Florkin and H. S. Mason (Eds.), *Comparative Biochemistry*, Vol. 1. Academic Press, New York, London, pp. 243–309.

Trinci, A. P. J., and G. H. Banbury (1967). A study of the tall conidiophores of *Aspergillus giganteus*. *Trans. Br. mycol. Soc.*, **50**, 525.

Trinci, A. P. J., and G. H. Banbury (1968). Phototropism and light-growth responses of *Aspergillus giganteus*. *J. gen. Microbiol.*, **54**, 427.

Trione, E. J., C. M. Leach, and J. T. Mutch (1966). Sporogenic substances isolated from fungi. *Nature*, **212**, 163.

Turian, G., and E. C. Cantino (1959). The stimulating effect of light on nucleic acid synthesis in the mould *Blastocladiella emersonii*. *J. gen. Microbiol.*, **21**, 721.

Uebelmesser, E. R. (1954). Uber den endonomen Tagesrhythmus der Sporangienträgerbildung von *Pilobolus*. *Arch. Mikrobiol.*, **20**, 1.

Wager, H. (1913). Life-history and cytology of *Polyphagus euglenae*. *Ann. Bot.*, **27**, 173.

Walkey, D. J. A., and R. Harvey (1966). Spore discharge rhythms in pyreno-mycetes. 1. A survey of the periodicity of spore discharge in pyrenomycetes. *Trans. Br. mycol. Soc.*, **49**, 582.

Weinhold, A. R., and F. F. Hendrix (1963). Inhibition of fungi by media previously exposed to light. *Phytopathology*, **53**, 1280.

Welty, R. E., and C. M. Christensen (1965). Negatively phototropic growth of *Asperigillus restrictus*. *Mycologia*, **57**, 311.

Zalokar, M. (1954). Studies on the biosynthesis of carotenoids in *Neurospora crassa*. *Archs Biochem. Biophys.*, **50**, 71.

Zalokar, M. (1955). Biosynthesis of carotenoids in *Neurospora*. Action spectrum of photoactivation. *Archs Biochem. Biophys.*, **56**, 318.

Zankel, K. L., P. V. Burke, and M. Delbrück (1967). Absorption and screening in *Phycomyces*. *J. gen Physiol.*, **50**, 1893.

CHAPTER 12

Photoperiodic effects in microorganisms

M. J. DRING

Department of Botany, The Queen's University of Belfast, Northern Ireland

Introduction

The study of photoperiodism since its first demonstration over 50 years ago (Tournois, 1912), has shown that daylength is widely used by both plants and animals as an indicator of the time of year. Processes as diverse as flowering and leaf-fall in plants, or diapause and migration in animals, may be under photoperiodic control, and the subject has consequently received a considerable amount of intensive investigation. Studies of photoperiodism in plants have been concentrated almost exclusively on responses in flowering plants, and lower plants and microorganisms have received little critical photoperiodic investigation. The purpose of the present paper is to review the reported effects of changes in daylength on microorganisms, especially algae, and to assess the potential for further work in this field.

The extensive work on the physiology of photoperiodism in the control of flowering (see reviews by Hillman, 1962; Lang, 1965; Salisbury, 1963) has shown that photoperiodic responses can be characterized by the following facts:

(1) Short-day (SD) plants respond if the period of uninterrupted darkness in a light–dark cycle of 24 hours exceeds some critical length.

(2) Long-day (LD) plants respond if the period of uninterrupted darkness in a light–dark cycle of 24 hours is less than some critical length.

(3) If a long dark period, alternating with a short light period in a 24-hour cycle, is interrupted by a short light break, the response of SD plants is inhibited and that of LD plants is promoted.

(4) Plants can be induced to respond by a relatively short exposure to cycles of the appropriate daylength, and will then remain in the induced state for some time after their return to a non-inductive daylength.

These characteristics provide a convenient definition of a genuine photoperiodic effect, which will be used to test all daylength effects for photoperiodic properties.

Algae

Daylength effects in algae

It is well known that the rates of growth, the phases of reproduction, and even the morphology and chemical composition of algae may vary from season to season. These seasonal variations have been extensively studied both in fresh-water and in marine habitats, and have most often been correlated with, and attributed to, similar variations in light intensity, temperature and nutrient supply. The influence of daylength has rarely been invoked as a causative factor in this seasonal variation, in spite of its recognized importance in the control of seasonal changes in flowering plants, and the possible existence of photoperiodic responses among the algae has been largely ignored. Indeed, it has been argued (for example, see Bünning, 1964), that in aquatic habitats, the temperature is sufficiently stable to serve as a reliable indicator of the time of year (in contrast to the situation in terrestrial habitats), and that the ability to respond to changes in daylength will carry little ecological advantage. However, this argument can only be applied satisfactorily to the phytoplankton, and to sublittoral or benthic algae in large well-mixed bodies of water. There are many other aquatic habitats in which the temperature is far from stable. All littoral marine algae, for instance, are subjected to extreme diurnal and tidal temperature fluctuations, which must rule out the use of temperature as a seasonal indicator. Diurnal and 'unseasonable' temperature fluctuations may also occur in small bodies of fresh water and in the surface layers of stratified lakes, which are probably only slightly less extreme than those of the surrounding terrestrial habitats. There seems little reason to suppose, therefore, that photoperiodic responses would be ecologically unnecessary or superfluous, at least in those algal species which are to be found in such habitats.

As a result of culture work in controlled conditions of light intensity, daylength and temperature, a few examples of seasonal variations in growth or reproductive activity have been attributed to changes in daylength; a list of these daylength effects is given in Table 12.1. Some of the responses listed were the result of casual observations made during the course of work quite unrelated to photoperiodism (for example, see Føyn, 1955; Fries, 1963; Provasoli, 1964), and few details are, therefore, available. Other responses were the main object of the investigation, and some have been described as photoperiodic (for example, see Hygen, 1948; Karling, 1924), although the strict test of photoperiodism has rarely been applied.

The life cycle of the edible red seaweed, *Porphyra tenera*, which is extensively cultivated in Japan (Kurogi, 1961) includes a macroscopic thallus phase, known as 'laver' or 'nori', and a microscopic filamentous phase, the conchocelis phase, which ramifies through the calcareous shells of oysters and other molluscs on the sea-bed. The thallus phase grows most luxuriantly during the winter months, and disappears almost completely by the end of May. Its reappearance in early autumn (mid-September to early November) is due to the germination of conchospores liberated by the conchocelis phase at this season. The formation of conchosporangia and the liberation of conchospores are stimulated by SD conditions (Kurogi, 1959). This response to SD can be inhibited by a short light break given in the middle of a long dark period (Dring, 1966, 1967a; Rentschler, 1967), and can also be induced by a relatively short exposure to SD conditions (Dring, 1967a). This was the first clear demonstration of a photoperiodic response in an alga, and has been the subject of detailed physiological investigation, the results of which will be considered in the next subsection. Two subsequent demonstrations of algal photoperiodic responses have also involved species of red algae. *Bangia fuscopurpurea* appears to exhibit a SD response which is remarkably similar to that of *Porphyra tenera*. The life history of this alga also includes a conchocelis phase, and, again, it is the formation of conchosporangia (or 'fertile cell rows', as they were originally called by Drew, 1954) by the conchocelis phase which appears to be under photoperiodic control (Richardson and Dixon, 1968). Conchosporangia were only observed when the conchocelis phase was grown in daylengths of 12 hours or less, but their formation was completely inhibited by 1 hour of light given in the middle of a 15-hour dark period.

A different type of SD response has been reported in the sublittoral red alga *Constantinea subulifera* (Powell, 1969, and personal communication). The thallus of this alga consists of an erect stipe, which is terminated by a

Table 12.1. The responses of algae to changes in daylength

Alga	Response to SD	Response to LD	Reference
Chlorophyta			
Acetabularia cremulata	Increased growth in high light Increased chlorophyll content	Increased rate of cap formation	Terborgh and Thimann (1964, 1965)
Chara fragilis	Vegetative	Sexual reproduction	Karling (1924)
Pediastrum duplex	—	Increased growth and development	Ermolaeva (1960)
Stigeoclonium amoenum	—	Zygote formation	Abbas and Godward (1963)
Ulothrix flacca	Asexual reproduction	Sexual reproduction	Hygen (1948)
Ulva lactuca	Germination of zygotes	Dormancy of zygotes Inhibition of growth of Mediterranean form	Føyn (1955)
Bacillariophyta			
Biddulphia aurita	—	Inhibition of growth in high light	Castenholz (1964)
Coscinodiscus concinnus	Auxospore formation	—	Holmes (1966)
Fragilaria striatula	Disruption of filaments	Increased growth	Castenholz (1964)
Melosira moniliformis	—	Inhibition of growth in high light	Castenholz (1964)
Melosira nummuloides	Auxospore formation	(intermediate daylengths)	Bruckmayer-Berkenbusch (1955)
Stephanopyxis palmeriana	Resting spores	Sexual reproduction	Steele (1965)
Synedra tabulata	—	Increased growth	Castenholz (1964)

Table 12.1. (*continued*)

Alga	Response to SD	Response to LD	Reference
Xanthophyta			
Vaucheria sessilis	—	Gametangium formation	League and Greulach (1955)
Phaeophyta			
Ectocarpus siliculosus	—	Increased ratio of unilocular to plurilocular sporangia	Müller (1962)
Scytosiphon lomentaria	Unilocular sporangium formation	—	Tatewaki (1966)
Rhodophyta			
Acrochaetium pectinatum	Tetrasporangium formation		West (1968)
Acrochaetium sp.	'Spore' formation		Fries (1963)
Bangia fuscopurpurea	Conchosporangium formation		Richardson and Dixon (1968)
Constantinea subulifera	Initiation of new thallus blade		Powell (1969)
Goniotrichum elegans	'Spore' formation		Fries (1963)
Porphyra tenera	Conchosporangium formation		Kurogi (1959)
		Carpospore formation; inhibition of thallus growth	Iwasaki (1961)
Porphyra unbilicalis	—	Conchosporangium formation	Kurogi and Sato (1967)
Porphyra sp.	Conchosporangium formation		Kurogi and Sato (1962b)
Rhodochorton purpureum	Tetrasporangium formation		West (1967, 1969)
Rhodochorton tenue	Tetrasporangium formation		West (1967, 1969)
Spyridia sp.	Inhibition of growth	Germination of tetraspores	Provasoli (1964)

horizontal circular blade. Growth is continued by the development of a new cylindrical stipe from the centre of the old blade, which in turn expands into a new horizontal blade. In natural conditions, each stipe initiates one new blade in the autumn of each year, and this response can be induced artificially by exposing plants to daylengths of less than about 13 hours. In daylengths greater than this, or in continuous light (even at light intensities as low as 100 lux), the elongation of the new stipe continues indefinitely, but no blade expansion occurs. Interruption of a long dark period by a 15-minute light break will suppress the response to SD.

Although the property of induction—(4) in the introductory definition—has not yet been demonstrated for these two SD responses, the inhibitory effect of light break treatment justifies their provisional inclusion as genuine photoperiodic responses, and both are receiving detailed investigation.

The ability of a light break given during a long dark period to stimulate a LD response has been tested for three of the LD responses listed in Table 12.1, but a negative result was obtained in each case. The species involved were the unicellular green alga *Acetabularia crenulata*, the xanthophyte *Vaucheria sessilis* and the filamentous brown alga *Ectocarpus siliculosus*. It was concluded that the responses to daylength in these algae resulted from the different total energies of light given in the various treatments, rather than to the different periods of the light treatments. Thus, an apparent stimulation of growth or reproductive activity by LD could be attributed to the increased availability of food reserves resulting from increased photosynthesis. This interpretation was confirmed directly for the LD stimulation of gametangium formation in *Vaucheria sessilis*, since the addition of glucose and meat peptone to the medium of plants grown in SD conditions resulted in earlier and more abundant gametangium formation, similar to that obtained in LD conditions (League and Greulach, 1955). Similar interpretations probably also apply to the LD responses of *Pediastrum* and *Stigeoclonium*, and to the diatom responses reported by Castenholz (1964), although none of these examples have been investigated in detail.

The results obtained by Karling (1924) in his early work on the induction of sexual reproduction in *Chara fragilis* by LD conditions, indicate that this response is not dependent on photosynthetic activity. In one of his experiments, plants exposed to continuous artificial light of about 100 lux produced sex organs after only 4 days of treatment, whereas comparable plants grown in natural SD in a greenhouse remained vegetative for several weeks. Other results of his also suggest that this may be a genuine, and rapid, photoperiodic response, but the experimental conditions were not sufficiently well controlled to establish this in the original investigation.

Morphological or reproductive responses to SD cannot be explained away quite so simply as can LD responses, but high light intensities over long periods may cause inhibition of photosynthesis in certain organisms, thus producing an apparent SD response. Castenholz (1964) observed such inhibition in the diatoms *Biddulphia* and *Melosira moniliformis*, and this interpretation could, therefore, account for the SD stimulation of auxospore formation in the centric diatoms *Coscinodiscus* and *Stephanopyxis*. The results for *Coscinodiscus* (Holmes, 1966), for example, indicate that SD conditions extend the range of temperatures and light intensities in which auxospore formation can occur, rather than producing a specific inductive effect. This suggests that high light for long periods has an inhibitory effect similar to that observed on the growth of *Biddulphia* and *Melosira*. The apparent response of *Melosira nummuloides* to intermediate daylengths could also be explained in this way. Auxospore formation in this diatom was observed to be maximal in 15-hour days, minimal in continuous light and intermediate in 5- and 10-hour days (Bruckmayer-Berkenbusch, 1955). These results suggest that auxospore formation is proportional to daylength, but that continuous conditions are inhibitory.

The SD stimulation of tetrasporangium formation in the tetrasporophytes of the red algae *Acrochaetium pectinatum* (West, 1968) and *Rhodochorton purpureum* (West, 1967, 1969, and personal communication) cannot be interpreted on the basis of the total light energies received, since a range of light intensities was investigated at each daylength. The total light energy given in some of the SD treatments which resulted in tetrasporangium formation was as great as, or greater than, that given in some of the LD treatments in which no sporangia were formed. These results indicate that the SD responses of these plants are dependent solely upon the length of the light period, and that they should therefore be classified as photoperiodic responses. However, the effects of interrupting a long dark period by a 15-minute light break (intensity of 1250 lux) were also investigated, and sporangium formation was found to be unaffected by this treatment. This suggests that the responses are not photoperiodic. However, *Porphyra tenera* has been found to require a light break of one hour at 1000 lux before the response to SD is completely inhibited (see page 356), and, thus, a possible explanation of this apparent paradox is that the light breaks used in the investigation were insufficiently long or insufficiently intense to cause a complete inhibition of the response to SD. The use of light breaks of greater total light energy might, therefore, confirm the photoperiodic nature of these responses.

SD-reproductive responses similar to that in *Porphyra tenera* have been observed in the conchocelis phases of several other species of *Porphyra* (*P. kuniedai, P. yezoensis, P. pseudolinearis* and *P. angusta*; Kurogi and Sato,

1962b), which show a similar seasonal behaviour. In *P. umbilicalis*, however (the thallus phase of which grows most luxuriantly during the summer months in Japan), conchosporangium formation by the conchocelis phase appears to be stimulated by LD conditions (Kurogi and Sato, 1967). None of these responses has been tested for the properties of induction or light-break activity, although rough values for the critical daylengths have been determined. Conchosporangium formation is reported to occur in day-lengths shorter than 10 hours in *P. pseudolinearis*, 12 hours in *P. yezoensis* and *P. angusta*, and 13 hours in *P. kuniedai*, and in daylengths longer than 12 hours in *P. umbilicalis*. The conchocelis phases of three American species of *Porphyra* were cultured in 10-hour days by Krishnamurthy (1969). The reproductive response was found to vary from species to species, but since no other daylengths were investigated it cannot yet be concluded that these species vary in their photoperiodic behaviour. It is probable that these responses will prove to be genuine photoperiodic responses of the same type as that in *P. tenera*, but they cannot be accepted as such until the appropriate experiments have been done. It has also been reported (Iwasaki, 1961) that LD conditions inhibit the growth of the leafy thallus phase of *Porphyra tenera* and induce carpospore formation. The latter observation appears to be at variance with the observed behaviour of the plant in nature, since spore production may occur at any time between November and May (Kurogi, 1961), and the poor stunted growth of the thalli observed in this investigation suggests that this may have been an artefact produced by the culture conditions. An attempt to reproduce the response and subject it to light-break treatment was unsuccessful since it resulted in similarly poor growth of the thalli and little or no spore production (Dring, unpublished). The photoperiodic response of the thallus phase of *Porphyra tenera* thus requires further confirmation.

The filamentous green alga *Ulothrix flacca* forms biflagellate zoospores, whose morphology is unaffected by daylength. In LD conditions, however, the zoospores fuse in pairs to form globular zygotes, whereas in SD they behave as asexual spores and develop directly into new filaments (Hygen, 1948). The germination of the zygotes formed in LD also appears to be affected by daylength, since they remain dormant for as long as LD conditions continue, but germinate as soon as they are transferred to a SD regime. These responses are difficult to attribute to quantitative differences in the light treatments, and so could also prove to be genuine photoperiodic responses. The remaining SD responses listed in Table 12.1—in the red algae *Acrochaetium*, *Goniotrichum* (Fries, 1963) and *Spyridia* (Provasoli, 1964), and in the brown alga *Scytosiphon* (Tatewaki, 1966)—are all the result of casual observations made during the course of culture work on these plants.

Consequently, there is little data available on which to base an assessment of their photoperiodic status, and no useful conclusions can be drawn at present.

Only three of the reported effects of daylength on algae can, therefore, be accepted as genuine photoperiodic responses at present. These are the responses of *Porphyra tenera, Bangia fuscopurpurea* and *Constantinea subulifera*. There is evidence for most of the effects reported for other species of red algae which suggests that these too may be photoperiodic effects, and the responses of the various species of *Porphyra, Acrochaetium* and *Rhodochorton* especially deserve attention. In the other algal groups the most promising organisms would appear to be *Chara* and *Ulothrix*, and a reinvestigation of these two responses would be valuable.

The photoperiodic response of PORPHYRA TENERA

The stimulatory effect of SD conditions on conchospore production in the conchocelis phase of *Porphyra tenera* was first observed by Kurogi (1959), and has since been confirmed by several workers using a variety of culture conditions. The response has been obtained when the conchocelis is grown in its natural shell substrate, either in natural sea water with small additions of nitrate and phosphate only (Kurogi, 1959) or in an enriched sea water medium (Provasoli's SWII, 1964) buffered with 'tris' (trihydroxymethyl aminomethane) and chelated with EDTA (ethylene diamine tetraacetic acid; Rentschler, 1967). Conchocelis has also been grown free of a shell substrate (*in vitro*) in a variety of enriched sea water media and completely artificial media (Iwasaki, 1961), and the SD response was observed in all media which permitted the growth of the conchocelis. The work of Iwasaki (1961) and Rentschler (1967) was largely qualitative, but quantitative methods have been employed in the investigation of the response by Kurogi and co-workers, and by the present author. The extent of the reproductive response to various conditions of daylength and temperature has been estimated in shell cultures by the number of spores falling to the bottom of the culture vessel (Kurogi, 1959), and also, in *in vitro* cultures, by the number of conchosporangial branches produced per unit volume of conchocelis filaments (Dring, 1967a). Accurate estimation of the amount of conchocelis material in a culture, and hence of the rate of growth, has only been possible in the latter studies using *in vitro* cultures. The results described here are mainly those obtained by these methods (Dring, 1967a, 1967b, 1967c), but they are in broad agreement with those of other workers, to which reference is made where appropriate.

The original observations of Kurogi (1959) showed that, although spore counts in the thousands were recorded in SD conditions, small numbers of spores were also discharged in 15-hour days, and sporangial counts have confirmed that a few sporangia are formed in LD conditions and even in continuous light (Dring, 1967a). These results indicated that the response to SD was a quantitative stimulation of sporangium production, rather than an

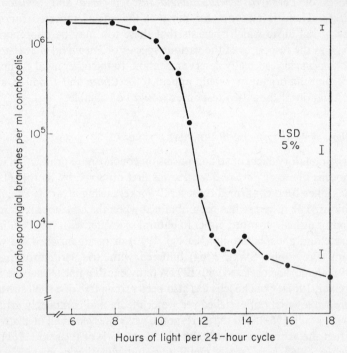

Figure 12.1. Effects of daylength on sporangium production at 20°c. Cultures received 25 cycles of 15 regimes from 6 : 18 to $\overline{18}$: $\overline{6}$ hours. Each point represents the mean of five samples from each of three cultures

'all or nothing' induction; the full extent of the response over a wide range of daylengths is shown in Figure 12.1. The quantitative nature of the response makes it impossible to define an exact critical daylength in excess of which no induction occurs, but the steepest section of the daylength-response curve (Figure 12.1) serves to indicate its range. There is a twenty-fold decrease in sporangium production when the daylength is increased from 11 to 12 hours at 20°c, and this range can be regarded as the most useful estimate of the critical daylength. This value can be compared with 10 to 12 hours at 20°c (Kurogi and Sato, 1962a) and 10 to 11 hours at 16 to

18°C (Rentschler, 1967). The apparent sharpness of the critical daylength is of interest since the thalloid liverwort *Lunularia* was found to have a poorly defined critical daylength (Schwabe and Nachmony-Bascombe, 1963); this suggested that lower plants might have a slightly less accurate timing mechanism than flowering plants.

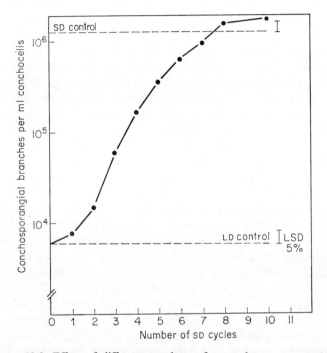

Figure 12.2. Effect of different numbers of SD cycles on sporangium production at 20°C. Cultures received 0–10 cycles of 8 : $\overline{16}$, and were then transferred to LD conditions (8 : $\overline{7\frac{1}{2}}$: 1 : $\overline{7\frac{1}{2}}$ hours). Controls received continuous SD or LD treatment. All cultures were sampled 25 days after the beginning of SD treatment. Each point represents the mean of five samples from each of two cultures

The ability of light-break treatment to simulate LD conditions was demonstrated by comparing sporangium production in cultures exposed to 8-hour days, with or without a further hour of light in the middle of the dark period. After 25 days of treatment, sporangia were found to be 200 times more abundant in the SD regime than in the interrupted night regime (control values, see Figure 12.2). The property of induction of the response is also illustrated in Figure 12.2. Cultures received between one and ten cycles of an 8-hour day before being returned to LD conditions. Seven to eight SD cycles induced the formation of as many sporangia as were formed after 25 days of continuous SD treatment, although no sporangia could be detected at the end of the short inductive treatments. Although seven to eight SD cycles are needed for 'full' induction—a state which is possibly reached when all the receptive sites have been converted to sporangial initials—significant induction will occur after only two SD cycles. *Porphyra* thus compares favourably with all but the few exceptional flowering plants (e.g. *Xanthium, Pharbitis*) which will respond to a single inductive cycle.

Since this demonstration, that the SD response of *Porphyra* was a genuine photoperiodic response of the type common among flowering plants, further work has attempted to determine the extent of the similarity between the physiology of this response and that of flowering plant responses. The photoreceptor system for the response was characterized by investigating the amount of light energy which is required to saturate the plant's response to a light break, and by determining the action spectrum of the response to light-break treatment. The system was found to be saturated by about $1 \cdot 8 \times 10^7$ ergs/cm^2 (i.e. 1000 lux for 1 hour, 1 lux $= 6$–8×10^6 cal/cm^2 minute; Westlake, 1965) of incandescent white light, and a highly significant linear correlation was obtained between the logarithm of the light energy used and the degree of inhibition of the response. Thus, reciprocity—the dependence of the response on the energy, rather than on the intensity or duration, of illumination—appeared to hold even for hour-long light treatments. This result agrees with those obtained from similar experiments on flowering plants (for example, Parker, Hendricks and Borthwick, 1950) and provides a further example of reciprocity in photobiological systems. Action spectrum experiments on the SD response of *Porphyra tenera* have been reported independently by Rentschler (1967) and by Dring (1967b). Both authors found that maximum inhibition was caused by red light (the peak wavelengths used by the two authors were 674 nm and 662·5 nm respectively), and that the effects of red light could be reversed by subsequent exposure to far-red light (at 753 and 737·5 nm respectively). It was inferred from these results that phytochrome is active as the photoreceptor pigment in this response, as in all of the flowering plant responses which have been investigated. This was the first demonstration of the presence of phytochrome in a red alga, or in any plant in which the photosynthetic pigments are not predominantly green. The implications of this discovery are discussed by Dring (1967b, 1967c) and the role of phytochrome in aquatic plants is considered more fully in the next subsection.

If photoperiod is to be more reliable than temperature as an indicator of the time of year it is essential that the timing mechanism involved should be unaffected by the changes in temperature which occur under natural conditions. This requirement implies that the critical daylength for any photoperiodic response should be little changed when measured in the full range of temperatures to which the organism may be subjected during those seasons when the daylength is close to the critical daylength. If the critical daylength of the *Porphyra* response is redefined as that daylength below which a rapid increase in sporangium production occurs with decreasing daylength (see Figure 12.1), a reasonable estimate for the critical daylength at 20°c would be 12·5 hours. The critical season is, therefore, that in which the daylength is reduced from 14 to 12 hours. Environmental data for Matsushima Bay, Japan (Iwaski, 1961; Kurogi, 1961), which is in the region from which the material used by the present author was originally obtained, indicate that when the daylength is 14 hours, the sea temperature ranges from 23·5°c at 0700 hours to 27°c at 1700 hours, and when the daylength has descreased to 12 hours, the temperature range is 20 to 21·5°c. A comparison of sporangium production at 20° and 25°c in a range of daylengths from 10 to 14 hours showed that there was no significant difference in either total sporangium production or critical daylength at the two temperatures (Dring, 1967c). When the same daylength range was investigated at 15°c, however, a ten-fold reduction in total sporangium production was observed, and the critical daylength was apparently reduced by about 1 hour. The observed reduction in sporangium production at 15°c agrees with the results obtained by Kurogi and Hirano (1956) and by Rentschler (1967) with this alga. It is difficult, however, to distinguish between the effects of low temperature on the reproductive process as a whole and on the timing mechanism of the photoperiodic response; the observed reduction in critical daylength requires confirmation. A low value for the critical daylength was also reported by Rentschler (1967), but since quantitative methods were not employed it is possible that this was due to a failure to observe small numbers of sporangia in the cultures. It is clear, therefore, that the critical daylength for the response is stable at the normal ecological temperatures of 20 to 25°c, which are also those optimal for sporangium production (Kurogi and Hirano, 1956), but below 20°c there is possibly a slight decrease in critical daylength with decreasing temperature. If endogenous rhythms are invoked as the basis for the measurement of time in photoperiodic responses, this observation would imply that the period of the rhythm increased with increasing temperature and, therefore, possessed a Q_{10} of less than unity. *Porphyra* would thus provide an exception to the general rule that endogenous rhythms have a Q_{10} slightly in excess of unity, but it is of interest that the rhythms of two other algae, *Gonyaulax* and *Oedogonium*, show a similar property (Bünning, 1964).

The possible involvement of endogenous rhythms in the timing mechanism of photoperiodic responses has been the subject of considerable controversy among investigators of flowering responses to daylength, and the experimental evidence is still somewhat confused (see the discussions in Cumming and Wagner, 1968; Hillman, 1967). This latter phrase can be applied equally well to the *Porphyra* response. The experimental approach to the problem has been to investigate the effects of light breaks given at different times during long dark periods (Dring, 1967c). When the total length of the dark period was 16 hours, maximum inhibition was obtained after 11 to 12 hours of darkness, and the resultant asymmetrical curve was comparable with similar curves obtained for *Xanthium* (Salisbury and Bonner, 1956), *Kalanchoë* (Bünning, 1964) and *Lemna* (Purves, 1961), which are best interpreted on the basis of a rhythmic response to light breaks. When a 40-hour dark period was used, however, maximum inhibition was observed after 4 and after 36 hours of darkness, with a marked stimulation of sporangium production by light-breaks given in the middle of the dark period (after 13 to 27 hours of darkness). These results are very similar to those obtained by Claes and Lang (1947) for *Hyoscyamus* and by Wareing (1954) for *Glycine*, which were convincingly interpreted on a non-rhythmic basis. Further experiments are required to elucidate the situation in *Porphyra*.

The effects of indole-3-acetic acid (IAA) and gibberellic acid on the reproduction of the conchocelis phase of *Porphyra tenera* have also been investigated (Dring, 1967c). Sporangium production in LD conditions was unaffected by IAA within the range of concentrations investigated (10^{-7} to 10^{-3} M), but in SD conditions high concentrations of IAA (10^{-3} M) caused a substantial inhibition of the reproductive response. This result agrees with many previous reports of the inhibition by auxin of the response of SD plants to inductive conditions (Doorenbos and Wellensiek, 1959), but appears to be the first report of an inhibitory effect of auxin on the reproduction of an alga (see Conrad and Saltman, 1962). Gibberellic acid was found to have little effect on the response of the conchocelis to either LD or SD conditions, apart from a slight inhibition of sporangium production in SD. This, again, is in agreement with results obtained with SD flowering plants (for example, see Harder and Bünsow, 1958; Lona, 1956).

Conclusions and implications

The investigation of the photoperiodic response of *Porphyra tenera* indicates that the physiological mechanisms involved in this response are remarkably similar to those which are thought to be involved in flowering plant responses. The characteristics which the *Porphyra* response appears to have in common with flowering plant responses include: the response to light break treatment, the property of induction, the clear-cut critical day-length, reciprocity and the activity of phytochrome in the photoreceptor system, independence of temperature, and the response to auxin treatment —and, even, an equivocal response to rhythmic experiments! These many similarities encourage the assumption that the physiology of photoperiodism in *Porphyra* is basically identical to that in the flowering plants. If the mechanism of photoperiodism can be so similar in plants which are as distinct, morphologically and phylogenetically, as a filamentous red alga and a flowering plant, the phenomenon seems likely to be of general occurrence throughout the algae, rather than being confined to the 'higher' or even to 'green' plants. This conclusion is supported by the results obtained in the preliminary investigations of the SD responses of *Bangia fuscopurpurea* and *Constantinea subulifera*. This neat and acceptable situation is in danger of being upset, however, if the inability of light-break treatment to inhibit the SD responses of *Acrochaetium pectinatum* and *Rhodochorton purpureum* is confirmed. In the present state of our knowledge about the physiology of photoperiodism, it is difficult to conceive of a mechanism which would not respond to light-breaks. Any mechanism mediated by phytochrome is likely to respond to short exposures to light because of the low light saturation of the pigment, its photoreversible properties, and because of the apparent metabolic activity of P_{FR}. Also, it is generally postulated that any mechanism involving an endogenous rhythm will respond to light-breaks as a result of an interaction between the light-break and the phase of the rhythm at the time of the light-break. It is clear, then, that a photoperiodic response that does not respond to light-breaks must involve a very different mechanism from that in the flowering plants—and, apparently, that in *Porphyra*—and it is profitless to speculate further on the possibilities until more detailed results are available.

The role of phytochrome in algae deserves some comment here. The activity of the pigment has been inferred, as a result of physiological studies, in only four species of algae: the green algae *Mougeotia* (Haupt, 1959) and *Mesotaenium* (Haupt and Thiele, 1961), the dinoflagellate *Gyrodinium dorsum* (Forward and Davenport, 1968) and the red alga *Porphyra tenera* (Dring, 1967b; Rentschler, 1967). The pigment has so far only been extracted from one alga, *Mesotaenium* (Taylor and Bonner, 1967). The small number of such reports of phytochrome activity may be due to the fact that, as pointed out by West (1968), the ecological significance of the pigment must be rather limited in aquatic plants by the absorption characteristics of water. As is well known, phytochrome absorbs most strongly at the long wavelength end of the visible spectrum, but it is also in this region

Table 12.2. Sensitivity of phytochrome system in various photoperiodic responses

Plant	Response	Energy of 660 nm light required (ergs/cm^2)	Reference
Xanthium	50 per cent inhibition of flowering	$3 \cdot 0 \times 10^4$	Parker and coworkers, 1946
Hordeum	50 per cent promotion of flowering	$4 \cdot 0 \times 10^4$	Borthwick and coworkers, 1948
Chenopodium	50 per cent inhibition of flowering	$1 \cdot 5 \times 10^3$	Kasperbauer and coworkers, 1963
Lunularia	Approximately 50 per cent inhibition of growth	$5 \cdot 4 \times 10^4$	Wilson and Schwabe, 1964
Porphyra	50 per cent inhibition of conchosporangium production	$3 \cdot 24 \times 10^5$	Dring, 1967b

that water absorbs most strongly. Although phytochrome is notoriously sensitive to low energies of red light, there is a limit below which it appears not to respond. The magnitude of this limit is indicated by the figures for a variety of plant species given in Table 12.2. The average energy of 660 nm radiation required to produce a 50 per cent response in these species is $5 \cdot 5 \times 10^4$ ergs/cm^2, so that the minimum energy for stimulation of a phytochrome-controlled response is probably of the order of $1 \cdot 0 \times 10^4$ ergs/cm^2. Solar radiation at zero air mass contains $0 \cdot 159$ w/cm^2 μ at 660 nm (Johnston, 1954). This energy will be subject to a 20 per cent attenuation by passage through the atmosphere (Gates, 1962). Assuming that this energy will be available for 10 hours per day, and that phytochrome will absorb all wavelengths in a 50 nm band on either side of 660 nm equally, the maximum energy available for absorption by phytochrome at the surface of the sea under a zenith sun will be in the order of $5 \cdot 0 \times 10^9$ ergs/cm^2. This energy will be reduced to $1 \cdot 0 \times 10^4$ ergs/cm^2 by a transmission of $2 \cdot 0 \times 10^{-4}$ per cent.

According to the transmission data given by Jerlov (1968), insufficient radiation at 660 nm—even to satisfy these low energy requirements—will penetrate to the limits of the photic zone in oceanic waters. Since oceanic water is only found outside the continental shelf, where the depth exceeds 200 m, however, benthic algae will only be found in coastal waters, in which variable quantities of suspended matter and 'yellow substance' will restrict the penetration of all wavelengths. The lower limit of the photic zone will, therefore, occur at much shallower depths than in oceanic waters, and small energies of 660 nm radiation will penetrate to this limit (e.g. $0 \cdot 75$ per cent of surface energy in type 9 water; $0 \cdot 37$ per cent in type 7; $0 \cdot 12$ per cent in type 5; $0 \cdot 0082$ per cent in type 3; $2 \cdot 9 \times 10^{-4}$ per cent in type 1; Jerlov, 1968). In all but the clearest coastal waters (type 1), therefore, there will be sufficient energies of 660 nm radiation throughout the photic zone to activate the phytochrome system. In littoral environments, however, the situation will be complicated by tidal variations in the depth of water overlying an individual plant. These variations will result in considerable fluctuations in the ratio of red to far red radiation (e.g. the ratio of 660 to 730 nm radiation, on a quantum basis, will be $1 \cdot 07$ in full sunlight, $114 \cdot 1$ at 5 m, and $12 \; 148 \cdot 0$ at 10 m; data from Jerlov, 1968). Since this red/far red ratio is important in determining the photostationary state of phytochrome (Cumming, Hendricks and Borthwick, 1965), fluctuations in the ratio may affect the physiological response of the plant. The photoperiodic responses of littoral algae are, however, unlikely to be influenced by these fluctuations, since the ratio can never decrease below its value in full sunlight; the fluctuations will always serve, therefore, to reinforce the light–dark rhythm of the terrestrial environment.

West (1968) argues that a red/far red reversible photoperiodic process is unlikely to be operative in sublittoral algae because neither type of light is present at depths where they grow, and that, if phytochrome systems are present, they must operate through the absorption of blue light. The calculations outlined above indicate that in most littoral or sublittoral environments there is sufficient radiation at 660 nm for the phytochrome system to operate by a mechanism similar to that in flowering plants. Benthic algae growing in relatively deep, clear waters, however, will be exposed to much higher energies of blue than of red radiation (e.g. below 20 m type 1 coastal water, the ratio of 450 to 660 nm radiation, on a quantum basis, will exceed 250), and it is possible that, here, the minor absorption bands, which both forms of phytochrome exhibit in the blue region of the visible spectrum (Butler, Hendricks and Siegelman, 1965), may assume a greater significance than in terrestrial or shallow water environments. It is these bands which are now thought to be responsible for the effects of prolonged or high-intensity radiation with blue (450 nm) light, known as the high energy reaction (HER: Hartmann, 1966; Hillman, 1967). There is evidence in the action spectrum results obtained by Rentschler (1967) and by Dring (1967b) for the inhibitory effects of light-breaks on sporangium production in *Porphyra tenera*, that red algal phytochrome is also sensitive to high energies of blue radiation. Dring used light breaks of low total energy (about 3.24×10^5 ergs/cm^2) and observed 50 per cent inhibition at 662·5 nm and no inhibition at 447 nm, whereas Rentschler, using higher total energies of up to 1.44×10^7 ergs/cm^2, observed almost 100 per cent inhibition at 674 nm and a secondary peak of 75 per cent inhibition at 447 nm. The latter treatments may have involved light energies high enough to activate an HER-type response, whereas the former clearly did not. The energy of red light which is required to achieve a 50 per cent inhibition of the photoperiodic response in *Porphyra* is compared with similar published values for the responses of higher plants in Table 12.2. Although the value for *Porphyra* is an order of magnitude higher than that for *Xanthium*, the difference between them is no greater than the differences between the values for different flowering plant responses. The system measured by Dring would seem, therefore, to be a 'low energy response', and a response to blue light would not be expected. Further detailed investigations of the action spectra of the known photoperiodic responses of red algae are clearly required in order to test the hypothesis that the phytochrome system of benthic algae is able to respond to high energies of blue radiation in their natural habitats.

The simple morphology of the conchocelis phase of *Porphyra*—the thallus is little more than an irregularly branched mass of filaments, in which sporangia may appear apparently at random—indicates that the high level of organization of the flowering plants is not essential for a photoperiodic response. This conclusion is supported by the observation that small fragments of the thallus of the liverwort *Lunularia* appear to respond as readily to photoperiod as do whole plants (Schwabe and Nachmony-Bascombe, 1963). These two responses, therefore, suggest that both the perception of, and the response to, photoperiod can be accomplished at the cellular level of organization, and that flowering responses are merely a particularly obvious expression of a more basic response of plant cells to photoperiod.

The current, almost general, acceptance of the hypothesis that circadian rhythms are implicated in photoperiodic time measurement (Cumming and Wagner, 1968; Hillman, 1967; Sweeney, 1969; Ehret and Wille, Chapter 13) carries the corollary that organisms which do not exhibit circadian rhythms cannot respond to photoperiod. No circadian rhythms have as yet been reported for any prokaryotic organisms (Chapter 13, p. 371), and this fact must cast some doubt on the possible existence of photoperiodic responses in blue-green algae. There was, however, no evidence for the existence of rhythms in red algae before the discovery of the photoperiodic responses discussed above, and a clear demonstration of a circadian rhythm in a red alga is still awaited. In conclusion, there seems little reason to suppose that photoperiodic responses will not be found in most of the major groups of algae, with the possible exception of the Cyanophyta.

The discussion of daylength responses (see page 346) has already indicated the most promising organisms for investigation—*Chara*, *Ulothrix*, the littoral species of red algae and possibly the diatoms—but reports of seasonal responses abound in the older algal literature, and many of these might repay photoperiodic investigation. Indeed, if such reports were followed up, and if critical screening of new algal cultures for photoperiodic effects were to become standard practice, the rate of discovery of algal photoperiodic responses might well be as rapid as that of flowering plant responses following the publication of Garner and Allard's classic paper in 1920.

Among the general advantages of algae as experimental organisms in physiological research, several features are particularly valuable in the study of photoperiodism. The first and most important of these is the simplicity of the responses which are likely to be found relative to the massive and fundamental change from a vegetative to a floral apex, which is being studied in flowering responses. Secondly, the simple morphology of most algae enables the physiological complications, which are involved in such processes as translocation in flowering plants, to be avoided, and also

makes it easier to ensure that all parts of the plant receive uniform treatment. Finally, the quantitative nature of the response in *Porphyra*, which
can also be expected in other algal responses, means that the effects of
marginally inductive treatments can be assessed more accurately than they
can with flowering responses, since the problems involved with the measurement of flowering (see the discussions in Hillman, 1962; Salisbury, 1963) are
largely avoided. Photoperiodic studies on algae could, therefore, make a
valuable contribution to our general understanding of the phenomenon of
photoperiodism, and it is hoped that their potential will soon be more
widely realized and exploited.

Fungi

Various species of the larger fungi, particularly among the Basidiomycetes
and Discomycetes, produce their fruiting bodies at distinct seasons of the
year, and the appearance of individual species may be limited to short
periods. The morel season, for example, only lasts for about a month
(Alexopoulos, 1952). Several important plant parasites have complex life
histories, the different stages of which occur exactly in phase with the corresponding developmental stages of their host plants. These seasonal cycles
of reproduction and development are usually thought to be controlled by
the temperature and humidity of the environment, although parasitic or
mycorrhizal species may obtain additional seasonal information from their
host by intercepting the translocation pathways of endogenous growth
substances. So far there is no evidence that any fungal processes are under
photoperiodic control. As far as the author is aware, only one effect of
daylength has been reported for a fungus. When cultures of *Phymatotrichum
omnivorum* were grown on a defined medium containing calcium in white or
blue light, sporulation was observed in 16-hour days, but not in 8-hour days
(Woods, Bloss and Gries, 1967). However, subsequent more detailed
investigation has shown that maximum sporulation occurs in 20- and 24-
hour days, and it is thought that the fungus has a requirement for a specific
amount of light energy, rather than a strictly photoperiodic requirement
(Baniecki, 1968). Light requirements of this type are common among the
fungi, and the specific requirement may even be for a period of light followed
by a period of darkness (see Carlile, Chapter 11 of this treatise). However,
specific requirements for a regular alternation of light and dark periods of
well-defined length appear to be lacking.

Two features of the general biology of the fungi suggest that photoperiodic responses will not be common in these organisms. A large number
of fungal species have a cosmopolitan distribution and are not confined to

specific latitudes, as are the majority of autotrophic plants. Distributions of this type are incompatible with photoperiodism. In addition, the majority of fungal responses, including complex reproductive responses, occur at great speed and may be complete within 24 hours of receiving the appropriate stimulus. They are thus ideally adapted to respond to and exploit a suddenly favourable situation, and need not rely on accurate seasonal information which merely indicates the average conditions which should prevail over an extended period.

Bacteria

There is little indication that bacteria are susceptible to changes in season. In the present state of our knowledge of bacterial ecology it seems unnecessary to postulate the existence of photoperiodic responses among the bacteria.

Acknowledgements

I thank Drs. H. E. Bloss, J. Powell and J. A. West for their permission to refer to material which is unpublished or in the press; Dr. M. J. Carlile for valuable discussion about fungal responses; and Professor E. W. Simon for criticism of the manuscript.

REFERENCES

Abbas, A., and M. B. E. Godward (1963). Effects of experimental culture in *Stigeoclonium*. *Br. Phycol. Bull.*, **2**, 281.

Alexopoulos, C. (1952). *Introductory Mycology*. Wiley, New York. 482 pp.

Baniecki, J. F. (1968). Influence of chemical and physical factors upon sporulation of *Phymatotrichum omnivorum*. Ph.D. Diss., University of Arizona.

Borthwick, H. A., S. B. Hendricks, and M. W. Parker (1948). Action spectrum for photoperiodic control of floral initiation of a long-day plant, Wintex barley (*Hordeum vulgare*). *Bot. Gaz.*, **110**, 103.

Bruckmayer-Berkenbusch, H. (1955). Die Beeinflussung der Auxosporenbildung von *Melosira nummuloides* durch Aussenfaktoren. *Arch. Protistenk.*, **100**, 183.

Bünning, E. (1964). *The Physiological Clock*, Springer-Verlag, Berlin. 145 pp.

Butler, W. L., S. B. Hendricks, and H. W. Siegelman (1965). Purification and properties of phytochrome. In T. W. Goodwin (Ed.), *Chemistry and Biochemistry of Plant Pigments*, Academic Press, London, New York. pp. 197–210.

Castenholz, R. W. (1964). The effect of daylength and light intensity on the growth of littoral marine diatoms in culture. *Physiologia Pl.*, **17**, 951.

Claes, H., und A. Lang (1947). Die Blutenbildung von *Hyoscyamus niger* in 48-stündigen Licht-Dunkel-Zyklen und in Zyklen mit aufgeteilten Lichtphasen. *Z. Naturforsch.*, **2b**, 56.

Conrad, H. M., and P. Saltman (1962). Growth substances. In R. A. Lewin (Ed.), *Physiology and Biochemistry of Algae*, Academic Press, London, New York. pp. 663–671.

Cumming, B. G., S. B. Hendricks, and H. A. Borthwick (1965). Rhythmic flowering responses and phytochrome changes in a selection of *Chenopodium rubrum*. *Canad. J. Bot.*, **43**, 825.

Cumming, B. G., and E. Wagner (1968). Rhythmic processes in plants. *Ann. Rev. Pl. Physiol.*, **19**, 381.

Doorenbos, J., and S. J. Wellensiek (1959). Photoperiodic control of floral induction. *Ann. Rev. Pl. Physiol.*, **10**, 147.

Drew, K. M. (1954). Studies in the Bangioideae. III. The life-history of *Porphyra umbilicalis* (L.) Kütz. var. *laciniata* (Lightf.) J. Ag. *Ann. Bot.*, N.S., **18**, 183.

Dring, M. J. (1966). Preliminary studies on photoperiodism in *Porphyra tenera*. *Br. Phycol. Bull.*, **3**, 153.

Dring, M. J. (1967a). Effects of daylength on growth and reproduction of the conchocelis-phase of *Porphyra tenera*. *J. Mar. Biol. Ass.*, *U.K.*, **47**, 501.

Dring, M. J. (1967b). Phytochrome in red alga, *Porphyra tenera*. *Nature*, **215**, 1411.

Dring, M. J. (1967c). Photoperiodic studies on algae. *Ph.D. Thesis*, University of London.

Ermolaeva, L. M. (1960). The importance of daylength for the development of *Pediastrum*. *Bot. Zh.*, *S.S.S.R.*, **45**, 1069.

Forward, R., and D. Davenport (1968). Red and far-red light effects on a short-term behavioural response of a dinoflagellate. *Science*, **161**, 1028.

Føyn, B. (1955). Specific differences between northern and southern European populations of the green alga, *Ulva lactuca* L. *Pubbl. Staz. Zool. Napoli*, **27**, 261.

Fries, L. (1963). On the cultivation of axenic red algae. *Physiologia Pl.*, **16**, 695.

Garner, W. W., and H. A. Allard (1920). Effects of the relative length of day and night and other factors of the environment on growth and reproduction in plants. *J. Agric. Res.*, **18**, 553.

Gates, D. M. (1962). *Energy Exchange in the Biosphere*, Harper, New York, 151.

Harder, R., und R. Bünsow (1958). Uber die Wirkung von Gibberellin auf Entwicklung und Blütenbildung der Kurztagpflanze *Kalanchoë blossfeldiana*. *Planta*, **51**, 201.

Hartmann, K. M. (1966). A general hypothesis to interpret 'high energy phenomena' of photomorphogenesis on the basis of phytochrome. *Photochem. Photobiol.*, **5**, 349.

Haupt, W. (1959). Die Chloroplastendrehung bei *Mougeotia*. I. Uber den quantitativen und qualitativen Lichtbedarf der Schwachlichtbewegung. *Planta*, **53**, 484.

Haupt, W., und R. Thiele (1961). Chloroplastenbewegung bei *Mesotaenium*. *Planta*, **56**, 388.

Hillman, W. S. (1962). *The Physiology of Flowering*, Holt, Rinehart and Winston, New York. 164 pp.

Hillman, W. S. (1967). The physiology of phytochrome. *A. Rev. Pl. Physiol.*, **18**, 301.

Holmes, R. W. (1966). Short-term temperature and light conditions associated with auxospore formation in the marine centric diatom, *Coscinodiscus concinnus* W. Smith. *Nature*, **209**, 217.

Hygen, G. (1948). Fotoperiodiske reaksjoner hos alger. *Blyttia*, **6**, 1.

Iwasaki, H. (1961). The life-cycle of *Porphyra tenera in vitro*. *Biol. Bull. Mar. Biol. Lab.*, *Woods Hole*, **121**, 173.

Jerlov, N. G. (1968). *Optical Oceanography*, Elsevier, Amsterdam. 194 pp.

Johnson, F. S. (1954). The solar constant. *J. Meteorol.*, **11**, 431.

Karling, J. S. (1924). A preliminary account of the influence of light and temperature on growth and reproduction in *Chara fragilis*. *Bull. Torrey Bot. Club*, **51**, 469.

Kasperbauer, M. J., H. A. Borthwick, and S. B. Hendricks (1963). Inhibition of flowering of *Chenopodium rubrum* by prolonged far red irradiation. *Bot. Gaz.*, **124**, 444.

Krishnamurthy, V. (1969). The conchocelis phase of three species of *Porphyra* in culture. *J. Phycol.*, **5**, 42.

Kurogi, M. (1959). Influence of light on the growth and maturation of conchocelis-thallus of *Porphyra*. I. Effect of photoperiod on the formation of monsporangia and liberation of monospores (1). *Bull. Tohoku Reg. Fish. Res. Lab.*, **15**, 33.

Kurogi, M. (1961). Species of cultivated *Porphyras* and their life histories. (Study of the life history of *Porphyra* II). *Bull. Tohoku Reg. Fish. Res. Lab.*, **18**, 1.

Kurogi, M., and K. Hirano (1956). Influences of water temperature on the growth, formation of monosporangia and monospore liberation in the conchocelis-phase of *Porphyra tenera* Kjellm. *Bull. Tohoku Reg. Fish. Res. Lab.*, **8**, 45.

Kurogi, M., and S. Sato (1962a). Influences of light on the growth and maturation of conchocelis-thallus of *Porphyra*. II. Effects of different photoperiods on the growth and maturation of conchocelis-thallus of *Porphyra tenera* Kjellman. *Bull. Tohoku Reg. Fish. Res. Lab.*, **20**, 127.

Kurogi, M., and S. Sato (1962b). Influence of light on the growth and maturation of conchocelis-thallus of *Porphyra*. III. Effect of photoperiod in the different species. *Bull. Tohoku Reg. Fish. Res. Lab.*, **20**, 138.

Kurogi, M., and S. Sato (1967). Effect of photoperiod on the growth and maturation of conchocelis-thallus of *Porphyra umbilicalis* (L.) Kütz. and *Porphyra pseudocrassa* Yamada et Mikami. *Bull. Tohoku Reg. Fish. Res. Lab.*, **27**, 111.

Lang, A. (1965). Physiology of flower initiation. In W. Ruhland (Ed.), Encyclopedia of Plant Physiol., Vol. 15–1, Springer, Berlin, 1380.

League, E. A., and V. A. Greulach (1955). Effects of daylength and temperature on the reproduction of *Vaucheria sessilis*. *Bot. Gaz.*, **117**, 45.

Lona, F. (1956). L'azione dell'acido gibberellico sull'accrescimento caulinare di talune piante erbacee in condizioni esterne controllate. *Nuovo G. Bot. Ital.*, **63**, 61.

Müller, D. (1962). Uber jahres- und lunarperiodische Erscheinungen bei einigen Braunalgen. *Botanica Mar.*, **4**, 140.

Parker, M. W., S. B. Hendricks, and H. A. Borthwick (1950). Action spectrum for the photoperiodic control of floral initiation of the long-day plant *Hyoscyamus niger*. *Bot. Gaz.*, **111**, 242.

Parker, M. W., S. B. Hendricks, H. A. Borthwick, and N. J. Scully (1946). Action spectrum for the photoperiodic control of floral initiation in short-day plants. *Bot. Gaz.*, **108**, 1.

Powell, J. (1969). *J. Phycol.* (In the press.)

Provasoli, L. (1964). Growing marine algae. *Proc. 4 Int. Seaweed Symp.*, 9.

Purves, W. K. (1961). Dark reactions in the flowering of *Lemna perpusilla* 6746. *Planta*, **56**, 684.

Rentschler, H. (1967). Photoperiodische Induktion der Monosporenbildung bei *Porphyra tenera* Kjellm. (Rhodophyta–Bangiophyceae). *Planta*, **76**, 65.

Richardson, N., and P. S. Dixon (1968). Life-history of *Bangia fuscopurpurea* (Dillw.) Lyngb. in culture. *Nature*, **218**, 496.

Salisbury, F. B. (1963). *The Flowering Process*, Pergamon Press, Oxford. 234 pp.

Salisbury, F. B., and J. Bonner (1956). The reactions of the photoinductive dark period. *Pl. Physiol.*, **31**, 141.

Schwabe, W. W., and S. Nachmony-Bascombe (1963). Growth and dormancy in *Lunularia cruciata* (L.) Dum. II. The response to daylength and temperature. *J. Exp. Bot.*, **14**, 353.

Steele, R. L. (1965). Induction of sexuality in two centric diatoms. *Bioscience*, **15**, 298.

Sweeney, B. M. (1969). *Rhythmic Phenomena in Plants*, Academic Press, London. 147 pp.

Tatewaki, M. (1966). Formation of a crustaceous sporophyte with unilocular sporangia in *Scytosiphon lomentaria*. *Phycologia*, **6**, 62.

Taylor, A. O., and B. A. Bonner (1967). Isolation of phytochrome from the alga, *Mesotaenium*, and liverwort. *Sphaerocarpus*. *Pl. Physiol.*, **42**, 762.

Terborgh, J., and K. V. Thimann (1964). Interactions between daylength and light intensity in the growth and chlorophyll content of *Acetabularia crenulata*. *Planta*, **63**, 83.

Terborgh, J., and K. V. Thimann (1965). The control of development in *Acetabularia crenulata* by light. *Planta*, **64**, 241.

Tournois, J. (1912). Influence de la lumière sur la floraison du houblon japonais et du chanvre. *Compt. Rend. Acad. Sci. (Paris)*, **155**, 297.

Wareing, P. F. (1954). Experiments on the 'light-break' effect in short-day plants. *Physiologia Pl.*, **7**, 157.

West, J. A. (1967). The life-histories of *Rhodochorton purpureum* and *Rhodochorton tenue* in culture. *J. Phycol.*, **3** (Suppl.), 11.

West, J. A. (1968). Morphology and reproduction of the red alga *Acrochaetium pectinatum* in culture. *J. Phycol.*, **4**, 89.

West, J. A. (1969). The life histories of *Rhodochorton purpureum* and *R. tenue* in culture. *J. Phycol.*, **5**, 12.

Westlake, D. F. (1965). Some problems in the measurement of radiation under water: a review. *Photochem. Photobiol.*, **4**, 849.

Wilson, J. R., and W. W. Schwabe (1964). Growth and dormancy in *Lunularia cruciata* (L.) Dum. III. The wavelengths of light effective in photoperiodic control. *J. Exp. Bot.*, **15**, 368.

Woods, R., H. E. Bloss, and G. A. Gries (1967). Induction of sporulation of *Phymatotrichum omnivorum* on a defined medium. *Phytopath.*, **57**, 228.

CHAPTER 13

The photobiology of circadian rhythms in protozoa and other eukaryotic microorganisms

CHARLES F. EHRET

Division of Biological and Medical Research, Argonne National Laboratory, Argonne, Illinois

JOHN J. WILLE

Department of Biological Sciences, University of Cincinnati, Cincinnati, Ohio

(*This work was supported under the auspices of the U.S. Atomic Energy Commission.*)

Prologue

In a billion years of evolution no single factor has been more influential in shaping life's options of design than the light of the sun. In water, on land, and wherever life exists the rays of sunlight have left their mark. Yet, as we scan the macromolecular consequences in our mind's eye, the marvellous catalogues of photosynthetic and photodynamically active pigments, of energy levels and excited states, and of photolabile and stable structures, another obvious and interestingly complementary factor, *darkness*, comes to mind, demanding now (just as before, in the late Cryptozoic eon) equal time with light. Peer now into a chic salon at midnight where a tiny kitten purrs a claw-sheathed sleep upon her weary mistress' lap and when every prodigal human has found his own abode for slumber, and see awesome evidence become commonplace by its familiarity for this daily complementarity. Thus, have cosmic algorithms of darkness and light shaped earth's diurnal landscape into precisely bounded temporal niches to be richly exploited in our own Phanerozoic eon by creatures and their codes.

The eukaryotic–circadian principle

Biological data of especially the last twenty years permit and encourage the following assertion: that circadian clocks are limited to eukaryotes, that all eukaryotes have regulatory capacities for circadian time-keeping, that these regulatory capacities are invariably turned-on in the slow (infradian) growth mode, and that they can be set by light. In other words one can, and we do, claim these properties as inescapable and general ones that manifest a fundamental underlying principle of existence (the eukaryotic–circadian principle) for all living systems of higher order than the prokaryotes.

When they were first discussed as early as thirty years ago (Jennings, 1939; Sonneborn, 1938), the daily rhythms of mating in protozoa under light–dark cycling were generally regarded as amusing curiosities which were exceptions to ordinary functional behaviour, rather than as indicators of a general organismic principle. The realization that the rhythms continued and were endogenous (free-running) in darkness under constant conditions, and that they were light settable was a significant step forward. This was established by studies on the phototactic response of the green phytoflagellate *Euglena gracilis* (Pohl, 1948) and on the mating reaction in the chlorella-less (non-green) ciliated protozoon *Paramecium bursaria* (Ehret, 1951, 1953). The general biological significance of an endogenous daily rhythm at the cellular level of organization was dimly glimpsed even then in the light of similar phenomena described at higher levels in Man (Kleitman, 1949), in invertebrates (Brown and Webb, 1949) and in the higher plants (Bünning, 1936). Next, it was found that the length of the period, 'τ', for the free-running circadian cycle was only slightly temperature dependent, and much more important, that free-running τ was significantly different from a day, but *about* a day in length (*circa dies*) for the phototactic response in *Euglena* (Bruce and Pittendrigh, 1956) and for the cell division cycle in the green dinoflagellate *Gonyaulax* (Sweeney and Hastings, 1958).

As late as 1958, the control of synchrony by (circadian) biological clocks was regarded in a distinguished review (Zeuthen, 1958) as a special case for those cells equipped with such clocks; however, by the time of the Cold Spring Harbor Symposium in 1960, the sweep of evidence from many fields led most investigators to agree that natural selection has endowed all higher organisms with the attribute of capacity for circadian time-keeping (Ehret, 1960a; Ehret and Barlow, 1960; Pittendrigh, 1960).

The physical basis for the eukaryotic–circadian principle remains unknown. Ideas for its composition range from

(1) the notion that optimal design for feed-back could have been newly invented and naturally selected from the components of each representative of a class in the higher taxa in turn and on its own terms (one might suppose cell–cell, organ–organ, system–systems, and intra- and inter-hierarchal interactions) to

(2) that eukaryotic structural organization implies and inevitably leads to circadian–infradian temporal organization.

A special case of this latter view is in the replicon–chronon theory (Ehret and Trucco, 1967), in which polycistronic replicons of optimal length (chronons) have become genetically indispensable and are the rate-limiting elements common to all eukaryotes.

In the present chapter we place particular emphasis not only upon validating the generality of the eukaryotic–circadian principle, but also upon pointing out signal experiments that distinguish between the two fundamental population growth modes: the fast, or ultradian mode (in which cells divide rapidly), and the slow, or circadian–infradian mode. Several excellent reviews have thoroughly covered the essential early data; these are Hastings, 1959; Bruce, 1965 and Bünning, 1967. Our plan is to discuss earlier studies only wherever in the present context the older data are especially illuminated by the new. In particular we desire not simply to condense each work under consideration into a mere resumé, but instead to show in each case its relevance to the broad subject of eukaryotic rhythms.

The *Gonyaulax–Euglena–Tetrahymena* effect (*G–E–T* effect) and circadian–infradian rule

When we first decided to test experimentally the replicon–chronon theory of circadian time-keeping, it was clear to us that we required large quantities of DNA and RNA from well-synchronized populations of cells. Our initial choice was *Paramecium* (Ehret and Trucco, 1967), until we ourselves took to heart the principle that the ciliated protozoan *Tetrahymena*, being a eukaryote, should also have a circadian rhythm. It was after all a prime choice for growth in defined or axenic medium, for synchronizability in the ultradian mode (Zeuthen, 1958) and for cell density at peak growth. However, no reliable reports existed for a circadian component to growth. Then we reasoned that even the circadian mating reaction rhythm in *Paramecium* does not appear during rapid exponential growth (the ultradian mode), but rather it shows up later during the so-called stationary phase—when the rate of cell division has diminished markedly. The time corresponding to this in *Tetrahymena* had hardly ever been studied—yet it appeared that this 'stationary phase' (in our terms, the infradian mode) would be the most likely to show light inducibility and circadian rhythmicity. As detailed elsewhere (Wille and Ehret, 1968a, 1968b) and later in this chapter, this proved to be the case, and in retrospect it relates well to what we should already have realized emphatically from other studies with *Gonyaulax* and with *Euglena*: circadian frequencies are seen in cells *after* rapid exponential growth (after the ultradian mode), at which time the cells of a population are light synchronizable. The population density following the ultradian mode rarely remains stationary; even batch cultures show some increase, and integral values for generation time may be of the order of 40 hours to 5 or more days. It is during this infradian growth mode that the cells are light synchronizable, and a composite description of this effect for *Gonyaulax*, for *Euglena* and for *Tetrahymena* (the *G–E–T* effect) is plotted in Figure 13.1. Turning from batch cultures to continuous cultures with *Tetrahymena* bore out these expectations fully: the circadian component of an infradian culture is readily revealed by light synchronization, just as the ultradian component of ultradian and late-ultradian cultures is revealed by heat synchronization. In the sections that follow the experiments are considered in greater detail. We regard then, as a corollary of the eukaryotic–circadian principle, the rule that a cell in switching its state from the ultradian to the infradian mode is invariably capable of circadian outputs in the latter mode.

Reflections on the older literature (mostly pre-1965)

The circadian clock in EUGLENA

A circadian rhythm of phototactic sensitivity was first demonstrated in non-growing populations of the autotrophic phytoflagellate *Euglena gracilis* by Pohl (1948). The rhythm exhibits a maximal response at midday and a minimal response at night. Bruce and Pittendrigh (1956, 1957, 1958) explored the circadian rhythm of phototaxis in *Euglena* in detail. They found that the rhythm of phototaxis persists for weeks with a free-running period τ close to 24 hours. The period is essentially temperature independent, can be entrained by light–dark cycles, and the phase may be changed by a single exposure to light.

None of the authors cited above emphasized the necessity for employing

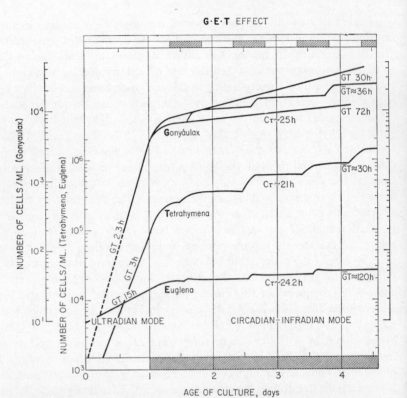

Figure 13.1. The *G–E–T* effect. After a day or so in small batch cultures, cells switch from the ultradian mode to the circadian–infradian mode and population densities continue to increase impressively, even though slowly. Ultradian generation times (GT) are given at the left and infradian generation times at the right. During the infradian mode the cells are light-synchronizable; although infradian GTs are long and variable, in each case the consequence is a circadian output: the circadian period (cτ) is ≈1 day (by definition). *Gonyaulax* is shown in continuous light (top open bar) as an asynchronous population with average infradian GTs in the 30–70 hr range (from the data of Hastings, Sweeney and Mullin, 1962), and in light–dark cycles of 12 hrs each as a synchronized population with an integral generation time of ≈ 36 hrs (from the data of Sweeney and Hastings, 1958). The curves for *Euglena* and *Tetrahymena* might also have included asynchronous infradian slopes (viz. Figures 13.3 and 13.11), but have been omitted here for clarity; the synchronized populations represented in each of these cases show free-running endogenous circadian rhythms following photoinduction by switch-down from light (white open bar, bottom of graph) to darkness (black bar) at a critical transition point between the two modes of growth. (From the data of Edmunds, 1966; Wille and Ehret, 1968a)

non-growing populations of *Euglena* for the phototatic assay. Pohl (1948) merely comments that during the course of the experiment there was no change in the number of cells in the experimental apparatus. This apparent disregard of the significance of the nutritive condition of the cells in the manifestation of the circadian rhythm can be understood on methodological grounds. The design of the experimental apparatus employed in the photo-tactic assay requires relatively dense cell populations. They were obtained by allowing the culture to reach a steady-state cell density in batch culture, in which case the growth dynamics of batch cultures were of little interest, except in as far as they provided sufficient numbers of cells for the photo-tactic assay. Nevertheless, the important fact remains: circadian rhythmi-city in *Euglena* was observed only when the cells were in the infradian mode of growth.

The circadian clock in PARAMECIUM

(a) *Paramecium bursaria*

The ciliated protozoan *Paramecium bursaria* deprived of its symbiotic chlorellae was the first non-green cell shown to possess a circadian clock; the clock regulates mating reactivity (Ehret, 1951, 1953). Mating between cells of different mating types occurs at midday on an *LD* : *12*,12 cycle (that is to say, *Light 12* hours, Darkness 12 hours) in populations of non-growing cells. The phase of the rhythm can be reset by brief exposures to light and is predominantly effective 16 to 18 hours after removal to continuous darkness, or corresponding to middle of the night phase of the subjective day. The rhythm persists as long as one week in continuous dark.

In a subsequent study, Ehret (1959a) found that far ultraviolet light is highly efficient in inducing a phase shift, which is reversed by a subsequent exposure of the cells to white light. This discovery of photoreactivation and other observations on the slight but significant endogenous metabolic rhythm of RNA nucleotides led to the suggestion that nucleic acids were involved in the clock mechanism (Ehret, 1959b, 1960a; Ehret and Trucco, 1967). In the visible region an action spectrum was obtained for phase shifting (phase advance) the *P. bursaria* mating reaction rhythm: it has peaks in the red (650 nm), blue (440 nm) and the near ultraviolet (350 nm) (Ehret, 1960a).

Earlier studies on the genetics of the mating type system in *P. bursaria* had revealed a number of interesting observations concerning the optimum conditions for mating between various stocks (Jennings, 1939). Jennings found that the nutritional state of the cells was one of the important variables affecting mating reactivity. He pointed out that 'clotting and mating do not occur when individuals to be mixed are taken from a rich nutrient medium in which they are plump and *rapidly dividing*, nor, usually, when the mixture is made in such a medium'. He further adds that starvation is not a necessary condition, and that 'well-fed, well-grown individuals mate rapidly, provided their plumpness is not excessive'. Again, Jennings notes that 'mature strong clones show little dependence on nutritive conditions unless excessively over-fed and plump'. We may only guess what criterion Jennings had in mind here for 'excessive', but it seems certain that manifestation of the mating reaction, which is clock controlled, is suppressed in rapidly dividing or excessively over-fed cultures, i.e., when *P. bursaria* is in the ultradian mode of growth. On the contrary, mating reactivity is strongest when the cells are slow growing, i.e., when the cultures are in the infradian mode of growth. On the basis of these observations, *P. bursaria* belongs to the *G–E–T* group of organisms displaying the circadian-infradian rule.

(b) *Paramecium multimicronucleatum*

Sonneborn and Sonneborn (1958) reported a fascinating example of daily rhythm involving a change of mating type on an *LD* : *6,18* entrainment cycle. For those unfamiliar with the jargon of mating type, this amounts to a kind of sex reversal, in which one sex is manifest in the cells of a population during the daytime, and the complementary sex is present at night! The rhythm persists for 3 days in darkness. The same cycle of mating type switch-overs is observed regardless of the relationship between solar time and time of schedule of the artificial lights. A phase shift of the rhythm occurs following an appropriate exposure to light; brighter light is more effective in this regard than dimmer light.

The circadian clock in GONYAULAX

A number of endogenous circadian rhythms have been demonstrated in the autotrophic dinoflagellate, *Gonyaulax polyedra*. It possesses a persistent diurnal rhythm of bioluminescence, cell division and photosynthesis, each of which continues for a period of approximately 24 hours, when the temperature and light intensity are held constant (Hastings and Astrachan, 1959; Hastings and Sweeney, 1959). The three rhythms have their own characteristic phase relationships which are maintained under constant conditions.

(a) *Rhythm of bioluminescence in* GONYAULAX

Like a number of other circadian systems, the rhythm of bioluminescence in *G. polyedra* damps out under continuous bright illumination. However, according to Hastings (1959), under dim light conditions (120 foot-candles, 21°c) 'no apparent damping occurs'. This observation remains unexplained, and raises an interesting possibility. If no apparent damping occurs under dim light conditions, but does under bright light conditions, there must be something keeping the cells in phase in the former case. Presumably in bright light the individual cells drift out of phase with each other, and the population loses synchrony because the cells receive no new phase information. On the contrary, in dim light the cells do receive new phase information which we suggest results from cells 'talking to one another'. An obvious means whereby *Gonyaulax* cells in the population communicate phase information to one another is by their capacity to emit light. By this means, pacemaker cells in the culture which are in-phase with the endogenous period of the rhythm photoentrain the out-of-phase cells and thereby sustain the synchrony of oscillation in the population.

The problem of possible cross-talk among cells as an explanation of undamped oscillations exists for all of the cases of persistent rhythms in unicellular organisms, and has not been adequately discussed in previous treatments of the subject, although the converse question has been raised and experimentally dealt with by Sweeney (1960). Evidence for the existence of cross-talk as a *Zeitgeber* for other eukaryotic cell populations has not been looked for, but should be sought when evidence for undamped oscillations exists, as it does in the above example in *Gonyaulax*. This conclusion may not be immediately obvious. Indeed, most workers in the field of circadian rhythms associate undamped oscillations with the endogenicity of the rhythm. This may be because they deal not with uniform cell populations, but with a complex behavioural output derived from the operation of a multicellular organism, where internal *Zeitgebers*, such as hormones, exert an overriding control on the circadian response, and where there is evidence for considerable cross-talk among tissues of the whole organism.

(b) *Rhythm of cell division in* GONYAULAX

The rhythm of cell division in *G. polyedra* (Sweeney and Hastings, 1958) demonstrates beautifully the *G–E–T* circadian–infradian rule. The integral generation time is characteristically always greater than 24 hours, e.g. 4·2 days, and it can be any non-multiple of 24 hours. The cell division event occurs in only a small fraction of the population (less than 5 per cent), but is gated and occurs only at circadian intervals (see Figure 13.2).

13

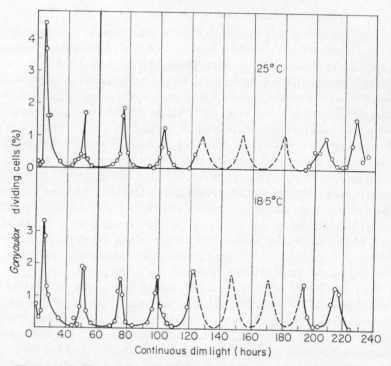

Figure 13.2. The per cent dividing cells present in a cell suspension (*Gonyaulax*) transferred from *LD* to continuous dim light (200 foot-candles) at the beginning of a light period (zero time) at 18·5°c (lower curve) and at 25°c (upper curve). The average length of the period of cell division: 23·9 hours at 18·5°c and 25·4 hours at 25°c. The average generation time: 4·2 days at 18·5°c and 3·8 days at 25°c. (From Sweeney and Hastings, 1958)

On an *LD* : *12*,12 entrainment cycle, cell division occurs at the end of the dark period. The period of the cell division rhythm is nearly temperature independent, while the average generation time is temperature dependent.

Temperature dependence revisited

In most circadian oscillations the circadian period, τ, shows a remark-ably small temperature dependence (Sweeney and Hastings, 1960), and a number of alternative hypotheses have been advanced to account for this seemingly unchemical aspect of biological clocks. The *Gonyaulax* clock displays yet another paradox, inasmuch as the measured values for the temperature coefficient (Q_{10}) of the circadian period are less than 1·0. Hastings (1959) has interpreted this result as 'over-compensation' of the clock mechanism. This thinking is in line with the view expressed by Bruce and Pittendrigh (1956). In their model the clock mechanism is *temperature compensated*, so that the effect of temperature on the period of the free-running rhythm operates mainly through a series of transient reactions which act in turn to adjust the clock mechanism. A specific cellular mechanism to account for temperature compensation has not been found, but one possibility frequently advanced is the coupling of two or more temperature-sensitive enzyme reactions, one of which is operative as a high temperature cut-out circuit and the other as a low temperature cut-out circuit. Such a metabolic scheme would require the clock mechanism to be coupled to steady-state enzyme reactions as an environmental buffer. At the close of a scholarly review in which he discussed the possible mechanisms underlying comparable compensatory changes in response to thermal conditions in intertidal organisms, Newell (1969) emphasized on the one hand the low dependency upon temperature of physical processes (e.g. diffusion) and on the other hand, especially in cases of long term acclimation, probable changes in specific physical properties of proteins. Ehret and Trucco (1967) had already considered each of these categories as likely mechanisms for eukaryotic regulation in the infradian mode. However, given that the mean-free path of an informational macromolecule in an infradian eukaryote can be considerably longer than that in a well-fed prokaryote, and that diffusion processes are the rate-limiting and temperature-independent steps involved in the clock mechanism, the finding of temperature coefficients with values of less than 1·0 still requires additional explanation.

A possible explanation consistent with the chronon–replicon model proposed by Ehret and Trucco (1967) is that the infradian–circadian state of cell differentiation is not a unique state, but represents a number of possible alternative states each of which is associated with the 'switched-on' condition of a particular chronon–replicon; for example, under one set of conditions (GT—60 hours, 25°C) the cell is specified by the operation of one particular chronon–replicon, while other mutually exclusive sets are switched off, whereas, under other conditions (GT—60 hours, 18·5°C) the cell is specified by the operation of a different chronon–replicon set, while the other sets are then repressed. So long as the condition 'infradian generation time' exists, 'the switched-on' chronons will set the pace for the whole cell. On this hypothesis, the eukaryotic cell is characterized by a panoply of differentiated states which are possible because of the multi-replicon nature of euchromosomes. A Q_{10} less than 1·0 is therefore an evolutionary accident in the case of *Gonyaulax* and in *Oedogonium* (Bühne-mann, 1955), and comes about as a result of a natural selection which endows these cells with an assortment of chronon–replicons, one set of which, when switched on at lower temperatures, requires a shorter duration to complete its transcription sequence before recycling than it would at a higher temperature. In this context one would predict that the *m*-RNAs transcribed from the chronon-characterizing 18·5°c clock control of the cell division rhythm in *Gonyaulax* would not be identical to the set of *m*-RNAs transcribed from the chronon which characterizes the 25°c circadian rhythm of cell division. Evidence for the existence of temperature induction of specific differentiated states has been amply documented for the serotype system in *Paramecium* (Beale, 1954); inspection of the experimental methods in each instance reveals that induction quite probably occurred during the infradian mode. The antigen type expressed on the cilia of the cell is dependent on the temperature range to which the *Paramecium* cell has been acclimated. Since each of these antigen types is the expression of a different genic potentiality, thermal induction of different antigens represents the induction and repression of genes. It is not known whether these different genes reside on different replicons, as the above hypothesis anticipates, but it is fully consistent with the absence of linkage between antigen genes.

Further reflections by Pavlidis (1969) on the older observations of small temperature dependence in the circadian values for τ result in his very interesting demonstration that the oscillation can be simulated by inter-action between a number of high frequency oscillators; temperature compensation in this Pavlidis model occurs when a rise in temperature increases the number of active oscillators, thereby assuring the relative constancy of τ over a remarkably broad range.

Photoentrainment revisited

In *Gonyaulax*, *Euglena* and *Paramecium* the endogenous circadian rhythm damps out after only a few cycles under conditions of continuous bright illumination. We have already commented on the possibility that on-time pacemaker cells may communicate with other cells in the popula-tion by light flashes in *Gonyaulax*, thereby acting as a *Zeitgeber* in the absence of entraining light–dark cycles. Such signals would be effective by virtue of the fact that phase resetting requires only one change in light intensity, serving to give phase information. Damping out of the endo-genous rhythm then reflects the normal course of events to be expected when no *Zeitgeber* influences are present.

Photoentrainment with light–dark cycles other than *LD* : *12*,12, while the sum of $L + D$ is ~ 24 hours, is a common property of circadian systems. However, in *Gonyaulax* Hastings (1959) found that the cell divisions and luminescence rhythms can be maintained on an $LD : 8,8$ regime, with the maximum of each rhythm occurring once every 16 hours. Even so, after 7 months of this regime, the cells free-run at their 'natural' circadian period. With regard to the *G–E–T* circadian–infradian rule, it is of further interest to note that no more than 70 to 100 cell divisions occurred during this 7-month entrainment period. This means that the average cell generation time of entrained cultures was between 50 and 72 hours, and hence the population was in the infradian mode of growth throughout the experiment. Photoentrainment during the ultradian mode of growth has never been clearly shown in either *Gonyaulax* or *Paramecium bursaria*. Photoentrainment of unicellular organisms undergoing more than one doubling per day has been reported for *Euglena* (Bruce and Pittendrigh, 1956), *Chlamydomonas* (Bernstein, 1964; Bruce, 1968) and *Paramecium* (Barnett, 1966). These exceptional cases will be discussed in a later section.

New observations consistent with the circadian–infradian rules

The circadian clock in TETRAHYMENA

Because it was expected, the discovery of a circadian rhythm in the non-green ciliated protozoan *Tetrahymena pyriformis* (W) was especially gratifying. It was expected not only because of the eukaryotic–circadian principle, but also because the action spectra for phase shift in other protozoa suggesting a porphyrin photoreceptor (Ehret, 1960a) seemed in harmony with certain observations in the literature on porphyrin content in light-grown and dark-grown cells. Thus the observations of high protoporphyrin IX concentrations in dark-grown *Tetrahymena* (Rudzinska and Granick, 1953; Kneuse and Shorb, 1966) were added factors in predicting the presence of a photoinducible circadian rhythm in *Tetrahymena* (Wille and Ehret, 1968a, 1968b). The utility of this organism for biochemical studies in cell physiology is well known. Therefore, the finding that it possesses a bona fide circadian rhythm has opened the way for an intense investigation on the molecular basis of the biological clock. In this section we shall review the recent observations on the circadian clock in *Tetrahymena*, paying particular attention to the establishment of the circadian–infradian rule.

(a) *Photoentrainment in batch cultures*

Batch cultures of *T. pyriformis* (W) can be characterized by a variable but short (1 to 3 hours) lag period, followed by an exponential phase of

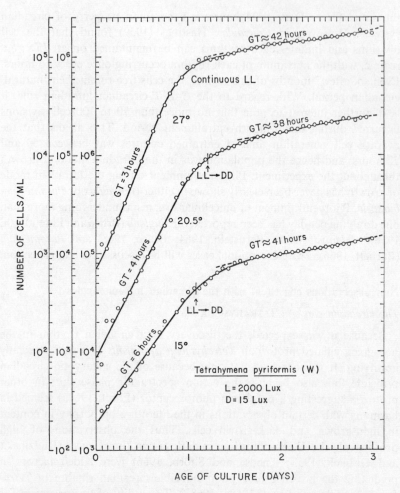

Figure 13.3. Growth curves of batch cultures of *T. pyriformis* (W) at three temperature conditions. The outermost ordinate is for the growth curve at 27°c, the middle ordinate is for the 20·5°c growth curve, and the innermost ordinate pertains to the 15°c growth curve. Note particularly the conspicuous temperature dependence during the ultradian mode. (Szyszko and coworkers, 1968)

growth, the *ultradian* mode, and followed by a period of extended slow growth, the *infradian* mode (Szyszko and coworkers, 1968). Typical batch culture growth curves obtained with three separate cultures at three different temperatures are shown in Figure 13.3. Basically, the growth

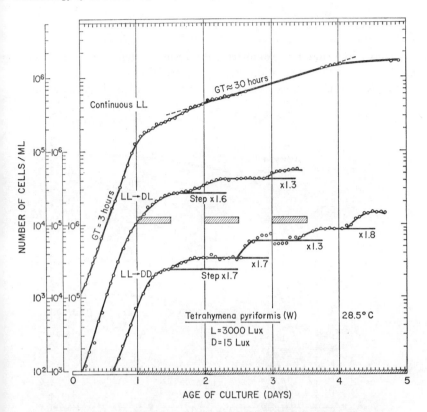

Figure 13.4. Growth of *T. pyriformis* (W) batch cultures under three different conditions of illumination. The outermost ordinate is for LL; the middle ordinate is for LL → DL; and the innermost ordinate is for LL → DD. LL is continuous bright illumination (3000 lux); LL → DD is continuous dim illumination (15 lux) following exponential growth in continuous bright illumination; and LL → DL is an *LD* : *12,12* cycle of dim and bright illumination following exponential growth in continuous bright illumination. The average generation time (GT) of the population during continuous illumination (LL) after exponential phase of growth is indicated. The increase in cell number or step following a division burst is given as a ratio. The abscissae index marks denote the sampling times, which occurred at 90 minute intervals

curve data demonstrate the strong temperature dependence that is characteristic of the ultradian mode of growth. The infradian mode of growth is characterized by a monotonically non-decreasing function, rather than by the stationary phase frequently seen in bacterial batches. Wille and Ehret (1968a) found that photoentrainment of a circadian rhythm of cell division

occurred in batch cultures of *T. pyriformis* (W) when cells had reached the
end of ultradian exponential growth, but not before then.

Figures 13.4 and 13.5 summarize the results obtained for the photoentrainment of the cell division rhythm in *Tetrahymena*. On the lower curve
in Figure 13.4 an endogenous circadian rhythm was initiated by a single

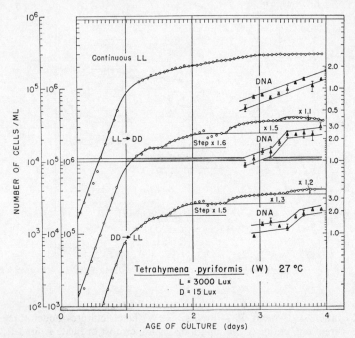

Figure 13.5. LL and LL → DD are the same as in Figure 13.4. DD → LL
is growth in continuous bright light following exponential growth in
dim light. The step in the amount of DNA per culture is shown in
relative units. The double bars indicate the magnitude of the spread in
DNA values

change in light intensity, a sudden decrease in the level of irradiance and
a 'switch-down' (LL → DD); Figure 13.5 shows that 'switch-up"
(DD → LL) is also effective. The rhythm persists for 3 or 4 days in either
constant dim or constant bright light. On an *LD* : *12*,12 entrainment
cycle the maximum in the cell division rhythm occurs at the end of the
light phase. The free-running period, τ, is approximately 21 hours. The
relationship between time of the onset of the rhythm and the time when a
'switch-down' or 'switch-up' occurs is not entirely clear, but seems to
follow the rules for 'on' and 'off' rhythms discovered in *Kalanchöe* and

Drosophila pseudoobscura by Englemann (1966). Further support for this notion was obtained from experiments on photoentrainment of *Tetrahymena pyriformis* (W) in a continuous culture apparatus.

(b) *Photoentrainment in continuous cultures*

As mentioned above, no more than 3 to 4 days of circadian–infradian growth can be supported by the batch culture technique before the decline phase sets in. In an attempt to overcome this difficulty, *T. pyriformis* (W) cells were grown in an electronically controlled nephelostat (Eisler and Webb, 1968). In this apparatus continuous culture growth of *Tetrahymena* cells was maintained on the chemostat principle of controlled flow rate of nutrient medium. The cell titre is kept at a constant value by keeping the amount of scattered light at a fixed level through periodic interrogation of the culture and subsequent feeding events if the amount of scattered light has changed upward with cell growth. If cell growth is characterized by a random exponential increase in cell numbers then the record of feeding events will show no signs of periodicity. If on the other hand, the cell number increases periodically, the frequency of feeding events will be periodic. Cultures grown in the nephelostat were subjected to different light regimes. Photoentrainment of an endogenous circadian rhythm of cell division was achieved on an *LD* : *10*,14 entrainment cycle. Inspection of Figure 13.6 shows that the cell division rhythm has a free-running period of 21 hours under constant dim red ($\lambda > 670$ nm) conditions for 5 days. The rhythm is characterized by increased cell division during the dark phase. Figure 13.6 also points out the sensitivity of the rhythm to such subtle changes as the rate of aeration. An increase in the aeration rate results in a decrease in the average generation time.

The nephelostat type of experiment indicates the feasibility of designing a continuous culture device which does not need to operate on the nephelostat principle, but which controls growth rate by a programme of feeding events. Such a device was constructed and employed for the study of circadian rhythms in large volumes (2 to 3 litres) of *Tetrahymena* cultures. A description of this programmed feeder has been published elsewhere (Wille and Ehret, 1968a). It consists of a growth flask and a Mariotte bottle connected by a siphon tube. An interval timer controls a solenoid-operated valve which allows nutrient to flow from the Mariotte bottle to the growth flask according to the schedule of feeding events prescribed by an interval timer programme. The programmed feeder can be arranged to yield a *symmetrical* programme, i.e. feeding events occur at equal time intervals, or an *asymmetrical* programme, i.e. unequal time

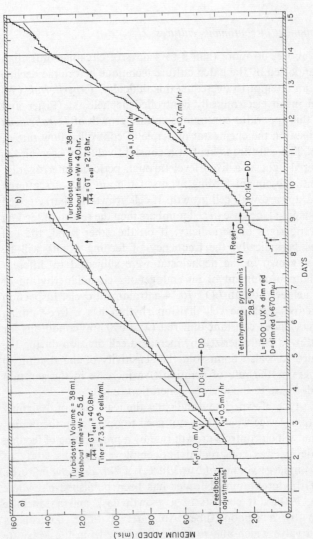

Figure 13.6. (A) Growth rate of *T. pyriformis* (W) in a nephelostat on a *LD* : *10,14* light–dark regime, and in 'constant dark' DD (dim red light, λ = 670 nm). (B) Growth rate of *T. pyriformis* (W) in a nephelostat of same culture as shown in A with the introduction of one new *LD* : *10,14* followed by continuous dark (DD). The nephelostat volume, washout time (*W*), cell generation time (GT) and titre are indicated. The slope lines were derived by extending the linear portions of the feeding curve and by fitting those which were approximately of the same slope to a common parallel slope.

intervals between feeding events. Figure 13.7 illustrates an unexpected finding, strongly suggestive of 'stationary-phase' synchronization in prokaryotes (Cutler and Evans, 1966), when the continuous culture programmed-feeder device was employed. On an LL light regime synchroniza-

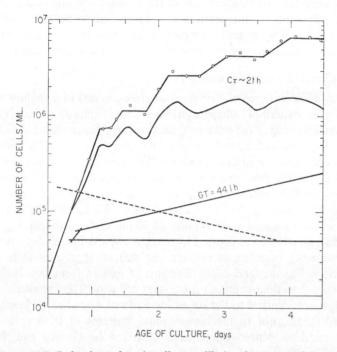

Figure 13.7. Induction of a circadian oscillation by onset of feeding during the late ultradian mode of growth. The cells were under continuous illumination throughout the experiment. The culture was grown as a batch culture until onset of feeding (symmetrical, 96 equal aliquots per day, and equivalent to an integral generation time of 44·1 hours); onset occurred at about 10^5 cells per ml. The sinusoidal curve is a direct tracing of the data from an electronic cell counter; the step function is a derivative of the washout rate and the cell counts (see the text)

tion of an endogenous circadian rhythm of cell division was achieved when batch cultures at the end of the ultradian exponential were converted to continuous cultures by the onset of a *symmetrical* feeding programme. If, on the other hand, *symmetrical* feeding is initiated in batch cultures which are in the early ultradian exponential phase of growth, synchronization of an endogenous circadian rhythm of cell division does not occur.

However, synchronization can then be brought about by photoentrainment with an LD : 12,12 light–dark entrainment cycle. In each case the washout rate of the continuous cultures gives an average generation time (GT—44 hours) that is clearly infradian. Hence it is apparent that the critical condition for manifestation of the circadian rhythm of cell division in *Tetrahymena* is the transition from the ultradian to the infradian mode of growth. Following that transition either light or food can act as a *Zeitgeber* for the induction of circadian rhythmicity in the entire population.

(c) *Circadian rhythm of autotaxis*

When dense cultures of *Tetrahymena* are examined in a shallow dish a very regular pattern of cell aggregations is seen (Figure 13.8, Plate V). The hexagonal packing of the cell aggregates, the strong edge effects seen during the development of the autotactic patterns, the requirements for a minimal cell number and critical fluid depth led Wille and Ehret (1968b) to suggest that this phenomenon was due to physical forces akin to those which produce similar patterns among inert particles, commonly known as Bénard-cell patterns. At a concentration of 3.5×10^5 cells per ml, a depth of 3 mm and a total volume of 10 ml, each autotactic aggregate (the white clouds seen in Figure 13.8) contains approximately 5×10^4 cells. The frequency (number of clusters per cm) of aggregates is inversely proportional to the fluid depth. The rate of pattern formation is directly proportional to fluid depth, at a constant cell titre. The autotactic aggregates are not identical to the streaming patterns described by Loefer and Mefferd (1952), nor to the bioconvective patterns of Platt (1961), but are identical to Bénard-cell patterns described by Gmitro and Scriven (1966). The direct proof of the Bénard-cell basis of the autotactic phenomenon in *Tetrahymena* was accomplished by the simple expedient of mixing inert carmine particles (selected for a size roughly comparable to *T. pyriformis* (W) by sedimentation under gravity) with live *T. pyriformis* cells under the conditions suitable for demonstration of the autotactic phenomenon. When this was done the inert carmine particles (which are red) took up the hexagonal edges of the Bénard-cell domains, while the motile cell aggregates occupied the centres of the domains.

Further evidence on the factors which influence pattern formation point to a strong stabilizing effect on the pattern by gaseous carbon dioxide; nitrogen and oxygen do not appear to exert any obvious effects. Light has a destabilizing effect on the pattern when the experimental set-up is illuminated from one side only. A series of experiments was carried out to examine the possibility of a daily rhythm in the rate of autotactic aggregation. Figure 13.9 shows that there is an endogenous circadian

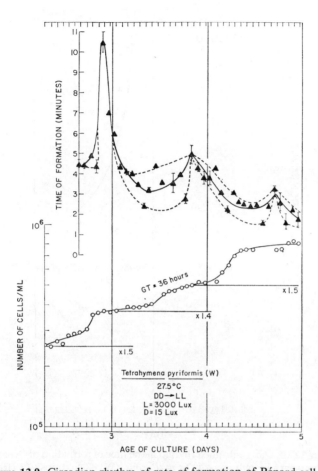

Figure 13.9. Circadian rhythm of rate of formation of Bénard-cell-like patterns and cell division in *T. pyriformis* (W), following a *switch-up* experiment. The overall generation time (GT) during the circadian–infradian growth phase is 36 hours. The step sizes of the circadian synchronized cell divisions are shown under the extended lines as ×1·5, ×1·4 and ×1·5. The increase in irradiance (DD → LL) occurred at 10^5 cells per ml. Dotted lines show the range and the solid line shows the best fit (by eye, and after Figure 13.10). Despite the scatter of points, they are clustered in such a way as to make any application of a uniformly descending monotonic function extremely improbable; instead there is clear evidence for at least two and probably three circadian peaks, each one resembling the form of Figure 13.10 (LD) in showing an abrupt rise and a precipitious plunge around the time maximum for pattern formation. (Wille and Ehret, 1968b)

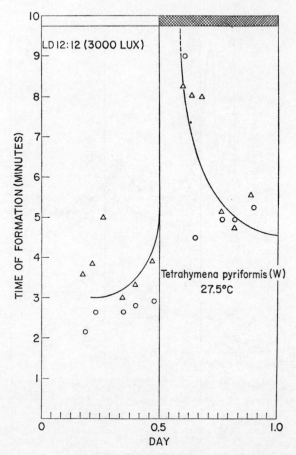

Figure 13.10. Daily entrained rhythm in the time of formation of Bénard-cell patterns in cultures of *T. pyriformis* (W). The transition from continuous light (3000 lux) to dim light (15 lux) is shown at the top of the figure by open and cross-hatched bars respectively. Open circles and triangles are data from two different experiments. (Wille and Ehret, 1968b)

rhythm in the rate of pattern formation. The rhythm can be brought about by a single change in light intensity sufficient to initiate the circadian rhythm. The minimum rate is primarily dependent on cell concentration, hence the oscillations appear to damp out after a few cycles in batch culture. Cultures which have been entrained by an *LD* : *12*,12 light–dark regime and which were kept at roughly constant cell density by periodic feeding in the infradian phase of growth, showed a daily rhythm in the time of pattern formation, as shown in Figure 13.10. The maximum in the time of pattern formation occurs at the change-over from bright light (LL) to dim light (DD). The pattern formation rhythm has a free-running period of 21 hours, and the maximum in the time of pattern formation has its own characteristic phase relationship to the cell division rhythm (see Figure 13.9). The requirement for a minimal cell density of at least $1\cdot5 \times 10^5$ cells/ml means that all these experiments were carried out with cells in the infradian mode of growth.

Circadian-infradian rules in EUGLENA

(a) *Photoentrainment of a circadian rhythm of cell division*

Over the past few years considerable evidence has accumulated to show that light–dark cycles can synchronize the cell cycle in *Euglena* (Bernstein, 1964; Cook, 1961; Petropulos, 1964). Edmunds (1965) has shown that a synchronous cell division rhythm can be attained in populations of *Euglena gracilis* Klebs strain Z by subjecting the culture to alternating light–dark cycles with varying photofractions (e.g. *LD* : *16*,8, *LD* : *12*,12 and *LD* : *14*,10).

On either of these entrainment cycles, synchronization of the division bursts during any given cycle began 13 to 14 hours after the onset of the light period. Edmunds (1965) points out that this result is reminiscent of that obtained for the release of autospores from the wall of the parent cell in *Chlorella pyrenoidosa* reported by Pirson and Lorenzen (1958). Both results can be taken as evidence for the involvement of a biological clock that regulates cell division; both results also relate to the 13 to 15 hours' lag in onset of the cell division rhythm in *Tetrahymena* (Wille and Ehret, 1968a, and see Figures 13.4 and 13.5). The 13-hour lag preceding synchronous and circadian rhythmic behaviour in *Tetrahymena*, *Chlorella* and *Euglena* following a change in the level of irradiance is too common an occurrence to be accidental. It is suggested here that it reflects a general rule regarding the phase of an emergent rhythm following either a switch-up or switch-down in the level of irradiance, as postulated for the 'on' and 'off' rhythms by Engelmann (1966).

Edmunds (1965) further noted that *Euglena gracilis* which is grown autotrophically on an *LD* : *16*,8 entrainment cycle can be desynchronized by the addition of 0·006 M ethanol to the culture. Following the ethanol addition, growth is delayed for 8 hours and then resumes in an exponential mode with an average generation time of 13 hours. Likewise, continuous bright light brings about desynchronization by inducing rapid exponential growth, if the culture is placed in continuous light (3500 lux) at the end of the normal dark period (10 hours in *LD* : *14*,10). In terms of the circadian–infradian rule, these desynchronizing treatments have one thing in common: they cause a return from an infradian to an ultradian mode of growth.

Some peculiarities of the growth kinetics of *E. gracilis* populations following the transition from continuous bright light (LL$_b$, 3500 lux) to

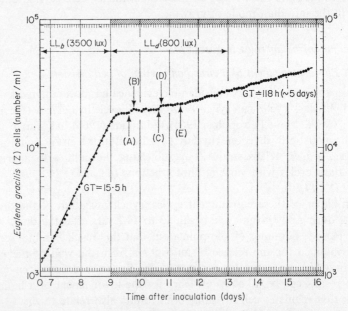

Figure 13.11. Growth of autotrophic *E. gracilis* (Z) in continuous dim light (800 lux) following exponential growth in continuous bright illumination (3500 lux). An initial plateau includes two small division bursts (A–B and C–D) at 14 and 38 hours following the transition from bright to dim light. The infradian logarithmic growth that takes over (point E) may contain circadian steps. Number of cells/ml is given as a function of time. Average generation times (GT) of the population are indicated. (From Edmunds, 1966)

continuous dim light (LL$_d$, 800 lux), led Edmunds (1966) to seek the proper autotrophic growth conditions whereby a single switch in the level of irradiance would synchronize an endogenous circadian rhythm of cell division, under constant temperature and illumination. Inspection of Figure 13.11 reveals that at least two transient and endogenous bursts of cell division occur following a switch-down from LL$_b$ to LL$_d$. Two things are immediately apparent. The time elapsed between the two endogenous bursts of cell division has a circadian period ($\tau \simeq 24$ hours), and the endogenous circadian rhythm is confined to the infradian slow growth mode of the growth curve (i.e. the average generation time, GT, was 118 hours). In contrast the average GT during the period of growth in continuous bright light was 12 to 13 hours, i.e. the ultradian mode. Although no specific evidence is available to decide whether the transition from the ultradian to infradian mode of growth is not itself sufficient to

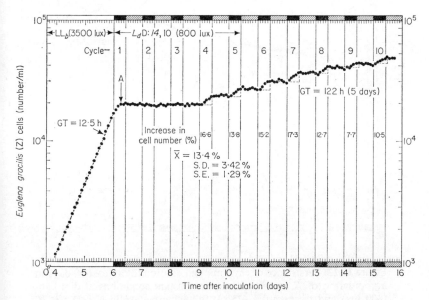

Figure 13.12. Synchronous growth of *E. gracilis* (Z) in a $L_dD : 14,10$ (800 lux) light–dark cycle for 10 days following exponential growth in continuous bright light (3500 lux). The increase in cell number (per cent) is given for each division burst. Average generation times (GT) of the population are indicated. (From Edmunds, 1966)

synchronize cell division independently of the switch-down event, there is little doubt that the endogenous circadian bursts of cell division do not appear when cells are in the ultradian mode of growth.

The rhythm of cell division bursts can be maintained by imposing a photoentrainment rhythm of $L_dD : 14,10$ on the *Euglena* culture growing in the infradian mode (see Figure 13.12). In Figure 13.12 note the average generation time of 122 hours, contrasting to the 118 hours without photoentrainment. This finding would seem to indicate that the sustained cycles of entrained cell division bursts do not derive this capacity from requirement for light in autotrophic growth. The slopes of the infradian growth phase in the entrained and non-entrained cultures are nearly the same although the two cultures received vastly different amounts of light. The average generation time during infradian growth does not have to be a multiple of 24 hours in order for entrainment to occur. A circadian cell

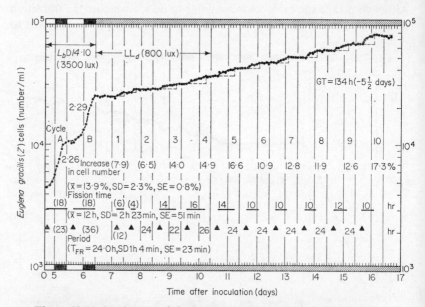

Figure 13.13. Persistence of rhythmic cell division for at least 10 days in a population of *E. gracilis* (Z) under constant conditions of dim light (800 lux) and temperature (25°c), following 6 days in a $L_bD : 14,10$ (3500 lux) light–dark cycle. For each division burst, the increase in cell number (per cent), fission time (hour), and period (hour) are given. Step size is shown for cycles A and B. The generation time (GT) of the population in dim light is indicated. (From Edmunds, 1966)

division rhythm was obtained when the average generation time of the culture had seven different absolute values, none of which was a multiple of 24 hours. In each of these seven experiments the period of the cell division rhythm was close to 24 hours. In no case, however, did the cell division rhythm appear when the absolute value of the average GT was not infradian.

Comparison of Figures 13.12 and 13.13 illustrate the dependence of the onset of the *Euglena* cell division rhythm on the Engelmann (1966) 'on' and 'off' rhythm rules. If we first compare Figures 13.11 and 13.13 we note that there is a quantitative difference in the time of onset between a 'switch-up' (DD → LL) event (16 to 18 hours, Figure 13.13), and a 'switch-down' (LL → DD) event (14 hours, Figure 13.11). The intreaction of 'switch-up' and 'switch-down' events as in the *LD* : *14*,10 entrainment cycle (Figures 13.12 and 13.13) invariably leads to a longer delay prior to onset of the rhythm than either the 'switch-up' or 'switch-down' generated rhythms. Apparently several cycles of light and darkness are necessary to obtain the photoentrained rhythm, as the photoentrainment regime consists of these two conflicting sets of rules: the 'switch-up' and 'switch-down' rules.

(b) *The action of heavy water and of cycloheximide on the phototactic rhythm*

(1) *Lengthening of the circadian period of the phototactic rhythm.* The work of Bühnemann (1955) on the green alga *Oedogonium* established that most metabolic poisons have little or no effect on the period of the circadian clock; later attempts with metabolic poisons have been largely unsuccessful. This dearth of known chemical agents affecting the period of the clock has recently been supplanted by two new observations. Bruce and Pittendrigh (1960) reported that heavy water (D_2O) dramatically increased the period of the circadian rhythm of phototaxis in *Euglena* from 23·5 to 28 hours. The exact reason for this lengthening effect is now known, but could involve conformational changes in macromolecules as well as diffusion-limited reactions.

More spectacular results on the lengthening of the period of the phototactic rhythm in *Euglena* was obtained by Feldman (1967) with the protein-synthesis inhibitor, cycloheximide. A dose of 4·0 mg/ml of cycloheximide lengthened the free-running period in continuous darkness from 23·7 hours to 36 hours (Figure 13.14). This is the first definitive evidence that protein synthesis is involved in the clock mechanism. Feldman (1968) was able to confirm the involvement of protein biosynthesis in circadian rhythms by showing that maxima in the rate of protein synthesis occurs once each 24 hours in free-running circadian cultures of *Euglena*.

(2) *Dark motility rhythm.* A circadian rhythm of dark motility in *Euglena* was reported by Brinkmann (1966). The rhythm is observed in dense cultures of *Euglena* in the stationary phase, and is probably akin to the 'bioconvective' pattern formation observed in dense cultures of free-swimming eukaryotic microorganisms (Platt, 1961; Robbins, 1952). The dark motility rhythm persists for months in a closed system, and damps out in continuous light. Induction of the rhythm by a single transition from LL to DD is independent of light intensity above 10 lux, i.e. the

level of irradiance of the previous continuous light regime is not a factor
influencing the induction of the endogenous rhythm. Of considerable
interest was the observation that the type of nutrition, mixotrophic or
autotrophic, has an influence on the degree to which the circadian period
is influenced by temperature. Under autotrophic conditions the circadian

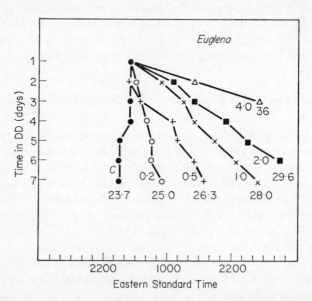

Figure 13.14. Period lengthening of the free-running rhythm by cycloheximide. The clock hours at which successive minima occurred are
plotted for successive days for each of the cultures tested. The numbers
next to the data lines indicate the concentrations (in μg/ml) of drug
added; C = control. The numbers below the data lines indicate the
period length (in hours). There is some slight variability (up to about
an hour) in the exact amount of lengthening of the period induced by
the highest concentrations of the drug (2 and 4 μg/ml). (From Feldman,
1967)

period, τ, is relatively independent of temperature (15° to 35°c), while
under mixotrophic conditions the frequency of the rhythm decreases with
increasing values of constant temperature. Nevertheless, the response
curves of autotrophic and mixotrophic cultures of a green wild type and
of a colourless mutant of the same strain are almost identical (Brinkmann,
Schnabel and Kirschstein, 1968). It would appear from the latter evidence
that the clock-resetting mechanism is influenced by some parameter other
than the type of energy supply.

Circadian–infradian rules in PARAMECIUM

Several recent advances have been made in the genetic and biochemical understanding of the clock mechanism in *Paramecium*.

(a) *Rhythm of mating-type reversal in* P. MULTIMICRONUCLEATUM

Barnett (1961, 1965, 1966) has made an extensive study of the rhythm of mating-type reversals in *P. multimicronucleatum*. As in the case of mating-type reactivity in *P. bursaria*, the conditions which allow expression of mating-type reversal in *P. multimicronucleatum* include the infradian mode of growth. Three different stocks (11, 17 and 19) were examined. Each stock was found to display different but characteristic phase relationships on an *LD* : *8,16* entrainment cycle, with regard to the time of switch-over from one mating type to the opposite type (e.g. mating type III switches-over to mating type IV). Genetic experiments with these three stocks and another stock which never cycles provided evidence for the direct genic control of the ability of the stock to cycle. The acyclic condition was found to be recessive to the cyclic condition. The phase relationships which characterize a particular stock were found to be inherited caryonidally, i.e. all sublines of a clone have the same phase, if the progenitor of that clone derived its macronucleus from one of the segregating macronuclear anlage formed at fertilization. The genetic evidence is consistent with the view that nuclear differentiation determines the phase of the mating-type switch-over times. Thus the fact that the nucleus controls the phase was known as early as 1961 from Barnett's work, and was affirmed later by Schweiger, Wallraff and Schweiger (1964) in *Acetabularia*.

Except for the most extreme photofractions (*LD* : *20,4*), photoentrainment of the mating-type reversal rhythm could be achieved with a number of different light–dark regimes which have varying photofractions and employ varying light intensities.

Barnett (1966) brings another new observation to the literature of circadian rhythms; this is that photoentrainment can be achieved when cells are reproducing asexually at a rate greater than one fission a day and when no rhythm is being expressed. This observation may be inconsistent with the circadian–infradian rule; it therefore merits special attention. It will be discussed again in a later section dealing with observations which challenge the rule.

(b) *Photoentrainment of circadian rhythm in* PARAMECIUM AURELIA

Photoentrainment of a circadian rhythm of mating type reactivity has been demonstrated in *P. aurelia* syngen 3 by Karakashian (1965, 1968).

The peak in the rhythm of mating-type reactivity in the three stocks examined was found to have a maximum during the light phase. One of the stocks becomes arhythmic when transferred to continuous darkness, while the other two show a persistent circadian rhythm. The free-running period of the circadian rhythm is 22 hours at 17°c, and the period of the rhythm is relatively temperature independent; however, temperature cycles (17° and 23°c) can replace light as the *Zeitgeber*.

The mating-type reactivity rhythm in *P. aurelia* syngen 3 adheres to the circadian–infradian rule. Manifestation of the rhythm is restricted to cells where the average generation time was reported to be 56 hours for the entire duration of the experiment.

(c) *Biochemistry of mating-type behaviour and physiology of cell division in circadian populations of* PARAMECIUM BURSARIA

The involvement of protein synthesis in the circadian rhythm of mating-type reactivity in *P. bursaria* has been shown by Cohen (1965). Levels of puromycin (160 mg/ml) and actinomycin D (20 to 40 mg/ml) which are sufficient to inhibit cell division led to the rapid loss of mating. In an earlier study Cohen (1964) had found that the circadian rhythm of mating was expressed by a cycle variation in the ability of 'excised cilia' to react with cells of the opposite mating type. These observations and new observations on the difference in stability of mating activity of cilia from antibiotic treated and untreated cells led Cohen (1965) to conclude that loss of mating reactivity is a programmed event, due to the cyclic metabolic turnover of the mating-type substance. The conditions for photoentrainment of the mating-type reactivity rhythm, and for maintenance of the persistent rhythm are consistent with the circadian–infradian rule. According to Cohen (1965), the average generation time for the cultures was one fission per week. Volm (1964) has reported a circadian rhythm of cell division in *P. bursaria*. The rhythm is photoentrainable with varying light–dark regimes and persists for at least one week. The maximum in the cell division rhythm occurs at the end of the light phase on an *LD* : *12*,12 cycle. As in *Gonyaulax*, *Euglena* and *Tetrahymena*, not every cell divides each day, which is consistent with the circadian–infradian rule.

Circadian–infradian rules in GONYAULAX

Recent observations on the circadian clock in the dinoflagellate *Gonyaulax* pertain to experiments aimed at discovering the biochemical basis of the biological clock. Karakashian and Hastings (1962) were the first to gain positive evidence for the clock-stopping effect of actinomycin D, an inhibitor of RNA synthesis. This evidence, in conjunction with earlier studies by Ehret (1959a) on the effect of ultraviolet light and its photo-reactivation on the circadian rhythm of mating type reactivity in *P. bursaria* and also by Sweeney (1963) on ultraviolet action in *Gonyaulax*, lend support to the direct involvement of DNA and DNA-directed RNA synthesis in the clock mechanism. Further work on the effect of protein-synthesis inhibitors on the circadian rhythms in *Gonyaulax* (McMurry, 1968) has shown that the characteristic phase relationships between the four known circadian rhythms cannot be uncoupled from each other following various phase-shifting treatments.

The circadian clock in the alga PHAEODACTYLUM

The marine diatom *Phaeodactylum tricornutum* possesses a circadian rhythm in the rate of photosynthesis (Palmer, Livingston and Zusy, 1964). The rhythm of photosynthetic capacity was observed by measuring the rate of uptake of labelled carbon dioxide into whole algae. The rhythm of photosynthetic capacity was brought about by a photoentrainment cycle of *LD* : *12*,12. The rhythm is not expressed in continuous darkness, and damps out after one cycle under constant bright illumination (600 foot-candles). At 40 or 80 foot-candles, or dim light the rhythm will persist for at least three cycles, but is characterized by a gradual damping out. Photoentrainment of the rhythm and subsequent circadian rhythmicity were observed in cultures approaching the stationary phase of growth, as predicted by the circadian–infradian rule.

The circadian clock in ACETABULARIA

The existence of a circadian rhythm of photosynthesis in the unicellular alga *Acetabularia crenulata* was first demonstrated by Sweeney and Haxo (1961). In their investigation they were able to demonstrate a persistent rhythm of the maximal rate of photosynthesis in enucleate fragments of the algae. Richter (1963) confirmed these observations and found that algae grown in continuous light till enucleation could be photoentrained following enucleation by a *LD* : *12*,12 light-dark cycle.

These experiments appear to rule out the necessity of a nucleus for photoentrainment. However, the results cannot be interpreted as ruling out DNA-dependent RNA synthesis in general, as chloroplasts contain DNA and can carry out DNA-dependent RNA synthesis on their own (for a discussion of this point see Vanden Driessche, 1967). Vanden Driessche (1966) has shown that *A. mediterranea* also has an endogenous rhythm of photosynthetic capacity and a circadian rhythm of chloroplast shape. Correlations between the rhythm of chloroplast shape and the functional capacity of the chloroplasts *vis-à-vis* photosynthetic capacity indicate that the changes in chloroplast shape are related to its photosynthetic state.

In a still later study Vanden Driessche (1967) has investigated the role of the nucleus in the circadian rhythms of *A. mediterranea*. Actinomycin D inhibits the rhythm of photosynthetic capacity in intact algae; it also prevents the rhythmic variation in chloroplast shape, at the same concentration and for the same length of treatment as used for the inhibition of the rhythm in photosynthetic capacity in enucleate algae at the concentration effective on whole algae (0·27 to 2·7 mg/ml). Likewise, the rhythm of variation of chloroplast shape in enucleate fragments is not sensitive to these concentrations of actinomycin D. On the other hand, ribonuclease was reported to abolish the rhythm of photosynthesis in the enucleate but not the nucleate fragments. Sweeney, Tuffli and Rubin (1967) also found that the rhythm of photosynthesis in enucleate fragments of *A. crenulata* was insensitive to actinomycin D. To complicate matters further, Schweiger, Wallraff and Schweiger (1964) clearly demonstrated in grafting experiments that the phase of the nucleus determines the phase of the rhythm. Hence, when a nucleus is present in this alga, phase information is primarily controlled by the nucleus.

To account for these apparent discrepancies Vanden Driessche (1967) proposed that in the absence of the nucleus, a nuclear messenger-RNA is stable in enucleate fragments, whereas the same messenger-RNA molecules undergo rapid turnover in intact algae. Such an interpretation is supported by the experiments of Cohen (1965) on *Paramecium bursaria*, discussed above, where active destruction of the rhythmic appearance of the mating-type substance is prevented by inhibition of protein synthesis. If the above interpretation is correct, enucleate fragments of *P. bursaria* prepared from cells exhibiting mating reactivity should keep their mating reactivity beyond the time when intact cells have already lost theirs.

The circadian clock in the phytoflagellate CHLAMYDOMONAS

Chlamydomonas reinhardii (RS) strain has been shown to possess a biological circadian clock which regulates the phototactic response and 'cell hatching' (Bruce, 1969). Demonstration of these two rhythms was accomplished in a continuous culture device operating on a chemostat principle. In contrast to the nephelostat experiments on *Tetrahymena* (Wille and Ehret, 1968a), there was no direct feedback of culture growth on the dilution rate. Steady-state growth was achieved in the Bruce continuous-culture apparatus when the cell density reached a level high enough for the mutual shading to reduce the growth rate. Under steady-state conditions the cell count remains constant when the cells are dividing randomly. If, however, cell hatching occurs synchronously, the log of the cell counts will periodically increase and decrease in a saw-tooth fashion. Thus a population of *C. reinhardii* can be synchronized by an *LD* : *12*,12 entrainment cycle, and a rhythm of cell-hatching will free-run in continuous light. The average period of the cycle was 24 hours 10 minutes. It is important to note that the dilution rate at the steady state was 50 hours. In this and in all of the other experiments on circadian rhythmicity in continuous cultures of *C. reinhardii*, the dilution rate always gave a culture with an average generation time that was infradian. In this same study Bruce found that there was endogenous circadian variation in the total protein per unit volume of culture. The maximum in total protein coincides with the maximum in cell numbers and occurs at the time in the daily cycle corresponding to the end of the dark phase.

The phototactic responses are similar to those obtained in non-dividing cultures of *Euglena*, and show the characteristic day-time responsiveness and night-time inactivity on an *LD* : *12*,12 cycle. Although the phototactic rhythm persists in continuous light, it is less pronounced than on the light–dark entrainment cycle.

Microscopic examination of samples removed from a culture on an *LD* : *12*,12 light–dark cycle indicate that most of the cells divide synchronously to produce two daughter cells, although a small fraction divide into four cells. Similar observations on a free-running circadian population suggested a greater degree of heterogeneity of cell stages which precluded attempts to establish the pattern of synchrony from direct microscopic examination.

It is noteworthy that in both the *Chlamydomonas* and *Tetrahymena* experiments circadian rhythms of cell division were achieved in continuous culture when the steady-state conditions of growth were such as to impose the infradian mode of growth on all cells of the population. A possible exception to this was noted by Bruce (personal communication). He observed the persistence of the cell division rhythm in *Chlamydomonas* cultures maintained in continuous light despite a ten-fold dilution of the culture. In this case, as in the previously described experiment of Barnett (1966), the cells would have had to retain the phase information through rapid ultradian growth. Bruce interprets this result as showing that the circadian clock is only very loosely coupled to the mitotic cycle. An alternative hypothesis that explains these exceptions is put forward in the next section.

New and old observations challenging some of the circadian–infradian rules

Is the circadian oscillation the sum of a large number of high-frequency oscillations?

The idea of a chemical clock regulating oscillatory behaviour in the cells of an organism was first suggested by Hoagland (1933). On the basis of some observations on the effect of fever-induced elevation of the body temperature on the time-sense of human subjects, Hoagland suggested that chemical clocks were pacemakers in oscillatory phenomena. The leading proponents of the high frequency chemical oscillator model for 'ticks' of the circadian biological clock base their notion on the recent work of Chance (1954). Chance found that during the transition to anaerobiosis in the yeast cells, *Saccharomyces carlsbergensis*, the higher level of reduced pyridine nucleotides (RPN) was often reached in an oscillatory manner. The periods of the high-frequency oscillations range from 5 to 90 minutes. Betz and Chance (1964) observed similar oscillations of RPN in a cell-free system, and Pye and Chance (1966) were able to obtain sustained oscillation in glycolysis by addition of the substrate trehalose. The oscillations of RPN can be terminated by dinitrophenol, arsenate or deoxyglucose, as well as by acetaldehyde and acetate. These results were interpreted as showing the involvement of both the ATP–ADP and DPNH–DPN systems, for which control functions in glycolysis have been postulated, in the generation of the oscillations (Betz and Chance, 1965). The frequency of the cycles has a large temperature coefficient which is fairly constant between 20° and 40°C (2·4), but is distinctly large in the range below 20°C (4·0). Betz (1966) has concluded that the source of the observed oscillations in yeast glycolysis may be due to alternating phases of inhibition and activation at phosphofructokinase and glyceraldehyd-phosphatedehydrogenase, two steps interconnected by the retarding aldolase step and further connected by feedback in the ATP–ADP system. In view of the extreme lability of the yeast cell oscillations to metabolic poisons, and of the large positive temperature coefficients for the effect of temperature on the frequency of the oscillation, it appears very unlikely that the sum of a large number of such temperature-dependent high-frequency oscillations would be the 'ticks' of the relatively temperature-independent circadian clock (Ehret, Wille and Trucco, 1968).

It is unfortunate that to date the high-frequency techniques (viz. measures of high-frequency outputs) have not been applied to classical circadian organisms, and reciprocally that circadian rhythms have not been systematically investigated in those eukaryotic microorganisms (especially *Saccharomyces*) so commonly used in the high-frequency experiments. Moreover, the converse strategy of regarding the consequences of diverse high-frequency inputs on circadian outputs are cogent here. Thus, Edmunds and Funch (1969) observed free-running circadian rhythms in *Euglena* under a variety of high-frequency LD regimes whose periods were integral submultiples of 24 hours, and in some whose periods were *not* submultiples of 24 hours (viz. LD : 5,5 or LD : 8,8). In most regimes

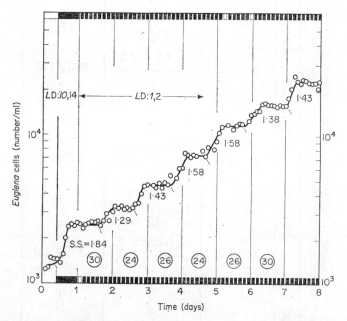

Figure 13.15. Effects of the high-frequency LD : *1*,2 cycle on the cell division rhythm in cultures of *Euglena* following initial synchronization by a LD : *10*,14 regime. The successive periods (hours) of an apparently free-running rhythm observed in the population are encircled at the bottom of the figure. Step-sizes (S.S.) are indicated after each division burst. (From Edmunds and Funch, 1969)

(LD : $\frac{1}{4}\frac{1}{2}$; LD : $_{\bar{2}}$,1; LD : *1*,2; LD : *1*,3; LD : *2*,4; LD : *2*,6; LD : *4*,4) synchronous cell division occurred in the culture with an average period of 26 to 27 hours, although only a fraction of the cells divided during any one burst. The results, derived from growth curves like that shown in Figure 13.15 are summarized in Table 13.1. They point out that the imposed high-frequency LD cycles might *impose* the rhythm, but find this hypothesis 'unattractive' because one would have to accept some sort of 'slipping clutch' mechanism to account for the synchronization to a

Table 13.1. Characteristics of the cell division rhythm in cultures of *Euglena* grown under six high-frequency *LD* cycles having periods that were integral submultiples of 24 hours. (After Edmunds and Funch, 1969)

Parameter	LD : 4,½	LD : ½,1	LD : 1,2	LD : 1,3	LD : 2,4	LD : 2,6
T = period of LD cycle (hour)	0·75	1·5	3	4	6	8
Light received/24 hour	8	8	8	6	8	6
Initial $\Delta\phi$ (hour)	−10	−30	−6	−10	−8	−4
Successive step-sizes (SS) No. 1	1·58	1·19	(1·29)	(1·14)	(1·41)	(1·38)
2	1·58	1·27	1·43	1·15	1·54	1·50
3	1·83	1·22	1·58	1·25	1·50	1·50
4	1·73	1·19	1·58	1·24	1·49	1·41
5	1·47	1·23	1·38	1·16	1·41	1·36
6	1·43	—	1·43	1·20	1·44	1·29
	$\overline{SS}$ = 1·60	$\overline{SS}$ = 1·22	$\overline{SS}$ = 1·48	$\overline{SS}$ = 1·20	$\overline{SS}$ = 1·47	$\overline{SS}$ = 1·41
τ = period of successive division cycles (hour) No. 1	28	26	24	22	24	26
2	24	26	26	24	32	28
3	32	26	24	26	24	26
4	32	24	26	28	26	30
5	30	26	30	22	29	26
	$\bar\tau$ = 29·2	$\bar\tau$ = 25·6	$\bar\tau$ = 26·0	$\bar\tau$ = 24·4	$\bar\tau$ = 27·0	$\bar\tau$ = 27·7

variable period which never equals either T or 24 hours and because 'random' illumination regimes (*LD* high-frequency) will also elicit a free-running rhythm having almost identical characteristics.

Photoentrainment of endogenous rhythms having non-circadian periodicities

Pirson and Schön (1957) discovered a rhythm in growth, photosynthetic capacity and respiration in the alga *Hydrodictyon*. The three rhythms are endogenous and persist in either continuous light or darkness. The peculiar feature of the rhythms in this alga is that photoentrainment cycles other than *LD* : *12*,12, such as *LD* : *6*,6 or *LD* : *9*,9, are able to induce a persistent rhythm with the same cycle length, as the entrainment cycle, instead of free-running at the usual circadian period. As Bünning (1967) points out the 12-hour rhythm can be argued away as an effect of desynchronization, but the cycle length of 18 hours cannot be explained in this way.

Another peculiar feature in the alga *Hydrodictyon* reported by Pirson and Schön (1957) is that intermittent carbon dioxide supply can substitute for light as a *Zeitgeber* for the induction of an endogenous rhythm. This result is reminiscent of the previously alluded to induction of an endogenous rhythm of cell division in *Tetrahymena* by intermittent feeding. In *Tetrahymena* entrainment of the rhythm by onset of feeding occurs under continuous light conditions just before the cells have switched from the ultradian to the infradian mode of growth. In both *Tetrahymena* and *Hydrodictyon* entrainment may be operating through the limitation of an essential metabolite whose threshold concentration within the cell determines the switch from the ultradian to the circadian–infradian mode of growth.

The occurrence of an endogenous rhythm of growth zonation in several fungi with infradian free-running periods of, say, 56 hours has been demonstrated by Jerebzoff (1965) and by Jerebzoff and Lambert (1967, 1968). The rhythm is photoentrainable and has a cycle length of 24 hours on an *LD* : *12*,12 light–dark regime. The period of the rhythm is controlled by the type of nutrition. In addition, Jerebzoff (1965) has shown that in the fungi certain amino acids promote periodicity, while others do not. An infradian τ of 56 hours in *Myrothecium* may be accompanied by a minor periodic component which is approximately 24 hours, and hence circadian. Since the longer period is not a multiple of the circadian period it requires some explanation, as does the 18-hour period length in the algae.

Examination of the process of growth and cellular multiplication in both groups has provided an interesting challenge to the concept of *average* cell generation times. *Hydrodictyon* is not a unicellular alga but a coenobial mass, i.e. a network of cells fused together and apparently restricted to addition of new material at fixed growing points in the mass. Likewise, the fungi are frequently multinucleate masses of filaments in which growth is restricted to the terminal segments of the filaments. Hence, from the viewpoint of the circadian–infradian rule the *average* cell generation time as measured from heterogeneous multinucleate masses covers a multitude of sins (in *Hydrodictyon*, as well as in the moulds and fungi, this average time is derived from the doubling time of the mass!). Since growth is not uniform with respect to each nucleus in the mass, the assignment of ultradian and infradian modes of growth on the basis of mass doubling time does not reflect the actual average nuclear condition in the multinucleate mass. It is possible for one part of the organism (the terminal segments of the hyphae) to have nuclei which are in the ultradian mode, and for another part of the organism to have nuclei which are in the infradian mode. If the circadian–infradian rule were to hold, only those nuclei in the infradian mode would be capable of manifesting a circadian rhythm in either the mould or algae. In addition the common cytoplasm permits internuclear communication and hence the synchronization of some physiological activities over a limited range of the filament. Since the period of the zonation rings in the mould is a periodicity in the linear differentiation of the mass of hyphae, the period length may be the result of the rate of propagation of a wave of transitions from the ultradian to the circadian–infradian mode of growth. Sporulation events are then triggered by the circadian output of nuclear products which are only produced in the region where nuclei are in the infradian state, and are seen as periodic linear growth zones. The length of the period is then a consequence of the rate of propagation of the circadian–infradian transition, and would be highly dependent on the nutritional state and the geometry of the race tube. The data of Chevaugeon and Nguyen Van (1969) and especially of Nguyen Van (1968) lend further support to these interpretations.

Absence of circadian rhythmicity in prokaryotes

It is now generally agreed that the circadian temporal organization is a property shared by all eukaryotes, i.e. in cells having a nucleus. The full significance of this sharing lies especially in the structural properties that distinguish the circadian–eukaryotes from the prokaryotes. Ehret and Trucco (1967) have postulated that regulation of the temporal aspects of the gene action system at a refined enough level to permit a circadian oscillation requires the multienvelope condition (Ehret, 1960b; Ehret and Barlow, 1960) found only in eukaryotes.

We would like to point out that perhaps an even more basic difference between prokaryotes and eukaryotes is the number of replicons per cell. Recent developments in the molecular biology of the DNA-replication process in prokaryotes have led to the knowledge that the entire genomic equivalent is a single DNA molecule, which during replication has but one growing point per molecule which proceeds in only one direction during the duplication of the entire genophore (Cairns, 1963). A unit of DNA replicated in this manner is called a replicon. Similar evidence has been adduced for the process of DNA replication in eukaryotes (Huberman and Riggs, 1968; Okada, 1968; Plaut, 1963). The number of replicating segments (Huberman and Riggs, 1968) or replicons is estimated to be in the order of 10^3 to 10^4 per vertebrate nucleus. On the basis of comparative values of the total amount of DNA per nucleus (see Table 1 in Markert, 1968), all eukaryotic cells must be characterized by multirepliconic nuclei. From an entirely different kind of observation, namely, redundancy of families of nucleotide sequences in all eukaryotic cells but not prokaryotes (Britten and Kohne, 1967), the presence of many similar DNA units has been inferred.

Prokaryotes and eukaryotes differ in one other respect—their modes of growth. By and large, the prokaryotes are restricted to the ultradian mode of growth or to 'non-growth'—the stationary phase. Eukaryotes, on the other hand, can express either the ultradian or the circadian–infradian mode, as well as a non-growing phase. In summary, the eukaryotic level of organization is associated with the multienvelope, multireplicon, circadian–infradian mode of growth, and represents the distinctive biotype of all higher plants and animals; the prokaryotic level of organization is restricted to the one-envelope, one-replicon, ultradian mode of growth, representative of blue-green algae, bacteria and actinomycetes. Are there any exceptions to this rule? In an attempt to resolve this question, Halberg and Conner (1961) reexamined the data of Rogers and Greenbank (1930) for evidence of circadian periodicity in the intermittent growth and/or the motility pattern seen in race-tube cultures of the bacterium *Escherichia coli*. By the application of a number of computational procedures they were able to demonstrate the existence of a cyclic component which in a periodogram analysis had a period of 21 hours. A minor component in the periodogram analysis with a period of about 16 hours was also present. As Halberg and Conner (1961) admit, the original data which were analysed are no longer available. This and a considerable number of questionable experimental procedures which have to do with what was actually measured, preclude ready acceptance of this case as an example of circadian rhythmicity in a prokaryote.

Circadian periodicity in cells undergoing more than one doubling in a 24-hour period

Bernstein (1964) made a study of the characteristics of light-induced cell division synchrony in the unicellular flagellate *Chlamydomonas moewusii*. In continuous bright light (600 foot-candles) growth is logarithmic and has an ultradian generation time of 5–6 hours. A daily rhythm of synchronous cell division occurs when the cells are subjected to an entrainment cycle of LD : *12*,12. Unlike *Euglena*, the synchronized bursts of cell division yield an eight-fold increase in cell number in each 24-hour cycle. The response of a cell population entrained on an LD : *12*,12 cycle to continuous light is as follows: after the switch-up in level of irradiance there is one more cycle of synchronized cell division; it occurs 24 hours later. The synchronized burst of cell division is not a doubling, but an eight-fold increase in cell number as in the entrained cycle. The rhythm damps out after only one cycle in continuous light. It is succeeded by asynchronous logarithmic growth 12 hours before the next expected endogenous rhythmic burst. The damping out in continuous bright light is not unexpected in consideration of the autotrophic condition of growth, and the results are similar to those found for the desynchronization of the *Euglena* cell division rhythm (Edmunds, 1965).

In an earlier report, Bernstein (1959) found that placing cells removed from the late logarithmic or stationary phase (infradian?) in new medium brought about a spontaneous synchronized burst of cell division, even under constant bright light conditions. In a series of experiments on the effect of adverse growth conditions on the capacity of cells to be synchronized, Bernstein (1964) concluded that photoentrainment of the cell division rhythm was not due to endogenous oscillations in *Chlamydomonas* cells, but to a gating of the cell division event by adverse physiological conditions.

This conclusion is not supported by more recent experiments, particularly those by Bruce (1969) discussed above, but may serve to supplement the fact that photoentrainment is not the only way to induce endogenous rhythmicity. The spontaneous synchronized cell division bursts and the eight-fold increase in cell number on a circadian period appear paradoxical and are at first difficult to explain. However, an explanation that encompasses the circadian–infradian rules is found in the rules which govern the

PLATE V

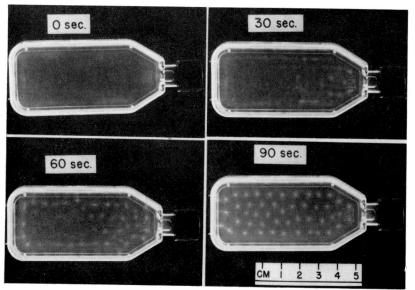

Figure 13.8. Photographs illustrating the kinetics of development of the Bénard-cell pattern in a dense motile culture of *Tetrahymena pyriformis* (W). The culture chambers are plastic tissue-culture flasks. (Wille and Ehret, 1968b)

rate of genophore replication, and its coupling to the cell cycle recently discovered for the prokaryotes. Cooper and Helmstetter (1968) have proposed a model whereby the rate of DNA replication is completely independent of the rate of cell division in the bacterium *Escherichia coli.* They studied the rate of DNA synthesis during the cell cycle in *E. coli* B/r growing with doubling times between 20 and 60 minutes, and concluded that the replication process of the entire genophore of the bacterium could be described by two constants, C, the time for a replication point to traverse the entire replicon, and D, the time between the end of a round of replication and cell division. In bacterial cells when the sum, C plus D, is equal to or greater than the cell doubling time, then there is only one growing point per replicon; when the doubling time is less than the sum, C plus D, the number of growing points per replicon increases. Furthermore, Helmstetter (1967) and Helmstetter and Cooper (1968) found that initiation of new rounds of DNA replication do not coincide with the beginning of the cell cycle, but can occur at increasingly earlier times in the cell cycle as the average generation time of the population decreases.

On the basis of this model of DNA replication, and by extending it to the multirepliconic eukaryotic chromosomes, the eight-fold increase in cell number once each 24-hour cell cycle in *Chlamydomonas moewusii* on an *LD* : *12*,12 entrainment cycle is understood as a result of growth rate control of the number of growth points per chronon–replicon. The time elapsed for one growing point to traverse the longest replicon in the set of multireplicons is such that the value of the constant corresponding to C is close to 24 hours. Hence, when cells are in the infradian mode of growth, one growing point per chronon is sufficient to just duplicate the entire complement of nuclear DNA. One growing point per chronon–replicon would not provide a sufficient rate of DNA synthesis to complete one round of replication if the cell doubling time were ultradian; in which case multiple growing points per replicon would be needed. Consider for example a replicon that contains four growing points: at the end of the 24-hour cycle the eight genomic equivalents produced are packaged up into eight new cells. At the start of each cell division cycle each cell receives a twice branched (three branch points) chronon–replicon with four growing points. This accounts for the production of eight genomic equivalents in one cell cycle whose period is circadian. The model predicts that the transition from the ultradian to the circadian–infradian mode of growth will be accompanied by a reduction from multiple growing points per replicon to a single growing point per replicon. Loss of circadian periodicity in cell division in rapid logarithmic growth occurs when the numbers of growing points become variable, not all replicons saturating

14

at the same number of growing points. The spontaneous occurrence of synchronized cell division bursts when cells are transferred from infradian to ultradian conditions can be understood as a consequence of the simultaneous initiation of new rounds of DNA replication, caused by a growth rate triggered accumulation of initiator substance (Maaløe and Kjeldgaard, 1966). The Cutler and Evans (1966) stationary phase technique for inducing cell division synchrony in bacteria may be another example of the coupling of the cell division cycle to the initiation of new rounds of DNA replication.

Inheritance of phase information through ultradian asexual reproduction

Barnett (1966) reported experiments in which phase information was inherited asexually through many cycles of *rapid* growth in *Paramecium multimicronucleatum*. In this instance, however, *expression* of the circadian rhythm was restricted to cells in the infradian state. For example, mating-type reversals could be entrained in cells dividing on the average more than once a day, and reappear in phase at least thirteen cell generations after growth in continuous darkness, during which time no rhythm is expressed.

This appears to be a clear case of entrainment occurring during the ultradian mode of growth. We could suggest an explanation involving branched-replicating chronons in the macronucleus, similar to that given in the discussion of the *Chlamydomonas* autospore release. The cilate macronucleus which divides amitotically might permit binary fission to occur even while DNA replication continues, thus allowing circadian chronons to function with ultradian division times. The challenge remains, however, as to why *Tetrahymena*, which also possesses a macronucleus, should not also be entrainable under ultradian conditions. Perhaps the experimental opportunity to test whether the $G-E-T$ effect is a mere consequence of the physiological system under study and the experimental design applied to it, or a general property of circadian systems, lies in a concentrated effort to entrain cells during ultradian growth.

Epilogue

Protozoa and other free-living eukaryotic microorganisms represent the lowest level of organization to manifest the general biological phenomenon of circadian rhythmicity. Because of their organizational simplicity they provide the best material with which to test the various theories for regulation at cellular and molecular levels on matters that range from the photo-induction of phase shift to the physical basis for the circadian clock escapement. The introduction of continuous culture devices in automated systems and the wider use of non-green protistan cells should stimulate a new breed of investigators to launch new waves of badly needed experiments (such as on the molecular aspects of action spectra for phase advance and phase delay), until now neglected because of technical limitations.

Acknowledgements

We are pleased to acknowledge the very valuable advice and assistance of Dr. Audrey Barnett and Arlene Zadylak.

REFERENCES

Barnett, A. (1961). Inheritance of mating type and cycling in *Paramecium multimicronucleatum*, syngen 2. *Am. Zoologist*, **1**, 341.

Barnett, A. (1965). A circadian rhythm of mating type reversals in *Paramecium multimicronucleatum*. In J. Aschoff (Ed.), *Circadian clocks*, North Holland Publ. Co., Amsterdam. pp. 305–308.

Barnett, A. (1966). A circadian rhythm of mating type reversals in *Paramecium multimicronucleatum*, syngen 2, and its genetic control. *J. Cell. Physiol.*, **67**, 239.

Beale, G. H. (1954). *The Genetics of Paramecium aurelia*, Cambridge University Press, Cambridge.

Bernstein, E. (1959). The synchronization of cell division in cultures of *Chlamydomonas moewusii*. *J. Protozool.*, **6**, 15.

Bernstein, E. (1964). Physiology of an obligate photoautotroph (*Chlamydomonas moewusii*). I. Characteristics of synchronously and randomly reproducing cells and an hypothesis to explain their population curves. *J. Protozool.*, **11**, 56.

Betz, A. (1966). Metabolic flux in yeast cells with oscillatory controlled glycolysis. *Physiologia Plantarum*, **19**, 1049.

Betz, A., and B. Chance (1964). Influence of inhibitors and temperature on the oscillation of reduced pyridine nucleotides in yeast cells. *Arch. Biochem. Biophys.*, **109**, 579.

Betz, A., and B. Chance (1965). Phase relationship of glycolytic intermediates in yeast cells with oscillatory metabolic control. *Arch. Biochem. Biophys.*, **109**, 584.

Britten, R. J., and D. E. Kohne (1965). Nucleotide sequence repetition in DNA. *Carnegie Institution of Washington Year Book*, **65**, 78.

Brinkmann, K. (1966) Temperatureinflüsse auf die Circadiane Rhythmik von *Euglena gracilis* bei Mixotrophic und Autotrophic. *Planta*, **70**, 344.

Brinkmann, K., G. Schnabel, and M. Kirschstein (1968). Control of circadian rhythm of light–dark cycles capable of circadian entrainment in green and colorless strains of *Euglena gracilis*. *Fifth International Congress on Photobiology*, Hanover, New Hampshire. p. 43.

Brown, F. A., and H. M. Webb (1949). Studies of the daily rhythmicity of the fiddler crab, Uca. modifications by light. *Physiol. Zool.*, **22**, 136.

Bruce, V. G. (1965). Cell division rhythms and the circadian clock. In J. Aschoff (Ed.), *Circadian Clocks*, North Holland Publ. Co., Amsterdam. pp. 125–138.

Bruce, V. G. (1969). The biological clock in *Chlamydomonas reinhardii*. *J. Protozool.* (In the press.)

Bruce, V. G., and C. Pittendrigh (1956). Temperature independence in a unicellular 'clock'. *Proc. Natl. Acad. Sci. U.S.*, **42**, 676.

Bruce, V. G., and C. Pittendrigh (1957). Endogenous rhythms in insects and microorganisms. *Amer. Nat.*, **91**, 179.

Bruce, V. G., and C. Pittendrigh (1958). Resetting the *Euglena* clock with a single light stimulus. *Amer. Nat.*, **92**, 295.

Bruce, V. G., and C. Pittendrigh (1960). An effort of heavy water on the phase and period of the circadian rhythm in *Euglena*. *J. Cell. Comp. Physiol.*, **56**, 25.

Bühnemann, F. (1955). Die Rhythmische Sporenbildung von *Oedogonium cardiacum* Wittr. *Biol. Zentr.*, **74**, 1.

Bünning, E. (1936). Die Endogene Tagesrhythmik als Grundlage der Photoperiodischen Reaktion. *Ber dtsch. bot. Ges.*, **54**, 590.

Bünning, E. (1967). *Physiological Clock*, Springer-Verlag, New York.

Cairns, J. (1963). The chromosome of *Escherichia coli*. *Cold Spring Harbor Symp. Quant. Biol.*, **28**, 43.

Chance, B. (1954). Enzyme mechanisms in living cells. In W. D. McElroy and B. Glass (Eds.), *A Symposium on the Mechanism of Enzyme Action*, Johns Hopkins Press, Baltimore, Maryland. p. 399.

Chevaugeon, J., and H. Nguyen Van (1969). Internal determinism of hyphal growth rhythms. *Trans. Br. Mycol. Soc.*, **53** (I), 1.

Cohen, L. W. (1964). Diurnal intracellular differentiation in *Paramecium bursaria*. *Exptl. Cell Res.*, **36**, 398.

Cohen, L. W. (1965). Biochemical basis for the circadian rhythm of mating in *Paramecium bursaria*. *Exptl. Cell Res.*, **37**, 360.

Cook, J. R. (1961). *Euglena gracilis* in synchronous division. I. Dry mass and volume characteristics. *Plant Cell Physiol.*, **2**, 199.

Cooper, S., and C. E. Helmstetter (1968). Chromosome replication and the division cycle of *Escherichia coli* B/r. *J. Mol. Biol.*, **31**, 519.

Cutler, R. G., and J. E. Evans (1966). Synchronization of bacteria by a stationary-phase method. *J. Bacteriol.*, **91**, 469.

Edmunds, L. N., Jr. (1965). Studies on synchronously dividing cultures of *Euglena gracilis* Klebs (Strain Z). I. Attainment and characterization of rhythmic cell division. *J. Cell Physiol.*, **66**, 147.

Edmunds, L. N., Jr. (1966). Studies of synchronously dividing cultures of *Euglena gracilis* Klebs (Strain Z). III. Circadian components of cell division. *J. Cell Physiol.*, **67**, 35.

Edmunds, L. N., Jr., and R. Funch (1969). Effects of 'skeleton' photoperiods and high frequency light–dark cycles on the rhythm of cell division in synchronized cultures of *Euglena*. *Planta.*, **87**, 134.

Ehret, C. F. (1951). The effects of visible, ultraviolet, and X-irradiation on the mating reaction in *Paramecium bursaria*. *Anat. Rec.*, **111**, 112.

Ehret, C. F. (1953). An analysis of the role of electromagnetic radiations in the mating reaction of *Paramecium bursaria*. *Physiol. Zool.*, **26**, 274.

Ehret, C. F. (1959a). Induction of phase shift in cellular rhythmicity by far ultraviolet and its restoration by visible radiant energy. In R. B. Withrow (Ed.), *Photoperiodism and Related Phenomena in Plants and Animals*, Washington: A.A.A.S. pp. 541–550.

Ehret, C. F. (1959b). Photobiology and biochemistry of circadian rhythms in non-photosynthesizing cells. *Fed. Proc.*, **18**, 1232.

Ehret, C. F. (1960a). Action spectra and nucleic acid metabolism in circadian rhythms at the cellular level. *Cold Spring Harbor Symp. Quant. Biol.*, **25**, 149.

Ehret, C. F. (1960b). Organelle systems and biological organization. *Science*, **132**, 115.

Ehret, C. F., and J. S. Barlow (1960). Toward a realistic model of a biological period-measuring mechanism. *Cold Spring Harbor Symp. Quant Biol.*, **25**, 217.

Ehret, C. F., and E. Trucco (1967). Molecular models for the circadian clock. I. The chronon concept. *J. Theoret. Biol.*, **15**, 240.

Ehret, C., J. Wille, and E. Trucco (1968). The circadian oscillation: an integral and undissociable property of eukaryotic gene-action systems. In abstracts of *Fed. European Biochem. Societies, The Czechoslovak Biochemical Society*, Prague, p. 235.

Eisler, W. J., Jr., and R. B. Webb (1968). Electronically controlled continuous culture device. *Appl. Microbiol.*, **16 (9)**, 1375.

Engelmann, W. (1966). Effect of light and dark pulses on the emergence rhythm of *Drosphila pseudoobscura*. *Experientia*, **22**, 606.

Feldman, J. F. (1967). Lengthening the period of a biological clock in *Euglena* by cycloheximide, an inhibitor of protein synthesis. *Proc. Natl. Acad. Sci. U.S.*, **57**, 1080.

Feldman, J. F. (1968). Circadian rhythmicity in amino acid incorporation in *Euglena gracilis*. *Science*, **160**, 1454.

Gmitro, J., and L. E. Scriven (1966). A physiochemical basis for pattern and rhythm. In J. F. Warren (Ed.), *Intracellular Transport*, Academic Press, London, New York. pp. 221–255.

Halberg, F., and R. Conner (1961). Circadian organization and microbiology: variance spectra and a periodogram on behaviour of *Escherichia coli*, growing in fluid culture. *Proc. Minn. Acad. Sci.*, **29**, 227.

Hastings, J. W. (1959). Unicellular clocks. *Ann. Rev. Microbiol.*, **13**, 297–312.

Hastings, J. W., and L. Astrachan (1959). A diurnal rhythm of photosynthesis. *Fed. Proc.*, **18**, 65.

Hastings, J. W., and B. M. Sweeney (1959). The *Gonyaulax* clock. In R. B. Withrow (Ed.), *Photoperiodism and Related Phenomena in Plants and Animals*, Washington: A.A.A.S. pp. 567–584.

Hastings, J. W., B. M. Sweeney, and M. M. Mullin (1962). Counting and sizing of unicellular marine organisms. *Ann. N.Y. Acad. Sci.*, **99**, 280.

Helmstetter, C. E. (1967). Rate of DNA synthesis during the division cycle of *Escherichia coli* B/r. *J. Mol. Biol.*, **24**, 417.

Helmstetter, C. E., and S. Cooper (1968). DNA synthesis during the division cycle of rapidly growing *Escherichia coli* B/r. *J. Mol. Biol.*, **31**, 507.

Hoagland, H. (1933). The physiological control of judgment of duration: evidence for a chemical clock. *Jour. Gen. Psychol.*, **9(2)**, 267.

Huberman, J. A., and A. D. Riggs (1968). On the mechanism of DNA replication in mammalian chromosomes. *J. Mol. Biol.*, **32**, 327.

Jennings, H. (1939). Genetics of *Paramecium bursaria*. I. Mating types and groups, their interrelations and distribution; mating and self sterility. *Genetics*, **24**, 202.

Jerebzoff, S. (1965). Manipulation of some oscillating systems in fungi by chemicals. In J. Aschoff (Ed.), *Circadian Clocks*, North Holland Publ. Co., Amsterdam. pp. 183–189.

Jerebzoff, S. et E. Lambert (1967). Détection par la methode du spectre de variance de deux rythmes régissant simultanément la sporulation de *Myrothecium verrucaria* (Alb. et Schw.), Ditm. et Fréres. *C. R. Acad. Sc. Paris*, **264**, 322.

Jerebzoff, S. et E. Lambert (1968). Caractéres particuliers d'un rythme journalier de zonations chez *Penicillium* expansum link. *C. R. Acad. Sci. Paris*, **266**, 1269.

Karakashian, M. W. (1965). The circadian rhythm of sexual reactivity in *Paramecium aurelia*, syngen 3. In J. Aschoff (Ed.), *Circadian Clocks*, North Holland Publ. Co., Amsterdam. pp. 301–304.

Karakashian, M. W. (1968). The rhythm of mating in *Paramecium aurelia*, syngen 3. *J. Cell Physiol.*, **71**, 197.

Karakashian, M. W., and J. W. Hastings (1962). The inhibition of a biological clock by actinomycin D. *Proc. Natl Acad. Sci. U.S.*, **48**, 2130.

Kleitman, N. (1949). Biological rhythms and cycles. *Physiol. Rev.*, **29**, 1–30.

Knuese, W. R., and M. S. Shorb (1966). Light induced changes in *Tetrahymena pyriformis*. *Proc. Soc. of Protozool.*, Abstract no. **29**.

Loefer, J. B., and R. B. Mefferd, Jr. (1952). Concerning the pattern formation by free-swimming microorganisms. *Am. Naturalist*, **86**, 325.

Maaløe, O., and N. Kjeldgaard (1966). *The Control of Macromolecular Synthesis*, Benjamin Press, New York.

Markert, C. L. (1968). Panel discussion: present status and perspectives in the study of cytodifferentiation at the molecular level. *J. Cell Physiol.*, **72**(1), 213.

McMurry, L. (1968). Coupling among circadian rhythms in *Gonyaulax*. *Fifth International Congress on Photobiology*, Hanover, New Hampshire. p. 43.

Newell, R. C. (1969). Effect of fluctuations in temperature on the metabolism of intertidal invertebrates. *Amer. Zool.*, **9**, 293.

Nguyen Van, H. (1968). Étude de rythmes internes de croissance chez le *Podospora anserina*. *Annls. Sci. Nat.* (Sér. 12), **9**, 257.

Okada, S. (1968). Replicating units (replicons) of DNA in cultured mammalian cells. *Biophys. J.*, **8**, 650.

Palmer, J. D., L. Livingston and D. Zusy (1964). A persistent diurnal rhythm in photosynthetic capacity. *Nature*, **203**, 1087.

Pavlidis, T. (1969). Populations of interacting oscillators and circadian rhythms. *J. Theoret. Biol.*, **22**, 418.

Petropulos, S. F. (1964). Automatic sampling device for the study of synchronized cultures of microorganisms. *Science*, **145**, 268.

Pirson, A. und H. Lorenzen (1958). Ein endogener Zeitfaktor bei der Teilung von *Chlorella*. *Z. Bot.*, **46**, 383.

Pirson, A., and W. J. Schön (1957). Experiments on the analysis of the metabolic periodicity in *Hydrodictyon*. *Flora*, **144** (3), 447.

Pittendrigh, C. S. (1960). Circadian rhythms and the circadian organization of living systems. *Cold Spring Harbor Symp. Quant. Biol.*, **25**, 159.

Platt, J. R. (1961). Bioconvection patterns in cultures of free-swimming organisms. *Science*, **133**, 1766.

Plaut, W. (1963). On the replication organization of DNA in the polytene chromosomes of *Drosophila melanogaster*. *J. Mol. Biol.*, **7**, 632.

Pohl, R. (1948). Tagesrhythmus im phototaktischen Verhalten der *Euglena gracilis*. *Z. Naturforsch.*, **3b**, 367.

Pye, K., and B. Chance (1966). Sustained sinusoidal oscillations of reduced pyridine nucleotide in a cell-free extract of *Saccharomyces carlsbergensis*. *Proc. Natl Acad. Sci. U.S.*, **55**, 888.

Richter, G. (1963). Die Tagesperiodik der Photosynthese *Acetabularia* und ihre Albangigkeit von Kernaktiritat, RNS- und Protein-Synthese. *Z. Naturforsch.*, **18b**, 1085.

Robbins, W. J. (1952). Patterns formed by motile *Euglena gracilis* var. *bacillaris*. *Bull. Torrey Botan. Club*, **79**, 107.

Rogers, L. A., and G. R. Greenbank (1930). The intermittent growth of bacterial cultures. *Jour. Bact.*, **19**(3), 181.

Rudzinska, M. A., and S. Granick (1953). Protoporphyrin production of *Tetrahymena geleii*. *Proc. Soc. Exptl Biol. Med.*, **83**, 525.

Schweiger, E. H., H. Wallraff, and H. Schweiger (1964). Endogenous circadian rhythm in cytoplasm of *Acetabularia*: influence of the nucleus. *Science*, **146**, 658.

Sonneborn, T. M. (1938). Mating types in *Paramecium aurelia*: diverse conditions for mating in different stocks; occurrence, number and interrelations of the types. *Proc. Amer. Phil. Soc.*, **79**, 411.

Sonneborn, T. M., and D. Sonneborn (1958). Some effects of light on the rhythm of mating type changes in stock 236-6 of syngen 2 of *P. multimicronucleatum*. *Anat. Rec.*, **131**, 601.

Sweeney, B. M. (1960). The photosynthetic rhythm in single cells of *Gonyaulax polyedra*. *Cold Spring Harbor Symp. Quant. Biol.*, **25**, 145.

Sweeney, B. M. (1963). Resetting the biological clock in *Gonyaulax* with ultraviolet light. *Plant Physiol.*, **38**(6), 704.

Sweeney, B., and J. Hastings (1958). Rhythmic cell division in populations of *Gonyaulax polyedra*. *J. Protozool.*, **5**, 217.

Sweeney, B. M., and J. W. Hastings (1960). Effects of temperature upon diurnal rhythms. *Cold Spring Harbor Symp. Quant. Biol.*, **25**, 87.

Sweeney, B. M., and F. T. Haxo (1961). Persistence of a photosynthetic rhythm in enucleated *Acetabularia*. *Science*, **134**, 1361.

Sweeney, B. M., C. Tuffli, and R. Rubin (1967). The circadian rhythm in photosynthesis in *Acetabularia* in the presence of actinomycin D, puromycin, and chloramphenicol. *J. Gen. Physiol.*, **50**, 647.

Szyszko, A. H., B. L. Prazak, C. F. Ehret, W. J. Eisler, and J. J. Wille (1968). A multi-unit sampling system and its use in the characterization of ultradian and infradian growth in *Tetrahymena*. *J. Protozool.*, **15**, 781.

Vanden Driessche, T. (1966). Circadian rhythms in *Acetabularia*: photosynthetic capacity and chloroplast shape. *Exptl Cell Res.*, **42**, 18.

Vanden Driessche, T. (1967). The role of the nucleus in the circadian rhythms of *Acetabularia mediterranea*. *Biochem. Biophys. Acta*, **126**, 456.

Volm, M. (1964). Die Tagesperiodik der Zellteilung von *Paramecium bursaria*. *Z. fur vergl. Physiol.*, **48**, 157.

Wille, J. J., Jr., and C. F. Ehret (1968a). Light synchronization of an endogenous circadian rhythm of cell division in *Tetrahymena. J. Protozool.*, **15**, 785.

Wille, J. J., Jr., and C. F. Ehret (1968b). Circadian rhythm of pattern formation in populations of a free-swimming organisms, *Tetrahymena. J. Protozool.*, **15**, 789.

Zeuthen, E. (1958). Artificial and induced periodicity in living cells. In *Advances in Biological and Medical Physics*, Vol. 6. Academic Press, London, New York. pp. 37–73.

CHAPTER 14

Bioluminescence

R. L. AIRTH, G. ELIZABETH FOERSTER AND R. HINDE*

Department of Botany and the Cell Research Institute, The University of Texas at Austin, Austin, Texas, U.S.A.

Research cited in this review which originated in the authors' laboratory was supported in part by the United States Public Health Service and The National Science Foundation.

* Presently at the School of Biological Sciences, MacQuarie University, North Ryde 2113, N.S.W., Australia.

Introduction

An exhaustive detailed review resulting in an annotated bibliography will not be attempted. It is hoped that the extent of the citation will be sufficiently broad to enable detailed information on specific systems to be located. The biological implications of bioluminescence can lead the investigator into such areas as anatomy, behaviour, biochemistry, ecology, genetics and physiology. Hence, the areas covered in this chapter reflect the authors' preferences.

The emission of light by a living organism is known as bioluminescence. The wavelength of maximum emission is a function of the system under consideration or, for *in vitro* studies, the reaction conditions (Johnson, 1967). Peak emissions have been measured (*in vitro*) from approximately 460 nm in *Cypridina* (ostracod) and *Apagon* (fish) to 580 nm in *Photinus* (firefly). The implication of light emissions in this spectral region is that 40–75 kcal per einstein must be made available with the oxidation of a mole of substrate (luciferin). One of the unique characteristics of bioluminescence is that this energy must become available in a single step; hence an intermediate in a highly excited electronic state must be formed. This excited intermediate will emit light when it returns to the ground level. Generally, biochemical studies of bioluminescence have been directed toward the identification of the intermediates involved in the enzymatic oxidation of luciferin, which is catalysed by the enzyme luciferase. The terms luciferin and luciferase are generic terms, in that cross-reactivity between different systems usually does not occur. If the luminescent systems are derived from different families, one does not expect cross-reactivity of the respective luciferins and luciferases. However, there are exceptions to this (Cormier, Crane and Nakano, 1968; Johnson, 1967).

Certain generalizations concerning type reactions in bioluminescence were compiled by Cormier and Totter (1964, 1968). Table 14.1 represents a modification of their original classification system. Hastings (1966) has added the fifth-type reaction, which has been found to take place in certain hydromedusae and dinoflagellates. This chapter will be concerned with organisms that carry out type reactions 1, 3 and 5. Organisms that carry out the other type reactions are considered in various reviews (Cormier and Totter, 1964, 1968; Hastings, 1966; Johnson, 1967). This classification should be considered tentative in that

(1) the biochemistry of some bioluminescent organisms is not known, and

(2) the details of those systems which have been investigated are sufficiently vague, in most instances, that reclassification may be necessary.

Table 14.1. Bioluminescent reactions categorized according to type of reactions

Type of reaction	Organism		BL_{max} (mμ)
1. Simple enzyme-substrate reactions			
$LH_2 + O_2 + LH_2ase^a \rightarrow light$	*Cypridina*	(crustacean)	460
	Apogon	(fish)	460
	Parapriacanthus	(fish)	460
	Pholas	(clam)	480
	Gonyaulax	(protozoan)	470
	Odontosyllis	(annelid)	507
	Latia	(limpet)	520
2. Peroxidation reaction			
$2LH_2 + H_2O_2 + LH_2ase \rightarrow light$	*Balanoglossus*	(acorn worm)	Blue
	Diplocardia	(annelid)	500
3. Pyridine-nucleotide linked			
(1) $\begin{cases} a) & NADH + H^+ + FMN + reductase \rightarrow FMNH_2 + NAD^+ \\ b) & FMNH_2 + RCHO + O_2 + LH_2ase \rightarrow light \end{cases}$	Bacteria		475–505
(2) $\begin{cases} a) & NADH + X + reductase \rightarrow XH_2 + NAD^+ \\ b) & XH_2 + O_2 + LH_2ase \rightarrow light \end{cases}$	Fungi		530
4. Adenine-nucleotide linked			
(1) $LH_2 + ATP + O_2 + LH_2ase \xrightarrow{Mg^{2+}} light$	*Photinus*	(firefly)	552–558
(2) $LH_2 + DPA^b + O_2 + LH_2ase \xrightarrow{Ca^{2+}} light$	*Renilla reniformis*	(sea pansy)	485
5. Ion-activate: 'pre-charged' systems			
(1) precharged particle $\xrightarrow[O_2]{H^+} light$	*Gonyaulax*	(protozoan)	470
(2) precharged protein $\xrightarrow{Ca^{2+}} light$	*Aequorea*	(hydromedusid)	460
6. Unclassified systems			
	Chaetopterus	(annelid)	460
	Octochaetus	(annelid)	Orange-yellow
	Hoplophorus	(shrimp)	Blue

a LH_2ase = luciferase.
b DPA = 3',5'-diphosphoadenosine.

When the biology of bioluminescence is considered, a constantly recurring question is: 'What, if any, is its significance?' There is a seemingly random distribution of bioluminescent species ranging from the protozoa to primitive fish in the animal kingdom, and the bacteria and fungi in the plant kingdom. This suggests that bioluminescence is a variation of some biochemical reaction that is fundamental to all intervening organisms. The answer to this question for the firefly is apparent, in that light emission is an attractant mechanism in mating (Buck, 1938, 1948; Seliger and coworkers, 1964). However, when bioluminescence is considered with reference to other organisms such as bacteria or fungi, its biological significance is far from obvious. Seliger and McElroy (1965a) have proposed that bioluminescence is possibly a 'vestigial' detoxification system which was essential during the conversion from an anaerobic to an aerobic environment. Thus, one is forced to argue that either bioluminescence still has some selective value or it has no biological significance. If the latter, selective pressures have not eliminated all light-emitting organisms. Keynan and Hastings (1961) found that in cultures (in a medium supporting active growth) of *Photobacterium fischeri* (*Achromobacter fischeri*) consisting of a mixture of luminous wild type and 'dark mutants', the latter outgrew the former. But, in a slow or non-growing environment (sea water) which closely approximated that of the natural habitat of the luminous bacterium, the death rate of the mutants was three times greater than that of the wild type (Keynan, Veeder and Hastings, 1963). One interpretation of these results is that under conditions of active growth, the expenditure of energy via light emission by the wild type limits its growth. In the non-growing environment bioluminescence provides the cell with a better energy balance for survival. These results would indicate that bioluminescence, at least in the bacteria, is not a vestigial process but may be, or has been, involved in detoxification.

It is rather interesting that of 51 'dark mutants' of *P. fischeri* isolated by Keynan and Hastings (1961) all emitted light that was detectable photometrically. The dimmest mutant emitted 10^{-5} less light than the parental strain. This suggested that the dark mutant had a concentration of excited-state species 10^{-5} less than that of the wild type. The reduced light emission could represent a reduction in the concentration of precursors to the excited-state species itself; or possibly the cell has another method of utilizing the energy of this species. In either case, the mutant may carry out a dark reaction (or a series of them) which may partially replace the former light reaction. Dark reactions are known to occur under the proper conditions in the bioluminescent systems of bacteria, ostracods, fungi and fireflies (Harvey, 1952). Thus, one could speculate that those organisms that are not bioluminescent have evolved dark reactions.

The luminous bacteria

Requirements for bacterial bioluminescence

Bacterial bioluminescence is a pyridine nucleotide-linked reaction (type 3). The steps involved in light emission have recently been reviewed (Cormier and Totter, 1964, 1968; Hastings, 1968; Hastings, Vergin and De Sa, 1966; Hastings and coworkers, 1966; Johnson, 1967; McElroy and Seliger, 1963; Seliger and McElroy, 1965a).

In 1953, Strehler demonstrated that NADH was the first constituent to become rate-limiting in an *in vitro* reaction using extracts of *P. fischeri*. In rapid succession, a series of publications established that a factor from hog kidney cortex, which was identified as palmitaldehyde, was found to stimulate light emission (Cormier and Strehler, 1953; Strehler and Cormier, 1953, 1954). This requirement was shown to be nonspecific in that any long-chain aliphatic aldehyde (from C_7 to C_{18}) was active (Strehler and Cormier, 1954).

The stimulatory effect of FMN (McElroy and coworkers, 1953; Strehler and Cormier, 1953) was due to its enzymatic reduction in the reaction:

$$\text{NADH} + \text{H}^+ + \text{FMN} \xrightarrow[\text{oxidase}]{\text{NADH}} \text{FMNH}_2 + \text{NAD}^+$$

(Cormier and Totter, 1957; Green and McElroy, 1956; Totter and Cormier, 1955).

The NADH + FMN requirement may be replaced by the addition of $FMNH_2$ (Strehler and coworkers, 1954). The final reactant necessary for light emission is oxygen. Thus, the minimal requirements for bioluminescence in the bacteria are oxygen, luciferase, $FMNH_2$ and possibly aldehyde.

Reactions leading to light emission

The basic question to be considered is: 'What is the chemical nature of the light-emitting species?' As a result of their elegant experiments, Hastings and coworkers (1966) suggested a series of possible steps involved in bacterial bioluminescence. Figure 14.1 is a modification of the reaction sequence this group currently proposes for microbial bioluminescence (Hastings, 1968). Addition of $FMNH_2$ to luciferase results in the stoichiometric oxidation of the cofactor in less than one second to form the reduced enzyme—intermediate I. $FMNH_2$ is also autooxidized to form H_2O_2 and FMN. Intermediate I, which has a half-life of 5–10 seconds at

10°C, may, under anaerobic conditions, be oxidized without light emission by ferricyanide. Under aerobic conditions, the reduced enzyme reacts with oxygen to form intermediate II (a reduced enzyme–oxygen complex) which is quite stable and detectable over a period of several minutes. In the absence of aldehyde, intermediate II decays relatively slowly with a low light yield. In the presence of aldehyde, it is converted to an excited-state

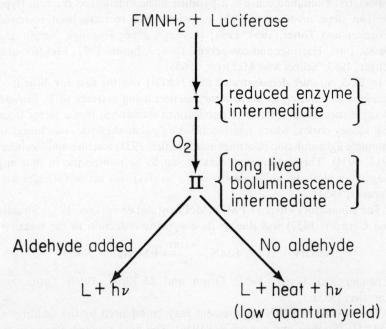

Figure 14.1. Hypothesized pathways in bacterial bioluminescence

species which emits light in high yield. While this reaction sequence is the most detailed yet proposed, it has raised several questions regarding the exact role(s) of $FMNH_2$ and aldehyde.

Hastings and Gibson (1963) originally proposed that one molecule of $FMNH_2$ was utilized to reduce luciferase. This interpretation was subsequently altered (Hastings, Gibson and Greenwood, 1964) to include the participation of two reduced flavin molecules. The latter scheme conforms to the earlier findings of Totter and Cormier (1955) and McElroy and Green (1955). The former workers (Totter and Cormier) suggested that one flavin was tightly bound to the enzyme and the other was freely dissociated. This led McElroy and Green to suggest that during luminescence one molecule of aldehyde combined with one molecule of $FMNH_2$

to form an aldehyde–FMNH₂ compound and the second FMNH₂ combined with oxygen to form a highly reactive organic peroxide. The peroxide could then act as an oxidant for the aldehyde–FMNH₂ compound to give an excited-state species. These reactions would account for the energetics of the bioluminescent reaction, but are difficult to reconcile with its spectral requirements.

A second observation has been made which would appear to invoke FMNH₂ at a level other than that indicated by Hastings and his colleagues. Seliger and McElroy (1965a) reported that a greater amount of light could be obtained from a given amount of FMNH₂ if the reaction was carried out in the presence of reduced glutathione (10^{-3} M). In the reaction system used by Hastings and associates the luciferase had virtually no turnover. It is possible that in the experiments of Seliger and McElroy enzyme turnover occurred, which might account for the greater amount of light. However, the rate of autooxidation of FMNH₂ would argue against this. Thus, the exact role of flavin in the bioluminescent reaction is far from understood.

The role of aldehyde in the luminous reaction is also unknown. An enumeration of some of the relevant facts are:

(1) total light emission is a function of the aldehyde concentration (McElroy and Green, 1955);

(2) one quantum of light is emitted for every twenty aldehyde molecules utilized (Cormier and Totter, 1957);

(3) the initial and decay rates, but not the quantum yield, are a function of aldehyde chain length (Hastings and coworkers, 1966); and

(4) the time to reach peak light intensity is about the same for all aldehydes tested.

Strehler (1961) and Terpstra (1963) contend that aldehyde is not a stoichiometric reactant and that no direct evidence of acid production (which would be produced on the basis of the reactions proposed) has been submitted. Much of the seemingly conflicting data would fall into place if the aldehyde were acting in a catalytic fashion as suggested by Strehler.

Hastings and coworkers (1966) have proposed a light reaction mechanism which would proceed in the absence of aldehyde. However, the criticism may be raised that the latter component may occur endogenously bound to the luciferase. The aldehyde concentration of the enzyme preparation was lowered by exhaustive dialysis against 2,4-dinitrophenyl-hydrazine, which resulted in the reduction of the endogenous light level to 6 per cent of the control. Since hydroxylamine also reacts with aldehyde, endogenous light emission may be reduced to 0·1 per cent of that obtained with decanal by the addition of 5×10^{-3} M hydroxylamine to the reaction. Eley and Cormier (1968) have recently carried out some interesting experiments on the role of aldehyde using luciferase preparations from both *P. fischeri* and *P. phosphoreum*. The results demonstrated that the quantum yield in the presence and absence of aldehyde is a function of the particular luciferase preparation being used. The greater the concentration of 'natural aldehyde' associated with the enzyme, the less the stimulation of light emission on aldehyde addition to the reaction. These workers propose that aldehyde may form part of a light-emitting complex.

Bacterial luciferase

McElroy and Green (1955) carried out the first extensive purification of this enzyme and reported on many of its properties. Luciferase has been crystallized in the laboratories of both Hastings (Hastings, Riley and Massa, 1965) and Cormier (Kuwabara and coworkers, 1965). On the basis of sedimentation and diffusion analysis, a molecular weight of 76 000 was reported, or if the molecule were spherical, the calculated molecular weight would be 58 000 (Hastings, Riley and Massa, 1965). Cormier and his group estimated the molecular weight of the protein to be 58 000 as determined by sedimentation equilibrium. One of the difficulties in this estimation is that bacterial luciferase tends to be present in a monomer–polymer equilibrium. The best estimate for the molecular weight of the monomer (2S) is 19 000 ± 3 000 (Kuwabara and coworkers, 1965). From these experiments the luciferase would appear to consist of a tetramer or possibly a trimer. Both the monomer (2S) and polymer species (4S and 5S) had luciferase activity, with the 4S species having a ten-fold greater specific activity. It is interesting that the 2S and not the 4S species had NADH oxidase activity. Friedland recently modified the purification procedure for bacterial luciferase in which the pure enzyme contained about 0·001 per cent of the oxidase (dehydrogenase) (see Hastings, 1968). The interpretation of the results of Kuwabara and coworkers (1965) has been questioned on the basis that even though the enzyme was recrystallized several times, it was impure (Hastings, 1968). The specific activity of the luciferase prepared by Cormier's group was only about 30 per cent of that prepared in Hastings' laboratory. However, it should be pointed out that purity as judged by specific activity in this instance may be hazardous. To use specific activity as an index for comparison of purity, the following minimal conditions should be met:

(1) the preparations under consideration should be made from the same isolate;

(2) the cells should be grown under identical conditions; and

(3) the assay conditions must be identical.

Friedland and Hastings (1967) dissociated luciferase with 8 M urea, and by acrylamide gel electrophoresis obtained two bands (α and β), neither of which had luciferase activity. When the α and β species were incubated in a 'recovery solution' (2×10^{-3} M dithiothreitol; 0·2 per cent bovine serum albumin; and 0·25 M, pH 7·0 sodium–potassium phosphate buffer) for several hours, luciferase activity was partially restored. Presumably, the α and β species reassociate to form an active enzyme. From these experiments they concluded that native luciferase is either a dimer or tetramer, but not a trimer as hypothesized by Kuwabara and coworkers (1965). Thus, in summary:

(1) the molecular weight of native luciferase has been estimated to be somewhere within the range of 58 000 to 76 000;

(2) luciferase may be dissociated to a monomer with a molecular weight of 19 000; and

(3) using gel electrophoresis, two non-identical subunits are obtained which may be reassociated to form a protein that is functional in light emission.

The problem of whether the native enzyme is a dimer (molecular weight of 40 000 if the monomer with a molecular weight of 20 000 is acceptable), a trimer (molecular weight of 60 000) or a tetramer (molecular weight of 80 000) is difficult to resolve at this time. The dimer does seem to be excluded in view of its molecular weight. The choice between a trimer or tetramer is more difficult. If the molecular weight of the monomer is 19 000 and that of the native enzyme somewhere between 58 000 and 76 000, then either a trimer or tetramer may be postulated. The findings of Friedland and Hastings (1967) would tend to suggest the latter, but this would place the molecular weight of the native enzyme at about 80 000, and they quote a value of 60 000. Further sedimentation analysis and electrophoretic separations at various pH values may resolve this question.

Seliger and McElroy (1965a) report that different strains of luminous bacteria may have *in vivo* peak light emissions that vary by as much as 200 Å. Corresponding *in vitro* shifts in peak emission were, in some instances, also noted. This would suggest that an alteration of the structure of the enzyme (at what level is unknown) is partly responsible for the shift in emission maximum. A similar observation has been made for various firefly species (*in vivo*) in which peak emission intensities depended upon the source of the enzyme. The substrate, luciferin, from all species studied was the same (McElroy and Seliger, 1966). These experiments with the firefly are of particular interest since the molecule undergoing excitation is chemically defined, as is the end-product of the reaction. The conclusion is reached that an alteration in luciferase structure is functional in the change in colour of the light emitted.

Bacterial mutants

It is rather surprising that the bacterial luminescent system has not been used in the development of the current concepts of protein biosynthesis. Hastings, Riley and Massa (1965) indicated that the quantum efficiency of bacterial luciferase is $2\cdot9 \times 10^{14}$ quanta per mg protein. If a molecular weight of 60 000 for luciferase is used and the lower limit of light detection is set at $1\cdot65 \times 10^7$ quanta sec^{-1} ml^{-1} (which is realistic for the instruments currently in use), then one should be able to readily detect the biosynthesis of 10^8 luciferase molecules per ml. A comparison between the sensitivity of the light assay and radioactive tracer methods used for the detection of protein biosynthesis, indicates that the two methods are approximately equal. In the reticulocyte haemoglobin system, the biosynthesis of approximately 10^8 haemoglobin molecules should be detectable (B. Hardesty, personal communication). Another advantage of the bacterial bioluminescent system for the study of protein biosynthesis is that the assay is specific for a single protein, luciferase. A final advantage of luminous bacteria as an experimental tool is that 'non-luminous' mutants are easily screened.

Several 'dark mutants' have been isolated which may be categorized as follows:

(1) nutritional mutants, in which growth and bioluminescence may or may not be associated, depending upon the mutant (McElroy and Farghaly, 1948);

(2) aldehyde mutants in which *in vivo* and *in vitro* light emission is greater upon the addition of aldehyde to the culture medium or reaction mixture (Rogers and McElroy, 1955); and

(3) luciferase mutants, which should not emit light or have reduced quantum efficiencies in the presence of aldehyde and $FMNH_2$.

A study of a series of luciferase mutants with respect to quality and quantity of light and amino acid composition would be most informative. Data concerning primary, secondary and tertiary protein structure and its effect on the nature of the excited state might also be obtained from such a series of mutants. Nealson and Markovitz (1968) have isolated a series of bacterial mutants using several mutagens. The isolates fell into three classes with reference to light emission:

(1) high *in vitro*, yet low *in vivo*;
(2) absent or very low *in vitro* with up to 1 per cent *in vivo*; and
(3) low values (10^{-3} per cent or less) for both *in vivo* and *in vitro*.

They suggest that there may be a relationship between bioluminescence and colony morphology.

Coffey (1967) carried out a study of the induction of luciferase by arginine using a nitrate-utilizing strain of *P. fischeri*. The induction was fairly specific for this amino acid and required a period of 12 minutes (27·5°c). The kinetics of induction suggested that during the first 2 minutes after addition of the inducer derepression occurred, followed by *m*-RNA synthesis. The latter reached a steady-state level of synthesis in about 4 minutes. Precursor luciferase polypeptide was synthesized in 2 to 4 minutes after the addition of the inducer. The remaining time, 8 to 10 minutes, was required for the conversion of luciferase precursors to active enzyme.

Light-induced luminescence

Photoinduction of the cell-free system of *P. fischeri* was possible and this was similar to that brought about by $FMNH_2$ (Gibson, Hastings and Greenwood, 1965). It was energy dependent (up to 300 J per flash under the conditions used), and the optimal wavelength for excitation occurred around 280 nm. Subsequent to photoinduction the quantum yield for bioluminescence was aldehyde dependent, in that the decay constant was a function of the aldehyde chain length. The photoinduced emission was similar to that obtained by $FMNH_2$ induction with peak emission occurring at 490 nm. Hastings and Gibson (1966) found light-induced bioluminescence to be oxygen dependent. Oxygen need not be present at the time of photo-induction, but it must be added shortly thereafter. There was light emission following anaerobic photoinduction of the enzyme, but it differed from the usual bioluminescent emission with respect to spectral quality, aldehyde dependence and oxygen quenching.

These experiments offer a very convincing argument for the role of $FMNH_2$ in bioluminescence as proposed by Hastings and coworkers (1966). In the case of photoinduction the quantum yield is independent of added flavin. The excitation spectrum differs from that of flavin. Photoinduction is maximal in the spectral region of optimal aromatic amino acid absorption. There is, however, a second minor photoinduction peak at approximately 400 nm (Gibson, Hastings and Greenwood, 1965). It is of interest that the absorption spectrum of flavin-free luciferase has a shoulder at 415 nm (Kuwabara and coworkers, 1965) and a fluorescence excitation band at approximately 400 nm (Cormier and Kuwabara, 1965). Whether these various optical properties of luciferase near 400 nm have any relevance to bioluminescence is unknown.

The luminous fungi

General features of in vivo light emission

The luminescence of decaying wood and occasionally of living plants intrigued many of the early scientists. It is now known that this phenomenon is due generally to luminous species of Basidiomycetes, with the possible exception of the Ascomycete *Xylaria hypoxylon*. Whether the latter is luminescent has been questioned by Wassink (1948). Lists and some descriptions of these light-emitting species have been published by Wassink (1948), Harvey (1952) and Haneda (1955).

The luminosity of the fungi generally occurs in the mycelium (*Omphalia flavida*), occasionally in the fruiting body, or in part of it (*Pleurotus japonicus*), or sometimes it is present in both (*Pleurotus olearius*) (Harvey, 1952). The light emitted is continuous. The intensity of the luminescence in *Collybia velutipes* was found to be dependent upon several factors such as the age of the culture (Foerster, Behrens and Airth, 1965) and its nutrition (Airth and Foerster, 1965). For *Armillaria mellea*, Berliner and Hovnanian (1963) demonstrated that, within the limits of resolution of the method of autophotography followed by photomicrography of the negative, light emission occurred throughout the entire cell. The *in vivo* emission spectra of several species of fungi show a maximum for each around 526 nm (Airth and Foerster, 1960; van der Berg, 1943). This emission maximum (blue-green) is one of the characteristics which distinguishes the fungi from the luminous bacteria (maximum at approximately 490 nm). The similarity of the emission maxima for the luminous fungi suggests that the constituents involved in each case are the same.

Another factor which will affect the intensity of the emission is exposure to ultraviolet radiation (Airth and Foerster, 1960). The effects of monochromatic ultraviolet irradiation on both *Panus stipticus* and *Armillaria mellea* were studied by Berliner and Brand (1962) and Berliner (1963). They demonstrated that the effects varied with the wavelength of the incident radiation, the time elapsed subsequent to treatment and with the species concerned. Although the effects on bioluminescence were multiple it was concluded that a constituent essential for emission, which absorbed at 280 nm, was destroyed. Inhibitor(s) of *in vivo* luminescence were also inactivated by ultraviolet radiation of wavelengths 245–265 nm and 366 nm.

Diurnal periodicity of the luminescence of *Armillaria mellea, Panus stipticus* and *Mycena polygramma* has been observed and was shown to be independent of the prior dark or dark–light (12-hour/12-hour) regimen in which the cultures were grown (Berliner, 1961). This interesting finding of bioluminescent rhythms warrants further investigation.

The biochemistry of fungal bioluminescence

The first fungal cell-free extracts capable of luminescence were obtained in 1959 (Airth and McElroy, 1959). Subsequently, this system has been elucidated by Airth and Foerster (1962) and Airth, Foerster and Behrens (1966).

The reaction mixture routinely used contains NADH (or NADPH), a hot-water extract of *Armillaria mellea* 1/11 and a soluble enzyme—reduced-pyridine nucleotide oxidase (RPN oxidase—from *Collybia velutipes*) and luciferase (a membrane-associated fraction from *C. velutipes*). The current concept is that fungal light emission involves more than one enzymatic reaction, viz (a type 3 reaction):

$$\text{NADH} + \text{H}^+ + \text{'X'} \xrightarrow{\text{RPN oxidase}} \text{'XH}_2\text{'} + \text{NAD}^+$$
$$\text{(NADPH)} \qquad\qquad\qquad\qquad \text{(NADP)}$$
$$\text{'XH}_2\text{'} + \tfrac{1}{2}\text{O}_2 \xrightarrow{\text{luciferase}} \text{'X''} + \text{H}_2\text{O} + \text{light}$$

The soluble RPN oxidase of *C. velutipes*, which has been purified some fourteen- to twenty-fold, utilized both NADH and NADPH. The enzyme has a molecular weight of approximately 25 000 (Hinde and Airth, unpublished, 1968).

It was the particulate, or 'membrane', fraction which appeared to function as the luciferase, since oxygen apparently was required only in the reaction catalysed by this enzyme (Airth, Foerster and Behrens, 1966). Purification of this fraction from homogenates has been obtained using discontinuous and linear sucrose-gradient fractionation. By this method a ten-fold increase in specific activity over that of the first discontinuous gradient fraction has been achieved (approximately 5 per cent of the total protein was recovered). Electron microscopy of the purified particles revealed the fraction to be composed of smooth surfaced vesicles, linear membranes and amorphous material. Which of the three, or combination of these, contained the luciferase was not determined. Of special interest is the origin of this fraction: 'Is it derived from mitochondria, the endoplasmic reticulum or some other membrane system within the cell?'

An investigation in progress in this laboratory is attempting to answer this question using 'marker' enzymes and the characteristics of membrane components.

Neither the chemical nature of the electron acceptor 'X' is known nor has its relationship to the product 'X" been established. Attempts to extract this acceptor from mycelial powder and lyophilized aqueous extracts of *Armillaria mellea* 1/11 with a number of organic solvents were unsuccessful (Airth, Foerster and Behrens, 1966). However, Kuwabara and Wassink (1966) did crystallize a component, which was possibly a dehydroluciferin, from a hot-water extract of *Omphalia flavida*. This compound was reported to be active in the production of light with the particulate enzyme, or to be chemiluminescent with the addition of hydrogen peroxide and sodium hydroxide. The emission peak in the enzyme-mediated reaction occurred at 524 nm, which compared favourably with that at 530 nm for the intact mycelium of *O. flavida* (Cormier, 1956). The emission maximum at 542 nm for the chemiluminescent system was shifted slightly toward the red end of the spectrum. The absorption spectrum of the purified compound at pH 6·5 showed a maximum at approximately 324 nm and a shoulder at 270 nm. Unfortunately, the structure of the compound is at present unknown. A sample of this crystalline compound shipped to the authors' laboratory from Kuwabara in England was inactive in the cell-free system from *Collybia velutipes* in both the presence and absence of NADH or NADPH. Since the possibility of inactivation during transport has not been disproved, the question as to whether Kuwabara and Wassink (1966) did crystallize fungal luciferin remains unanswered.

A substance with an apparent phosphorescence has been found in association with the mycelium of the luminous fungus, *Omphalia flavida* (Cormier, 1956; Cormier and Totter, 1966). Dissolved in water-free acetone, this compound, when excited with ultraviolet radiation, phosphoresced with an emission maximum identical to that of the *in vivo* bioluminescence of the mycelium. It is tempting to speculate that the phosphorescence and the bioluminescence are due to a molecular species common to both.

The genetics of fungal bioluminescence

The genetics of luminosity has been studied using luminous (North American) and non-luminous (European) strains of *Panus stipticus*. The North American and the European forms are intrafertile and completely interfertile (Macrae, 1937). After pairing the luminous and non-luminous forms and examining the mycelia from the resulting haploid spores,

Macrae (1942) concluded that light emission is 'governed by a single pair of Mendelian factors in which luminosity is dominant over non-luminosity'. Airth and Foerster (1964) have demonstrated that a North American strain contains both RPN oxidase and luciferase and that light emission could be obtained from these fractions using a hot-water extract of *Armillaria mellea* 1/11 as the electron acceptor. However, both enzymes were either absent or inactive in the European strain examined. No reliable evidence could be obtained for the presence or absence of an electron acceptor in either strain. The enzymes of the luminous strains of *Panus stipticus* were interchangeable with those obtained from *Collybia velutipes*.

These findings together with those of Macrae (1942) would appear to be at variance with the simple one gene–one enzyme concept. However, the genetic and biochemical studies described above were carried out on different isolates of the fungus and this might account for the discrepancy. The strain used for enzymatic study might have been a double-mutant.

The apparently conflicting genetic and biochemical observations might be interpreted as the result of multiple-protein changes occurring with a single-gene mutation similar to that found in *Neurospora crassa* (Mitchell, Mitchell and Tissieres, 1953) or in yeast (Ephrussi, 1953).

The luminous marine Protozoa

Taxonomy and ecology

The taxonomic distribution of bioluminescent organisms has been published by Harvey (1920, 1952). To remain within the scope of the term 'microorganism', this review will consider only the Phylum Protozoa.

Of the six classes of Protozoa, bioluminescent forms are present in only two, each of which contains only one luminous order. Class Mastigophora (Flagellata), Order Dinoflagellata is represented by the following luminous genera: *Ceratium, Peridinium, Prorocentrum, Pyrodinium, Gonyaulax, Blephorocysta, Gymnodinium, Noctiluca, Pyrocystis, Leptodiscus, Craspodetella* and *Glenodinium*. The last three are questioned by Harvey (1952) with respect to the validity of the luminosity observations. Only marine dinoflagellates have been observed to be luminous. Within Class Sarcodina (Rhizopoda), Order Radiolaria (strictly marine), the following genera have luminous forms: *Thalassicolla, Myxosphaera, Collosphaera, Collozoum* and *Sphaerozoum*.

Ryther (1955) has reviewed the general ecology of the dinoflagellates, considering temperature, salinity, nutrient requirements and the ecological significance of motility. These organisms, in addition to being the major contributors to marine bioluminescence, constitute a considerable percentage of individuaᵗs composing the population of plankton.

In vivo *protozoan luminescence*

Harvey (1952) published a comprehensive review considering the observations of the *in vivo* bioluminescence of various radiolarians and dinoflagellates.

(a) *Order Radiolaria* (*Class Sarcodina*)

The early investigations made by Brandt in 1885 on the physical location of bioluminescence in living *Sphaerozoum punctatum* have been discussed by Harvey (1952). The light emission of *Myxosphaera coerula* was also observed with respect to the response of the cells to agitation and the onset of fatigue. Harvey (1926) observed that *Collozoum inerme* and *Thalassicolla nucleata* cells would luminesce in the absence of free oxygen. That these earlier investigative methods for an oxygen requirement are currently still acceptable was briefly commented upon by Shimomura, Johnson and Saiga (1962). In view of more recent hypotheses, this might still be interpreted as a lack of free oxygen requirement rather than light emission in the complete absence of oxygen (McElroy and Seliger, 1963; Seliger and McElroy, 1965a). Oxygen may be bound, being released only when required for the luminescent reaction.

Combined light intensity data for individual specimens (approximately 12 mm in diameter) of *Cytocladus major* and *Autosphaera tridon* has been reported to range from $0·2 \times 10^{-6}$ to $1·7 \times 10^{-6}$ $\mu W/cm^2$ receptor surface at a distance of 5·6 cm in air (Nicol, 1958b). Assuming that the specimens were uniformly diffusing spheres, the total radiant flux from the entire surface of the animals was estimated to range from $0·69 \times 10^{-4}$ to $6·71 \times 10^{-4}$ $\mu W/4\pi$ steradians. No spectral emission was given, but the light was observed to be blue. Nicol used the spectral emission curve of *Chaetopterus* (marine polychaete annelid) with an emission peak at approximately 465 nm as a substitute in the calculation for entire surface luminosity. It is of interest that neither of these radiolarian species was reported to be luminous by Harvey (1952).

(b) *Order Dinoflagellata* (*Class Mastigophora*)

Systematic investigation of dinoflagellate bioluminescence (*in vivo* and *in vitro*) has been carried out on three genera, the armoured photosynthetic *Gonyaulax* and *Pyrodinium*, and the unarmoured holozoic *Noctiluca*. Recent reviews have discussed endogenous periodicity of bioluminescence and flash intensity with respect to specific periods of time during the circadian cycle of *Gonyaulax* (Chase, 1964; Johnson, 1967; McElroy and Seliger, 1963; Seliger and McElroy, 1965a, 1965b). The studies on *in vivo* bioluminescence of *Noctiluca* have been cited by Chase (1964).

In *Gonyaulax polyedra*, four systems which exhibit circadian cycles have been investigated: photosynthesis, cell division, flashing luminescence and bioluminescent glow. These observations have been reviewed by Hastings and Keynan (1965). Peak activities for the last three have been observed to occur during or toward the latter part of the dark period of a light–dark regimen, while the greatest photosynthetic capacity occurred during the light cycle. The endogenous diurnal periodicity of flashing luminescence in *Gonyaulax* was shown to be modified by continuous exposure of cells to 'white' fluorescent light and was reversed by removal to darkness (Sweeney and Hastings, 1957). A phase-shift of the time cycle of the *Gonyaulax* bioluminescent clock was obtained by changing the time of cell illumination (entrainment) (Hastings and Sweeney, 1959). Seliger and McElroy (1968) have noted a similar shift in *Pyrodinium* luminescence.

Maximal *Gonyaulax* photosynthetic activity along with photoenhancement and photoinhibition of bioluminescence was controlled by the intensity and wavelength of the incident light (Sweeney, Haxo and Hastings, 1959). The action spectra of all three phenomena, although distinct, had a similar shape and response with maxima occurring around 440 nm and 700 nm. These were very similar to the absorption maximum of the pigment system(s) for intact *Gonyaulax* cells. Under the conditions used, light of 800 foot-candle intensity gave maximal photoenhancement, as well as maximal photoinhibition of luminescent rhythm.

The amount of *Gonyaulax* stimulable bioluminescence obtained during the circadian cycle differed depending upon whether the cells were subjected to mechanical or acid stimulation (Sweeney, personal communication, 1968). Light inhibition of luminescence was not observed with acid-stimulated organisms (concentration below 0·05 M acetic acid), but it was readily demonstrated in mechanically stimulated cells. During the day phase more light may be produced by acid-stimulated cells than fro n a comparable population which was mechanically stimulated.

Several inhibitors of macromolecular biosynthesis have been used to investigate the intracellular control of the circadian clock mechanism in the *Gonyaulax* glow rhythm (Karakashian and Hastings, 1962, 1963). Actinomycin D (0·33 μg/ml) nearly abolished the rhythmic glow, this inhibition being observed after one normal cycle of luminescence periodicity subsequent to the addition of the inhibitor. Cell division also ceased under these experimental conditions. These data were interpreted as inhibition of RNA synthesis (possibly messenger-RNA). However, when inhibitors of protein synthesis—chloramphenicol $(3 \times 10^{-4} \text{ M})$ and puromycin $(2 \times 10^{-5} \text{ M})$—were tested, the results indicated that rhythmicity was not controlled by protein synthesis exclusively. Thus, the newly synthesized RNA might function not only as messenger for protein synthesis, but also as a 'control RNA' in another biochemical system. These initial studies of circadian rhythms at the molecular level represent an exciting new approach to what is a most complex biological problem.

Seliger and McElroy (1968) described a light-influenced bioluminescent cycle with a high dark/light ratio in *Pyrodinium bahamense* observed *in situ*. Once attained, the maximal stimulable luminescence occurred during the entire night, resulting in a square-wave pattern of diurnal light emission. These results were in contrast to the sinusoidal pattern of periodicity of luminescence obtained for laboratory cultures of *Gonyaulax polyedra* (Hastings and Sweeney, 1958; Sweeney, Haxo and Hastings, 1959). Seliger and McElroy (1968) suggested that possibly the differences in the results obtained with these organisms may be due to variations in both diurnal stimulability and the synthesis of luciferin and luciferase. The endogenous rhythm of stimulable bioluminescence in *Pyrodinium bahamense* (observed in nature) was light influenced, since the time of increase in luminescence was affected by seasonal differences in the time of sunset. On the basis of changes in the ratio of blue to red incident light, diurnal variation of *Pyrodinium* could involve spectral as well as light intensity changes (Seliger and Fastie, 1968). Although the stimulable intensity decreased in laboratory cultures of *P. bahamense* after approximately 12 hours of continuous darkness, the decrease was very slow and small compared to that noted *in situ*. The decrease in bioluminescence during the daylight period could result from the combined effects of endogenous rhythm and photoinhibition.

Experimental recording and analysis of the flash kinetics of stimulated *Noctiluca* was first undertaken by Nicol (1958a) and followed by Eckert (Eckert, 1965a, 1965b, 1966a, 1966b, 1966c, 1967; Eckert and Reynolds, 1967). Eckert (1966a) investigated subcellular microsources of luminescence, emission from single microsources and the relationships between the microflash (from individual microsources) and the macroflash (over the entire *Noctiluca miliaris* cell). From these results, a sequence of events leading to microsource light emission in *Noctiluca* was proposed:

(1) cell membrane displacement is caused by mechanical stimulus;
(2) a graded, slow potential develops across an excitable membrane;
(3) at threshold amplitude a regenerative potential change is initiated by the graded potential;

(4) the action potential thus initiated propagates non-decrementally from the site of initiation over the remainder of the periplast to the opposite side of the cell;

(5) undetermined events concomitant with the recorded action potential act at the microsource level to permit reactions which are maintained unreactive prior to triggering; and

(6) light is emitted by the microsource with a characteristic and stable time course of intensity (whereas flash intensity exhibits considerable lability).

A preliminary hypothesis proposed that the basic feature of the excitation-flash coupling mechanism is a controlled release of a cosubstrate which diffuses into microsources and acts as the rate-limiting component of the subsequent luminescent reaction (Eckert, 1966a). Evidence for substrate control of the luminescent flash in *Noctiluca* has been given (Eckert, 1966c).

Noctiluca microsources and their luminescence in live specimens were investigated by Eckert and Reynolds (1967). Fluorescence microscopy revealed that illumination with 365 and 404 nm lines showed numerous sources of fluorescence in the perivacuolar cytoplasm. Phase contrast used together with ultraviolet microscopy revealed that fluorescence originated in certain spherical phase-retarding bodies (SPRB). Not all SPRBs are identical with the luminescent microsources. Only those that fluoresced were capable of producing luminescence. The green fluorescence was consistent with the luminosity spectrum which had a maximum around 470 nm and extended beyond 540 nm (Nicol, 1958). Using image-intensification autophotography, the average number of microsources in an average 500 μ *Noctiluca* specimen was about $(3 \cdot 6 \pm 1 \cdot 5) \times 10^4$. The average number of photons emitted by unfatigued microsources, using two different methods of measurement, was $(4 \cdot 4 \pm 1 \cdot 5) \times 10^4$ per microsource. Each organelle responded repetitively with a reproducible time course to a succession of triggering potentials. Gradations in intensity of the micro-flashes caused reversible changes in the intensity of the macroflash. The shapes of the micro- and macroflashes were similar, but shorter rise and decay times occurred in the former. The authors concluded that the macroflash was the result of a somewhat asynchronous, but otherwise parallel, summation of microflashes.

In vitro *dinoflagellate bioluminescence*

Biochemical investigation of the dinoflagellate bioluminescent system was made possible after isolation and pure culture maintenance of

Gonyaulax was achieved (Hastings and Sweeney, 1957; Haxo and Sweeney, 1955). There are two separately isolatable systems: the particulate (scintillon) system and the soluble system. Many reviews concerning *Gonyaulax* bioluminescence *in vitro* have been published (Chase, 1964; Cormier and Totter, 1964, 1968; Hastings, 1966, 1968; Hastings and Keynan, 1965; Johnson, 1967; McElroy and Seliger, 1963; Seliger and McElroy, 1965a, 1965b).

(a) *The particulate (scintillon) system*

Ion-activated (H⁺), crystal-like birefringent particles capable of producing light in the presence of oxygen have been isolated from *Gonyaulax* (De Sa, Hastings and Vatter, 1963). *In vitro* flash from these particles (scintillons) did not occur until the suspending medium was rapidly acidified to an optimum pH of 5·7. The reaction may be represented by the equation:

$$\text{Particle} + O_2 \xrightarrow{\text{H}^+} \text{light}$$

Scintillon purification was achieved by density gradient centrifugation, and good correlation was obtained between particle numbers and the amount of light emitted from the gradient fraction. Examination of active particles in the electron microscope revealed rhombohedral and chevron shapes.

Identification and location of crystal-like particles in *Gonyaulax polyedra* cells have been described by Sweeney and Bouck (1966), who also reported the presence of membrane-enclosed scintillon-like particles in non-luminous as well as other luminous species. However, these were absent in the luminous *Pyrodinium bahamense* (Sweeney and Bouck, personal communication, 1968). The absence of scintillons attributable to fixation procedures for electron microscopy was ruled out by processing *Pyrodinium* and *Gonyaulax* cells simultaneously.

Guanine has been identified as a component of the scintillon. The differences have been determined between the buoyant densities of actively bioluminescent scintillons and the inactive heavier crystals of pure guanine isolated from *Gonyaulax* cells (De Sa, 1964; De Sa and Hastings, 1968; Hastings, Vergin and De Sa, 1966). From these experiments, it was proposed that the scintillon consists of a guanine crystal core with an enzyme–substrate-associated–lipid coat. It was pointed out, that although there is no morphological evidence for a lipid coat, the latter may be lost during preparation for electron microscopy.

Active scintillons have recently been isolated from *Noctiluca* (Hastings, Vergin and De Sa, 1966), using the same procedures as for *Gonyaulax*.

These particles have been found to be very similar to *Gonyaulax* scintillons in that they:

(1) are stable at pH 8·2 and are activated by rapidly lowering the pH to 5·7;
(2) have nearly identical *in vitro* flash kinetics;
(3) are markedly inhibited by copper and zinc at concentrations inhibitory to *Gonyaulax* scintillons; and
(4) have approximately the same buoyant density (1·23 g/cc) as those from *Gonyaulax*.

No morphological identification was made of the preparation.

(b) *The soluble system*

The reaction carried out by the soluble system may be represented by a simple type I reaction:

$$LH_2 + O_2 \xrightarrow[\text{salt (nonspecific)}]{\text{luciferase}} \text{light } (\lambda_{max} = 470 \text{ nm})$$

Hastings and Sweeney (1957) prepared the first *Gonyaulax* cell-free extract capable of catalysing the *in vitro* light reaction when salt and a heat stable substrate were supplied. The reaction had the following characteristics: optimal pH, 6·6; optimal temperature, 24°C; oxygen dependence; and spectral emission maximum at approximately 470 nm. A seventeen-fold purified luciferase preparation (Bode, 1961; Hastings and Bode, 1961) was studied with respect to the effect of salt and the addition of albumin to the reaction mixture. It was proposed that salt may be effective in the production of *in vitro* light by one or more of the following:

(1) a modification of the luciferin to an active form in the presence of salt;
(2) salt may have an effect on the excited state of luciferin, hence increasing quantum yield; or
(3) salt may act directly on the enzyme.

When the amount of total extractable luminescent activity from *Gonyaulax* is considered, the relative contribution by the soluble compared with that of the scintillon system appears to be unsettled. Completely different percentages for soluble system contribution were obtained depending upon the pH of the soluble assay reaction. The ratio of scintillon/soluble system total light was greater when the soluble system was assayed at pH 8·2, with a reversal when the assay was carried out at pH 6·6 (Hastings,

Vergin and De Sa, 1966; Sweeney and Bouck, 1966). Lee and Winans (1968) considered this, and also the fact that, since only a small percentage of the maximum stimulable *in vivo* light was extracted, misinterpretation of the data could result. With a modification of the preparation procedure, they were able to extract over 30 per cent of the maximum *in vivo* light activity obtainable from intact *Gonyaulax polyedra*. The total *in vivo* light emission was found to be a function of preillumination time (400 foot-candles of 'cool white' fluorescent light). The greater the length of the preillumination period, the less the total *in vivo* light emission. The reverse was found for the *in vitro* total extractable (soluble + scintillon) light systems. The yield of extractable light activity increased with preillumination, but the soluble/scintillon ratio decreased. *In vitro* luminescence was measured at pH 6·5 for the soluble system and 5·7 for the scintillon assay. These investigators suggested that the scintillon system might be an artifact of extraction, but indicated that other interpretations of their data were possible. Again, it is pointed out that although *Noctiluca miliaris* was not examined by electron microscopy for the presence of scintillons, *per se*, the cell does contain an organelle directly responsible for light emission (Eckert and Reynolds, 1967). The luminescent *Pyrodinium* did not contain scintillons (Sweeney and Bouck, personal communication, 1968). Thus, the question of the universal role of scintillons in dino-flagellate bioluminescence is unresolved at this time.

The ambiguity with respect to the optimal pH for the assay of the soluble system appears to have been clarified by the work of Krieger and Hastings (1968). Using molecular sieve chromatography on crude *Gony-aulax* extracts, two active luciferase components were isolated. A 35 000 mol. wt. component was functional in the bioluminescent reaction over a pH range of 6·0 to 9·0, while a 150 000 mol. wt. component was active in a much narrower range (approximately pH 6 to 7). A partial conversion of the 150 000 mol. wt. species to the smaller component was obtained by pretreatment at pH 6·0 followed by gel filtration. Krieger and Hastings (1968) suggested that the earlier observations of a narrow pH response reported for a crude luciferase preparation (Hastings and Sweeney, 1957) was the result of the 150 000 mol. wt. component. They also suggested that De Sa (1964) in assaying the soluble system at pH 8, was observing the activity of the partially purified 35 000 mol. wt. com-ponent. The apparent lack of pH 8 activity in a crude preparation of luciferase was attributed to the inhibitory effect of the 150 000 mol. wt. component. The source of the inhibitory action has not been identified, and interpretation of the effect was complicated by the presence of luciferin bound to the 150 000 mol. wt. species.

(c) *Luciferin*

Gonyaulax luciferin, although not yet chemically characterized, has been substantially purified by Bode (1961) and Bode and Hastings (1963). It was highly unstable at all degrees of purity, but was rendered more stable by the addition of sulphydryl reagents and bovine serum albumin or by anaerobic conditions. Oxidized luciferin could be reduced by hydrogen gas plus a platinum catalyst or with sulphydryl reagents (cysteine or mercaptoethanol).

Effect of circadian clock rhythmicity on the extractable in vitro *dinoflagellate bioluminescent activity*

The optimal times of harvest for bioluminescent activity of *Gonyaulax* luciferin and luciferase were controlled by an endogenous clock mechanism, and these components showed a daily periodicity cycle of luminescent activity (Bode, De Sa and Hastings, 1963). *Gonyaulax* luciferin extracts prepared from cells harvested late in the 12-hour dark period (12 hours dark/12 hours light artificial day) were about four times more active than those harvested from the light cycle. The optimal time of harvest with respect to maximal luciferin activity was shown only after pretreatment of the cells to inhibit spontaneous flashing in response to agitation during collection. Inhibition of substrate utilization was achieved by either rapidly raising the temperature of the culture from 22 to 34°c or by exposing the cells to bright light just prior to harvest. However, the yield of luciferin activity from light-cycle cells was not affected by heat.

The activity versus time-of-day plot for *Gonyaulax* luciferase exhibited a sinusoidal periodicity independent of the method of harvest, with its maximum occurring in the middle of the night cycle (Hastings and Keynan, 1965).

Optimal yield of scintillon activity was found to be affected not only by the method used for cell disruption, but also by the time in the light–dark cycle during which the cells were harvested. The optimal time of harvest was similar to that for luciferin, but whether scintillon activity was observed to be cyclic was not reported (Hastings, Vergin and De Sa, 1966).

Relationships between in vitro *and* in vivo *dinoflagellate luminescence*

The intracellular origin and location of *Gonyaulax* bioluminescence is still unresolved (De Sa and Hastings, 1968; Hastings, 1966, 1968; Lee and Winans, 1968; Sweeney, 1968, personal communication). The present state of knowledge suggests two possible hypotheses. In the first, only the scintillon is functional and is composed of luciferin and luciferase, bound in some fashion via a low-density component to crystalline guanine. This complex unit is in close cytological association with an ion-activated membrane. In this model, the soluble system exists only as an *in vitro* preparatory artifact. The second hypothesis assumes the simultaneous intracellular existence of a crystal-like enzyme–substrate–guanine particle (responsible for the flash), as described above, and in addition, a soluble luciferin and luciferase system dispersed throughout the cytoplasm. The latter might be responsible for the bioluminescent glow noted periodically in the intact cell (Hastings, 1966).

Various attempts to release active luciferin and luciferase from scintillons have failed, ending in either no release from the particle or complete inactivation (De Sa and Hastings, 1968; Hastings, Vergin and De Sa, 1966; Sweeney and Bouck, 1966). Osmotic shock, sonication and treatment with various detergents were all tested.

The *in vivo* data of Eckert (1966a) and Eckert and Reynolds (1967) present a convincing case for an organelle alone as the sole source of luminescence in *Noctiluca*. A corresponding series of experiments has not been carried out for *Gonyaulax* cells. Of note is the close similarity of the bioluminescence spectral emissions of the soluble system, the scintillon system, living *Gonyaulax* and the fluorescence spectrum of isolated luciferin (activated at 400 nm). All had emission maxima at approximately 470 nm (De Sa and Hastings, 1968). It is of interest that the soluble system was activated by the presence of high salt concentration (approximately 1 M) and at this concentration the enzyme was precipitated. However, the scintillon activity was markedly inactivated and/or quenched by neutral salts in the 1 to 3 M range (Hastings, Vergin and De Sa, 1966).

Since the *in vitro* flash of the particle was triggered by lowered pH, Hastings (1966) suggested that for *in vivo* bioluminescence the flash results from a process(es) involving hydrogen-ion transport. Within a membrane the luciferin anion and luciferase may be fixed in a stable configuration in the structure of the scintillon at an appropriate pH. Eckert's (1966a, 1966b, 1966c, 1967) investigations on *Noctiluca* suggest that after cellular stimulation the membrane is involved in proton transport in a series of electrophysical events which culminate in luminescence. Hastings (1966) presented the following equations to represent the sequence of reactions involved. Gu represents the particle (guanine) to which the enzyme (E) and substrate (LH_2) are attached.

$$(1) \quad \text{Gu-E-LH}_2^- + \text{H}^+ \longrightarrow \text{Gu-E-LH}_2$$

$$(2) \quad \text{Gu-E-LH}_2 + \text{O}_2 \longrightarrow \text{Gu-E-L(O)}^* + \text{H}_2\text{O}$$

$$(3) \quad \text{Gu-E-L(O)}^* \longrightarrow \text{Gu-E-L(O)} + \text{light}$$

Acknowledgments

We should like to take this opportunity to thank our many colleagues who informed us of their unpublished work. We also appreciate their constructive review of this manuscript and their many suggestions.

REFERENCES

Airth, R. L., and G. E. Foerster (1960). Some aspects of fungal bioluminescence. *J. Cell. Comp. Physiol.*, **56**, 173.

Airth, R. L., and G. E. Foerster (1962). The isolation of catalytic components required for cell-free fungal bioluminescence. *Arch. Biochem. Biophys.*, **97**, 567.

Airth, R. L., and G. E. Foerster (1964). Enzymes associated with bioluminescence in *Panus Stypticus luminescens* and *Panus Stypticus non-luminescens*. *J. Bact.*, **88**, 1372.

Airth, R. L., and G. E. Foerster (1965). Light emission from the luminous fungus *Collybia velutipes* under different nutritional conditions. *Am. J. Bot.*, **52**, 495.

Airth, R. L., G. E. Foerster, and P. Q. Behrens (1966). The luminous fungi. In F. H. Johnson and Y. Haneda (Eds.), *Bioluminescence in Progress*, Princeton University Press, Princeton. pp. 203–223.

Airth, R. L., and W. D. McElroy (1959). Light emission from extracts of luminous fungi. *J. Bact.*, **77**, 249.

Berliner, M. D. (1961). Diurnal periodicity of luminescence in three basidomycetes, *Science*, **134**, 740.

Berliner, M. D. (1963). The action of monochromatic ultraviolet radiation on luminescence in *Armillaria mellea*. *Rad. Res.*, **19**, 392.

Berliner, M. D., and P. B. Brand (1962). Effects of monochromatic ultraviolet light on luminescence of *Panus stipticus*. *Mycologia*, **54**, 415.

Berliner, M. D., and H. P. Hovnanian (1963). Autophotography of luminescent fungi. *J. Bact.*, **86**, 339.

Bode, V. C. (1961). The bioluminescence reaction in *Gonyaulax polyedra*: a system for studying the biochemical aspects of diurnal rhythms. *Doctoral Thesis*, University of Illinois.

Bode, V. C., R. De Sa, and J. W. Hastings (1963). Daily rhythm of luciferin activity in *Gonyaulax polyedra*. *Science*, **141**, 913.

Bode, V. C., and J. W. Hastings (1963). The purification and properties of the bioluminescent system in *Gonyaulax polyedra*. *Arch. Biochem. Biophys.*, **103**, 488.

Brandt, K. (1885). Die Koloniebildenden Radiolarien (Sphaerozoen) des Golfes von Neapel. *Fauna u. Flora Neapel*, **13**, 1–276. *Phosphorescenz*. pp. 136–139.

Buck, J. B. (1938). Synchronous rhythmic flashing of fireflies. *Quart. Rev. Biol.*, **13**, 301.

Buck, J. B. (1948). The anatomy and physiology of the light organ in fireflies. *Ann. N.Y. Acad. Sci.*, **49**, 397.

van der Burg, A. (1943). Spektrale onderzoekingen over chemo-en biol-lumenescentic. *Ph.D. Thesis*, University of Utrecht.

Chase, A. M. (1964). Bioluminescence—production of light by organisms. In A. C. Giese (Ed.), *Photophysiology*, Vol. 2. Academic Press, London, New York. pp. 389–421.

15

Coffey, J. L. (1967). Inducible synthesis of bacterial luciferase: specificity and kinetics of induction. *J. Bact.*, **94**, 1638.

Cormier, M. J. (1956). Studies on the luminescence of the fungus *Omphalia flavida* and some related aspects of the problem of bacterial luminescence. *Doctoral Thesis*, University of Tennessee.

Cormier, M. J., J. M. Crane, Jr., and Y. Nakano (1968). Evidence for the identity of the luminescent systems of *Porichthys porosissimus* (fish) and *Cypridina hilgendorfii* (crustacean). *Biochem. Biophys. Res. Comm.*, **29**, 747.

Cormier, M. J., and S. Kuwabara (1965). Some observations on the properties of crystalline bacterial luciferase. *Photochem. Photobiol.*, **4**, 1217.

Cormier, M. J., and B. L. Strehler (1953). The identification of KCF: requirements for long-chain aldehydes for bacterial extract luminescence. *J. Am. Chem. Soc.*, **75**, 4864.

Cormier, M. J., and J. R. Totter (1957). Quantum efficiency determinations on components of the bacterial luminescence system. *Biochim. Biophys. Acta*, **25**, 229.

Cormier, M. J., and J. R. Totter (1964). Bioluminescence. In E. E. Snell, J. M. Luck, P. D. Boyer and G. MacKinney (Eds.), *Ann. Rev. Biochem.*, Vol. 33. Annual Reviews, Inc., Palo Alto. pp. 431–458.

Cormier, M. J., and J. R. Totter (1966). The apparent phosphorescence of a substance extracted from the mycelium of the luminous fungus, *Omphalia flavida*. In F. H. Johnson and Y. Haneda (Eds.), *Bioluminescence in Progress*, Princeton University Press, Princeton, N.J. pp. 225–231.

Cormier, M. J., and J. R. Totter (1968). Bioluminescence: enzymic aspects. In A. C. Giese (Ed.), *Photophysiology*, Vol. 4. Academic Press, London, New York. pp. 315–353.

De Sa, R. (1964). The discovery, isolation and partial characterization of a bioluminescent particle from the marine dinoflagellate, *Gonyaulax polyedra*. *Doctoral Thesis*, University of Illinois.

De Sa, R., and J. W. Hastings (1968). The characterization of scintillons. Bioluminescent particles from the marine dinoflagellate, *Gonyaulax polyedra*. *J. Gen. Physiol.*, **51**, 105.

De Sa, R., J. W. Hastings, and A. Vatter (1963). Luminescent 'crystalline' particles: an organized subcellular bioluminescent system. *Science*, **141**, 1269.

Eckert, R. (1965a). Bioelectric control of bioluminescence in the dinoflagellate *Noctiluca*. I. Specific nature of triggering events. *Science*, **147**, 1140.

Eckert, R. (1965b). Bioelectric control of bioluminescence in the dinoflagellate *Noctiluca*. II. Asynchronous flash initiation by a propagated triggering potential. *Science*, **147**, 1142.

Eckert, R. (1966a). Excitation and luminescence in *Noctiluca miliaris*. In F. H. Johnson and Y. Haneda (Eds.), *Bioluminescence in Progress*, Princeton University Press, Princeton, N.J. pp. 269–300.

Eckert, R. (1966b). Emission characteristics of microluminescent organelles in the cytoplasm of *Noctiluca*. *Science*, **151**, 349.

Eckert, R. (1966c). Evidence for substrate control of the luminescent flash of *Noctiluca*. *Fed. Proc.*, (Abstracts) 25: Abstract No. 2896.

Eckert, R. (1967). The wave form of luminescence emitted by *Noctiluca*. *J. Gen. Physiol.*, **50**, 221.

Eckert, R., and G. T. Reynolds (1967). The subcellular origin of bioluminescence in *Noctiluca miliaris*. *J. Gen. Physiol.*, **50**, 1429.

Eley, M., and M. J. Cormier (1968). On the function of aldehyde in bacterial bioluminescence: evidence for an aldehyde requirement during luminescence from the frozen state. *Biochem. Biophys. Res. Comm.*, **32**, 454.

Ephrussi, B. (1953). *Nucleo-cytoplasmic Relations in Microorganisms*, Clarendon Press, Oxford.

Foerster, G. E., P. Q. Behrens, and R. L. Airth (1965). Bioluminescence and other characteristics of *Collybia velutipes*. *Am. J. Bot.*, **52**, 487.

Friedland, J., and J. W. Hastings (1967). Nonidentical subunits of bacterial luciferase. Their isolation and recombination to form active enzyme. *Proc. Nat. Acad. Sci. U.S.*, **58**, 2336.

Gibson, Q. H., J. W. Hastings, and C. Greenwood (1965). On the molecular mechanism of bioluminescence. II. Light-induced luminescence. *Proc. Nat. Acad. Sci. U.S.*, **53**, 187.

Green, A. A., and W. D. McElroy (1956). Crystalline firefly luciferase. *Biochim. Biophys. Acta*, **20**, 170.

Haneda, Y. (1955). Luminous organisms of Japan and the Far East. In F. H. Johnson (Ed.), *The Luminescence of Biological Systems*, The American Association for the Advancement of Science, Washington, D.C. pp. 335–385.

Harvey, E. N. (1920). *The Nature of Animal Light*, J. B. Lippincott Co., Philadelphia.

Harvey, E. N. (1926). Oxygen and luminescence, with a description of methods for removing oxygen from cells and fluids. *Biol. Bull.*, **51**, 89.

Harvey, E. N. (1952). *Bioluminescence*, Academic Press, London, New York.

Hastings, J. W. (1966). The chemistry of bioluminescence. In D. R. Sanadi (Ed.), *Current Topics in Bioenergetics*, Vol. 1. Academic Press, London, New York. pp. 113–152.

Hastings, J. W. (1968). Bioluminescence. In P. D. Boyer, A. Meister, R. L. Sinsheimer and E. E. Snell (Eds.), *Ann. Rev. Biochem.*, Vol. 37. Annual Reviews, Inc., Palo Alto. pp. 431–458.

Hastings, J. W., and V. C. Bode (1961). Ionic effects upon bioluminescence in *Gonyaulax* extracts. In W. D. McElroy and B. Glass (Eds.), *Light and Life*, The Johns Hopkins Press, Baltimore, Md. pp. 294–306.

Hastings, J. W., and Q. H. Gibson (1963). Intermediates in the bioluminescent oxidation of reduced flavin mononucleotide. *J. Biol. Chem.*, **238**, 2537.

Hastings, J. W., and Q. H. Gibson (1966). The role of oxygen in the photoexcited luminescence of bacterial luciferase. *J. Biol. Chem.*, **242**, 720–726.

Hastings, J. W., Q. H. Gibson, J. Friedland, and J. Spudich (1966). Molecular mechanisms in bacterial bioluminescence. On energy storage intermediates and the role of aldehyde in the reaction. In F. H. Johnson and Y. Haneda (Eds.), *Bioluminescence in Progress*, Princeton University Press, Princeton, N.J. pp. 151–186.

Hastings, J. W., Q. H. Gibson, and C. Greenwood (1964). On the molecular mechanism of bioluminescence. I. The role of long-chain aldehyde. *Proc. Nat. Acad. Sci. U.S.*, **52**, 1529.

Hastings, J. W., and A. Keynan (1965). Molecular aspects of circadian systems. In J. Aschoff (Ed.), *Circadian Clocks*, North-Holland Publishing Co., Amsterdam. pp. 167–182.

Hastings, J. W., W. H. Riley, and J. Massa (1965). The purification, properties and chemiluminescent quantum yield of bacterial luciferase. *J. Biol. Chem.*, **240**, 1473.

Hastings, J. W., and B. M. Sweeney (1957). The luminescent reactions in extracts of the marine dinoflagellate, *Gonaulax polyedra. J. Cell. Comp. Physiol.*, **49**, 209.

Hastings, J. W., and B. M. Sweeney (1958). A persistent diurnal rhythm of luminescence in *Gonyaulax polyedra. Biol. Bull.*, **115**, 440.

Hastings, J. W., and B. M. Sweeney (1959). The *Gonyaulax* clock. In R. B. Withrow (Ed.), *Photoperiodism and Related Phenomena in Plants and Animals*, American Association for the Advancement of Science, Washington, D.C. pp. 567–584.

Hastings, J. W., M. Vergin, and R. De Sa (1966). Scintillons: the biochemistry of dinoflagellate bioluminescence. In F. H. Johnson and Y. Haneda (Eds.), *Bioluminescence in Progress*, Princeton University Press, Princeton, N.J. pp. 301–329.

Haxo, F. T., and B. M. Sweeney (1955). Bioluminescence in *Gonyaulax polyedra*. In F. H. Johnson (Ed.), *The Luminescence of Biological Systems*, American Association for the Advancement of Science, Washington, D.C. pp. 415–420.

Johnson, F. H. (1967). Bioluminescence. In M. Florkin and E. H. Stotz (Eds.), *Comprehensive Biochemistry*, Vol. 27. Elsevier, New York. pp. 79–136.

Karakashian, M. W., and J. W. Hastings (1962). The inhibition of a biological clock by Actinomycin D. *Proc. Nat. Acad. Sci. U.S.*, **48**, 2130.

Karakashian, M. W., and J. W. Hastings (1963). The effects of inhibitors of macromolecular biosynthesis upon the persistent rhythm of luminescence in *Gonyaulax. J. Gen. Physiol.*, **47**, 1.

Keynan, A., and J. W. Hastings (1961). The isolation and characterization of dark mutants of luminous bacteria. *Biol. Bull.*, **121**, 375.

Keynan, A., C. Veeder, and J. W. Hastings (1963). Studies on the survival of dark and bright mutants of luminescent bacteria in sea water. *Biol. Bull.*, **125**, 382.

Krieger, N., and J. W. Hastings (1968). Bioluminescence: pH activity profiles of related luciferase fractions. *Science*, **161**, 586.

Kuwabara, S., M. J. Cormier, L. S. Dure, P. Kreiss, and P. Pfuderer (1965). Crystalline bacterial luciferase from *Photobacterium fischeri. Proc. Nat. Acad. Sci. U.S.*, **53**, 822.

Kuwabara, S., and E. C. Wassink (1966). Purification and properties of the active substance of fungal bioluminescence. In F. H. Johnson and Y. Haneda (Eds.), *Bioluminescence in Progress*, Princeton University Press, Princeton, N.J. pp. 233–245.

Lee, J., and M. D. Winans (1968). Light yields from soluble versus insoluble extracts of the bioluminescent marine dinoflagellate, *Gonyaulax polyedra. Biochem. Biophys. Res. Comm.*, **30**, 105.

McElroy, W. D., and A. H. Farghaly (1948). Biochemical mutants affecting the growth and light production in luminous bacteria. *Arch. Biochem.*, **17**, 379.

McElroy, W. D., and A. A. Green (1955). Enzymatic properties of bacterial luciferase. *Arch. Biochem. Biophys.*, **56**, 240.

McElroy, W. D., J. W. Hastings, W. V. Sonnenfeld, and J. Coulombre (1953). The requirement of riboflavin phosphate for bacterial luminescence. *Science*, **118**, 385.

McElroy, W. D., and H. H. Seliger (1963). The chemistry of light emission. In F. F. Nord (Ed.), *Advances in Enzymology*, Vol. 25. Interscience Publishers (J. Wiley), New York. pp. 119–166.

McElroy, W. D., and H. H. Seliger (1966). Firefly bioluminescence. In F. H. Johnson and Y. Haneda (Eds.), *Bioluminescence in Progress*, Princeton University Press, Princeton, N.J. pp. 427–458.

Macrae, R. (1937). Interfertility phenomena of the American and European forms of *Panus Stypticus* (Bull.) Fries. *Nature*, **139**, 674.

Macrae, R. (1942). Interfertility studies and inheritance of luminosity in *Panus stypticus*. *Can. J. Res.*, **20**, 411.

Mitchell, M. G., H. K. Mitchell, and A. Tissieres (1953). Mendelian and non-Mendelian factors affecting the cytochrome system in *Neurospora crassa*. *Proc. Nat. Acad. Sci. U.S.*, **39**, 606.

Nealson, K., and A. Markovitz (1968). Effect of chemical mutagens on the bioluminescent system of *Achromobacter fischeri*. *Bact. Proc.*, p. 46.

Nicol, J. A. C. (1958a). Observations on luminescence in *Noctiluca*. *J. Marine Biol. Assoc. U.K.*, **37**, 535.

Nicol, J. A. C. (1958b). Observations on luminescence in pelagic animals. *J. Marine Biol. Assoc. U.K.*, **37**, 705.

Rogers, P., and W. D. McElroy (1955). Biochemical characteristics of aldehyde and luciferase mutants of luminous bacteria. *Proc. Nat. Acad. Sci. U.S.*, **41**, 67.

Ryther, J. H. (1955). The ecology of autotrophic marine dinoflagellates with reference to red water conditions. In F. H. Johnson (Ed.), *The Luminescence of Biological Systems*, American Association for the Advancement of Science, Washington, D.C. pp. 387–413.

Seliger, H. H., J. B. Buck, W. G. Fastie, and W. D. McElroy (1964). Spectral distribution of firefly light. *J. Gen. Physiol.*, **48**, 94.

Seliger, H. H., and W. G. Fastie (1968). Studies on a bioluminescent bay. III. Measurements of underwater sunlight spectra with a double beam photoelectric spectrometer. *J. Marine Research*, **26**, 273.

Seliger, H. H., and W. D. McElroy (1965a). *Light: Physical and Biological Action*, Academic Press, London, New York. pp. 168–205.

Seliger, H. H., and W. D. McElroy (1965b). *Light: Physical and Biological Action*, Academic Press, London, New York. pp. 258–268.

Seliger, H. H., and W. D. McElroy (1968). Studies on a bioluminescent bay. I. Bioluminescence patterns *in situ*. *J. Marine Research*, **26**, 244.

Shimomura, O., F. H. Johnson, and Y. Saiga (1962). Extraction, purification and properties of Aequorin, a bioluminescent protein from the luminous hydromedusan, *Aequorea*. *J. Cell. Comp. Physiol.*, **59**, 223.

Strehler, B. L. (1953). Luminescence in cell-free extracts of luminous bacteria and its activation by DPN. *J. Am. Chem. Soc.*, **75**, 1264.

Strehler, B. L. (1961). In W. D. McElroy and H. G. Glass (Eds.), *Light and Life*, Johns Hopkins Press, Baltimore. pp. 306.

Strehler, B. L., and M. J. Cormier (1953). Factors affecting luminescence of cell-free extracts of the luminous bacterium, *Achromobacter fischeri*. *Arch. Biochem. Biophys.*, **47**, 16.

Strehler, B. L., and M. J. Cormier (1954). Isolation, identification, and function of long-chain aldehydes affecting the bacterial luciferin-luciferase reaction. *J. Biol. Chem.*, **211**, 213.

15§

Strehler, B. L., E. N. Harvey, J. J. Chang and M. J. Cormier (1954). The luminescent oxidation of reduced riboflavin or reduced riboflavin phosphate in the bacterial luciferin-luciferase reaction. *Proc. Nat. Acad. Sci. U.S.*, **40**, 10.

Sweeney, B. M., and G. B. Bouck (1966). Crystal-like particles in luminous and non-luminous dinoflagellates. In F. H. Johnson and Y. Haneda (Eds.), *Bioluminescence in Progress*, Princeton University Press, Princeton, N.J. pp. 331–348.

Sweeney, B. M., and J. W. Hastings (1957). Characteristics of diurnal rhythm of luminescence in *Gonyaulax polyedra*. *J. Cell. Comp. Physiol.*, **49**, 115.

Sweeney, B. M., F. T. Haxo, and J. W. Hastings (1959). Action spectra for two effects of light on luminescence in *Gonyaulax polyedra*. *J. Gen. Physiol.*, **43**, 285.

Terpstra, W. (1963). Investigations on the identity of the light-emitting molecule in *Photobacterium phosphoreum*. *Biochim. Biophys. Acta*, **75**, 355.

Totter, J. R. and M. J. Cormier (1955). The relation of bacterial luciferase to alternative pathways of dihydroflavin mononucleotide oxidation. *J. Biol. Chem.*, **216**, 801.

Wassink, E. C. (1948). Observation on the luminous fungi, I, including a critical review of the species mentioned as luminescent in literature. *Rec. Trav. Bot. Néere.*, **41**, 150.

Author Index

Page numbers in *italics* where the author's work is quoted in full.

Taxonomic Index

Subject Index